Bireswar Banerjee (Ed.)

**Rubber Products**

## Also of interest

*Sustainability of Polymeric Materials*
Marturano, Ambrogi, Cerruti (Eds.), 2020
ISBN 978-3-11-059093-7, e-ISBN (PDF) 978-3-11-059058-6,
e-ISBN (EPUB) 978-3-11-059069-2

*Polymer Surface Characterization*
*Recycling of Polyurethane Wastes*
Alavi Nikje, 2019
ISBN 978-3-11-064102-8, e-ISBN (PDF) 978-3-11-064159-2
e-ISBN (EPUB) 978-3-11-064126-4

*Smart Rubbers.*
*Synthesis and Applications*
Polgar, van Essen, Pucci, Picchioni, 2019
ISBN 978-3-11-063892-9, e-ISBN (PDF) 978-3-11-063901-8,
e-ISBN (EPUB) 978-3-11-063931-5

*Rubber Analysis.*
*Characterisation, Failure Diagnosis and Reverse Engineering*
Forrest 2019
ISBN 978-3-11-064027-4, e-ISBN (PDF) 978-3-11-064028-1,
e-ISBN (EPUB) 978-3-11-064044-1

# Rubber Products

Technology and Cost Optimisation

Edited by
Bireswar Banerjee

DE GRUYTER

**Editor**
Bireswar Banerjee
Karunamoyee Estate B-12/3
700091 Kolkata Salt Lake
West Bengal
India
banerjee_bireswar@yahoo.co.uk

ISBN 978-3-11-066724-0
e-ISBN (PDF) 978-3-11-066853-7
e-ISBN (EPUB) 978-3-11-066740-0

**Library of Congress Control Number: 2023949754**

**Bibliographic information published by the Deutsche Nationalbibliothek**
The Deutsche Nationalbibliothek lists this publication in the Deutsche Nationalbibliografie; detailed bibliographic data are available on the internet at http://dnb.dnb.de.


Cover image: DmyTo/iStock/Getty Images Plus
Typesetting: Integra Software Services Pvt. Ltd.
Printing and binding: CPI books GmbH, Leck

www.degruyter.com

# Contents

**Contributing authors —— VII**

Samar Bandyopadhyay
**Chapter 1**
**Fundamentals of rubber compounding —— 1**

Soumen Chakraborty
**Chapter 2**
**Carbon black morphology and its application in elastomer and polymer matrix —— 53**

Dipak Kumar Setua
**Chapter 3**
**Elastomer blend and compatibility: a flow visualisation study —— 121**

Mrinmoy Debnath and Abhijit Adhikary
**Chapter 4**
**Use of graphene in rubber nanocomposite, its processability, and commercial advantages —— 151**

Asit Baran Bhattacharya and Kinsuk Naskar
**Chapter 5**
**Transmission rubber V-belt technology —— 197**

Timir Baran Bhattacharyya
**Chapter 6**
**The Science of Rubber Conveyor Belt: A Comprehensive Guide' —— 229**

Saikat Das Gupta, Hirak Satpathi, Tirthankar Bhandary, and Rabindra Mukhopadhyay
**Chapter 7**
**Reverse engineering: a tool for the chemical composition analysis of finished rubber products —— 279**

Shambhu Lal Agrawal and Abhijit Adhikary
**Chapter 8**
**Thermal and mechanical analysis study of different rubber applications —— 323**

Dipankar Mondal, Soumyajit Ghorai, Dipankar Chattopadhyay, and Debapriya De
**Chapter 9**
**Devulcanisation of discarded rubber: a value-added disposal method of waste rubber products — 365**

Kasilingam Rajkumar and Santosh C Jagadale
**Chapter 10**
**Cost of quality in rubber processing — 445**

Bireswar Banerjee
**Chapter 11**
**Lean productivity and cost optimisation for rubber processing industries — 469**

**Index — 501**

# Contributing authors

**Dr. Samar Bandyopadhyay**
Pukhraj Additives LLP
601, Acacia Apartment, Old Airport Road
Siddharthnagar H Block
Jaipur 302017
Rajasthan
India
bandyopadhyay.drsamar@gmail.com

**Dr. Soumen Chakraborty**
Business President – CBD
Himadri Speciality Chemical Ltd, India
8, India Exchange Place, 2nd Floor, Kolkata
700001
West Bengal
India
drsc@himadri.com

**Dr. Dipak Kumar Setua**
Retd. Associate Director/Scientist 'G' of DMSRDE,
Kanpur (DRDO),
Kanpur, Uttar Pradesh
India
and
Former Director of Advanced Centre of Research
on High Energy Materials (ACRHEM)
University of Hyderabad
Telangana
India
and
Defence Materials and Stores Research and
Development Establishment
DMSRDE Post Office, G.T. Road
Kanpur 208013
Uttar Pradesh
India
dksetua@rediffmail.com

**Mrinmoy Debnath**
DGM, Technical Service Product Development
Reliance Industries Limited, India
J-402 Shreenath Putam
Near PVR-Deep Multiplex
Old Chhani Road, Nizampura
Vadodara 390002, Gujarat
India
mrinmoy1.debnath@ril.com

**Dr. Abhijit Adhikary**
Asstt. Vice President – Carbon Black
BKT CARBON
M-903 Bakeri Swara, Besides ABB Complex
Makarpura Main Road
Vadodara 390013, Gujarat
India
1307abhijit@gmail.com

**Dr. Asit Baran Bhattacharya**
Rubber Technology Centre
IIT Kharagpur
Kharagpur 721302
West Bengal
India
asitrkmv94@gmail.com

**Professor Kinsuk Naskar**
Rubber Technology Centre
IIT Kharagpur
Kharagpur 721302
West Bengal
India
naskark73@gmail.com

**Dr. Timir Baran Bhattacharyya**
Sr. Vice President
Forech India Pvt. Ltd.
No. 1, SIPCOT Industrial Park, Mangal
Cholavaram, Cheyyar
Thiruvannamalai 631701
Tamil Nadu
India
timir1957@gmail.com

**Tirthankar Bhandary**
Hari Shankar Singhania Elastomer and Tyre
Research Institute, India
Raghupati Singhania Center of Excellence
Plot No. 437, Hebbal Industrial Area
Mysore 570016
Karnataka
India
tirthankar@jkmail.com

https://doi.org/10.1515/9783110668537-203

**Dr. Hirak Satpathy**
Hari Shankar Singhania Elastomer and Tyre Research Institute, India
Raghupati Singhania Center of Excellence
Plot No. 437, Hebbal Industrial Area
Mysore 570016
Karnataka
India
hirak@jkmail.com

**Dr. Rabindra Mukhopadhyay**
Hari Shankar Singhania Elastomer and Tyre Research Institute, India
Raghupati Singhania Center of Excellence
Plot No. 437, Hebbal Industrial Area
Mysore 570016
Karnataka
India
rm@jkmail.com

**Dr. Saikat Das Gupta**
Hari Shankar Singhania Elastomer and Tyre Research Institute, India
Raghupati Singhania Center of Excellence
Plot No. 437, Hebbal Industrial Area
Mysore 570016
Karnataka
India
saikat.dasgupta@jkmail.com

**Shambhu Lal Agrawal**
General Manager
Reliance Industries Limited
C352, Sector-1, Reliance Township, Near Undera Circle
Vadodara 391345
Gujarat
India
Shambhu.Agrawal@ril.com

**Dr. Abhijit Adhikary**
Asstt. Vice President – Carbon Black
BKT CARBON
M-903 Bakeri Swara, Besides ABB Complex
Makarpura Main Road
Vadodara 390013
Gujarat
India
1307abhijit@gmail.com

**Dr. Debapriya De**
Professor & Principal
Panihati Mahavidyalaya
Barasat Road, Sodepur
Kolkata - 700110
West Bengal, India
and
Sourav Apartment
Flat No. 19, 244 Dum Dum Park
Kolkata 700055
West Bengal
India
debapriyad2001@yahoo.com

**Prof. Dipankar Chattopadhyay**
Professor, Department of Polymer Science and Technology
University College of Science and Technology
University of Calcutta
Kolkata 700009, West Bengal
India
and
15, Ramkrishna Pally, Haltu
Kolkata 700078
West Bengal, India
dipankar.chattopadhyay@gmail.com

**Dipankar Mondal**
Senior Research Fellow
Chemistry Department
MCKV Institute of Engineering,
Liluah, Howrah 711204
West Bengal, India
and
6/7 Marconi Avenue, B Zone
Durgapur (M-Corporation)
Steel Town East
Bardhaman 713205
West Bengal, India
mrdipu87@gmail.com

**Soumyajit Ghorai**
Senior Research Fellow
Chemistry Department
MCKV Institute of Engineering
Liluah, Howrah 711204
West Bengal, India
and

3, Masjid Lane, P.O. Ghurni
Krishnagar, Kotwali
Nadia 741103
West Bengal, India
soumyajit.ghorai@gmail.com

**Dr. Kasilingam Rajkumar**
Director, Indian Rubber Manufacturers Research Association (IRMRA)
IRMRA
Plot No. 254/1b
Road No. 16 V
Wagale Industrial Estate
Thane West 400604
Maharashtra
India
rk@irmra.org

**Santosh Jagadale**
Jr. Sc. Officer, Indian Rubber Manufacturers Research Association (IRMRA)
IRMRA
Plot No. 254/1b
Road No. 16 V
Wagale Industrial Estate
Thane West 400604
Maharashtra
India
sj@irmra.org

**Bireswar Banerjee**
Rubber and Quality Consultant
B–12/3, Karunamoyee Estate, Salt Lake
Kolkata 700091
West Bengal, India
banerjee_bireswar@yahoo.co.uk

Samar Bandyopadhyay

# Chapter 1
# Fundamentals of rubber compounding

## 1.1 Introduction

Rubbers seldom provide anticipated properties in many of the rubber products. They do not meet the service properties. Many chemicals are sequentially mixed with natural and synthetic rubbers in required quantities to make rubber compounds and to achieve desired end use properties of rubber goods. Rubber compounds are obtained by homogeneous mixing of raw rubber and other compounding ingredients. Once the basic rubber is selected, selection of other chemicals is dependent on property requirements of the final product and on best processing of the compound, as well as cost competitiveness of the final product. Mixing of rubber and its ingredients is performed by using suitable mixing equipment so that the properties of most of the ingredients remain unchanged in the final product, thereby maintaining a defined set of mechanical, chemical, and other properties [1, 2]; whereas rubber compounding is a process to modify rubbers or elastomers, or a blend thereof along with other ingredients for optimisation of properties and to meet a required set of service application or performance properties [1–3].

To design rubber formulation, rubber and its ingredients are chosen very judiciously to obtain properties within specifications. It is most desirable that the ingredients are safe for health and environment and they should be efficiently and cost-effectively processable with effective use of available equipment. Thus, the most important factor in rubber compounding is to secure an acceptable balance of processability, properties, and most importantly, cost. Compounded rubber should provide required properties, which are best found by mixing, and should fulfil the service requirement of rubber goods such as tyres, inner tubes, conveyor belts, re-treaded tyres, footwear, rubber rolls, hoses, belts, weather stripping, and many more.

Mixing is the most important stage in the manufacturing of rubber products, which affects the dispersion and the homogeneity of the chemical ingredients in the mixed compound [4, 5]. An economical mixing process is also essential to begin with cost-effective products. Thus, the basic objective of mixing operation includes obtaining a uniform blend of all the mixing ingredients to attain satisfactory stable dispersion of fillers and to produce successive batches, which are uniform both in degree of dispersion and viscosity.

https://doi.org/10.1515/9783110668537-001

## 1.2 Evolution of rubber mixing process

Mixing is the most critical step in achieving properties of the rubber compounds. Initially, mixing was carried out in Hancock's Pickle machine, which was made up of a single rotor [5, 6]. During that time, generally, two roll mills were used in small factories and are still very often preferred for many special mixing requirements such as to develop coloured, tacky, and very hard compounds. Generally, the friction ratio was maintained between 1:1.05 and 1:1.2 to maintain friction against each other and to keep the compound on single roll. To obtain better mixing, double rotor internal mixers, which were invented in the mid-eighteenth century, were not particularly suitable for efficient mixing [7].

Nowadays, mixing operation is carried out mainly in internal mixers and two roll mills [8–11], where the latter is used mostly for cooling and sheeting of the dumped compounds out of internal mixer and sometimes to carry out mixing in the absence of an internal mixer [14, 15]. Occasionally, incorporation of vulcanising chemicals during the final batch mixing process is carried out in two roll mills. Two roll mills are also used as breakdown and warm-up equipment prior to processing compounds in calenderer or extruder. As continuous, economical, and very precisely controlled mixing process is hardly possible by two roll mills due to various physical forms and nature of rubber and rubber chemicals and their intermediate compounds, the internal mixer is predominantly used for such compound mixing in big rubber industries like tyre or conveyor belt manufacture [14].

The mixing in an internal mixer takes place between the rotors and rotor to chamber wall [12, 13]. The Banbury type internal mixer was developed in 1916 [16]. Figure 1.1 depicts the schematic diagram of a typical Banbury mixer [18]. Currently, in large-scale production units, the mixing processes have been extensively automated. Internal mixers are classified into three types based on chamber and rotor design. Tangential and intermix internal mixer are named based on tangential and intermeshing rotor design. Rotor design of tangential and intermeshing has been shown in Figure 1.2.

The tandem mixer has a primary mixer unit with either tangential or intermeshing rotor and a lower tandem unit. It has two counterrotating rotors of different diameters that rotate independently at different speeds. Dispersive type of mixing occurs between the nip of the rotor and internal sidewall of the chamber, whereas distributive mixing happens by material shifting from one rotor to another and between rotor and chamber wall. A tangential mixer generates high shear mixing primarily between rotor tip and mixer sidewall, dispersive mixing between rotor wings and sidewall, and distributive mixing by transferring material around the mixing chamber from one rotor to the other, whereas intermeshing rotors provide dispersive mixing in the nip between the rotors and facilitate transfer of material from rotor to rotor.

Banbury mixers permits approximately 20% mixing and intermeshing rotors permits nearly 80% mixing between rotors. The rotors are designed in such a way that the tips of the rotor blade do not touch one another. Most Banbury mixers still have two-wing rotors. Owing to lower productivity in Banbury with two-wing rotors, conceptualisation of four-wing rotors design began in the beginning of 1930 and it started to gain importance due to its higher output and improved mixing quality for high volume mixing [14, 15]. Temperature is more effectively controlled; and drive power is generally 10–20% higher in intermeshing type of mixer in comparison with tangential mixer. The fill factor is nearly 5% lower in intermix [16].

Nowadays intermix comes with many features like special rotor design, variable speed of rotor, good control on cooling during mixing, and control on ram position and ram pressure to carry out reactive mixing [18]. Due to such merits, the usage of intermix is continually gaining acceptance in big rubber goods manufacture. The tandem mixer is a recent development in the rubber mixing process. The tandem mixer is composed of an upper ram type (primary mixer) and a lower ramless (tandem) units. The primary mixer contains either an intermeshing or a tangential rotor, which can incorporate and disperse high-quality level filler and other ingredients. This is a single-stage process, where the master batch is traditionally mixed in ram-type internal mixer unit and is discharged into larger ramless internal mixer unit under the primary unit and followed by curatives addition. It is a continuous process and a new master can be simultaneously prepared to eliminate multiple mixing steps, storage time, and cooling time of the intermediate batches. Though a tandem mixer is costlier by approximately 20%, it is nearly 100% more efficient than a conventional mixer.

Apart from efficient mixing by an internal mixer, the consistent product quality depends on the accuracy of weighing systems, controlled addition of rubber and rubber chemicals, mixing time, and dump temperature of the batches. Batch-to-batch variation could also be improved by keeping a close look on energy and power curves of the mixing process and constantly monitoring quality parameters of each batch. A typical diagram of the power curve of a conventional single-stage mixing profile is shown in Figure 1.3.

## 1.3 Processing and properties

Rubbers are long-chain macromolecules, and at the initial stage of the rubber mixing process, the long chains are broken down mechanochemically, which is called mastication [19–21]. Mastication was fundamentally developed to bring down Mooney viscosity of rubber to an acceptable limit, so that it is receptive to the other compounding additives. As most of the synthetic rubbers are produced with uniform viscosity, they minimise or prevent the need for mastication.

Mooney viscosities of natural rubber (NR)s are quite high and vary from lot to lot, source, time of cultivation, and collection of latex [22–26]. It is required to bring down the Mooney viscosity of NR to a processable limit in order to obtain uniform viscosity

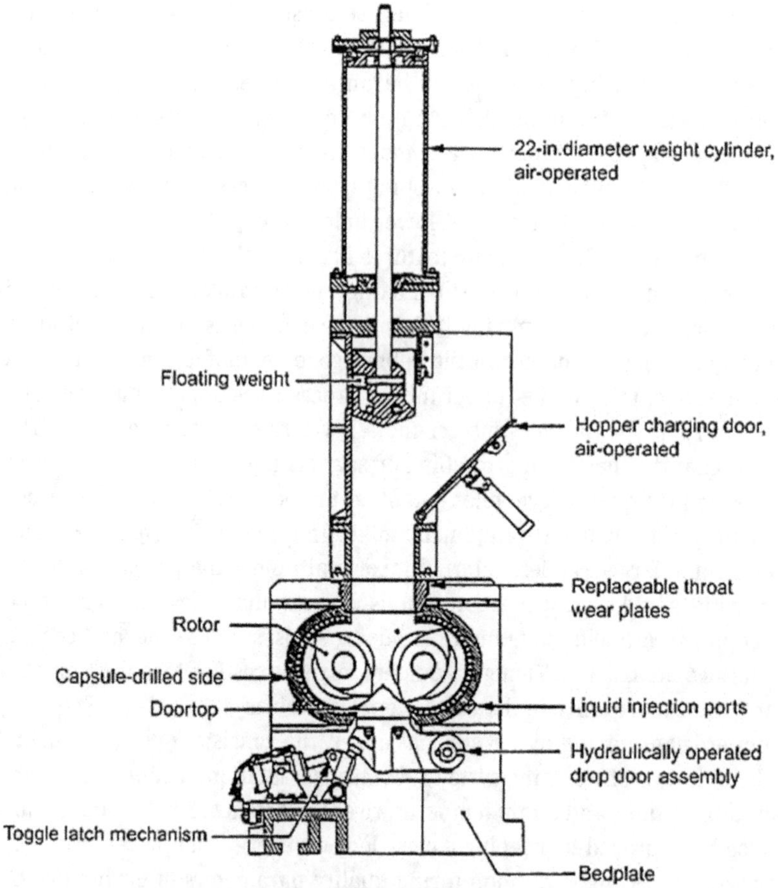


**Figure 1.1:** Schematic diagram of the Banbury mixer.

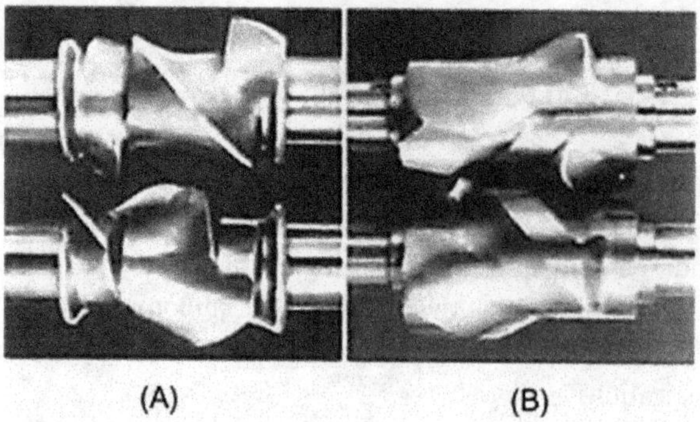


**Figure 1.2:** Design of tangential (A) and intermeshing (B) rotors.

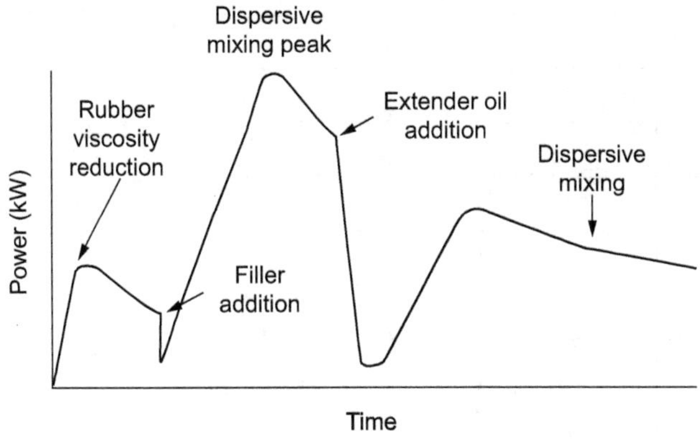


**Figure 1.3:** Power versus time profile of a conventional single-stage mixing process.

before further processing. Hence, this process is essentially employed in NR-based compounds or in a compound containing both natural and synthetic rubber.

Mooney viscosity is a measure of shearing torque obtained by using a Mooney viscometer and signifies the necessity for high energy of Mooney rubber or compound during mixing or processing and plays a critical role in designing the mixing process and subsequent processing of rubber compound. Figure 1.4 shows a typical Mooney viscosity and stress relaxation curve [23].

Measurement of Mooney viscosity of compounded rubber is also useful to predict the temperature generation and flow of compound during processing such as calendering, extrusion, and injection moulding or vulcanisation. Figure 1.4 also depicts the stress relaxation behaviour of rubber, where slow relaxation rate signifies higher elasticity and rapid rate of relaxation indicates a higher viscous response [23].

Scorching or premature vulcanisation of a rubber compound is detrimental for its further processing as it lessens the plastic property of the compound and makes it non-processable; therefore, measuring the scorching behaviour of rubber compounds is very important for process control and quality assurance. The scorching behaviour of rubber compound is measured by a Mooney viscometer, and a typical Mooney scorch curve is shown in Figure 1.5 [23]. It is one of the most useful tests to determine starting of cure or scorching characteristics of rubber compounds and provides essential data to design and control production processes as well as to check final compound consistency.

Scorch time ($t_5$) is the time interval, measured at rotor start-up, which corresponds to viscosity increase by 5 Mooney units over minimum Mooney viscosity and indicates the pre-vulcanisation behaviour of the rubber compound where higher value of $t_5$ signifies lower pre-vulcanisation tendency and therefore, the rubber compound can be more consistently processed by two roll mill, calender, or extruder. The vulcanisation time ($t_{35}$) is the time interval measured from rotor start-up and corre-

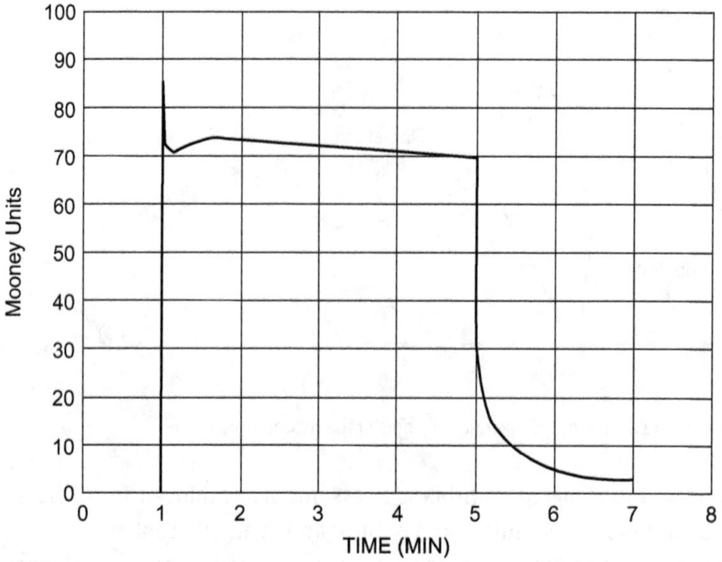


**Figure 1.4:** A typical Mooney viscosity and stress relaxation curve.

sponds to a viscosity increase of 35 units over the minimum Mooney viscosity value. Vulcanisation index ($t_{35} - t_5$), provides indications about the vulcanising ability of rubber compounds where a compound with a low vulcanisation index cures more rapidly than a compound with a higher vulcanisation index [23].

Tackiness of a rubber compound is its ability to stick to itself or to other rubber compounds with application of instant moderate pressure. Good tack of rubber compound is necessary to build rubber composite materials like uncured tyre or making conveyor belts by laying extruded and/or calendered plies one after another. NR (polyisoprene rubber) has high inherent tack, whereas many of the synthetic rubbers do not have enough tack; hence, tackifiers are often added to the formulations to build up tackiness in the compound [27, 28].

Many compounding ingredients like sulphur, accelerators, or antioxidants bloom out to the surface of the compounds on storage and create lesser tackiness of the compound [29]. On the other hand, stickiness signifies the ability of the rubber compounds to stick to nonrubber substances like metal or textile fabric surfaces.

Extreme stickiness to metal may result in poor release from the mill or internal mixer and may cause subsequent problems in processing of compounds as well as to the processing equipment. Low stickiness can also originate slippage of the compound against metal surface of rotors of internal mixture, calenderer, or extruder. Various compounding additives, such as mill release agents or external lubricants are used to control the level of stickiness. It is to be noted that the compound tackiness and stickiness are not same.

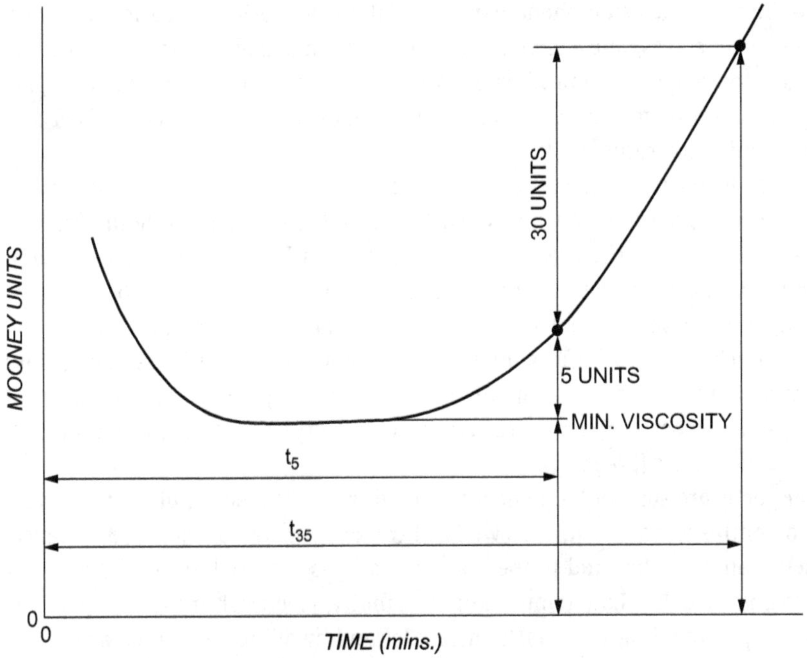


**Figure 1.5:** Illustrative Mooney scorch curve of rubber compound using a large rotor.
where MV is minimum viscosity; $t_5$ is scorch time to increase 5 units Mooney over minimum Mooney viscosity; $t_{35}$ is time to increase 35 units Mooney over minimum Mooney viscosity cure index = $t_{35} - t_5$.

The process of coating rubber compounds into textile fabric or tyre cord by mechanically squeezing it between pairs of counter rotating rolls is termed as calendering [30]. The calendering process, which is classified as fabric and steel cord calendering, consists of two to four more rolls. The surface compatibility and the effective adhesion of the rubber compound with the steel cords and fabric controls the effectiveness of calendering operations. Thus, the performance of calendered steel cord or fabric depends on the adhesion of the rubber compound and the rubber compound adjacent to the calendered sheet.

Rubber extrusion is carried out in an extruder machine at a prefixed temperature and under a specific pressure through a die to provide a specific shape to the product. Typically, good flow of the rubber compounds is one of the most important criteria. The uncured compounds are fed in the feeding section of the extruder and the rotating screw channels carry forward the compounds through the barrel to die. During the process, it generates additional temperature and pressure due to friction between the compound, die channel, and barrel.

After extrusion, swelling and shrinkage occur to the extruded profiles and depends on overall temperature and pressure exerted on the compounds, as well as on hardness of the compound. The design of the die is carried out accordingly to taking care of swelling and extrusion phenomena [31]. After calendering, extrusion, other essential processing stages, and assembling of all parts in sequential manner, it is required to give the shape of the product prior to start the vulcanisation process to give final shape to the products in curing press or in moulds where it is vulcanised at a desired time and temperature.

The tyre industry is the biggest consumer of worldwide production of natural as well as synthetic rubber and consumes nearly half of the rubbers, while the rest are used in mechanical mountings, seals, conveyor and other belts, hoses, gaskets, and in consumer products. Most of the products such as tyres are composed of different rubber compounds with different sets of properties. Many of the components of such composite material need to fulfil many desired properties like hardness, tensile and tear strength, elongation at break, modulus at various elongation levels, abrasion resistance, hysteresis, set properties, resistance to cut growth fatigue, flex cracking, and dynamic properties [32–35].

Rubber goods are susceptible to degradation due to metal poisoning (Cu and Mn), UV radiation of the sunrays, liquid, as well as heat, oxygen, and ozone [36, 37]. Hence, suitable selection of rubber and proper compounding is required to resist the degradation of rubber articles. Low temperature flexibility, electrical properties, gas permeability, proper bonding to metal and textile, and swelling resistance are a few critical property requirements [38–42]. Rubber products that remain in contact of food and drugs have additional requirement of properties, so that they do not harm human beings.

## 1.4 Compounding ingredients

Raw rubbers and their blends do not have sufficient inherent strength and cannot fulfil the requirement of most of the finished rubber products. Many compounding ingredients are mixed with the raw rubber to obtain specific desired properties. The compounding ingredients are added with natural or synthetic rubber to formulate rubber compounds as per required doses [2]. These are classified as fillers (carbon black and nonblack materials), antidegradants (antioxidants, antiozonants, and protective wax), plasticisers, softeners and tackifiers, processing aids (peptizers, lubricants, and release agent), colour pigments, vulcanising agents or curatives, accelerators, activators, retarders, resins, inert filler, and special-purpose materials like blowing agents and deodorants.

During mixing operation, rubber/rubbers and rubber chemicals are sequentially added. Rubbers and doses of chemicals are specifically chosen based on the required properties and performance of the product and often vary depending on the base elastomers, whereas performance of the products depends on consistent quality of raw materials, the mixing process, and subsequent process control [43].

Rubbers are selected depending on their applications, so that they do not degrade or lose basic properties during the service life. To attain anticipated properties, very often, different elastomers are blended. Fillers are chosen not only to cheapen the product but also to incorporate reinforcement and many other performance properties in the finished product. Carbon black, amorphous silica, clay, and so on are used as conventional fillers in compounding. Specific fillers are occasionally used to achieve very high abrasion resistance on the product. It is noteworthy to mention that no specific rubber formulation can fulfil all the expected properties.

It is observed that the improvement of one property can deteriorate the other. As rubber compounds are susceptible to degradation, various antidegradants such as antioxidants and antiozonants are used to arrest degradation of rubber compounds. Occasionally mixtures of antidegradants are used for better performance of the product.

Processing aid, normally oil, is added, which helps the mixing and processing. Selection of suitable curatives and curing system is also one of the most important criteria, which influences the properties of the final compound. It is important to note that a change by a small quantity of curing agent can substantially deviate the properties of the final compound; hence, sufficient care must be taken on addition of curatives. The curing process becomes faster in the presence of an accelerator, and it takes a longer time if accelerators are not used. Even if the compound formula is the same, there can still be differences in the properties of the final compound due to many variations such as source to source variation of raw materials, mixing process variation, unavoidable loss of chemicals, and many more. It is a trend that rubber materials manufacturers always create their own formulary, which are truly suited to their processing conditions and provide consistent application properties.

Natural as well as synthetic rubbers are used for compounding based on requirements. As per the American Society for Testing and Materials (ASTM), designation and chemicals composition of some common elastomers are shown in Table 1.1 [47–48].

**Table 1.1:** Designation and chemical composition of common elastomers.

| ASTM designation | Common name | Chemical composition |
|---|---|---|
| NR | Natural rubber | *cis*-Polyisoprene |
| IR | Synthetic rubber | *cis*-Polyisoprene |
| BR | Butadiene rubber | *cis*-Polybutadiene |
| SBR | SBR | Poly(butadiene-styrene) |
| IIR | Butyl rubber | Poly(isobutylene-isoprene) |
| CIIR | Chlorobutyl rubber | Chlorinated poly(isobutylene-isoprene) |
| BIIR | Bromobutyl rubber | Brominated poly(isobutylene-isoprene) |
| EPM | EP rubber | Poly(ethylene-propylene) |
| EPDM | EPDM rubber | Poly(ethylene-propylenediene) |
| CSM | Hypalon | Chloro-sulfonyl-polyethylene |
| CR | Neoprene | Polychloroprene |

**Table 1.1** (continued)

| ASTM designation | Common name | Chemical composition |
|---|---|---|
| NBR | Nitrile butadiene rubber | Poly(butadiene-acrylonitrile) |
| HNBR | Hydrogenated nitrile rubber | Hydrogenated poly(butadiene-acrylonitrile) |
| ACM | Polyacrylate | Poly(ethyl acrylate) |
| ANM | Polyacrylate | Poly(ethyl acrylate acrylonitrile) |
| T | Polysulfide | Polysulfides |
| FKM | Fluoroelastomer | Polyfluoro compounds |
| FVMQ | Fluorosilicone | Fluoro-vinyl polysiloxane |
| MQ | Silicone rubber | Poly(dimethylsiloxane) |
| VMQ | Silicone rubber | Poly(methylphenyl-siloxane) |
| PMQ | Silicone rubber | Poly(oxydimethyl silylene) |
| PVMQ | Silicone rubber | Poly(polyoxymethylphenylsilylene) |
| AU | Urethane | Polyester urethane |
| EU | Urethane | Polyether urethane |
| GPO | Polyether | Poly(propylene oxide-allyl glycidyl ether) |
| CO | Epichlorohydrin homopolymer | Polyepichlorohydrin |
| ECO | Epichlorohydrin copolymer | Poly(epichlorohydrin-ethylene oxide) |

## 1.5 Selection of rubbers

As discussed, the base rubber selection is the most important step in creation of rubber formulary. Therefore, choice of incorrect rubber may cause catastrophic failure of the product, which could be expensive; hence, for any desired application, rubbers with suitable properties should be chosen. While targeting the overall cost-effectiveness of the rubber compound, the selection of suitable rubber and rubber compounding ingredients depends on processing and performance properties of the rubber products [44–46].

To obtain the desired tensile and tear strength, elongation at break, modulus or stiffness, abrasion resistance, oil resistance, low temperature properties, fatigue properties, tack, compression strength, stress relaxation, dynamic properties such as hysteresis and damping characteristics, flammability, chemical resistance, and to withstand service temperature, compounds should comply with all the technical specifications. Tensile strength is the measure of stress that a rubber compound can withstand before it breaks. Primarily it depends on the molecular structure of rubber. The composition of rubber compounds also contributes highly to the tensile property. It also varies based on the application temperature. It signifies the interaction of fillers with rubber and the reinforcement. Modulus measures the stiffness of materials and higher modulus signifies a stiffer compound [46].

It is measured as the stress required to elongate the compound at various percent of elongations. Elongation at break of rubber compounds signifies the resistance to the change of shape of a rubber compound without generation of any crack and is measured as a ratio of length of the material at break to the initial length as percentage.

Hardness of a rubber compound is also an important parameter and is measured as resistance to indentation of a microprobe under specific load and time. It depends on elastic modulus and viscoelastic nature of the rubber compound [72]. Tear strength of a rubber compound is also an important parameter, which signifies its resistance to tear and is measured as the force required for initiating and propagating until the compound fails.

Many rubber materials are used in various types of oil atmosphere; hence, oil resistance of the rubber is an important criterion to select rubber. Many times, rubbery materials are deployed in sub ambient temperature. The glass transition temperature of rubber will be the main selection criteria in this case. When rubber compounds undergo constant cyclic deformation, they experience fatigue due to the exertion of mechanical force and cause initiation of crack and once crack is initiated, they start propagation. Hence, good resistance to fatigue is required for improved service life of rubber products.

NR, due to its strain-induced crystallisation, shows better crack propagation resistance than synthetic rubbers, whereas crack initiation resistance of polybutadiene and styrene butadiene rubber (SBR) is superior to NR. Tack is always required for processing of rubber and for adhesion between two different layers. NR and chloroprene rubber (CR) have inherent tack, whereas external tackifiers are required to improve tackiness of synthetic rubbers. When a rubber article undergoes constant compression stress, it shows compression set due to permanent deformation.

Lower compression set indicates higher elastic response of the rubber goods. Stress relaxation is a measure of time-dependent change of stress at a constant strain, whereas creep is a measure of change of strain under constant stress and a particular temperature in viscoelastic materials like rubber compounds. Stress relaxation of viscoelastic materials depends on both elastic and viscous components where a quick relaxation indicates better processing and less batch-to-batch variation of the compounds.

Lower value of creep is expected of viscoelastic materials for their storage stability. Synthetic rubbers show high heat generation compared to NR and limits their uses where less heat dissipation under dynamic conditions is required. These properties are dependent on service temperature of the products.

Rubber products are very often used in service where self-extinguishing property is a unique requirement to select rubber for the typical application. Similarly, chemical resistance of base rubber is to be considered before initiating compounding. Generally, rubber of more than 50% is used in a compound but for a few exceptions. Therefore, cost of the base rubbers is an important selection criterion for rubber compounds.

Besides cost, mixing of the rubber should be accounted; otherwise, it will require additional processing cost and therefore overall compound cost will increase. For example, if the Mooney viscosity of the rubber is selected on a higher side than the specified limit, energy to process the compound will increase, which will increase total cost of the compound. Ease of mixing also governs processability, cost, and compound quality. If blends of two or more rubbers are used, Mooney viscosity of the rubbers should be in the acceptable range, so that no phase separation happens, and the blending of the rubbers commences in the molecular level. Due to their higher molecular weight, NRs are properly masticated to bring down Mooney viscosity to the processing limit and higher molecular weight synthetic rubbers are conventionally oil extended to smoothen the mixing operation.

Molecular weight and its distribution are also significant measures to choose the right rubber. Broader molecular weight distribution of rubber eases the mixing operation, where as narrow molecular weight distribution improves the properties of the compound. The strength of raw rubbers solely depends on their molecular architecture, micro- and macrostructure.

NR, nitrile rubber, and hydrogenated nitrile rubber show excellent inherent strength and could be used alone to manufacture select products. Polybutadiene rubber does not have enough inherent strength and is used in blend with other rubbers like NR or SBR. Due to its intermediate strength, SBR could replace NR to a higher extent, which is also driven by availability of rubbers and cost benefits. The typical properties of a few elastomers are captured in Table 1.2 [46], and Figure 1.6 [48] depicts the classification system as per SAE J200 for IRM 903 oil (ASTM D 2000–03) for swelling in oil, heat, and temperature resistance [46–48].

**Table 1.2:** Relative properties of various elastomers.

| ASTM designation | NR | BR | SBR | IIR CIIR | EPM EPDM | CSM | CR | NBR | HNBR | ACM ANM | T | FKM | FVMQ | VMQ MQ, PMQ, PVMQ | AU EU | GPO | CO ECO |
|---|---|---|---|---|---|---|---|---|---|---|---|---|---|---|---|---|---|
| Durometer range | 30–90 | 40–90 | 40–80 | 40–90 | 40–90 | 45–100 | 30–95 | 40–95 | 35–95 | 40–90 | 40–85 | 60–90 | 40–80 | 30–90 | 35–100 | 40–90 | 40–90 |
| Tensile max, psi | 4500 | 3000 | 3500 | 3000 | 2500 | 4000 | 4000 | 4000 | 4500 | 2500 | 1500 | 3000 | 1500 | 1500 | 5000 | 3000 | 2500 |
| Elongation max., % | 650 | 650 | 600 | 850 | 600 | 500 | 600 | 650 | 650 | 450 | 450 | 300 | 400 | 900 | 750 | 600 | 350 |
| Compression set | A | B | B | B | B-A | C-B | B | B | B-A | B | D | B-A | C-B | B-A | D | B-A | B-A |
| Creep | A | B | B | B | C-B | C | B | B | B | C | D | B | B | C-A | C-A | B | B |
| Resilience | High | High | Med. | Low | Med. | Low | High | Med.-Low | Med. | Med. | Low | Low | Low | High-Low | High-Low | High | Med.-Low |
| Abrasion resistance | A | A | A | C | B | A | A | A | A | C-B | D | B | D | B | A | B | C-B |
| Tear resistance | A | B | C | B | C | B | B | B | B | D-C | D | B | D | C-B | A | A | C-A |
| Heat aging at 212 °F | C-B | C | B | A | B-A | B-A | B | B | A | A | C-B | A | A | A | B | B-A | B-A |
| $T_g$, °C | −73 | −102 | −62 | −73 | −65 | −17 | −43 | −26 | −32 | −24, −54 | −59 | −23 | −69 | −127, −86 | −23, −34 | −67 | −25, −46 |
| Weather resistance | D-B | D | D | A | A | A | B | D | A | A | B | A | A | A | A | A | B |
| Oxidation resistance | B | B | C | A | A | A | A | B | A | A | B | A | A | A | B | B | B |
| Ozone resistance | NR-C | NR | NR | A | A | A | A | C | A | B | A | A | A | A | A | A | A |
| Solvent resistance | | | | | | | | | | | | | | | | | |
| Water | A | A | B-A | A | A | B | B | B-A | A | D | B | A | A | A | C-B | C-B | B |
| Ketones | B | B | B | A | B-A | B | C | D | D | D | A | NR | D | B-C | D | C-D | C-D |
| Chlorohydrocarbons | NR | NR | NR | NR | NR | D | D | C | C | B | C-A | A | B-A | NR | C-B | A-D | A-B |
| Kerosene | NR | NR | NR | NR | NR | B | B | A | A | A | A | A | A | D-C | B | A-C | A |
| Benzol | NR | NR | NR | NR | NR | C-D | C-D | B | B | C-B | C-B | A | B-A | NR | C-B | NR | B-A |
| Alcohols | B-A | B | B | B-A | B-A | A | A | C-B | C-B | D | B | C-A | C-B | C-B | B | C | A |
| Water glycol | B-A | B-A | B | B-A | A | B | B | B | A | C-B | A | A | A | A | C-B | B | C |
| Lubricating oils | NR | NR | NR | NR | NR | A-B | B-C | A | A | A | A | A | A | B-C | A-B | D | A |

A, excellent; B, good; C, fair; D, use with caution; NR, not recommended. SOURCE: Seals Eastern, Inc. Common name and chemical composition of ASTM of elastomers in abbreviation are depicted in Table 1.1.

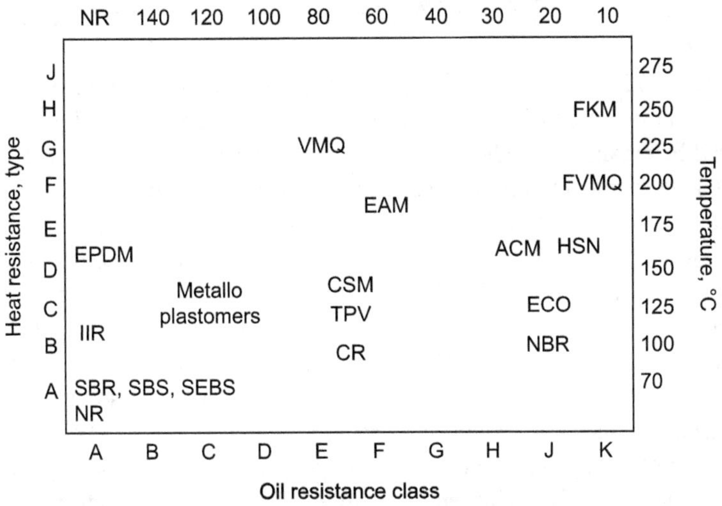


**Figure 1.6:** SAE J200 classification systems for IRM 903 (ASTM D 2000–03) oil for max. swell, %, heat resistance, and temperature resistance, where NR represents no requirement, and polymer abbreviations are defined in Table 1.1 [48].

## 1.6 Curing system

Curing is a process where an elastomer forms a cross-linked network and improves its strength. Selection of accelerator and activator types of curing system for the selective rubber compounds depends on the base rubber, processing, and curing conditions as well as service requirements of final product. Curing time and temperature are decided based on the method of curing and flow behaviours of the rubber compound, and so on. Use of processing equipment, temperature buildup during processing, extent of rework usage, and so on are also considered before constructing the curing system. It is also designed to provide stability and required specific properties such as mechanical, chemical, and dynamic property, ageing behaviour, and permanent set to the compounds during its service.

Sulphur is commonly used as a vulcanising agent to vulcanise diene-based elastomers, like NR, SBR, and butadiene rubber (BR) [50–53]. Preferably, rhombic sulphur or in-rubber soluble sulphur, which is crystalline in nature and made up of eight sulphur atoms in ring structure, is used in vulcanisation. If the doses of rhombic sulphur exceed its solubility limit in rubber matrix due to its crystalline nature, it blooms out to the rubber surface and the rubber compound loses its tackiness. Another allotrope of sulphur, which is amorphous and polymeric (in-rubber insoluble) in nature and does not bloom out to the rubber surface is also used in vulcanisation. It maintains tackiness of the rubber compound and does not reduce adhesion between two components [54, 55].

The sulphur-based vulcanisation system has few limitations as it can vulcanise diene elastomers.

Specialty elastomers like ethylene propylene diene monomer (EPDM) and butyl rubber have been chemically modified for sulphur vulcanisation and fully saturated elastomers are vulcanised by organic peroxides. Vulcanisation of rubber by only sulphur is a slow process and takes several hours at elevated temperature, which reduces the performance properties of vulcanisates. This process is accelerated by the presence of an accelerator, which also lowers the need for sulphur. To activate the efficiency of accelerators, accelerator-activators are also used. Table 1.3 lists a few common types of accelerators used in the rubber industry. Accelerators such as tetramethyl thiuramdisulfide (TMTD), dipentamethylenethiuramtetrasulfide (DPTT), tetrabenzylthiuramdisulfide (TBzTD) act both as effective accelerators and as sulphur donors. These accelerators also improve mono- and disulfide cross-links and are thermally more stable and add more heat and set resistance to the rubber compounds. These accelerators restrict their uses as they generate nitrosamine species, which is carcinogenic in nature. Peroxides are alternate curing agents and can cross-link both diene and non-diene rubbers.

Peroxide-cured compounds are thermally more stable than sulphur-cured compounds due to carbon-to-carbon cross-linkages [56]. The rate of peroxide vulcanisation reaction depends on temperature and half-life of the specific peroxide. Peroxide vulcanisations require coagents to improve cross-link efficiency while maintaining rate of curing reaction. Bismaleimides, triallyl isocyanurate, diallyl phthalate, and so on are a few coagents. Halogenated elastomers like polychloroprene and chlorosulfonated polyethylene are cured by metal oxides such as in combination of zinc oxide and magnesium oxide where zinc chloride, which is produced as a by-product of the reaction acts as auto-catalyst during the curing process [57, 58].

Polyacrylate elastomers are cured by hexamethylene diamine carbamate whereas fluoroelastomers are mostly vulcanised by amines and metal oxides [59–62]. Table 1.4 describes few organic accelerators and their typical uses in elastomer matrix.

**Table 1.3:** Commonly used accelerators for different types of rubbers.

| Accelerator type | Common use |
|---|---|
| Sulphur or sulphur -containing | Natural rubber, isoprene, SBR, butyl, polybutadiene, EPDM, nitrile, polynorbornenes |
| Organic peroxides | Urethane, silicone, chlorinated polyethylene, cross-linked polyethylene, ethylene acrylic elastomer, vinyl acetate-ethylene copolymer, nitrile-PVC |
| Metallic oxide | Polychloroprene, chlorosulfonated polyethylene, polysulfide |
| Organic amines | Acrylic, fluorocarbon, epichlorohydrin, ethylene acrylic elastomer |
| Phenolic resins | Butyl rubber |

**Table 1.4:** A few organic accelerators and their typical uses.

| Accelerator type | Example | Typical use |
|---|---|---|
| Aldehyde-amine | Reaction product of butyraldehyde and aniline | Fast curing accelerator for reclaim and hard rubber and self-curing cements |
| Amines | Hexamethylene tetramine | Delayed action slow accelerator for natural rubber |
| Guanidines | Diphenyl guanidine (DPG) | Secondary accelerator to activate thiazole-type accelerator |
| Thioureas | Ethylene thiourea (ETU) | Fast curing accelerator for neoprene, hypalon and epichlorohydrin |
| Thiazoles | Benzothiazyldisulfide (MBTS) | Safe processing, moderately fast curing accelerator for natural rubber, isoprene, SBR, nitrile, and EPDM |
| Thiurams | Tetramethylthiuram disulfide (TMTD) | Fast curing sulphur-bearing accelerator for SBR, nitrile, butyl, and EPDM rubbers |
| Sulfenamides | *N*-Cyclohexyl-2-benzothiazyl sulfenamide (CBS) | Safe processing, delayed-action accelerator for natural rubber, SBR, and nitrile rubbers |
| Dithiocarbamates | Zinc dimethyldithiocarbamate (ZDMC) | Fast curing accelerator for SBR and butyl rubbers |
| Xanthates. | Dibutylxanthogen disulfide | Fast curing, low-temperature accelerator for natural rubber and SBR |

SBR, styrene butadiene rubber; EPDM, ethylene propylene diene monomer.

## 1.7 Fillers

Fillers are particulate in nature and addition of fillers improves the mechanical properties, hardness, as well as stiffness of the compounds [63]. Addition of fillers also brings down the overall cost of the product. Based on the type of filler and its effect on the improvement of physical properties, they are classified as reinforcing, semi-reinforcing, and non-reinforcing types [64–66]. Use of reinforcing fillers in unfilled rubbers, which is known as rubber reinforcement, greatly improves the physical properties such as stiffness, modulus, tensile and tear strength, rupture energy and resistance to fatigue, and abrasion. Non-reinforcing fillers are used as diluents or extenders while semi-reinforcing fillers act intermittently between reinforcing and non-reinforcing fillers.

Improvements in physical properties of rubber compounds depend on particle size of fillers and fillers with smaller particle size provide higher reinforcement [66]. The shape of the particulate fillers and their distribution also greatly affect the reinforcement property of rubber compounds. Filler particles with higher aspect ratio contribute to higher reinforcement, whereas particles with broader distribution lower the viscosity of the compound owing to its compact interaction with rubber phase compared to narrow distribution of fillers. The morphology of filler particles, doses, and interfacial interaction with elastomers also significantly contribute to the property of compound [68, 69]. Wide grades of reinforcing, semi-reinforcing carbon black and silica fillers are used in rubber industries, based on property requirements. Apart from these, many mineral types of fillers such as carbonates, clays, silicas, silicates, and talc are used in rubber compounding. Various fillers are shown in the Figure 1.7, based on primary particle size of the fillers [70]. Quantity and type fillers are chosen based on their reinforcement capability, cost, ease of processing, colouration to the product, service, or any other special requirement to be incorporated in the rubber goods.

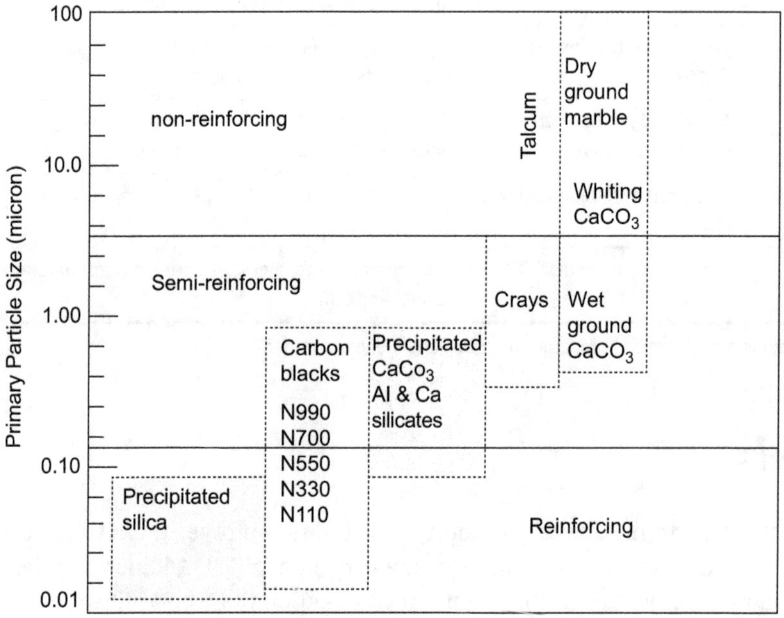


**Figure 1.7:** List of fillers used in rubber industry.

Hardness approximation is sometimes important and mostly depends on hardness of base rubber, and its contribution from fillers, process oil, and other liquid softeners. In general, Shore A hardness is followed to measure hardness of rubber compounds [71]. It measures the resistance to indentation where harder compounds show higher hardness and the softer ones, low hardness. Estimation of Shore A hardness in rubber compounds is discussed elsewhere [72].

## 1.8 Plasticisers and process aid

Processing aids [73, 74] selection also depends on the type of base polymer and solubility parameters of processing aid, anticipated Mooney viscosity of the compound, dispersion of fillers, and processing requirement. Table 1.5 briefly specifies a few common processing aids used in rubber compounding. Rubber plasticisers are basically used to reduce compound viscosity, to improve processability, to ease the incorporation and dispersion of fillers, to improve flexibility at low temperature, and to lower hardness of the rubber compound. Aromatic, naphthenic, and paraffinic oils, which are widely used in rubber as plasticisers are petroleum-based. Polar plasticisers are compatible with polar rubber, whereas nonpolar plasticisers are compatible with nonpolar rubber, based on solubility parameters.

The petroleum oil-based plasticisers are complex mixture of unsaturated aromatic, saturated naphthenic rings, and side chain paraffinic groups in various ratios. Aromatic oils mostly contain unsaturated single and multicomponent aromatic rings and paraffinic oils contain high level of saturated paraffinic molecules with side chains, whereas naphthenic oils mostly contain high levels of saturated rings than aromatic and paraffinic process oil. Aromatic oils are also denoted as distillate aromatic extract (DAE) oils. These oils are classified as carcinogenic as per EU legislation and contain high amount of polycyclic aromatic hydrocarbon (PAH). Alternative mineral oils like mild extract solvates and treated DAEs are not carcinogenic as PAH contents are within EU legislation.

As aromatic oils are highly polar, they have good compatibility with polar rubbers such as nitrile butadiene rubber (NBR) and CR. Being saturated and of low polar nature, paraffinic oils are mostly compatible with saturated rubber and used in EPDM and butyl rubber-based products. Naphthenic oils have intermediate polarity and are compatible with NR, PBR (pyridine–butadiene rubber), SBR, EPDM, and IIR rubbers. Fatty acids are primarily accelerator activators, but when used in large amounts, they work as process aid.

Tackifiers are used as processing additives in rubber compounds to improve its tackiness and adhesion by lowering molecular cohesion. These are low molecular weight resinous compounds or oligomers, and the glass transition temperature is higher than that of the base polymer. Tackifiers are chosen based on their compatibility with base polymer, effective low melt viscosity, and softening point of the rubber compounds. Tackifiers are generally classified into three groups which are rosin or resin, hydrocarbon resin, and terpene resins. Hydrocarbon resins are derived from petroleum feedstocks, whereas terpenes and rosin esters are from pine trees.

**Table 1.5:** Common processing aid used for rubber compounding.

| Processing aids | Function |
|---|---|
| Activated dithio-bisbenzanilide | Peptizer for natural rubber |
| Poly-*p*-dinitrosobenzene | Chemical conditioner for butyl rubber |
| Xylyl mercaptans | Peptizer for natural rubber, butyl rubber, SBR, and nitrile rubber; stabiliser for cement viscosity |
| Low-molecular-weight polyethylene | Release agent, lubricant |
| Calcium oxide | Desiccant |
| Aliphatic-naphthenic-aromatic resins | Homogenising agent for all elastomers |
| Paraffin wax | Release agent, lubricant |
| Polyethylene glycol | Activator for silica lubricant |

## 1.9 Antidegradants

Most rubber products are prone to degrade on exposure to oxygen and ozone gases, under application of heat and UV light, and due to heavy metal ions contamination. Degradation of rubber due to such factors alters the properties and performances of the rubber materials and shortens their service life. It is very much necessary to incorporate various antidegradants in rubber compounds to improve their storage and service life as well as performance properties.

Antidegradants are judiciously selected based on the nature and extent of protection required for the exposure of the rubber goods in the environment and nature of end use of the product. Compatibility and stability of antidegradants in rubber products, their staining or nonstaining characteristics, toxicological aspects, and cost are also taken in account to select an antidegradant [75–77]. Antidegradants are classified as antioxidants and antiozonants.

Antioxidants that prevent rubbers and rubbery materials from oxidative degradation are chemically categorised in two types: (i) amine type, which are staining or discolouring and (ii) phenolic type, which is nonstaining or non-discolouring. Staining varieties of antioxidants cause significant colour variation to the products under light and atmospheric conditions. Rubbers with high level of unsaturation are susceptible to react with atmospheric ozone and thus antiozonants are used to prevent them from ozone degradation. *p*-Phenylenediamine (PPD)-type antiozonants are most commonly used in unsaturated rubber compounds. Besides its antiozonant effects, PPDs also work effectively as primary antioxidants. To improve the protection efficiency against ozone degradation, blends of PPDs are often incorporated and the efficiency can be further enhanced by addition of waxes [78].

The efficiency of antiozonants depends on their rapid ozone-scavenging rate than the reaction rate of ozone with rubber. Antiozonants like waxes also ooze out from the compound and form a protective layer on the rubber surface that protects the surface against ozone degradation. Most antiozonants also work as an antioxidant but antioxidants cannot protect rubber compounds against ozonation. Figure 1.8 shows a typical structure of PPD.

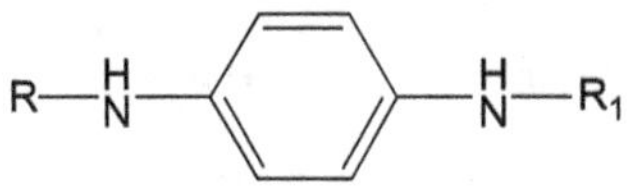


p-Phenylenediamine (PPD)
R, $R_1$ -alkyl or aryl group

**Figure 1.8:** Typical structure of *p*-phenylenediamine (PPD).

Rubber molecules form free radicals under heat, light, or mechanical stress during compound processing, in storage and service conditions that are extremely prone to react with atmospheric oxygen. Antioxidants are used to scavenge such free radicals and to stop their further propagation to more reactive peroxy radicals and hydroperoxide intermediates. Antioxidants are also classified as free radical scavengers or primary scavengers and peroxide scavengers or secondary antioxidants. Ironically, hindered phenols, which are nonstaining type, are most broadly used as free radical scavengers and are called primary antioxidants or radical chain terminators.

Secondary aromatic amines, which are staining types, react with the most reactive free radical hydro peroxides into nonreactive products before they further propagate and are called hydroperoxide decomposers. Trivalent phosphorus compounds are commonly used as secondary antioxidants. Table 1.6 shows a few examples of antidegradants and their staining behaviour.

**Table 1.6:** Examples of few antidegradants of elastomers.

| Antidegradants | Example | Staining |
|---|---|---|
| Hindered phenol | 2,6-Di-t-butyl-*p*-cresol | None to slight |
| Hindered bis-phenols | 2, 2′-Methylene-bis(4-methyl-6-tert-butyl) phenol | None to slight |
| Hindered thiobisphenols | 4,4′-Thiobis(6-tert-butyl-3-methylphenol | Slight |
| Hydroquinones | 2,5-Di(tert-amyl) hydroquinone | None to slight |
| Phosphites | Tri(mixed mono- and di-nonyl-phenyl) phosphite | None to slight |

**Table 1.6** (continued)

| Antidegradants | Example | Staining |
|---|---|---|
| Diphenylamines | Octylated diphenylamine | Slight to moderate |
| Naphthylamines | Phenyl-alpha-naphthylamine | Moderate |
| Quinolines | Polymerised 2,2,4-trimethyl-1,2-dihydroquinoline | Slight to moderate |
| Carbonyl-amine condensation product | Reaction product of diphenylamine and acetone | Considerable |
| *p*-Phenylene diamines | Mixed diaryl-*p*-phenylene diamines | Considerable to severe |

## 1.10 Other rubber chemicals

A good balance between rubber processing safety and faster curing rates is important for productivity and economic production. To improve productivity, high temperature of processing and faster accelerator are employed to decrease the vulcanisation time, which may originate scorch or premature vulcanisation, and to avoid such premature vulcanisation, retarders or pre-vulcanisation inhibitors (PVI) are often used to increase the induction time. *N*-(Cyclohexylthio) phthalimide (CTP) is found to be most effective at 0.1–0.2 phr dosages with sulphur quantities between 1.5 and 1.3 phr and is very sensitive with sufanamide accelerators. Excess CTP dosage has to be avoided as it will delay the cure time [79].

To improve quality and performance of rubber products and process rationalisation factices are often used as softener and processing aids. These are commonly a cross-linked solid form of vegetable oil containing sulphur and classified as sulphur factice, sulphur chloride factice, and sulphur-free factice based on the use of cross-linking agent. Factices are used to improve green strength and dimensional stability during extrusion, surface smoothness, plasticiser adsorption, handling of softer compound, and fatigue crack resistance. They also reduce mixing and milling energy by shortening mixing time, extrusion time and reducing calendering defects and density, hence cost per volume [80].

Blowing agents, which are often used to make open cell or microcellular sponge, are stable at room temperature but decompose to release nitrogen or carbon dioxide gases on and before vulcanisation. Generally organic blowing agents are suitable as dispersion of inorganic blowing agents is poor in rubber. The performance of good blowing agents is measured by large quantities of nontoxic, odorless gas generation, non-discolouring, ease of dispersion in the matrix of rubber, ease of decomposition well below curing temperature, and negligible effect of vulcanisation kinetics.

Inorganic blowing agents such as sodium bicarbonate in combination with tartaric, oleic, or stearic acid are often used in rubber as blowing agent. Azodicarbamide, hydrazine derivatives, benzene-1,3-disulfohydrazide, *N*-nitroso derivatives of secondary amines and *N*-substituted amides, N,N'-Dinitrosopentamethylenetetramine (DNPT) are good examples of organic blowing agents.

To improve dimensional stability and load-bearing capacity of tyres, v-belt, conveyor belt, hose, and so on, additional reinforcements are provided by using synthetic fibres, steel and wire, glass fibres, and aramid fibres, where adequate bonding between rubber and the fibre phases are carried out by using special adhesives known as bonding agents.

Resorcinol formaldehyde resin in vinyl pyridine polymeric latex is used to improve adhesion of various fabrics with rubber compounds. A dry bonding system composed of solid RF resin and a suitable methylene donor, like hexamethylene tetramine (HMT) or hexamethoxymethyl melamine (HMMM), or nitromethylpropanol (NMP) is often used. Silica is conventionally used with resorcinol/formaldehyde resins to promote adhesion to fabric and also been revealed to improve adhesion, if used in combination with cobalt salt. The bonding between rubber and metal is one of the most complex problems as such materials do not have any affinity to each other, and moreover, have modulus difference. Bonding agents, such as a combination of adhesion promoters and brass coating or rarely zinc or bronze coating are effectively used as bonding agent for rubber to metal adhesion [81].

## 1.11 Rubber compound development

The material researchers in the rubber industry continuously work to improve the quality of existing products to meet the expectation of end users, to sustain in the competitive market, and to improve the service life of the products. Therefore, a few basic frameworks followed for the improvement of existing products are:

a. Quality: The compound and the final product must meet their quality requirements.
b. Performance: Performance expectations must be met by the product.
c. Cost: The product cost must be competitive and offer value to the end user.
d. Other aspects: Products must comply with all regulatory requirements and should cause minimum environmental concern.

In this context, extensive knowledge is required for development of a new compound, as performance of a rubber product not only depends on its design and area of application, but also very importantly on composition of the rubber compound. Most importantly, failure of any product due to improper formulation may initiate loss of reputation to the organisation.

It may also cause monetary loss due to warranty claim and taking back the faulty materials. As regards rubber compounding, the failure of products may happen due to the selection of inappropriate type of rubber and inappropriate selection of rubber ingredients. There are more areas that are taken into consideration, while developing new rubber compounds.

The objective of rubber compounding is to be clearly defined initially, considering the expected properties along with its processing conditions and overall optimal price. Rubbers are subsequently selected and existing formulary reviews are carried out to select a preliminary formulation and initiate mixing and compound evaluation at laboratory scale. Once the compound meets the desired set of properties at laboratory, the cost is estimated and processability of the compound is evaluated in factory. Subsequently, products are made and tested to check their performance specifications.

### 1.11.1 Tyre compounding

A tyre is a circular or ring-type part fitted with the rim of a vehicle wheel, which makes contact with the ground. Tyres are composite materials made of rubbers, steel or metal fabric, and carbon black along with many other ingredients. They are filled with compressed air and are mechanically flexible [82]. A tyre possesses load-carrying capacity and absorbs cushioning and damping shock during moving the vehicle on the road surface. A tyre also transmits forces of acceleration and breaking of vehicle to the ground, delivers cornering force, generates steering response, and maintains directional stability of the vehicle.

The tyre is expected to resist abrasion to improve mileage, to minimise vibration, and minimise noise generation for ride comfort during travel. The safety of the vehicle on road is of prime importance and hence, a tyre must provide sufficient grip (or traction). A tyre also has to be cool while running on road to improve low rolling resistance and thus fuel efficiency. All the components of tyres are not made of the same compounds. The type of rubber, being the major ingredient of the compounds, greatly varies along with the other compounding ingredients based on the property requirement.

Rubber compounding entails proper mixing of different types of rubber and rubber chemicals at specific required amounts to produce different tyre compounds with specific features based on different types of materials of tyre components. Individual components of each tyre segment are separately designed to provide maximum specific benefit. The structure and function of important components of a tyre are briefly described below.

## 1.12 Tyre structure

Various rubber compounds are assembled during tyre building to make a green tyre, which is then cured to provide the shape where all the different components retain their original unique functions. Based on the use of the tyre, the types and construction vary; light truck tyres are produced to operate in very severe and even off-road conditions with high load capacity, and hence their construction differs from passenger car (PCR) tyres [83]. Light truck tyres are heavier than PCR tyres due to the use of extra belt, casing ply with stronger steel belt cord, bigger bead, and thicker sidewall [84]. Figure 1.9 shows a typical cross-section of both bias and passenger radial tyres.

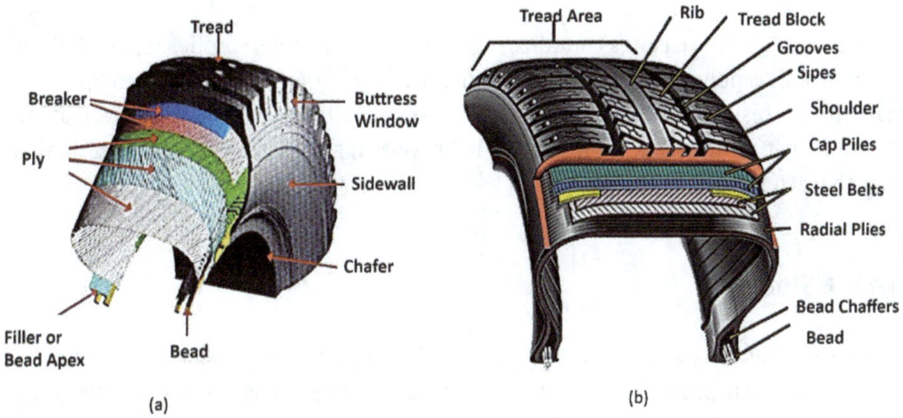


**Figure 1.9:** Typical components of (a) bias truck tyre and (b) radical passenger tyre.

A tyre is broadly divided into two segments: (i) the tread area and (ii) the casing. The tread area includes tyre tread compounds and belts, sub-tread, tread base, cushion, wedges, and overlay, whereas the casing area includes carcass plies, sidewall, bead, inner liner, chaffer, and apex.

A brief description of the different parts of tyre is given below.

### 1.12.1 The tread

The coextruded tread compound profile, which contains tread and tread base is positioned on the belt compound profile during building process. The belt compound is made up of calendered polyester cords called plies. The tread base compound meets the belt system and improves stability and durability of ply system of the cured tyre. Tread cap and tread base compounds are so designed to add more mileage, grip, and low rolling resistance to the vehicle. Sometimes, sub-tread, a cooler-running extruded compound is placed under the tread profile to typically lower hysteresis, which improves rolling resistance of the tyre and in order to meet the target fuel economy.

During the moulding operation, tread patterns are formed. The tread profiles are sometime fine-tuned to improve ride quality, less noise, and better handling. During the tyre-building operation, the belt system, made up of two plies of steel cords placed at opposite angles, which is mounted above the casing profile to improve strength, stability, and steadiness of tread area and to enhance mileage, rolling resistance, and grip of the tyre tread.

### 1.12.2 Body ply

Most of the PCR tyre casings contain several layers of calendered plies made of different fabric like polyester, nylon, or rayon cords, which enhance the strength of the tyre and its resistance to damage on road. Polyester fabric is preferred over others due to its better adhesion with rubber, light weight, good strength, and heat dissipation characteristics.

### 1.12.3 Sidewall

The tyre sidewall extends from bead area to the tread of a tyre and acts as protective covering of cord plies. The side wall compounds are designed to be flexible to improve riding comfort. Although the sidewall does not meet road surface, due to its use, it must be scraping, and flexing resistant and antidegradants are used in the rubber compounding to improve its weather resistance.

### 1.12.4 Bead

Tyre bead made of rubber-coated steel cords holds the tyre with the rim of wheel, so as to protect air loss. It includes loops of body plies, apex, and bead filler compounds. A hard and tough rubber compound is used as bead insulation outside the bead area, which forms an inextensible hoop to seal the tyre against the rim and provides tension hoops to prevent air leakage. It transfers load between the tyre and the rim.

### 1.12.5 Belt wedge

Belt wedge is a small trip of rubber compound specially used in the radial tyre belt edges to prevent tread separation by reducing inter-ply shear. Belt wedge compounds are formulated for high dynamic stiffness to restrict relative belt motion, to improve cushioning, and to absorb stress between belts and its neighboring area, good fatigue, and tear resistance.

### 1.12.6 Shoulder inserts

Shoulder inserts are small strips of rubber placed on the body ply, under the belt ends, which help maintain a smooth belt contour and insulate the body plies from the belt edges.

### 1.12.7 Belt skim compound

Belt skim compound is applied by calendering over the brass-plated steel cords. It is primarily formulated to resist fatigue and tear.

### 1.12.8 The inner liner

Typically, all the radial tyres are tubeless, and a special rubber compound made up of halobutyl rubber is used as an integral part in the inside of the tyre and acts like an inner tube. The inner liner along with the bead and bead filler works to hold air inside the tyre.

## 1.13 Tyre composition

A tyre is made up of different rubber compounds containing variety of rubbers such as NR, polybutadiene rubber, SBR, EPDM rubber, and halobutyl rubber. Carbon black, silica, clays, and so on are used as fillers. Other additives like rubber process aids: different types of process oil, weathering resistance additives, aliphatic and aromatic hydrocarbon resin along with curatives as sulphur, accelerators, and accelerator activator (zinc oxide and stearic acid), and different bonding promoters are used based on requirement. Several types of reinforcing fabric, steel cords and bead wires are also used for tyre manufacturing [85]. A brief comparison of tyre composition major ingredients used in passenger/light truck tyres and truck tyres is depicted in Figure 1.10.

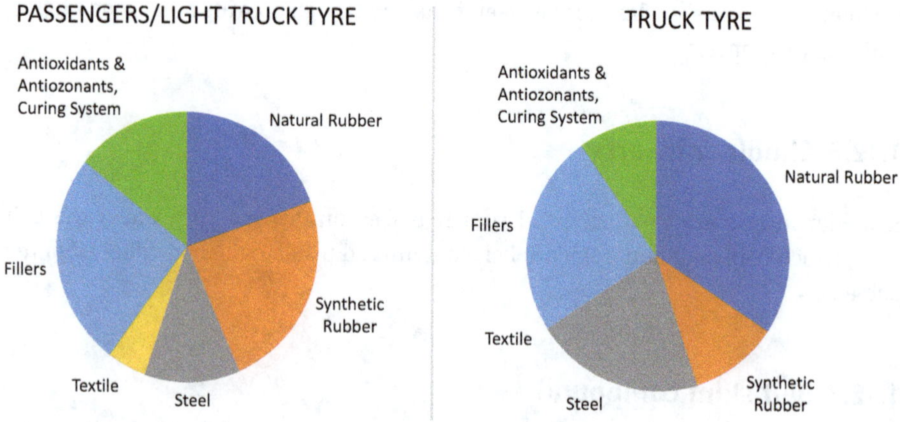


**Figure 1.10:** Typical composition of passenger/light truck tyres and truck tyres.

### 1.13.1 Truck tyre tread compound

Low cost per tyre mile is the basic requirement of the truck tyre, which is dependent on effective retreadability of tyre [86]. The type of tyre, its tread compound, and the design of casing contribute to the retreading efficiencies of the tyre. The tread compound is designed to comply with other basic requirements; for instance wet traction, ride comfort, fuel efficiency, and handling, along with meeting the overall service life of the tyre. Generally, natural or SBR along with carbon black as reinforcing filler is used in bias tyre tread compound. Table 1.7 shows a typical bias tyre tread recipe [72].

**Table 1.7:** Typical bias truck tyre tread formulation [72].

| Ingredients | Rib type | Lug type |
|---|---|---|
| Ribbed smoke sheet NR | 70 | 100 |
| Peptisers | 0.1 | 0.1 |
| High *cis*-BR | 30 | – |
| N234-carbon black | 65 | – |
| N330-carbon black | – | 50 |
| Aromatic oil | 15 | 10 |
| 6PPD | 2.75 | 2.65 |
| TMQ | 0.75 | 0.75 |
| Wax blend | 1 | 1 |
| Stearic acid | 1.5 | 1.75 |
| Zinc oxide | 5 | 5 |
| Sulphur | 1.4 | 1.2 |
| MBS | 1.4 | 0.7 |
| PVI | 0.2 | 0.2 |

**Table 1.7** (continued)

| Ingredients | Rib type | Lug type |
|---|---|---|
| Rheometric properties by MDR @ 150 °C | | |
| $t_{s2}$ (min) | 8.8 | 5.6 |
| $t_{c90}$ (min) | 13.2 | 13.9 |
| Maximum torque (lb in) | 58.5 | 70.1 |
| Unaged physical properties (cured at 150 °C, $t_{c90}$ + 2 min) | | |
| 300% modulus (MPa) | 10.8 | 15.0 |
| Tensile strength (MPa) | 22.5 | 25.5 |
| Elongation at break (%) | 560 | 480 |
| Hardness, IRHD | 72 | 75 |

N234-carbon black, intermediate super abrasion furnace (ISAF); N330-carbon black, high abrasion furnace (HAF); 6PPD, *N*-(1,3-dimethylbutyl)-*N'*-phenyl-*p*-phenylenediamine-quinone; TMQ, polymerised 2,2,4-trimethyl-1,2-dihydroquinoline; MBS, *N*-oxydiethylene -2-benzothiazole sulfenamide; PVI, *N*-(cyclohexylthio) phthalimide.

## 1.13.2 Passenger radial tyre tread compound

Precipitated silica is preferably used as a reinforcing filler in high-performance pneumatic PCR tyre tread by part or full replacement of carbon black. Solution styrene butadiene rubber (S-SBR) due to its controlled molecular weight distribution, branching, vinyl content, and functionalisation shows good interaction with silica fillers in presence of silane coupling agent [87]. As a result, a combination of precipitated silica and S-SBR as the main elastomer shows improvement of rolling resistance and wet grip along with comparable abrasion resistance of tyre [88–90].

Glass transition temperature ($T_g$) of S-SBR could be varied by altering the ratio of styrene and vinyl content. The vinyl content is almost constant and cannot be varied in emulsion styrene butadiene rubber (E-SBR) [91]. Low primary-chain molecular weight, branching, and soap residue of E-SBR rubber contributes to the hysteresis of the compound and is therefore not preferred for high-performance PCR tyre tread [92–95].

Physically, this is demonstrated during the energy loss function tan $\delta$-$G''/G'$ (or $E''/E'$) as a function of the temperature. Figure 1.11 displays tan $\delta$, measured at low frequency and low deformation amplitude of a carbon black-filled tread rubber with emulsion E-SBR as matrix polymer and for silica-filled S-SBR with equal filler content and compound stiffness. The silica compound has a lower loss factor for temperatures at 40–70 °C indicating lower contributions of the tread compound to the overall rolling losses of the rolling tyre.

Structural deformations of tyre are responsible for rolling resistance at low frequency. Simple sinusoidal lab testing yields an approximate estimate of the loss differences between different rubber compounds. This figure also displays the loss factor at

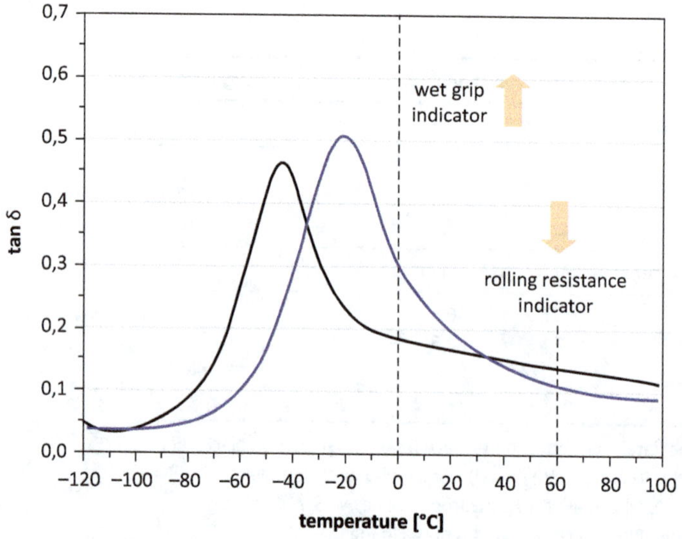


**Figure 1.11:** Viscoelastic response for rolling resistance and wet traction.

lower temperatures around 0 °C, where the loss factor of the silica compound is higher than the carbon black compound, which correlates with the surface deformation at high frequency region.

According to the time-temperature superposition principle, as frequency is inversely related to temperature, we can argue that higher energy dissipation at low temperatures correlates with grip and lower energy dissipation at high temperatures correlates with rolling resistance [57, 63, 67]. Table 1.8 depicts a typical PCR tyre tread recipe [115].

Generally, tyre sidewall is made with NR and polybutadiene rubber in 60:40 blend ratios and vice versa. The combination of natural and polybutadiene rubber provides tyre sidewalls a good flex-fatigue and cut initiation and crack propagation resistance along with high dynamic flexibility. Sometimes, EPDM rubber is also used in combination of NR/BR rubber blend to improve the ozone resistance, flex-fatigue properties, and cut growth resistance of sidewall compounds with a small sacrifice of tensile properties [95]. Table 1.9 shows a typical sidewall formulation.

### 1.13.3 Skim compound of steel cord

Steel cords are primarily used as reinforcing materials for carcass and belts of radial tyres. Most of the skim compounds are made of NR-based compounds with higher amounts of sulphur and sulphur-to-accelerator ratio for initial tack of the compound. As adhesion is a major concern in heterogeneous materials, the skim compounds are specially designed for good adhesion with brass-plated steel cords [96]. Cobalt com-

**Table 1.8:** Passenger car tyre tread formulation [115].

| Ingredients | Formulation-1 | Formulation-2 |
|---|---|---|
| Functionalised solution SBR | 75 | – |
| Solution SBR (S-SBR) | – | 75 |
| Hi-*cis*-polybutadiene rubber | 25 | 25 |
| Highly dispersible silica | 75 | 75 |
| N339- carbon black | 6 | 6 |
| Low PCA oil | 27.5 | 27.5 |
| X50S | 12 | 12 |
| Zinc oxide | 3 | 3 |
| Stearic acid | 2 | 2 |
| 6PPD | 1.5 | 1.5 |
| TBBS | 1.6 | 1.6 |
| DPG | 1.5 | 1.5 |
| Soluble sulphur | 1.7 | 1.7 |
| Compound Mooney viscosity | | |
| ML (1 + 4) at 100 °C, MU | 81 | 73 |
| Rheometric property by MDR, 160 °C for 20 min | | |
| Maximum torque MH (dN m) | 25.5 | 20.5 |
| Minimum torque ML (dN m) | 3.3 | 3.4 |
| Delta torque (dN m) | 22.2 | 17.1 |
| $t_{s2}$ (min) | 4.2 | 2.8 |
| $t_{c90}$ (min) | 13.5 | 11.7 |
| Cured at 160 °C for 2 × $t_{c90}$ min | | |
| 200% modulus (MPa) | 9.0 | 8.5 |
| Tensile strength (MPa) | 14.6 | 14.8 |
| Elongation at break (%) | 272 | 320 |
| Hardness (Shore A) | 71 | 69 |
| Cured at 141 °C for 60 min | | |
| Heat build-up (°C) | 14 | 17 |
| DIN abrasion loss ($mm^3$) | 82 | 92 |
| Cured at 160 °C for 2 × $t_{c90}$ min | | |
| tan $\delta$ at 0 °C | 0.47 | 0.42 |
| tan $\delta$ at 30 °C | 0.18 | 0.22 |
| tan $\delta$ at 70 °C | 0.10 | 0.14 |

N339-carbon black, rubber grade-hard particle furnace carbon black; X50 S, 50:50 blend of bis(3- triethoxysilylpropyl)-tetrasulfide (TESPT) and N330 carbon black; 6PPD, *N*-(1,3-dimethylbutyl)-*N'*- phenyl-*p*-phenylenediamine-quinone; TBBS, *N*-tert-butyl-2-benzothiazyl sulfenamide; PPG, polypropylene glycol.

**Table 1.9:** Tyre sidewall formulation [95].

| Ingredients | Sidewall-1 (phr) | Sidewall-2 (phr) |
|---|---|---|
| Natural rubber | 45.0 | 45.0 |
| Polybutadiene rubber | 55.0 | 35.0 |
| EPDM (high ENB grade) | – | 20.0 |
| N339 carbon black | 45.0 | 45.0 |
| Aromatic oil | 9.0 | 6.0 |
| TMQ | 0.75 | 0.25 |
| Stearic acid | 2.0 | 2.0 |
| Zinc oxide | 6.0 | 6.0 |
| MC wax | 2.0 | 0.5 |
| Phenol formaldehyde resin | 3.0 | 3.0 |
| DTPD | 0.75 | 0.25 |
| 6PPD | 3.0 | 1.25 |
| Sulphur | 1.6 | 1.6 |
| CBS | 0.1 | 0.1 |
| DCBS | 0.6 | 0.6 |
| PVI | 0.2 | 0.2 |
| Rheometer @ 160 °C/30 min | | |
| MH-ML (dN m) | 7.6 | 7.9 |
| $t_{c90}$ (min) | 12.6 | 11.4 |
| Stress–strain properties (cured time @ 160 °C for 20 min) | | |
| 300% modulus (MPa) | 3.1 | 3.7 |
| Tensile strength (MPa) | 18.3 | 17.5 |
| Elongation at break (%) | 875 | 800 |
| Shore A hardness | 49 | 54 |

N339 carbon black, rubber grade-hard particle furnace carbon black; TMQ, polymerised 2,2,4-trimethyl-1,2-dihydroquinoline; DTPD, *N,N'*-dixylene-*p*-phenylenediamine; 6PPD, *N*-(1,3-dimethylbutyl)-*N'*-phenyl-*p*-phenylenediamine-quinone; CBS, *N*-cyclohexyl-2-benzothiazyl sulfenamide; DCBS, *N,N*-dicyclohexyl-2-benzothiazolsulfene; PVI, *N*-(cyclohexylthio)phthalimide

pounds are added to change the morphology in the interfacial layers between the rubber compound and steel cords as an adhesion promoter and improve the pull-out force [97]. Cobalt being a transition metal is found to be detrimental for rubber compounds; recent studies on metallic coating where cobalt salts are not in use are found [98, 99]. A typical formulation of skim compound of steel cord is detailed in Table 1.10.

**Table 1.10:** A typical skim compound of steel cords.

| Ingredients | PHR |
|---|---|
| Ribbed smoked sheet (RSS-I) | 100.0 |
| N220-carbon black | 55.0 |
| Zinc oxide | 8.0 |
| Peptisers | 0.25 |
| PF resin | 2.5 |
| TMQ | 2.0 |
| 6PPD | 1.0 |
| Cobalt stearate | 1.5 |
| Granular silica | 0.35 |
| CTP | 0.3 |
| 20% oil-treated insoluble sulphur | 4.5 |
| DCBS | 1.0 |

N220-carbon black, intermediate super abrasion furnace black; TMQ, polymerised 2,2,4-trimethyl-1,2-dihydroquinoline; 6PPD, *N*-(1,3-dimethylbutyl)-*N*' -phenyl-*p*-phenylenediamine-quinone; CTP, *N*-(cyclohexylthio)phthalimide; DCBS, benzothiazyl-2-dicyclohexyl sulfenamide.

### 1.13.4 Bias truck tube compound

The basic requirement of a tube is to retain air inside and to maintain the structural support of tyre and provide suspension. A tube must be low permeable to air and resistant to heat to retain physical properties. It should have low growth during its service life for low tension set and good flex resistance. It should also provide splice durability, tear resistance, and weather and heat ageing resistance.

Tubes are generally made up of butyl rubber having very high air retention, heat ageing, and weather-resistant properties. Butyl rubber is sometimes blended with low unsaturation EPDM to improve heat and fatigue resistance while sacrificing air retention property considerably. Higher particle size blacks are normally used in tube compounds. Blends of blacks may also be used to improve processability and splice durability along with higher dosages of paraffinic oil. A typical recipe of tube formulation is shown in Table 1.11. Coumarone Indene Resin (CI) may be used for better green splice. Zinc oxide and stearic acid are added in normal dosage. Antioxidants are not used normally as unsaturated rubber and are inherently stable against degradation. A blend of TMTD and MBT is used for better processing and cure characteristics [100–101].

**Table 1.11:** Tyre inner tube formulation.

| Ingredients | Compound-1 | Compound-2 |
|---|---|---|
| Butyl rubber | 100 | 100 |
| N330-carbon black | – | 50 |
| N660-carbon black | 70 | – |
| Paraffinic oil | 25 | 20 |
| Stearic acid | 1.0 | 1.0 |
| Zinc oxide | 5.0 | 5.0 |
| PF resin | 4.0 | – |
| Aliphatic hydrocarbon resin | – | 3.0 |
| Sulphur | 2.0 | 1.0 |
| CBS | 1.5 | – |
| MBTS | – | 1.0 |
| Dithiocarbamate accelerator | 1.5 | 2.9 |
| Mooney viscosity, (ML 1 + 4), @ 100 °C | | |
| MU | 45 | 50 |
| Rheometric properties @ 160 °C, 60 min | | |
| Minimum torque (dN m) | 1.3 | 1.8 |
| Maximum torque (dN m) | 9.6 | 8.3 |
| $t_{s2}$ (min) | 3.6 | 3.9 |
| $t_{90}$ (min) | 25.5 | 20.4 |
| Stress–strain properties (cured @ 160 °C, $t_{90}$ + 2 min) | | |
| Shore A hardness | 50 | 48 |
| 100% modulus (MPa) | 1.3 | 1.2 |
| 300% modulus (MPa) | 3.1 | 3.2 |
| Tensile strength (MPa) | 9.5 | 14.0 |
| Elongation at break (%) | 665 | 725 |

N330-carbon black, high-abrasion furnace black; N660-carbon black, general-purpose furnace black; CBS, *N*-cyclohexyl-2-benzothiazyl sulfenamide; MBTS, 2–2′-dithiobis (benzothiazole).

## 1.14 Tyre inner liner for tubeless tyre

When a vehicle runs at high speed, a tube inside the tyre generates friction between the tube and tyre causing temperature rise in tube. Insertion of any sharp object may cause puncture in tube tyre and the pressurised air gets emitted immediately. These increase possibilities of explosion and devastating consequences such as that the vehicle may lose its control and may meet with an accident. Such instances are least possible in tubeless tyre where a low air-permeable calendered inner linear sheet compound made up of chlorobutyl (CIIR) or bromobutyl (BIIR) is applied as an innermost layer in the tyre.

A good quality inner liner retains air pressure inside the tyre besides contributing fuel economy of the tyre. If the tyre with inner liner gets punctured, the air comes out slowly and the driver may reach a safe place. Both the halobutyl elastomers are co-vulcanised with general-purpose elastomers and show good adhesion and vulcanisation kinetics [102]. Both the elastomers show excellent ageing as well fatigue resistance and improve tyre life. A typical recipe of tyre inner liner is depicted in Table 1.12 [103, 107].

**Table 1.12:** Tyre inner liner formulation.

| Ingredients | Light truck (phr) | Passenger car (phr) |
|---|---|---|
| Bromobutyl rubber | 100 | – |
| Chlorobutyl rubber | – | 100 |
| N660-carbon black | 60 | 60 |
| Naphthenic oil | 8 | 2 |
| Hydrocarbon resin | 7.5 | 10 |
| PF resin (tackifier) | 4 | 2 |
| Magnesium oxide | 0.2 | – |
| Stearic acid | 2 | 1.2 |
| Zinc oxide | 1 | 1 |
| Sulphur | 0.5 | 0.5 |
| 2–2′-Dithiobis(benzothiazole) | 1.25 | 1.25 |
| Mooney viscosity, ML (1 + 4) at 100 °C | | |
| MU, 100 °C | 56 | 70 |
| Rheological property by MDR (160 °C/30 min) | | |
| MH – ML (dN m) | 3.6 | 5.0 |
| $t_{c90}$ (min) | 12.9 | 6.3 |
| Stress–strain properties (cure at 160 °C for 2 × $t_{c90}$ min) | | |
| Modulus 300% (MPa) | 3.4 | 4.1 |
| Tensile strength (MPa) | 9.5 | 9.2 |
| Hardness (Shore A) | 47 | 50 |

N660-carbon black, general-purpose furnace black.

## 1.15 Conveyor belt

Figure 1.12 shows a typical conveyor belt system. The conveyor belt is a part of continuous transportation belt conveyor system used to transfer materials discharged on it at its loading to the release point and runs by driving pulley and through pulley idlers. The length of the conveyor belt may be in the range of kilometers depending on its use. It is widely used in manufacturing units where continuous transportation of materials is required. The conveyor belt should have sufficiently high strength, so that it can carry the load, along with low growth or slack, and should have high impact resistance to take care of high shock. Good toughness is required for conveyor belt to convey lumps or granular materials.

Other required key properties for a good performing conveyor belt are low bending resistance, low power consumption for economic transportation of the materials, effective mechanical splicing for initial splicing and to repair its damaged parts, good adhesion between components, resistance to environment – for durability, along with good flexibility and low creep, and high compression resistance [104].

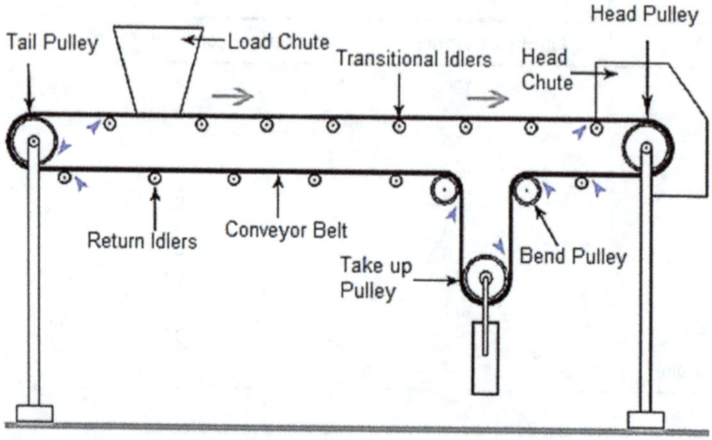


**Figure 1.12:** A typical conveyor belt system.

The prime requirement of a conveyor belt is to withstand specific types of abrasion caused by transporting different types of materials such as heavy and sharp objects, which can cut the cover compound aggressively and should have very high abrasion resistance properties compared to cover compounds carrying finer materials. Cover compounds is also used to protect the belt carcass from damage due to impact, abrasion loss, and cutting, and thereby extends its service life. The cover compound, which is exposed to the materials conveying side is called top cover and the bottom cover protects the belt carcass from abrasive wear generated by pulleys and distributes the shocks. Elastomers commonly used to make cover compounds include NR, SBR, CR, NBR, and EPDM as per specific application requirements.

The carcass of the belt provides structural strength and gives support to the pulling of load. Based on the type of carcass materials, conveyor belts are categorised as textile belt and steel belt conveyor belts. Typical cross sections of a textile belt conveyor belt and a steel belt conveyor belt are depicted in Figure 1.13(a) and (b), respectively. For textile-based conveyor belts, layers of fabric reinforcements are impregnated with rubber by means of latex dipping, friction, and skimming which are then laminated to obtain intended carcass to support, toughness, tensile strength, modulus, mechanical fastener holding and resistance to flex fatigue, impact, rip, and tear.

In a textile conveyor belt, the fabric reinforcements are normally woven-type consisting of warp threads, which provide strength and dimensional stability to the belt and weft threads impact strength, and the required toughness to the belt. Commonly used fabrics used for this purpose are nylon-polyester, polyester-polyester, nylon- nylon, aramid, cotton, and so on [105].

For steel cord-based conveyor belts, the steel cord cables are coated with brass or galvanised zinc. Steel cords have high strength, length stability, and toughing characteristics with high impact and fatigue resistance with less creep. These characteristics make the steel.

Cords useful in highly wear-resistant, heavy loads, high tension and for long-haul conveying conveyor belt applications.

Breaker, made up of loosely woven fabric, used between face cover and carcass, is often used to provide additional instant impact resistance by spreading it in all directions to the carcass and leaving slight force at the impact place. It improves additional anchorage through locking and resists cover from grouping, stripping, and tearing and finds application where impact due to fall of lumps from greater height is experienced.

The inner ply or skim rubber compounds are used to improve the bondage between plies, so that they do not get separated during service life of conveyor belt. It also delivers additional cushioning to the plies to protect them from shocks. They are normally soft compound. For textile belt, they are used for improvement of adhesion between plies and ply to cover. In steel cord belt, they are used to improve adhesion with metal and to improve dynamic flex resistance.

Overall, they improve adhesion, prevent damage from chafing and friction, provide load flexibility and load support, and impart resistance to impact failure. Materials used as skim compounds are NR, NR-SBR, NR-PB, NR-EPDM, and NR-BIIR-EPDM [106].

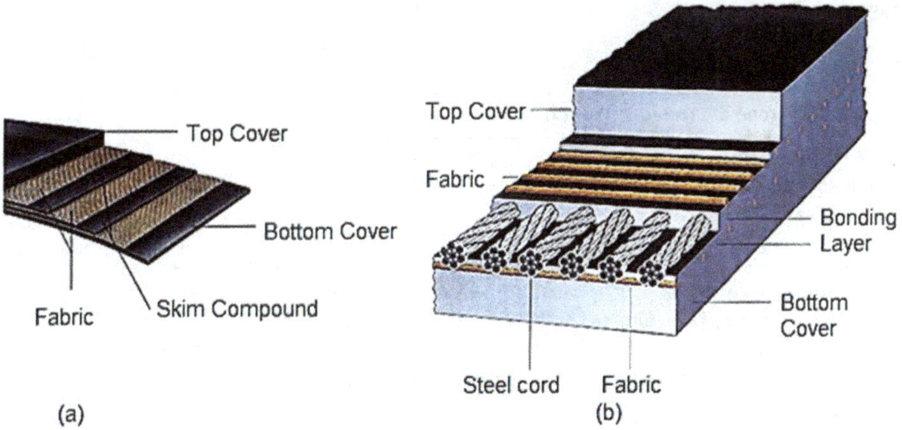


**Figure 1.13:** Typical section of conveyor belt: (a) Textile belt conveyor belt and (b) steel belt conveyor belt.

## 1.15.1 Heat-resistant conveyor belt cover compound

Heat-resistant conveyor belts are used in conveying very hot materials; its requirements are most demanding and applications are most detrimental. Accelerated ageing at high temperature makes rubber compound so hard that crack generates easily during the service of conveyor belt. The high heat also deteriorates adhesion between cover compound and fabric plies with carcass, which causes de-lamination of the belt. Owing to change of hardness at high temperature, tensile strength as well as elongation at break drop drastically, which reduce the flexibility and operative strength of the belt. Subsequently, abrasion loss also increases substantially.

Therefore, a heat-resistant conveyor belt has to be designed to withstand high heat and abrasion loss at high temperature ageing [106]. General-purpose as well as heat-resistant conveyor belt cover compounds are portrayed in Tables 1.13 and 1.14 [107].

**Table 1.13:** General-purpose rubber conveyor belt cover compound.

| Ingredients | PHR | PHR |
|---|---|---|
| Natural rubber | 80.0 | 10.0 |
| SBR 1502 | – | 70.0 |
| High-*cis*-polybutadiene | 20.0 | 20.0 |
| Aromatic oil | 15.0 | 11.0 |
| Stearic acid | 1.0 | 1.5 |
| Zinc oxide | 3.0 | 3.0 |
| 6PPD | 1.5 | 1.5 |
| Wax blend | 2.5 | 2.5 |
| N220-carbon black | 50.0 | – |
| N339-carbon black | – | 50.0 |
| NOBS | 1.5 | – |
| TBBS | 1.5 | – |
| Sulphur | 2.0 | 2.5 |
| TMTD | 0.2 | 0.2 |
| Stress–strain properties (cured at 150 °C/20 min) | | |
| 300% modulus (MPa) | 12.2 | 10.7 |
| Tensile strength (MPa) | 24.8 | 19.8 |
| Elongation at break (%) | 560 | 560 |
| Hardness (Shore A) | 61 | 62 |

6PPD, *N*-(1,3-dimethylbutyl)-*N'*-phenyl-*p*-phenylenediamine-quinone; N220-carbon black, intermediate super abrasion furnace black; N339-carbon black, rubber grade-hard particle furnace black; NOBS, *N*-oxydiethylene-2-benzothiazole sulfonamide; TBBS, *N*-tert -butyl-2-benzothiazyl sulfenamide; TMTD, tetramethylthiuram disulfide.

**Table 1.14:** Heat-resistant conveyor belt cover compound.

| Ingredients | PHR |
|---|---|
| Bromobutyl rubber | 100.0 |
| N550-carbon black | 45.0 |
| Naphthenic oil | 5.0 |
| Stearic acid | 1.0 |
| Polyethylene glycols | 1.5 |
| Magnesium oxide | 1.0 |
| 2-Mercapto benzimidazole (MBI) | 1.0 |
| Dicumyl peroxide | 2.0 |
| *N,N-m*-Phenylene-dimaleimide | 1.0 |
| Stress–strain properties (cured at 170 °C/21 min) | |
| Hardness (Shore A) | 55 |
| 100% modulus (MPa) | 2.8 |
| Tensile strength (MPa) | 9.4 |
| Elongation at break (%) | 890 |
| Stress–strain properties after ageing (at 150 °C/14 days) | |
| Hardness (Shore A) | 50 |
| 100% modulus (MPa) | 2.5 |
| Tensile strength (MPa) | 6.8 |
| Elongation at break (%) | 265 |

N550-carbon black, fast-extrusion furnace black.

### 1.15.2 Chemical-resistant conveyor belt cover compound

Chemical-resistant conveyor belts cover compounds are made up of oil and chemical-resistant rubbers such as nitrile rubber, PVC, and SBR which provid maximum protection to the conveying chemicals following international standards and customer requirements to provide long service life. Oil-resistant cover compounds are made up of NBR, CR, CSM, ECO, or ACM depending on the nature of oil contact and based on other specific requirements such as ozone and heat resistance. A few more typical formularies of conveyor belt cover compound, friction, and skim compound are shown in Table 1.15 [107].

## 1.16 Automotive weather strip compound

The weather-stripping compound in vehicles plays many important roles such as to protect water leakage inside the vehicle, to minimise wind and dampening noise to improve ride comfort and to seal its doors, windows, hood, and trunk from the environment. A good quality weather-stripping compound also protects it from ageing

**Table 1.15:** Typical high temperature-resistant transmission belt cover, friction, and skim compound.

| Ingredients | Cover | Friction and skim |
|---|---|---|
| EPDM rubber | 80.0 | – |
| Chloro-sulfonated polyethylene (CSM 40) | 20.0 | – |
| Natural rubber, RSS II | – | 100.0 |
| Processing aid | – | 2.0 |
| Stearic acid | – | 1.0 |
| Zinc oxide | 2.5 | 5.0 |
| TMQ | 1.0 | 1.5 |
| Diphenyl amine | – | 0.5 |
| Pine tar | – | 5.0 |
| Calcium carbonate | – | 20.0 |
| DCP 40 C | 7.0 | – |
| N330-carbon black | 50.0 | – |
| TMPTA | 1.5 | – |
| Paraffinic oil | 7.0 | – |
| Sulphur | – | 3.0 |
| MBS | – | 1.0 |
| Rheological properties (cured at 151 °C for 25 min) | | |
| 300% modulus (MPa) | 7.1 | 1.6 |
| Tensile strength (MPa) | 14.2 | 16.0 |
| Elongation at break (%) | 550 | 745 |
| Hardness (Shore A) | 60 | 43 |

TMQ, polymerised 2,2,4-trimethyl-1,2-dihydroquinoline; DCP 40C, 40% dicumyl peroxide on calcium carbonate; N330-carbon black, high-abrasion furnace black; TMPTA, trimethylolpropane triacrylate; MBS, 2-(4-morpholinyl-thio)-benzothiazole.

and adds aesthetics to the vehicle. As compounds remain exposed to the direct environment, very good weather resistance and ozone resistance are the primary requirements of the strip compound.

High hardness and moderate tensile strength are also required for good performance of the strip compound. EPDM as base rubber with 30–40 wt% of carbon black meets good performance properties of a weather strip compound. EPDM-based rubber compound has inherent weather resistance due to no unsaturation in backbone. EPDM-based rubber with medium diene content is suitable for faster cure rate. Blends of carbon black may be used for better cost performance. High dose of naphthenic oil in the compound is used as plasticiser. A typical formulation of weather stripper is illustrated in Table 1.16 [107].

**Table 1.16:** Weather stripper formulation.

| Ingredients | PHR |
|---|---|
| EPDM rubber | 100.0 |
| Kaolin clay | 35.0 |
| N990-carbon black | 50.0 |
| N550-carbon black | 30.0 |
| Whiting | 50.0 |
| Zinc oxide | 4.0 |
| Stearic acid | 1.5 |
| Naphthenic oil | 65.0 |
| Azodicarbonamide | 8.0 |
| Surface-treated urea | 1.0 |
| 2-Mercaptobenzothiazole (MBT) | 2.0 |
| Ethyl tellurac | 1.1 |
| Zinc di-*n*-butyldithiocarbamate | 1.9 |
| Dipentamethylene thiuram tetrasulfide | 2.1 |
| Sulphur | 1.9 |

N990-carbon black, medium thermal black; N550-carbon black, fast-extrusion furnace black.

## 1.17 Pharmaceutical stoppers

Pharmaceutical companies use glass vials primarily for storing injectable drug products and to ensure optimised storage and protection for an injectable drug, stoppers made of rubber compounds are used. The rubber stopper remains fastened into the glass container by an aluminum seal or other standard type of seal. Thus, the stopper works as a system of container closure and to keep the contents of vial sterile, to avoid any contamination inside the vial once the needle is injected, and allows instant resealing of the vial once the needle is taken out. An ideal stopper should be low gas- and moisture-permeable, highly resistant to ageing, and should be disinfected. The most widely used stoppers are made with elastomeric compounds.

As halobutyl rubbers possess very low levels of air and moisture permeability, are extractable by chemical and biological substances, have resistance to heat and ultraviolet light, and due to their self-sealing and low disintegration on needle insertion, they are widely used to make pharmaceutical stoppers. Vulcanisations of bromobutyl rubber by zinc oxide also find specific application in making stoppers. Due to low viscosity of this rubber, it can be formulated with less or no plasticiser, which lowers extractables of low molecular weight ingredients and reduces the air permeability.

Clays are used as non-reinforcing fillers and owing to their surface acidity, they may affect the vulcanisation, which is compensated using appropriate dosages of curatives and retarders. Calcined clay is mostly used due to its purity, low water absorption, and uniform particle size. Table 1.17 offers typical formulations of pharmaceutical closure compounds [103].

**Table 1.17:** Rubber pharmaceutical stopper formulations.

| Ingredients | Phr |
|---|---|
| Bromobutyl rubber | 100.0 |
| Zinc oxide | 5.0 |
| Stearic acid | 0.5 |
| Titanium dioxide | 10.0 |
| Calcined clay | 80.0 |
| Plasticiser | 10.0 |
| Paraffin wax | 1.0 |
| MBT | 0.5 |
| TMT | 1.0 |
| Sulphur | 1.5 |
| Colour | As desired |
| Press cure: 20 min @160 °C | |

MBT, 2-mercaptobenzothiazole; TMT, tetramethylthiuram disulfide.

## 1.18 Rice rubber polisher brakes

Rice rubber polisher brakes are an integral part of modern rice milling industry and are used to polish large quantities of rice and to get good quality rice on regular basis. They are mostly rectangular-shaped long sticks, made of NR, polybutadiene rubber, and SBR, which add a shining appearance to rice grains that do not break easily. They improve the quality of polished rice and add to the effectiveness of rice mills. Varied quality of the rice rubber polishers are available based on their uses. A typical recipe of rice rubber polisher brake is shown in Table 1.18 [108].

**Table 1.18:** Rice polisher brake formulation.

| Ingredients | PHR |
|---|---|
| Natural rubber (RSS I) | 70.0 |
| Polybutadiene rubber | 30.0 |
| Zinc oxide | 5.0 |
| Stearic acid | 1.0 |
| Ppt. silica | 90.0 |
| Diethylene glycol | 1.0 |
| Coumarone indene (CI) resin | 5.0 |

**Table 1.18** (continued)

| Ingredients | PHR |
|---|---|
| Rubber process oil | 6.0 |
| Nonstaining antioxidant | 1.5 |
| Accelerator CZ | 1.5 |
| Tetramethyl thiuram disulfide (TMT) | 0.5 |
| Sulphur | 3.0 |
| Salicylic acid | 0.5 |
| Cured 25 min @ 150 °C | |
| Hardness IRHD | 75 |

Accelerator CZ, *N*-cyclohexylbenzothiazole-2-sulfenamide.

## 1.19 O-rings

O-ring is a rubber gasket of circular cross section placed between grooves of two surfaces to mechanically seal under compression to prevent leakage of gaseous and liquid substances in static as well as dynamic applications. Various elastomers such as polytetrafluoroethylene, neoprene, EPDM, fluorocarbon, and silicon rubbers are used to make O-rings depending on the requirements. As nitrile rubbers have outstanding oil and good chemical resistance, O-rings made of nitrile rubber are used in fuel tank, automotive, and aerospace applications.

O-rings made of EPDM rubber are suitable in automotive, households, and medical sectors as EPDM rubber is extremely resistant to chemicals and heat and weathering conditions. As fluorocarbon elastomers can withstand temperature extremes and are stable in contact with organic and inorganic chemicals, these elastomers are used to make O-rings for the aerospace industry.

O-rings made of neoprene rubber are used for air-conditioning and refrigerating applications owing to its very good weather-, good heat-, and abrasion-resistance. Silicon rubber can withstand high to low temperature and have very good resistance to ultraviolet rays and ozone gases. It is used to make O-rings for automotive and electronics applications.

Fluorocarbon elastomers are very stable in heat, flame, oil, and acidic environment and these elastomers are used to make O-rings for use in chemical and automotive applications. Inorganic bases commonly are used as acid acceptors in compounding of fluoroelastomers such as calcium oxide, calcium hydroxide, magnesium oxide, and zinc oxide, which affect both the rheological and the physical properties. The best balance of properties is achieved by using a combination of calcium hydroxide and magnesium oxide [109]. A formulation of O-ring compound is shown in Table 1.19 [107].

**Table 1.19:** General-purpose O-ring formulation.

| Ingredients | Phr |
|---|---|
| Medium ACN nitrile rubber | 100 |
| Zinc oxide | 5 |
| Stearic acid | 1.5 |
| C. I. resin | 5 |
| High abrasion carbon black | 30 |
| Ppt. calcium carbonate | 20 |
| Sulphur | 1.5 |
| TDQ | 1.5 |
| 6PPD | 0.5 |
| DOP | 10 |
| MBTS | 1 |
| TMTD | 0.5 |
| **Typical properties** | |
| Cured @ 153 °C | 6 min |
| Hardness, IRHD | 70 |
| Tensile strength (MPa) | 15 |
| 300% modulus (MPa) | 12 |
| Elongation at break (%) | 350 |
| Tear strength (kN/m) | 45 |

C. I. resin, coumarone indene resin; TDQ, trimethyl dihydroquinoline; 6PPD, phenyl paraphenylenediamine; DOP, dioctyl phthalate; MBTS, dibenzothiazole disulfide; TMTD, tTetramethylthiuram disulfide.

## 1.20 Curing bladder

Curing bladder is used to apply pressure from inside to inflate green tyres mounted on mould, to increase flow of rubber compounds to take final shape and designs engraved inside the tyre mould, and provide heat to initiate curing reaction in between rubber compounds and materials. The bladder is filled with recirculation steam, hot water, or inert gases as heat transferring medium. Once the tyre is cured, the pressure is bled down to open the mould. Thus, tyre curing bladder has one of the most severe applications in terms of heat and flexing resistance.

Butyl rubber is generally used in curing bladder application. As sulphur-vulcanised compounds soften at high temperature (148–204 °C) applications, resin is preferably used to formulate a heat-resistant compound. Table 1.20 depicts a tyre curing bladder formulation [100]. In order to achieve optimum performance of the bladder compound, dispersion of the ingredients is essentially required. There is a trend towards injection moulding of bladders to improve production efficiency and uniformity of the tyres and better flow properties are therefore essential.

**Table 1.20:** Tyre curing bladder formulation.

| Ingredients | Phr |
|---|---|
| Butyl rubber | 100.0 |
| Chloroprene rubber | 5.0 |
| N330- carbon black | 50.0 |
| Castor oil | 5.0 |
| Stearic acid | 1.0 |
| Zinc oxide | 5.0 |
| Octylphenol formaldehyde resin | 10.0 |
| Compound Mooney viscosity | |
| ML (1 + 4) @100 °C, MU | 76 |
| Rheometric properties by MDR (@190 °C for 60 min) | |
| Minimum torque (ML) | 2.5 |
| Maximum torque (MH) | 12.5 |
| $t_{s2}$ (min) | 3.2 |
| $t_{c90}$ (min) | 23.7 |
| Stress–strain properties (cured @ 190 °C, $t_{c90}$ + 2 min) | |
| Hardness (Shore A) | 59 |
| 100% modulus (MPa) | 1.5 |
| 300% modulus (MPa) | 4.2 |
| Tensile strength (MPa) | 13.9 |
| Elongation at break (%) | 640 |
| Aged stress–strain properties (@ 125 °C for 7 days) | |
| Shore A hardness | 86 |
| 100% modulus (MPa) | 4.3 |
| 300% modulus (MPa) | 10.0 |
| Tensile strength (MPa) | 11.7 |
| Elongation at break (%) | 410 |

N330- carbon black, high abrasion furnace black.

## 1.21 Ball bladders

Ball bladders are used as inner layers in footballs, basketballs, volleyballs, and so on, where retention of air inside the balls is the most important criterion. It may either be made up of natural or synthetic rubber. Butyl rubber is preferably used to make bladders due to its best air retention property and better contact quality with the outer linings. Table 1.21 shows the standard butyl rubber recipe for ball bladders [100].

**Table 1.21:** Ball bladder formulation.

| Ingredients | Phr |
|---|---|
| Butyl rubber | 100.0 |
| N550-carbon black | 70.0 |
| Paraffinic oil | 23.0 |
| Stearic acid | 1.0 |
| Aliphatic hydrocarbon resin | 3.0 |
| Zinc oxide | 5.0 |
| MBTS | 0.5 |
| ZBEC | 2.0 |
| Sulphur | 1.5 |
| Compound Mooney viscosity | |
| Mooney viscosity, ML (1 + 4) @100 °C | 45 |
| Rheometric properties by MDR @ 160 °C, 60 min | |
| Minimum torque (ML, dN M) | 1.7 |
| Maximum torque (MH, dN M) | 8.9 |
| $t_{s2}$ (min) | 3.2 |
| $t_{c90}$ (min) | 22.3 |
| Stress–strain properties (samples cured @ 190 °C $t_{c90}$ + 2 min) | |
| Hardness (Shore A) | 54 |
| 100% modulus (MPa) | 1.6 |
| 300% modulus (MPa) | 4.3 |
| Tensile strength (MPa) | 11.5 |
| Elongation at break (%) | 710 |

N550-carbon black, fast-extrusion furnace black; MBTS, dibenzothiazole disulfide ; ZBEC, zinc dibenzyl dithiocarbamate.

## 1.22 Rubber hoses

Rubber hose is a flexible hollow tube used to transfer fluids from one place to other and keeps the conveying medium separated from the environment. The use of rubber hose depends on the application requirements, and components of reinforcing layers, cover compounds, and so on are accordingly modified. A hose can have more than one specially formulated rubber compound to add desired properties; irrevocably, very high levels of adhesion between compounds are essential. A cross section of a typical rubber hose is illustrated in Figure 1.14.

Oil and aromatic substance resistance, resistance to fatigue, ozone, and temperature are the key requirements of hoses. Fuel hose, radiator hose, industrial hose, garden hose, automotive hose, and so on are based on applications. Using a suitable rubber hose makes convenient dispersal of fluids like water, oil, and chemicals. Most hoses are made of three elementary parts like tube, carcass, and cover compound [110].

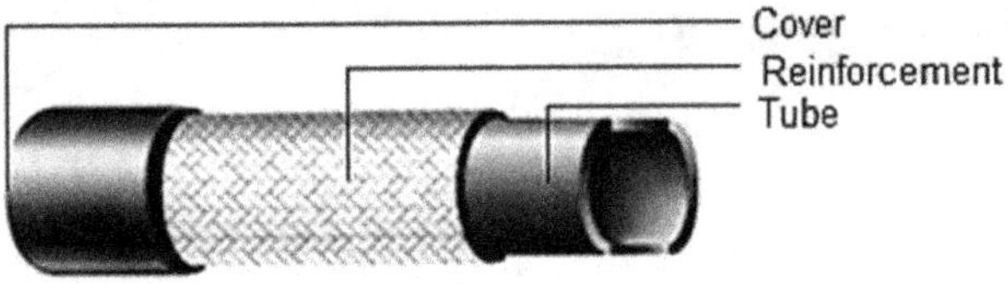


**Figure 1.14:** Cross section of a typical rubber hose.

The fluid contacts with rubber or plastic inner tube, which is reinforced with reinforcing carcass made of reinforcing textiles to withstand the associated high pressures and to enable the tube to be protected from internal pressure as well as outside force. The carcass is made of fabrics like cotton, rayon, polyester, nylon, aramid fibre and steel wire. The lining and outer layer are made from abrasion-, ozone-, and weather-resistant synthetic rubbers like natural, SBR, nitrile, butyl, and EPDM to protect the tube from external and environmental impacts [110–114].

To withstand the oil and grease resistance, the inner tube of automatic air brake rubber hoses is mostly made of nitrile, chloroprene, or blend of nitrile and SBR rubbers, whereas the cover compound is made with CR, SBR or CSM rubbers. Synthetic textiles are used as reinforcement parts. The inner tube of automatic fuel hoses is produced with NBR or sometimes, blend of NBR and PVC to get oil-resistant properties and the cover compound is made with CR or CSM rubbers. EPDM rubber is typically used to make inner tubes in radiator coolant hoses whereas CR, NBR, and sometimes SBR rubbers are used to make the top cover compound. Other uses of hoses include transferring liquid petroleum gases, air, oxygen, acetylene gases, and water.

Temperature and pressure rating are important performance criteria of hydraulic hoses as selection of hose against specifications may lessen the service life of the hose. Braided steel-wire hoses are manufactured based on their application to withstand pressure of liquid transport and layers of reinforcing fabric and steel wires are used as required. Few braided hoses are found to be very rigid while others are highly flexible. These are used to arrest elongation on air, gases, and liquids transportation under pressure extremes. A formulation of hydraulic hose tube and cover is shown in Table 1.22 and air-conditioning hose in Table 1.23 [72].

**Table 1.22:** Formulations of a typical hydraulic hose tube and cover compound.

| Ingredients | Hydraulic hose tube | Hydraulic hose cover |
|---|---|---|
| Acrylonitrile butadiene rubber (medium nitrile content) | 100 | – |
| Chloroprene rubber (W-grade) | – | 100 |
| Magnesium oxide | – | 4.0 |
| Zinc oxide | 5.0 | 5.0 |
| Stearic acid | 1.5 | 0.5 |
| 6PPD | 0.75 | 0.75 |
| TMQ | 1.75 | 1.0 |
| N550-carbon black | 85 | 45 |
| Precipitated silica | 20 | 20 |
| Calcium carbonate | – | 40 |
| Aromatic oil | – | 12 |
| DOP | 6.0 | – |
| Ethylene thiourea | – | 0.6 |
| MBTS | 1.1 | 0.25 |
| TMT | 1.1 | – |
| Sulphur | 1.1 | – |
| Rheometric properties by ODR at 150 °C for 60 min | | |
| $t_{s2}$ (min) | 2.3 | 3.1 |
| $t_{c90}$ (min) | 3.9 | 22.2 |
| Max. torque (dN m) | 16.5 | 6.3 |
| Stress–strain properties (cured @ 150 °C for $t_{c90}$ + 2 min) | | |
| Tensile strength (MPa) | 18.6 | 16.0 |
| Elongation at break (%) | 250 | 370 |
| 300% modulus (MPa) | – | 12.5 |
| Hardness, IRHD | 86 | 73 |
| Tear strength (kN/m) | 66 | 46 |

6PPD, *N*-(1,3-dimethylbutyl)-*N'*-phenyl-*p*-phenylenediamine-quinone; TMQ, polymerised 2,2,4-trimethyl-1,2-dihydroquinoline; N550-Carbon Black, fast extrusion furnace black; DOP, dDioctyl phthalate; MBTS, dibenzothiazole disulfide; TMT, tetramethyl thiuram disulfide.

**Table 1.23:** Air-conditioning hose compound.

| Ingredients | phr |
|---|---|
| Chlorobutyl rubber | 100.0 |
| N774-carbon black | 75.0 |
| Stearic acid | 1.0 |
| Zinc oxide | 5.0 |
| Magnesium oxide | 3.0 |
| Paraffin wax | 1.5 |
| Paraffinic oil | 10.0 |
| MBTS | 1.0 |

**Table 1.23** (continued)

| Ingredients | phr |
|---|---|
| TMTMS | 0.5 |
| Vultac 5 | 0.8 |
| Rheometric properties at 160 °C | |
| ML (dN m) | 2.6 |
| MH (dN m) | 20.9 |
| $t_{s2}$ (min) | 1.5 |
| $t_{c90}$ (min) | 3.6 |
| Stress–strain properties @ 160 °C for 2 × $t_{c90}$ min | |
| Hardness (Shore A) | 72 |
| 100% modulus (MPa) | 6.3 |
| Tensile strength (MPa) | 12.7 |
| Elongation at break (%) | 250 |

N774 carbon black, semi-reinforcing furnace black; MBTS, dibenzothiazole disulfide; TMTMS, tetramethylthiuram monosulfide; Vultac 5, alkylphenol polysulfides.

# References


[1] Sisanth, K. S., Thomas, M. G., Abraham, J., Thomas, S. General introduction to rubber compounding. In: Maria, H. J., Thomas, S. (eds) Progress in Rubber Nanocomposites, Netherlands: Elsevier Science 1–39 (2017). ISBN-*9780081004289*.

[2] Bhatia, S. C., Goel, A. Rubber Technology: Two Volume Set (Vol-1). India: Woodhead Publishing India PVT. Limited (2019).

[3] Rodgers, B., Waddell, W. The Science of Rubber Compounding, Science and Technology of Rubber, 3rd edition. Netherlands: Elsevier Academic Press 401–454 (2005).

[4] Wiedmann, W. M., Schmid, H. M. Optimization of rubber mixing in internal mixers. Rubber Chemistry and Technology 55(2):363–381 (1982).

[5] Funt, J. M. Rubber mixing. Rubber Chemistry and Technology 53(3):772–779 (1980).

[6] Visakh, P. M., Thomas, S., Chandra, A. K., Mathew, P. A. Advances in Elastomers I: Blends and Interpenetrating Networks. Germany: Springer Berlin Heidelberg, (2013).

[7] http://tirenews4u.wordpress.com

[8] Peterson, A. Rubber Mixing Technology Course. Milwaukee: Center for Continuing Engineering Education, Univ. Of Wisconsin (1999).

[9] Sommer, J. G. Molding of rubber for high performance applications. Rubber Chemistry and Technology 58:672 (1985).

[10] Wheelans, M. A. Injection Moulding of Rubber. London: Hastead Press (1974).

[11] Hofmann, W. Rubber Technology Handbook. Munich: Carl Hanser Publishers (1989).

[12] Christy, R. L. Rubber World 180:100 (1979).

[13] Lambright, J., Long, H. (ed). Basic Compounding and Processing of Rubber. Akron: ACS Rubber Division (1985).

[14] Takashi Moribe. Advanced intermeshing mixers for energy saving and reduction of environmental impact. Mitsubishi Heavy Industries Technical Review 49(4) 38–43 (December 2012).
[15] Wood, P. R. Rubber Mixing. Shawbery: Rapra Technology Ltd (1996).
[16] Grossman, R. F. The Mixing of Rubber. Netherlands: Springer (1997).
[17] Borzenski, F. J., Hartley, Q. Rubber World. United States: Lippincott and Peto Inc. (Jul 1 2013).
[18] Drobny, J. G. Fluoroelastomers Handbook, 2nd edition. (2016).
[19] Pike, M., Watson, W. F. Mastication of rubber, I. Mechanism of plasticizing by cold mastication. Journal of Polymer Science 9:229–251 (1952).
[20] Fries, H., Pandit, R. R. Mastication of rubber. Rubber Chemistry and Technology 55(2):309–327 (1 May 1982).
[21] Busse, W. F. Mastication of rubber an oxidation process. Industrial and Engineering Chemistry Research 24(2):140–146 (1932).
[22] Mooney, M. 0. A shearing disk plastometer for unvulcanized rubber. Rubber Chemistry and Technology 7(3):564–575 (1934).
[23] ASTM D1646-15. Standard Test Methods for Rubber – Viscosity, Stress Relaxation, and Pre-Vulcanization Characteristics (Mooney Viscometer). West Conshohocken, PA: ASTM International (2015).
[24] ASTM D6204-15. Standard Test Method for Rubber – Measurement of Unvulcanized Rheological Properties Using Rotorless Shear Rheometers. West Conshohocken, PA: ASTM International (2015).
[25] Kim, C., Morel, M., Beuve, J., Guilbert, S., Bonfils, F. Better characterization of raw natural rubber by decreasing the rotor speed of Mooney viscometer: Role of macromolecular structure. Polymer Engineering and Science 50(2):240–248 (2009).
[26] Pawlowski, H., Dick, J. Viscoelastic Characterization of Rubber with a New Dynamic Mechanical Tester. United States: Akron Rubber Group (1992).
[27] Gwin, L. E., Weaver, E. J. A new dimension in rubber compound tackifiers. Journal of Elastomers and Plastics 9(3):289–298 (July 1977). doi: 10.1177/009524437700900304.
[28] Smitthipong, W., Nardin, M., Schultz, J., Nipithakul, T., Suchiva, K. Study of tack properties of uncrosslinked natural rubber. Journal of Adhesion Science and Technology 18(12):1449–1463 (January 2004).
[29] Kuzminskiĭ, A. S., Feldshteĭn, L. S., Reĭtlinger, S. A. The Blooming of sulfur and other ingredients from compounded stocks. Rubber Chemistry and Technology 35(1):147–152 (1962).
[30] Drobny, J. G. Handbook of Thermoplastic Elastomers, 2nd edition. United Kingdom: Elsevier Science (2014).
[31] Limper, A., Schramm, D. Process description for the extrusion of rubber compounds–Development and evaluation of a screw design software. Macromolecular Materials and Engineering 287:824–835 (December 2002).
[32] Findik, F., Yilmaz, R., Köksal, T. Investigation of mechanical and physical properties of several industrial rubbers. Materials and Design 25(4):269–276 (2004).
[33] Arguello, J. M., Santos, A. Hardness and compression resistance of natural rubber and synthetic rubber mixtures. Journal of Physics: Conference Series 687:012088 (2016).
[34] Galimberti, M., Agnelli, S., Cipolletti, V. 11 – Hybrid filler systems in rubber nanocomposites. In: Thomas, S., Maria, H. J. (eds) Woodhead Publishing Series in Composites Science and Engineering, Progress in Rubber Nanocomposites. Netherlands: Woodhead Publishing, 349–414 (2017).
[35] Wisojodharmo, L. A., Fidyaningsih, R., Fitriani, D. A., Arti, D. K., Indriasari,, Susanto, H. The influence of natural rubber – Butadiene rubber and carbon black type on the mechanical properties of tread compound. Materials Science and Engineering 223:012013 (2017).
[36] Duh, Y. S., Ho, T. C., Chen, J. R., Kao, C. S. Study on exothermic oxidation of acrylonitrile-butadiene-styrene (ABS) resin powder with application to ABS processing safety. Polymers 2:174–187 (2010).

[37] Aguele, F. O., Idiaghe, J. A., Apugo-Nwosu, T. U. A study of quality improvement of natural rubber products by drying methods. Journal of Materials Science and Chemical Engineering 3:7–12 (2015).
[38] Tanaka, Y., Shûkambara, J. N. The oil resistance of rubber. I Study of the swelling of vulcanized rubber. Rubber Chemistry and Technology 9(1):70–73 (19361 March).
[39] Bukhina, K. Low-Temperature Behaviour of Elastomers, 1st edition. Netherlands: Taylor & Francis (2007).
[40] Sexsmith, F. H., Polaski, E. L. Mechanisms of adhesion in elastomer-to-textile bonding. In: Lee, L. H. (ed) Adhesion Science and Technology. Polymer Science and Technology. Boston: Springer, Vol. 9 259–279 (1975).
[41] Solomon, T. S. Bonding textiles to rubber. In: Skeist, I. (ed) Handbook of Adhesives. Boston, MA: Springer 583–597 (1990).
[42] Van Amerongen, G. J. The effect of fillers on the permeability of rubber to gases. Rubber Chemistry and Technology 28(3):821–832 (1 September 1955).
[43] McKeen, L. W. 1 – Introduction to plastics and polymers. In: McKeen, L. W. (ed) Plastics Design Library, Film Properties of Plastics and Elastomers, 4th edition. United Kingdom: William Andrew Publishing, 1–24 (2017).
[44] Gupta Verlag, D. Materials and compounds – Chapter 2. In: Deartment of Polymer Science. Ohio 44325-3909, USA: The University of Akron.
[45] Rodgers, B., Halasa, A. Compounding and processing of rubber/rubber blends. In: Isayev, A. I. (ed) Encyclopedia of Polymer Blends. Germany: Wiley, 163–206 (2011).
[46] Ronald, J. S. Chapter 33: Mechanical properties of rubber. In: Piersol, A. G. (ed) Harris' Shock and Vibration Handbook, 5th edition. New York, NY, USA: The McGraw-Hill Companies, Inc. 18 (2002).
[47] SAE J200. Classification System for Rubber Materials. Warrendale, PA: Society of Automotive Engineers (2000).
[48] ASTM D471-79. Standard test Method for Rubber Property – Effect of Liquids. ASTM D5964- 96, Standard Practice for Rubber IRM 902 and Replacement Oils for ASTM No. 2 and ASTM No. 3 Oils (1979).
[49] Das Gupta, S., Mukhopadhyay, R., Baranwal, K. C., Bhowmick, A. K. Reverse Engineering of Rubber Products: Concepts, Tools, and Techniques. United Kingdom: CRC Press, 241 (2014).
[50] Flory, P. J. Principles of Polymer Chemistry, 1st edition. United Kingdom: Cornell University (1953).
[51] Oae, A. Organic Chemistry of Sulfur. New York: Plenum Press (1977).
[52] Krejsa, M. R., Koenig, J. L. A review of sulfur crosslinking fundamentals for accelerated and unaccelerated vulcanization. Rubber Chemistry and Technology 66(3):376–410 (1993).
[53] Joseph, A. M., George, B., Madhusoosaban, K. N., Alex, R. Current status of sulphur vulcanization and devulcanization chemistry: Process of vulcanization. Rubber Science 28(1):82–121 (2015).
[54] Guo, R., Talma, A. G., Datta, R. N., Dierkes, W. K., Noordermeer, J. W. M. Solubility study of curatives in various rubbers. European Polymer Journal 44:3890 (2008).
[55] Guillaumond, F. X. The influence of the solubility of accelerators on the vulcanization of elastomer blends. Rubber Chemistry and Technology 49:105 (1976).
[56] Kruželák, J., Sýkora, R., Hudec, I. Vulcanization of rubber compounds with peroxide curing systems. Rubber Chemistry and Technology 90(1):60–88 (2017).
[57] Smejda-Krzewicka, A., Olejnik, A., Strzelec, K. The effect of metal oxide on the cure, morphology, thermal and mechanical characteristics of chloroprene and butadiene rubber blends. Polymer Bulletin 77:4131–4146 (2020).
[58] Kruželák, J., Hudec, I. Vulcanization systems for rubber compounds based on IIR and halogenated IIR: An overview. Rubber Chemistry and Technology 91(1):167–183 (1 January 2018).
[59] George Drobny, J. 11 – Fluoroelastomer applications. In: Drobny, J. G. (ed) Plastics Design Library, Fluoroelastomers Handbook, 2nd edition. United States: William Andrew Publishing, 435–438 (2016).

[60] Xiao, D., Chen, D., Zhou, Z., Zhong, A. Three-group type mechanism in the curing behavior of polyacrylate and blocked toluene diisocyanate. Journal of Applied Polymer Science 83:112–120 (2002).
[61] Chakraborty, S. K., De, S. K. Epoxy-resin-cured carboxylated nitrile rubber. Journal of Applied Polymer Science 27:4561–4576 (1982).
[62] Ghosh, P. Chapter 9: Rubbers-materials and processing technology. In: Polymer Science and Technology: Plastics, Rubbers, Blends and Composites, United States: McGraw-Hill Education LLC 429 (2001).
[63] Rothon, R. Particulate fillers in elastomers. In: Fillers for Polymer Applications, Germany: Springer International Publishing 125–146 (March 2017).
[64] Donnet, J.-B., Custodero, E. 8 – Reinforcement of elastomers by particulate fillers. In: Mark, J. E., Burak, E., Eirich, F. R. (eds) Science and Technology of Rubber, 3rd edition. Germany: Elsevier Science Academic Press, 367–400 (2005).
[65] Rothon, R. Particulate Fillers in Elastomers. In: Rothon, R. (ed) Fillers for Polymer Applications. Polymers and Polymeric Composites: A Reference Series. Cham: Springer 125–146 (2017).
[66] Pegoretti, A., Dorigato, A. Polymer composites: Reinforcing fillers. In: Mark, H. F., Seidel, A. (ed) Encyclopedia of Polymer Science and Technology. United Kingdom: Wiley 1–72 (2021).
[67] Roland, C. M. Reinforcement of Elastomers. Netherlands: Elsevier Inc (2016).
[68] Shao-Yun, F., Feng, X.-Q., Lauke, B., Mai, Y.-W. Effects of particle size, particle/matrix interface adhesion and particle loading on mechanical properties of particulate–polymer composites. Composites Part B: Engineering 39(6):933–961 (2008).
[69] DeArmitt, C., Rothon, R. Particulate fillers, selection, and use in polymer composites. In: Rothon, R. (ed) Fillers for Polymer Applications. Polymers and Polymeric Composites: A Reference Series. Cham: Springer 3–27 (2017).
[70] Internet page. www.rtvanderbilt.com/nonblackfillers.pdf, (July 15, 2012).
[71] Qi, H. J., Joyce, K., Boyce, M. C. Durometer hardness and the stress-strain behavior of elastomeric materials. Rubber Chemistry and Technology 76(2):419–435 (2003).
[72] Internet file. http://www.nocil.com/Downloadfile/CCompoundingFormulations&UsefulInfo-Dec2010.pdf; Starting Point Rubber Compounding Formulations, 28.
[73] Lloyd, D. G. Additives in rubber processing. Materials and Design 12(3):139–146 (1991).
[74] Leblanc, J. L. Assessing rubber processing aids effectiveness. In: Kleintjens, L. A., Lemstra, P. J. (eds) Integration of Fundamental Polymer Science and Technology. Dordrecht: Springer 394–397 (1986).
[75] Datta, R. N., Huntink, N. M., Datta, S., Talma, A. G. Rubber vulcanizates degradation and stabilization. Rubber Chemistry and Technology 80(3):436–480 (2007).
[76] Sharj-Sharifi, M., Taghvaei-Ganjali, S., Motiee, F. The effect of protecting waxes on staining antidegradant performance in tyre sidewall formulation. Journal of Rubber Research 23:111–124 (2020).
[77] Ferradino, A. G. Antioxidant selection for peroxide cure elastomer applications. Rubber Chemistry and Technology 76(3):694–718 (2003).
[78] Sulekha, P. B., Joseph, R., Prathapan, S. Synthesis and characterization of chlorinated paraffin wax-bound paraphenylenediamine antioxidant and its application in natural rubber. Journal of Applied Polymer Science 81: 2183–2189 (2001). 2. D.
[79] RaaKhimi, S., Pickering, K. L. A new method to predict optimum cure time of rubber compound, a new method to predict optimum cure time of rubber compound. Journal of Applied Polymer Science (2014. doi: 10.1002/APP.40008.
[80] Nai, Y. Function of factice (rubber substitute) in rubber compounds. Nippon GomuKyokaishi 89(3):80–84 (2016).

[81] Bobrov, Y. A., Kandyrin, K. L., Shmurak, I. L., Potapov, E. E. Bonding of rubbers to metal cord using compounds of variable-valency metals. International Polymer Science and Technology 32(11) 18–26 (2005).
[82] Mark, J. E., BurakErman, M. R. The Science and Technology of Rubber Netherlands: Elsevier Science (2013).
[83] Internet page. https://madhavuniversity.edu.in/automobile-tire.html
[84] Internet page. https://www.tiremarket.com/tiremantra/radial-vs-bias-play-comparison-revealed
[85] Internet page. https://www.ustires.org/whats-tire-0
[86] McDowell, W., Reynolds, C. Considerations in Truck Tire Retreading. SAE Technical Paper 650061 (1965).
[87] Sheridan, M. F., Vanderbilt, R. T. Inc (eds). The Vanderbilt Rubber Handbook, 4th edition. Norwalk, CT : R. T. Vanderbilt Company (2010).
[88] Schulze, T., Bolz, G., Strübel, C., et al. Tire technology in target conflict of rolling resistance and wet grip. ATZ Worldwide 112:26–32 (2010).
[89] Goerl, U., Hunsche, A., Mueller, A., Koban, H. G. Investigations into the silica/silane reaction system. Rubber Chemistry and Technology 70(4):608–623 (1 September 1997).
[90] Thaptong, P., Sae-oui, P., Sirisinha, C. Influences of sstyrene butadiene rubber and silica types on performance of passenger car radial tire tread. Rubber Chemistry and Technology 90(4):699–713 (1 December 2017).
[91] Dhanorkar, R. J., Mohanty, S., Kumar Gupta, V. Synthesis of functionalized stirene butadiene rubber and its applications in SBR–silica composites for high performance tire applications. Industrial and Engineering Chemistry Research 60(12):4517–4535 (2021).
[92] Thaptong, P., Sae-oui, P., Sirisinha, C. Influences of styrene butadiene rubber and silica types on performance of passenger car radial tire tread. Rubber Chemistry and Technology 90(4):699–713 (2017).
[93] Heinrich, G., Dresden, Vilgis, T. A., Mainz. Why Silica technology needs S-SBR in high performance tires? KGK. Kautschuk Gummi Kunststoffe, 368–276 (July/August 2008).
[94] Hardy, D., Albino, F., Steinhauser, N., Douglas, J., Gross, T.; Innovations in Solution SBR Technology, Lanxess Customer Seminar, Chengdu (June 11–15th, 2012).
[95] Ghosh, S., Bhattacharyya, S., Bandyopadhyay, S., Dasgupta, S., Mukhopadhyay, R. Development of a passenger-car radial-tyre sidewall compound by the reactive processing of an NR/BR/EPDM blend in a banbury mixer. Progress in Rubber, Plastics and Recycling Technology 28(1):15–26 (2012).
[96] Datta, R., Huntink, N., Made, M., Pierik, B. Improved hysteresis and adhesion to steel cord by using chemically activated aramid fiber. KGK Kautschuk Gummi Kunststoffe 61:580–583 (2008).
[97] Mohanty, T. R., Chandra, A. K., Bhandari, V., Chattopadhyay, S. Steel cord skim compound for radial tire based on natural rubber-carbon black-organoclay nanocomposites. Journal of Materials Science and Engineering 6 (2016).
[98] Stephen Fulton, W. Steel tire cord-rubber adhesion, including the contribution of cobalt. Rubber Chemistry and Technology 78(3): 426–457 (1 July 2005). doi: https://doi.org/10.5254/1.3547891
[99] Buytaert, G., Luo, Y. Study of Cu–Zn–Co ternary alloy-coated steel cord in cobalt-free skim compound. Journal of Adhesion Science and Technology 28 (2014). doi: 10.1080/01694243.2014.903567.
[100] Internet page. Exxon™ butyl rubber compounding and applications manual.
[101] Internet page. Exxon™ butyl rubber innertube technology manual.
[102] Hermenegildo, G., Bischoff, E., Mauler, R., Giovanela, M., Carli, L., Crespo, J. Development of chlorobutyl rubber/natural rubber nanocomposites with montmorillonite for use in the inner liner of tubeless ride tires. Journal of Elastomers and Plastics 49(1):47–61 (2017).
[103] Internet page. Exxon™ bromobutyl rubber compounding and applications manual.

[104] Lucas, J., Thabet, W., Worlikar, P. Using virtual reality (VR) to improve conveyor belt safety in surface mining, CIB W078 24th International Conference on Information Technology for Construction and 5th ITCEDU Workshop and 14th EG-ICE Workshop (2007).
[105] Internet page. http://www.meroller.com/index.php?m=Product&a=show&id=92&l=en
[106] Internet page. https://www.mining-technology.com/contractors/materials/phoenix/
[107] Ciullo, P. A., Hewitt, N. Plastics design library. In: Hewitt, N. (ed.) The Rubber Formulary. William Andrew Publishing Ukraine: Elsevier Science 74–772 (1999).
[108] Internet page. https://www.relflex.in/web-formulary.html
[109] Internet page. https://multimedia.3m.com/mws/media/1408677O/3m-dyneon-fluoroelastomers-compounding-guide.pdf; 3M™ Dyneon™ Fluoroelastomers, Compounding Guide.
[110] Internet page. http://www.arctechhydraulics.co.za/hydraulic-hose-construction/
[111] Mills, D. Chapter 8 – Pipelines and valves. In: Mills, D. (ed) Pneumatic Conveying Design Guide, 3rd edition. Netherlands: Butterworth-Heinemann, 183–195 (2016).
[112] Simmons, D. High performance environments: The case for polymeric flexible hoses. Polymers and Polymer Composites 5(8):563–571 (1997).
[113] Majumdar, S. R., Oil Hydraulic Systems: Principles and Maintenance. New Delhi: Tata Mc Graw-Hill. Thirteenth reprinting (2006).
[114] Pazur, R. J., Kennedy, T. A. C. Effect of plasticizer extraction by jet fuel on a nitrile hose compound. Rubber Chemistry and Technology 68(2):324–342 (2015).
[115] Mazumder, A., Chanda, J., Bhattacharyya, S., Dasgupta, S., Mukhopadhyay, R., Bhowmick, A. K. Journal of Applied Polymer Science 138(42):e51236 (2021). https://doi.org/10.1002/app.51236

Soumen Chakraborty

# Chapter 2
# Carbon black morphology and its application in elastomer and polymer matrix

## 2.1 Introduction

Carbon black is a collective term for a group of small particles made up of carbon nuclei that are mostly shapeless or have a static nature. These particles join together to create clusters of different sizes and shapes. Carbon black is generated through the heat-induced breakdown of poly-aromatic hydrocarbons in an oxygen-restricted setting, resulting in sub-stoichiometric amounts of nuclei. Industrial production yields numerous specific commercial grades, each differing in primary particle size, structure, aggregate size and shape, porosity, surface area, and chemical composition.

Carbon black finds widespread use as reinforcing fillers in rubber products. The reinforcement effect depends on the interaction between rubber molecules and the carbon black particles themselves. For elastomer reinforcement, the size of the primary particles (measured by specific BET surface area) and the surface activity of the carbon black types are crucial, along with their structure. Additionally, the achieved dispersion of carbon black and the amount used in the rubber composite significantly affect the outcome. The specific grade of carbon black chosen depends on various factors, including size and morphology, tailored to meet the specific requirements of different rubber goods and tyre components for various applications [1].

Although carbon black is expected to continue dominating the rubber market in the foreseeable future, they face stiff competition from alternative reinforcing fillers such as precipitated silica and regenerated silica. Silica has been reported to offer advantages in terms of lower rolling resistance and reduced heat buildup, contributing to improved fuel efficiency and lower emissions. This trend of silica's growing popularity is on the rise. However, carbon black remains superior in terms of wear resistance, especially in harsh climates. The emergence of low PAH (polyaromatic hydrocarbon) and low hysteresis, coupled with low rolling resistance, as per ASTM standards, closely aligns with the goals of improved fuel efficiency and reduced emissions [2].

Specifically formulated carbon black grades are developed for use as pigments in coatings, paints, inks, and plastic masterbatches. In some cases, they are utilised to provide conductivity in polymeric compounds for electrostatic dissipation and increased conductivity. These grades also act as UV stabilisers in polymeric compounds and serve as alternatives to harmful organic dyes [3].

https://doi.org/10.1515/9783110668537-002

## 2.2 History of carbon black

Carbon black, known as lamp black, has a long history as one of the earliest manufactured materials used as a pigment and reinforcement. Its early applications can be traced back to ancient China, where it was utilised in ink and cosmetics, and the ancient Egyptians used it for writing and painting, and in cosmetics as well. The demand for carbon black began to rise with the invention of the bicycle and the automobile in the nineteenth century, as well as by the invention of printing in the fifteenth century.

In the early nineteenth century, the discovery that carbon black could reinforce natural rubber and significantly enhance the durability of tyres propelled the material into the modern era. The first industrial application of carbon black took place in 1860. Today, carbon black is present in various aspects of modern life. It is used in inkjet printer ink as a reinforcing agent for natural and synthetic rubber, as an active component in electrically conductive plastics, and as a pigment and tinting aid in paints, coatings, newspaper inks, and cosmetics, among other applications. Carbon black has become ubiquitous in our daily lives.

## 2.3 Is carbon black soot?

Carbon black is often associated with soot, which is a commonly used term for any carbon-based substance or by-product produced through incomplete combustion in a low-oxygen environment. However, carbon black is distinct from soot in terms of its morphology, chemical composition, and physical characteristics. Carbon black is primarily composed of 97% elemental carbon, arranged in grape-like structures known as acini particles. It has a partially crystalline nature dispersed within the amorphous carbon phase.

In contrast, soot consists of 60% of the total particles responsible for black carbon. The shape, size, morphology, and heterogeneity of soot particles depend on the specific source from which it originates. Both soot and carbon black contain trace amounts of PAHs, which can be quantitatively identified through soxhlet distillation using specialised apparatus. However, the quantity of PAH present can vary, depending on the source of manufacture.

The morphological structure of carbon black and soot differs significantly, leading to variations in the properties exhibited when applied as reinforcements in rubber products. Each substance is produced through distinct processes, separate from those used for commercial carbon black production, resulting in unique physical and chemical properties. A detailed comparison between carbon black and soot is provided in Table 2.1.

**Table 2.1:** Carbon black versus soot.

| | Carbon black | Soot |
|---|---|---|
| 1 | Carbon black is manufactured under controlled conditions for commercial use, primarily in the rubber, plastic, and printing industries. | Soot is unwanted byproduct from the combustion of carbon-based materials for the generation of energy or heat, or for the disposal of waste. |
| 2 | Carbon black contains >90% elemental carbon, arranged as aciniform particulate. | Depending on the type of soot, the relative amount of carbon (<60% of the total particle mass), the type of particulate carbon, and particle characteristics (size, shape, and heterogeneity) can vary considerably. |
| 3 | Total inorganics (ash) represent <1% of the carbon black particle mass. Organic compounds can be extracted from particle surfaces (solvent extractable fraction), and for carbon black, total inorganics are <1% of the particle mass. | Soots have much greater percentages of ash, solvent extractable fraction, or both, than carbon black.<br>(solvent extractable fraction) |

According to current understanding, carbon black formation involves the presence of certain particles that possess a fullerene structure, specifically C60, which plays a crucial role in reinforcement. The formation of fullerene occurs through the synchronisation of PAH-containing dimers, resulting in the creation of 3D elemental 'C' and ultimately leading to the formation of C60 fullerene. Therefore, when these dimmers aggregate, they transform into soot [4]. Hence, it is important to recognise that carbon black and soot are distinct from each other, and when discussing potential health effects, it is essential to differentiate between these two types of carbon-based particles.

## 2.4 What is carbon black?

Carbon black is a commonly used term for a type of pure inorganic chemical that belongs to group IV in the periodic table. It is composed of small-sized carbon particles that are mostly para-crystalline (turbostatic) in nature. These particles come together to form aggregates of varying sizes and shapes, as shown in Figure 2.1.

It pertains to a category of industrial particles produced through channel, thermal, furnace, and plasma black processes. These materials are characterised by near-spherical particles of colloidal size that coalesce and form oligomers of various sizes, initially creating aggregates and then agglomerates [5].

Carbon black is primarily obtained through the furnace process, which involves the partial combustion or thermal decomposition of hydrocarbons in the gas phase, in the presence of oxygen in specific proportions. The Carbon black evaluation cycle is depicted in Figure 2.2.

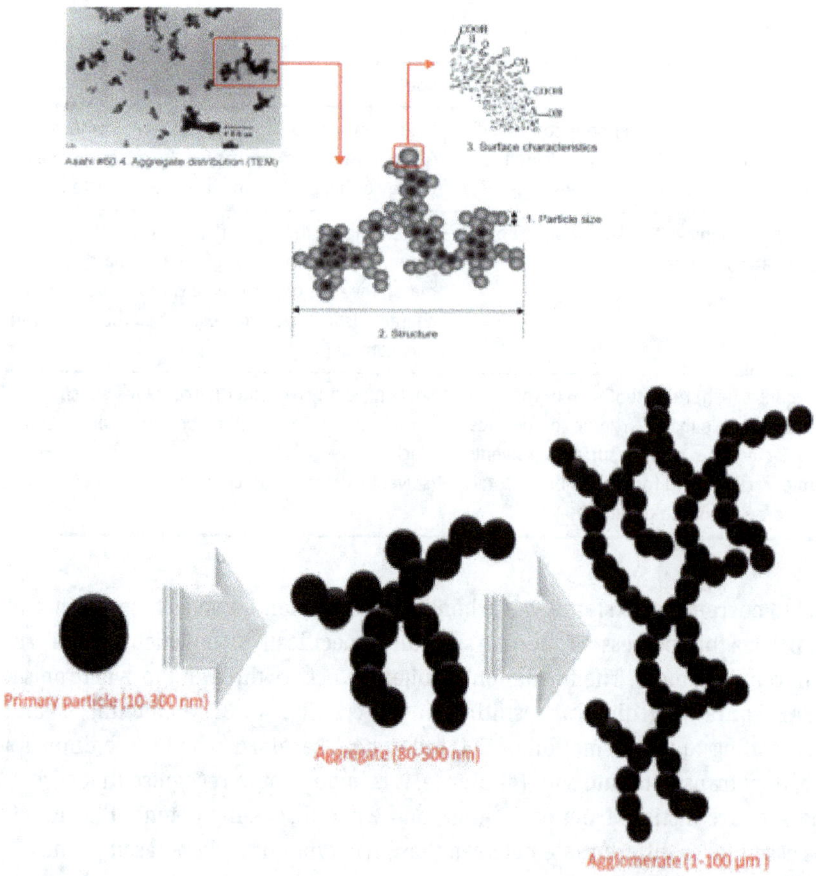


**Figure 2.1:** Carbon black aggregate, agglomerate, particle size, structure and surface characteristics.

Presently carbon black has more than 100 commercial grades, with differences in primary particle size, shape, primary and secondary structure, porosity, and surface chemistry.

### 2.4.1 Difference between carbon black and other fillers

Carbon black distinguishes itself from other reinforcing fillers such as silica, and semi-reinforcing china clay and calcined china clay. The reinforcing effect of carbon black primarily arises from its aggregate shape, primary particle size, and surface chemistry. Silica, on the other hand, achieves reinforcement through the presence of –OH groups and their specific arrangements within the surface structure. The reinforcement mechanism of semi-reinforcing clay primarily relies on the orientation of $Al_2O_3$ and $SiO_2$ within its multilayer structure [6].

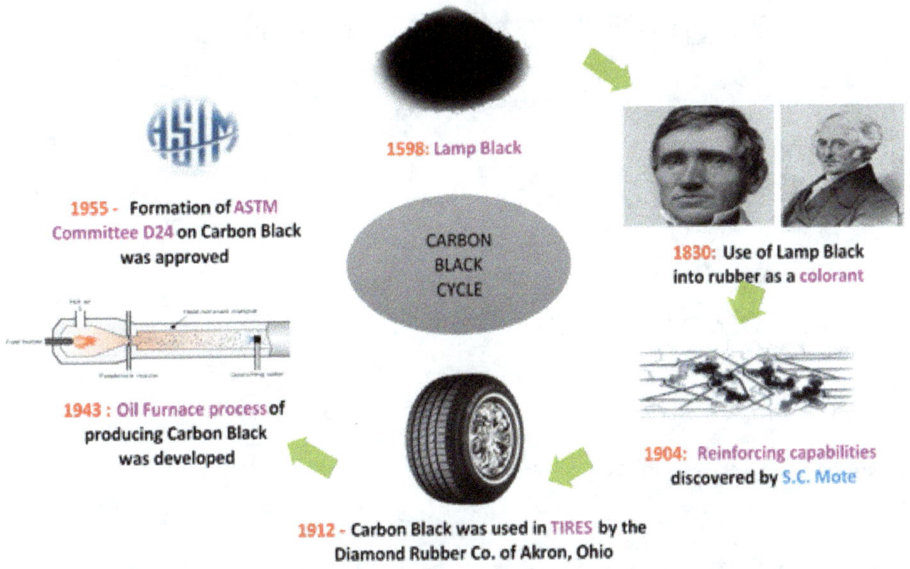


**Figure 2.2:** Carbon black evaluation cycle.

When incorporated into elastomers, carbon black generally imparts superior abrasion resistance and mechanical properties, compared to other reinforcing and semi-reinforcing fillers. Figures 2.3 and 2.4 illustrate the variations in surface groups between carbon black and silica.

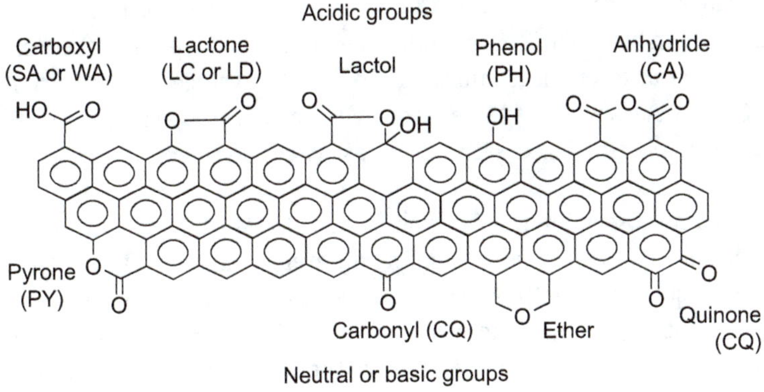


**Figure 2.3:** Surface group of carbon black.

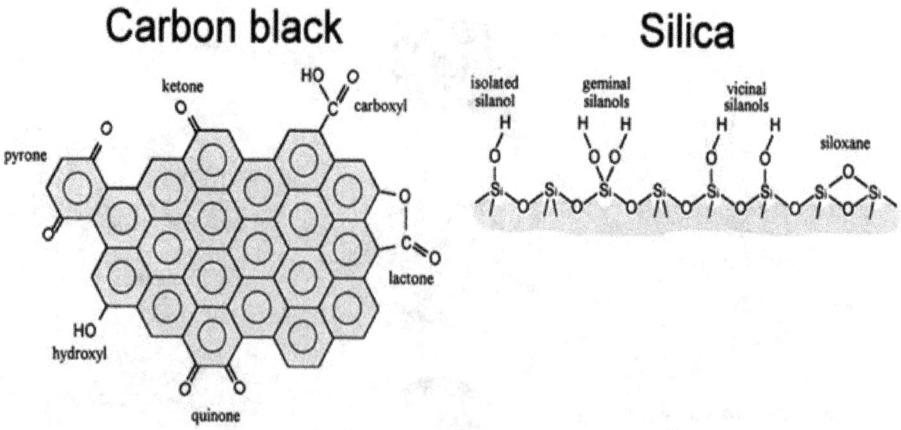


**Figure 2.4:** Surface characteristics carbon black and silica.

## 2.4.2 Carbon black nomenclature

The classification of different types of carbon black has evolved over many years, with each manufacturer assigning their own names. In 1967, the ASTM D1765 proposed a system for carbon black nomenclature, which has since been adopted worldwide. The International Standards Organization (ISO) also adopted this system in 1980. This ASTM system utilises a one-letter code, followed by three numerals.

For rubber applications, carbon black is typically identified by a four-character code, starting with either 'N' or 'S', such as NXXX or SXXX. The first character provides an indication of the carbon black's influence on the rate of cure in a typical rubber formulation. The second character provides information about the average surface area of the carbon black. Carbon black with the same second character is grouped into a series ending in '00', for example, the N200 series. The last two characters are assigned arbitrarily.

In the past, the iodine absorption number (iodine no.) (ASTM D1510, ISO 1304) was the primary measure of surface area used to define different grades of carbon black. However, the nitrogen surface area (NSA, ASTM D6556) and statistical thickness surface area (STSA, ASTM D6556) are now more commonly used for this purpose. The *N*-di-butyl phthalate absorption, which is now referred to as oil absorption number (OAN) (ASTM D2414), has traditionally served as the primary indicator of structure [7]. Table 2.2 describes the properties of carbon black, according to ASTM D1765-16.

**Table 2.2:** Carbon black properties (ASTM D1765 – 16).

| ASTM classification | Iodine adsorption no. D1510 (g/kg) | N2SA multipoint D6556, $10^3$ m$^2$/kg (m$^2$/kg) | STSA D6556 $10^3$ m$^2$/g (m$^2$/kg) | Oil absorption no. D2414, $10^{-5}$ m$^3$/g | Oil absorption no. compressed sample, D3493, $10^{-5}$ m$^3$/kg | Tint strength, D3265 |
|---|---|---|---|---|---|---|
| *Tread Black* | | | | | | |
| N110 | 145 | 127 | 115 | 113 | 97 | 123 |
| N115 | 160 | 137 | 124 | 113 | 97 | 123 |
| N120 | 122 | 126 | 120 | 114 | 99 | 129 |
| N121 | 121 | 122 | 114 | 132 | 111 | 119 |
| N134 | 142 | 143 | 137 | 127 | 103 | 131 |
| N219 | 118 | – | – | 78 | 75 | 123 |
| N220 | 121 | 114 | 106 | 114 | 98 | 116 |
| N231 | 121 | 111 | 107 | 92 | 86 | 120 |
| N234 | 120 | 119 | 112 | 125 | 102 | 123 |
| N326 | 82 | 78 | 76 | 72 | 68 | 111 |
| N330 | 82 | 78 | 75 | 102 | 88 | 104 |
| N339 | 90 | 91 | 88 | 120 | 99 | 111 |
| N347 | 90 | 85 | 83 | 124 | 99 | 105 |
| N351 | 68 | 71 | 70 | 120 | 95 | 100 |
| N375 | 90 | 93 | 91 | 114 | 96 | 114 |
| *Carcass black* | | | | | | |
| N539 | 43 | 39 | 39 | 111 | 81 | – |
| N550 | 43 | 40 | 39 | 121 | 85 | – |
| N582 | 100 | 80 | – | 180 | 114 | 67 |
| N650 | 36 | 36 | 35 | 122 | 84 | – |
| N660 | 36 | 35 | 34 | 90 | 74 | – |
| N683 | 35 | 36 | 34 | 133 | 85 | – |
| N762 | 27 | 29 | 28 | 65 | 59 | – |
| N772 | 30 | 32 | 30 | 65 | 59 | – |
| N774 | 29 | 30 | 29 | 72 | 63 | – |

Change in oil and air ratio in the manufacturing process of furnace black can alter the surface area and change in quench determines the secondary structure along with oil flow and additives. Carbon black morphology is described in Figure 2.5.

Iodine no./N2SA/STSA is a generally a measure of carbon black surface area. OAN and COAN are generally measures of the available structure of carbon black.

Tint is another important parameter; it is a single measure of uniformity of particle size distribution and also an important factor of polymer carbon black reinforcement.

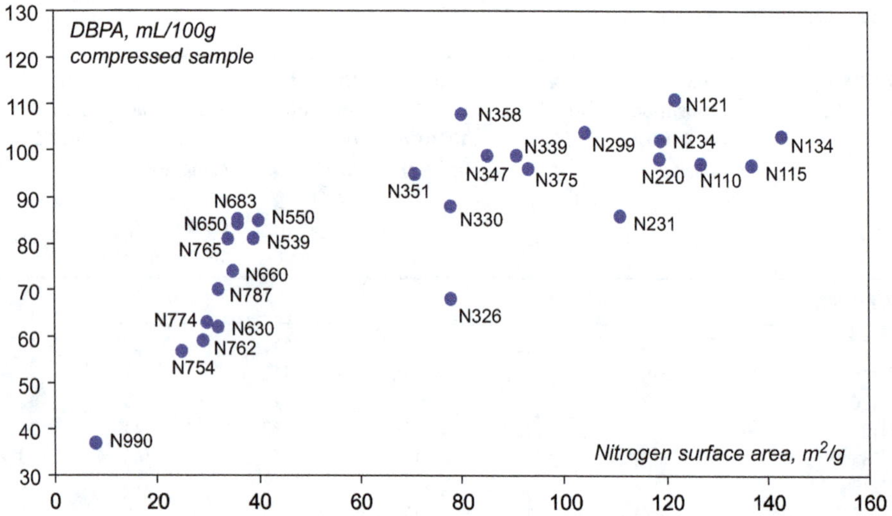


**Figure 2.5:** Carbon black morphology map.

## 2.5 Carbon black manufacturing process

Earliest from of carbon black was probably lamp black, used by the ancient Egyptian, Indians, and Chinese as a medium for writing, painting, and even as cosmetics. Lamp black, semicommercialised around 1598 is a pigment made as pure pine carbon, originally from the soot produced by burning vegetable oil/coal tar in lamps. It has been used as cosmetics – an oldest pigment known to mankind. In 1860, carbon black was first utilised for industrial application.

Carbon black manufacturing process and percentage of global production, with respect to feedstock, is explained in Table 2.3.

### 2.5.1 Lamp black

Lamp black is produced in a cast iron burner pan (where hydrocarbon fluid is being combusted) located under brick-lined walls. The carbon black is deposited in chambers or is separated by cyclone separators and bag filters. Detailed lamp black process is illustrated in Figure 2.6.

**Table 2.3:** Carbon black manufacturing process and percentage of global production, with respect to feedstock.

| Chemical process | Manufacturing process | Percentage of global production | Feedstock |
|---|---|---|---|
| **Partial combustion** | Furnace black process | >95% | Petrochemical oils, coal tar oils, and Natural gas |
| | Gas black process | <5% | Coal tar oils |
| | Channel black process | | Natural gas |
| | Lamp black process | | Petrochemical/coal tar oils |
| **Thermal cracking** | Thermal black process | | Natural gas, oil |
| | Acetylene black process | | Acetylene |

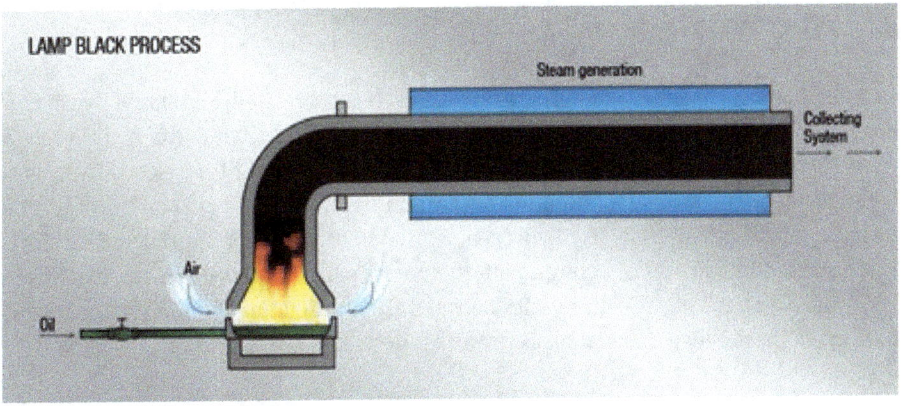


**Figure 2.6:** Lamp black process.

## 2.5.2 Channel black

The Channel black is multiple small diffusion of flames burned in air and the carbon black is deposited on cool surface. The fuel used here is generally natural gas fed from ceramic openings impinged upon the underside of water-cooled iron channels. The deposited carbon black is scraped into a funnel-shaped trough and collected in screw conveyor. The channel black process is described in Figure 2.7.

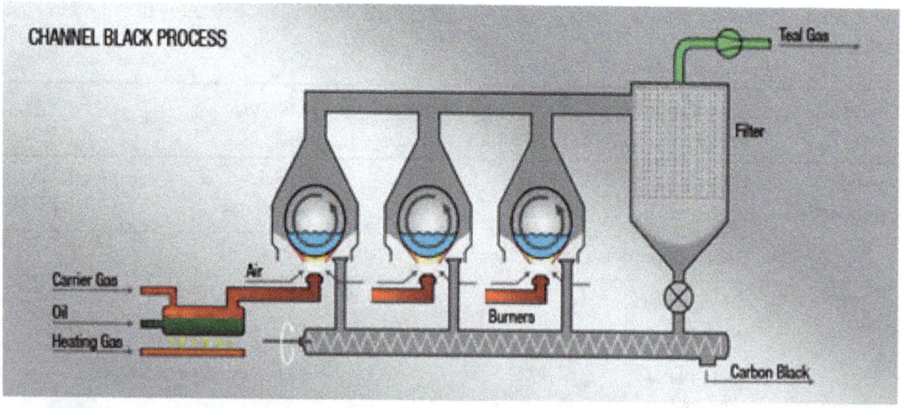


**Figure 2.7:** Channel black process.

Channel black process was developed in 1872 and the yield was only 5%; the rest was wasted.

## 2.5.3 Thermal black

Thermal black is made using natural gas, coke oven gas, shale oil, and liquid hydro carbons in the absence of air and flame. So, it is produced by cracking in a hydro carbon refractory-lined furnace, previously heated by the combustion of a gaseous hydrocarbon and its mixtures. As the liquid hydrocarbon decomposes to carbon black in one furnace, the off gas, particularly hydrogen, is used to heat a second furnace. Therefore, it is a continuous process, cycling to alternative furnaces. Notable thermal blacks are N990 and N880; the particle size varies between 150 and 200 nm and the surface area is 8–12 $m^2/g$. The detailed thermal black process is described in Figure 2.8.

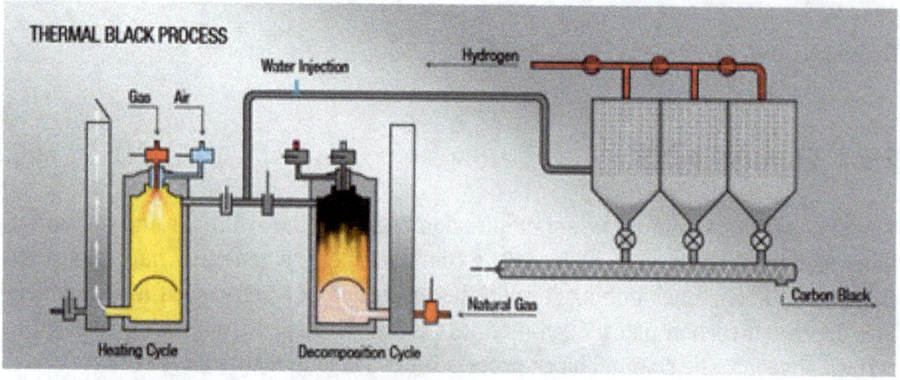


**Figure 2.8:** Thermal black process.

These blacks have unique properties, like low hardness, low compression set, high elasticity, low hysteresis, and very good processability.

### 2.5.4 Furnace black

Furnace black process is the most recently developed process and has become the most common in large-scale carbon black manufacturing. Above 95% of carbon black nowadays are produced by this method. This process was developed in the United States in the 1920s and since then, it has been greatly refined. This is a continuous process and uses liquid and gaseous hydrocarbons as feedstock and as heat source, respectively.

This process uses the principle of oxidative decomposition within the temperature range of 1,200–2,250 °C in closed reactors, where highly turbulent air flow prevails due to high velocity. Because it occurs at a very high temperature, the reaction is confined to a refractory-lined furnace; hence the name. Carbon black produced in the lower part of this range of temperature is used in 'carcass' or 'soft' black reactors, while those produced in the upper part of the range of temperature are used in 'tread' or 'hard' black reactors.

These descriptive terms derive from their application in rubber. The lower temperature carbon black gives soft compounds that are much used in tyre carcass compound and the higher temperature carbon black gives hard compound that is often used in tyre treads.

## 2.6 Description of the process

At the core of a furnace black plant is the furnace where the carbon black is generated. The primary feedstock is introduced, typically as a finely dispersed spray, into a zone of high temperature and high energy density. This is achieved by burning a secondary feedstock such as natural gas or oil, together with air.

Due to the excess oxygen, in relation to the secondary feedstock, complete combustion of the primary feedstock does not occur. Instead, the majority of it undergoes pyrolysis, resulting in the formation of carbon black at temperatures ranging from 1,200 to 2,250 °C.

The reaction mixture is then rapidly cooled by water and further brought to a lower temperature using heat exchangers. The carbon black is separated from the remaining gases through a filtration system. The gas that is released, following the production of carbon black, typically enters the combustor of a boiler. This process generates medium-pressure steam, which is then utilised to generate power through a steam turbine, as illustrated in Figure 2.9.

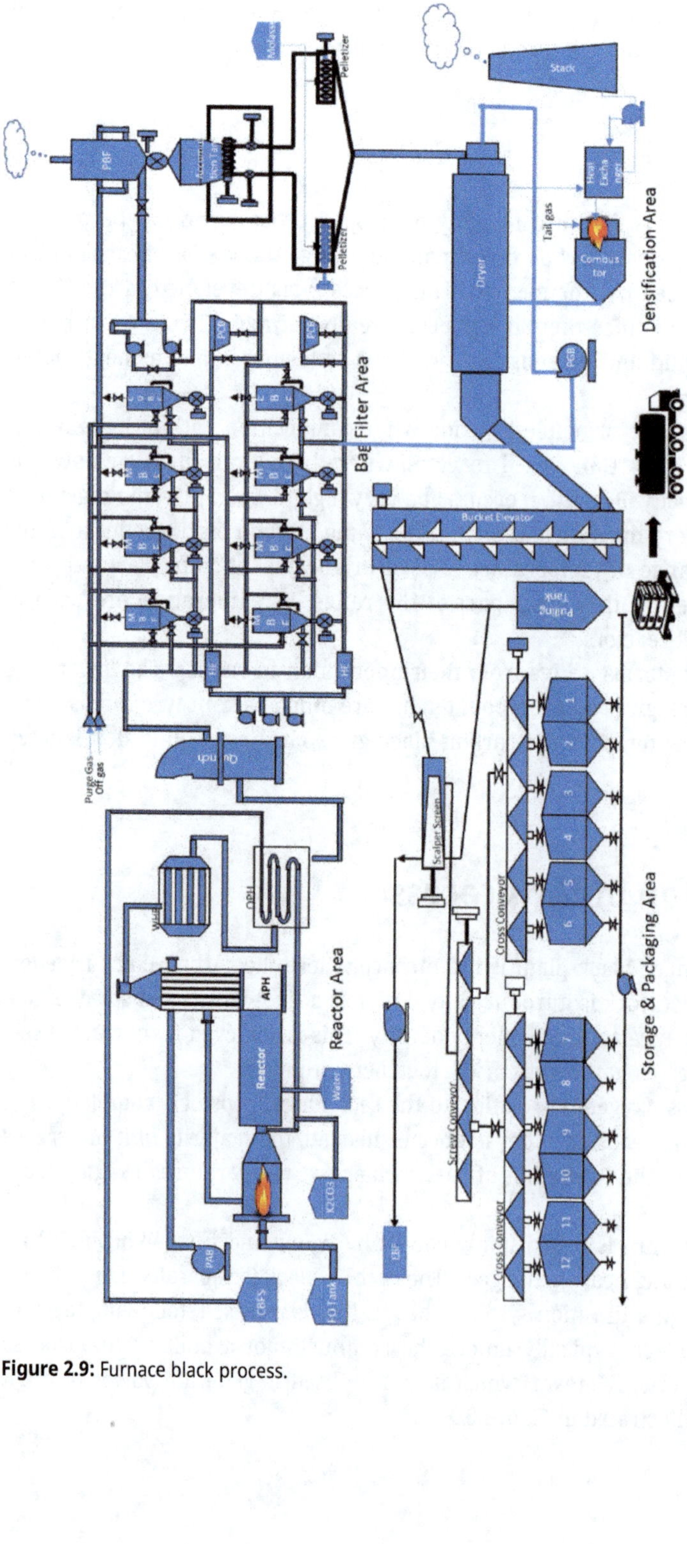


Figure 2.9: Furnace black process.

The carbon black feedstock (CBFS) is transferred from the storage tank to the reactor using an oil preheater. Gas and preheated process oils at temperatures around 200–250 °C are introduced into the reactor. Within the reactor's high-temperature section, known as the chock section, the hydrocarbon feedstock undergoes thermal oxidative cracking. This results in a phase change, from a liquid or gaseous state (CBFS) to a solid mist of soot. The reaction is halted by injecting water at approximately 950 °C, which leads to the formation of carbon black and the cooling of the reactor's off-gas.

Following further cooling in heat exchangers, with temperatures below 250 °C, the mixture of carbon black and off-gas is directed into a series of bag filters, where the carbon black is separated from the 'tail gas'.

The filtered carbon black is then pneumatically conveyed to the pelletising section. Any present grits are reduced using a pulveriser or a hammer mill. Once separated from the off-gas, the fluffy powdered black is densified in the pelletiser to facilitate transportation.

Subsequently, the pelleted product undergoes drying in a rotary kiln dryer at temperatures below 160 °C, which is heated by hot gases, preferably derived from the combustion of the tail gas. After leaving the dryer and prior to entering the silo, the pelletised black undergoes screening and passes through a magnetic separator to eliminate coarse particles and metallic impurities. The carbon black is then conveyed to storage silos using a bucket elevator or a screw conveyor.

From the silos, the final product is packaged in various bags or loaded directly onto trucks. The schematic diagrams of the furnace black reactor and carbon black formation mechanism are portrayed in Figures 2.10 and 2.11, respectively.

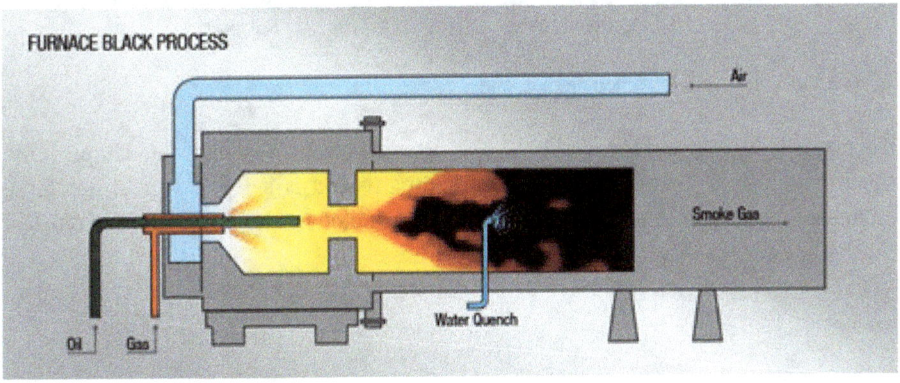


**Figure 2.10:** Schematic diagram of furnace black reactor.

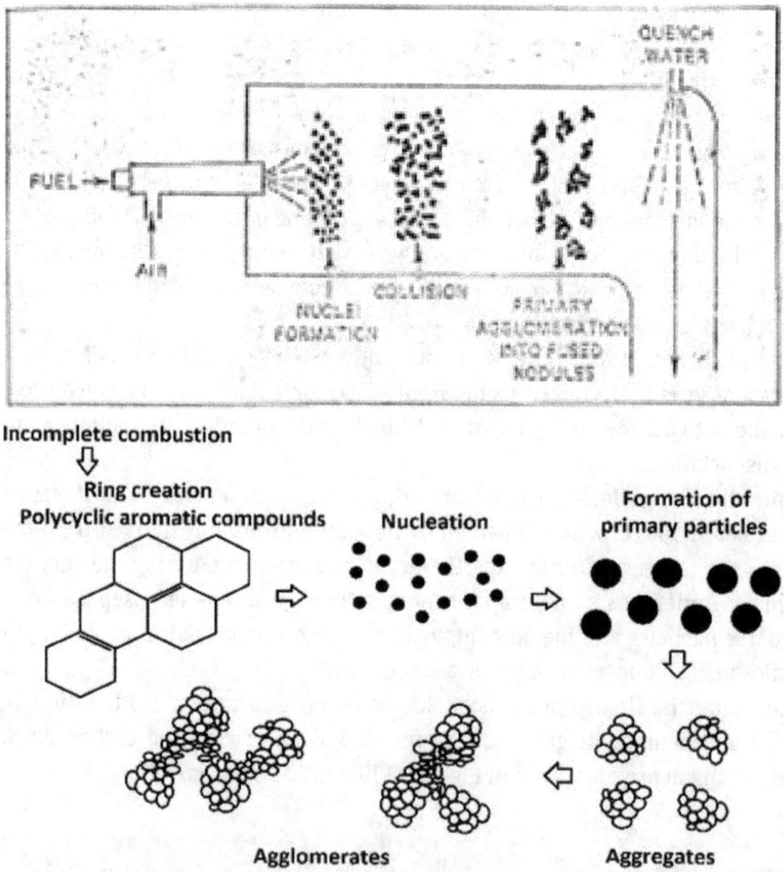


**Figure 2.11:** Carbon black formation mechanism.

The major technology of the above process depends on the geometry and shape of the chock, which is a proprietary of every company and is a trade secret. But for a broader understanding, it can be depicted that, fundamentally, there are two types of chocks, as shown in Figures 2.12 and 2.13.

Feedstock generally used are petroleum-based and coal-based, generally known as CBFS.

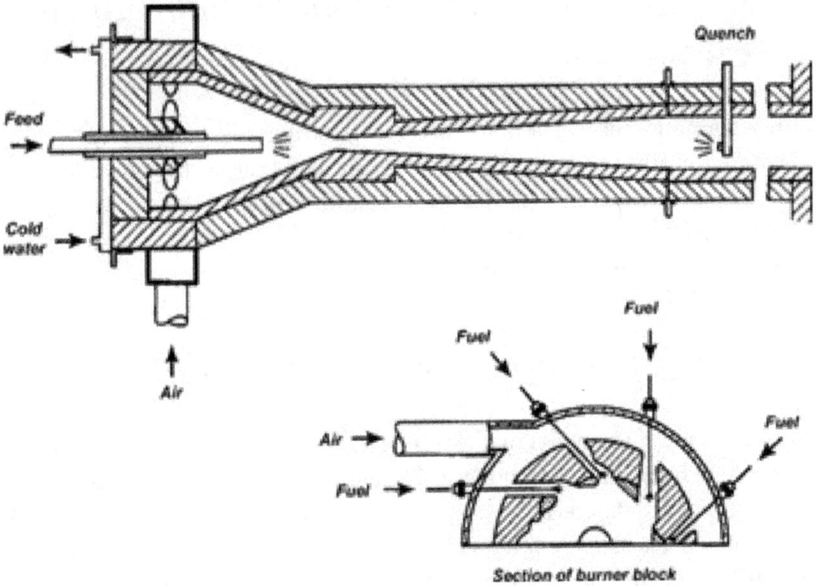


**Figure 2.12:** Venturi-type reactor.

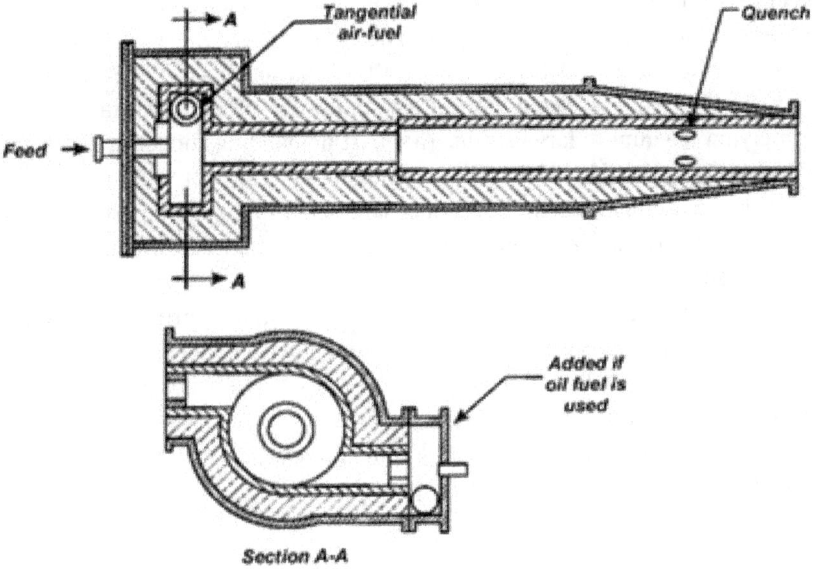


**Figure 2.13:** Square chock reactor.

## 2.7 Physical interpretation of rubber reinforcement with carbon black

Based on the information provided in the preceding sections, we have learned about the manufacturing process of carbon black, particularly by the furnace method. The terminology and morphological properties associated with carbon black are crucial, especially when considering particle size and its distribution in polymers. Understanding these aspects can be complex, as there are various types of size references involved.

### 2.7.1 Carbon black & its effect in rubber reinforcement

The reinforcement is generally defined as conversion of elastomer from its particle state to elastic state by adding carbon black. Without fillers, cross-linked rubbers are generally partially plastic materials, which do not have much nerve. Carbon black particles enhance the elasticity and the mechanical properties, which are defined by increasing tensile strength, tear resistance, and abrasion resistance of elastomers.

The degree of reinforcement varies with particle size, surface area, primary secondary structure, and nature of elastomer–filler interactions, which also depends on the shape of the primary particle and type of branching/agglomeration. This also determines the elastomer–filler interaction properties, and ultimately leads to change in static and dynamic characteristics. The quantity of carbon black determines the stress-strain curve, hardness, strain-induced crystallisation effect (in certain elastomers) by applying medium to large strain. The most demanding rubber applications such as automobile tyres and rubber goods employ many grades of different surface area and structure. Nanostructured materials such as graphene and carbon nanotubes, which are also carbon allotropes, impart critically balanced different properties for special application [22].

The global production of carbon black is approx. 21 million metric tons per year, and 93% of this material goes into rubber applications (automobile tyres (73%) and non-tyre rubber products (20%)), with the remaining 7% used in paints, coatings, inks, fibre, and conductive and plastics compounding. Present research reveals the importance of specific carbon black for specialty applications. Sometimes, dual carbon black is used for enhancement of mechanical properties.

### 2.7.2 Why is it called reinforcement?

Carbon black is utilised as filler in elastomers and polymers (plastics), as well as in paints, to modify their viscoelastic properties and enhance their mechanical, electrical, and optical characteristics.

In elastomers, the addition of carbon black enhances the strength and physical properties of the material. This process, known as reinforcement, involves the interaction between elastomer molecules and carbon black particles within the modified composite matrix.

The primary particle size (measured by NSA) and surface activity of the carbon black, along with its loading, are crucial factors in elastomer reinforcement.

Carbon black significantly impacts the properties of elastomers (rubber), affecting their fracture behaviour, abrasion resistance, and hardness. High-resolution transmission electron microscopy images, such as Figure 2.14(a), reveal the crystalline structure of carbon black. Figure 2.14(b) demonstrates the effect of elevated temperatures on the nanostructure of carbon black.

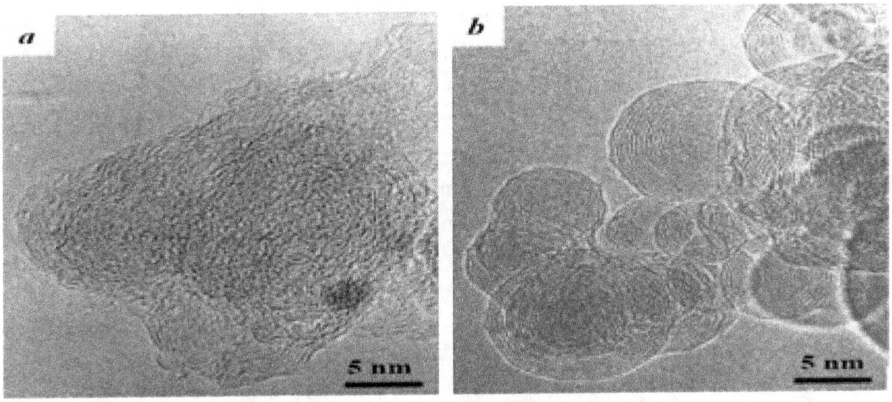


**Figure 2.14a:** Carbon black, as viewed under high-resolution transmission electron micrographs, generally display very low crystallinity.

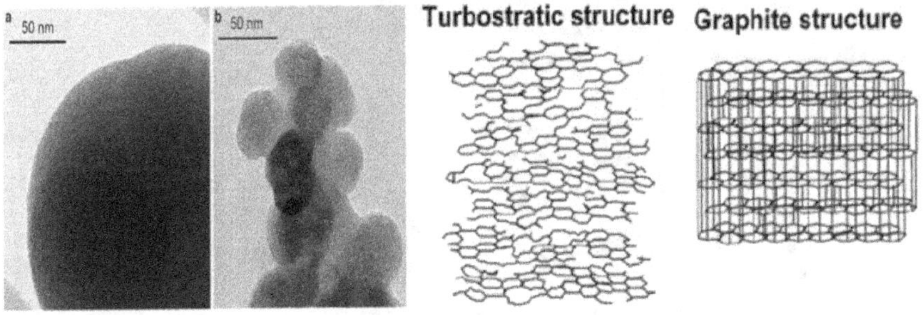


**Figure 2.14b:** Carbon black can be converted into carbon nano-onions, upon high-energy irradiation at a higher temperature.

## 2.7.3 Major parameters influencing the reinforcement

The major parameters influencing reinforcement are:
1. Carbon black microstructure
2. Primary particle size
3. Porosity
4. Aggregate size, shape, and structure of carbon black
5. Surface activity

### 2.7.3.1 Carbon black microstructure

Carbon black sets itself apart from other forms of carbon due to its predominantly spherical primary particles, which have diameters ranging from 10 nm to several hundred nanometers. It exhibits various morphologies and shapes that must be taken into account. The atomic structure within these primary particles is commonly described as a para-crystalline model, where carbon atoms are arranged in continuous layers in a hexagonal pattern, forming fundamental building blocks [21]. Within the primary particle, the carbon layers are concentrically organised around one or more growth nuclei, resulting in adjacent layers that are parallel to each other. This arrangement of parallel stacks is referred to as turbo static, as depicted in Figure 2.15.

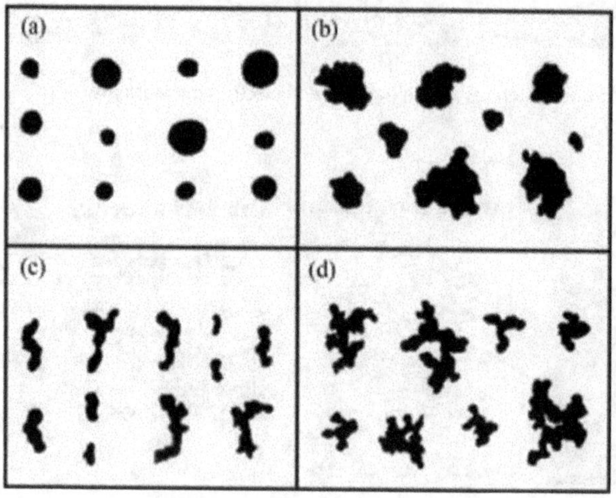


**Figure 2.15:** (a) Carbon black microstructure, (b) SEM images, (c) turbostratic structure, and (d) graphite structure.

These parallel stacks can be rotated around the crystallographic C atom.

#### 2.7.3.2 Primary particle size

The primary particle size refers to the diameter of carbon black particles and is commonly estimated based on the specific surface area. Assuming a spherical particle shape and disregarding particle porosity, the primary particle size (PMS) can be approximated using the specific surface area. Each primary particle is composed of concentric or irregularly arranged graphitic crystallites. In rubber, the adsorption of carbon black predominantly occurs at the edges of these graphitic crystallites or in the transitional zone between different graphitic crystallites.

#### 2.7.3.3 Porosity

Porosity is a critical parameter of carbon black, indicating the presence of small openings, cavities, or channels within the particles. It should not be confused with the space between particles. Porosity is often referred to as void volume, which provides a fundamental understanding of porosity.

Porosity can be classified into two categories: open pores and closed pores. Closed pores exist entirely within the solid phase, without any outlets on the surface, while open pores have outlets on the surface of the carbon black particles. Closed pores primarily affect the specific gravity but have minimal impact on other properties [4]. On the other hand, open pores significantly influence the surface area, resulting in intimate mixing with polymers, higher bound rubber content, lower volatiles, and improved reinforcement. The effect of open pores depends on the size of the pore openings, polymer molecular weight, and its flexibility.

Traditionally, pore sizes are divided into three types, based on their dimensions:
- Macropore size: >50 µm
- Mesopore size: >2–50 µm
- Micropore size: <2 µm

These pore size surfaces can be measured by differential pressure, technically in a BET apparatus using nitrogen gas.

#### 2.7.3.4 Aggregate size, shape, and structure of carbon black

Aggregates play a crucial role in the interaction between carbon black and polymers, where both their size and shape are important factors. The term 'structure' has two connotations. The first relates to the size and shape of the aggregates and agglomerates, which form a permanent structure. The second connotation refers to the carbon black network formed within the polymer, which is essential for the polymer–filler network.

Immediately after formation through the partial decomposition of hydrocarbon, small carbon black primary particles do not exist in isolated form. They quickly bind together to form discrete rigid colloidal entities of various shapes, including spheroidal, ellipsoidal, linear, and branched shapes. These different shapes contribute to reinforcement in distinct ways. These carbon black aggregates, which are the smallest dispersible units, separated from the aggregate only through rupture, are formed by the merging of continuous carbon layers and the connection of primary particles through strong $sp^2$ (covalent) bonds.

The size and shape of the aggregates can vary, ranging from individual spheroidal or oval particles to clusters of a few or multiple layers of primary particles, with irregular, three-dimensional, chain-like, or grape-like morphologies.

Aggregates can be categorised into four basic types: spheroidal, ellipsoidal, linear, and branched. Each carbon black sample contains a distribution of these types. Higher grade structures have more branched and linear aggregates, while lower grade structures have more ellipsoidal and spheroidal aggregates, as illustrated in Figure 2.16.

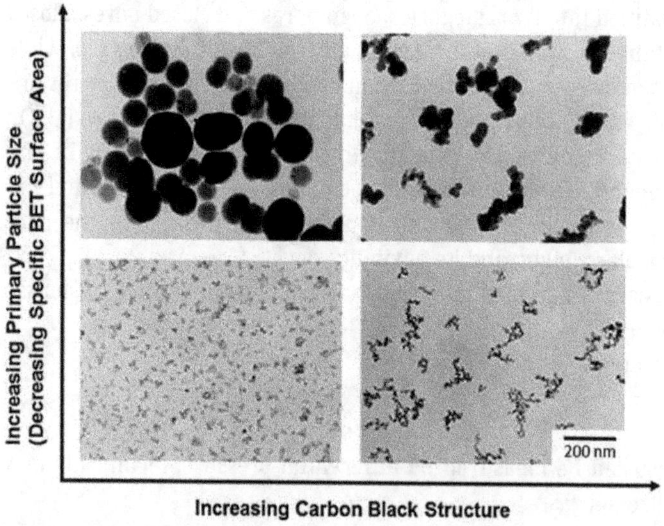


**Figure 2.16:** Different shapes of carbon black: (a) spheroidal, (b) elliptical, (c) linear, and (d) branched.

During intense mixing processes at higher shear rates and temperatures, the linear and branched aggregates have a tendency to undergo breakdown, resulting in smaller aggregates. However, this breakdown does not lead to the formation of primary particles or elemental carbon; it is limited to the secondary structure.

Recent research has indicated that aggregates possess an anisotropic morphology, exhibiting a somewhat more two-dimensional or planar nature than previously thought. Gaining a better understanding of these characteristics can provide valuable insights

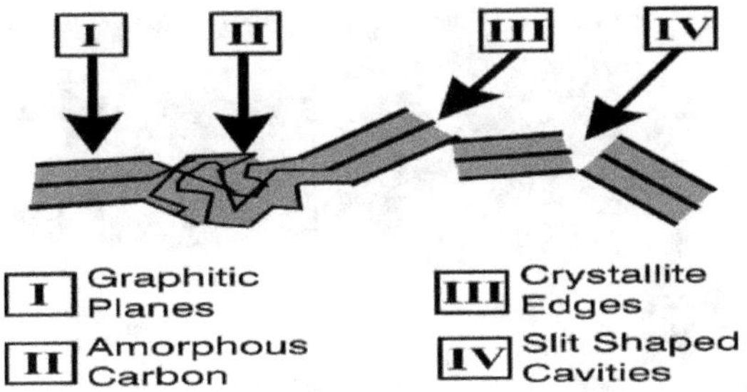


**Figure 2.17:** Different classes of carbon black aggregate.

into the formation and performance of carbon black. Figure 2.17 explains the different classes of aggregate morphology observed in carbon black.

#### 2.7.3.4.1 Morphology

Usually, the tendency to form aggregate decreases with increase in primary particle size.

### 2.7.3.5 Surface activity

The chemical behaviour at the surface refers to the chemistry of the surface groups and the structure of the surface at a microscopic level. The presence of different features such as graphitic planes, amorphous carbon, crystallite edges, and narrow openings indicates the existence of adsorption sites with varying energies. The interaction between the carbon black's surface and the surrounding polymer chains is influenced to some extent by the type and amount of functional groups attached to the surface. Figure 2.18 demonstrates the characterisation of the surface activity of carbon black and its connection to the interaction between the polymer and the filler.

Functional groups in the surface are: –H and oxygen-bearing groups as for example –carboxylic (–COOH), carbonyl (=C=O), hydroxyl (–OH), quinonic, lactonic, oximes (=N–OH) as well as oxygen (=O), and sulphur (–$SO_3$ and S). These surface oxides can react as Lewis base, which may act as the anchor atom for Lewis acid as the polymer.

Both adsorptive surface and functional groups are responsible for reinforcement. Adsorptive activity of carbon black is irregularly distributed on a surface. The adsorptive active surface in carbon black is only about 5–6% of the total available surface and is primarily concentrated on the graphitic crystallite edges and in the amorphous region between the crystals [24].

Very high structured carbon black with greater surface roughness allows more adsorption by relatively more graphitic edges and amorphous region crystals.

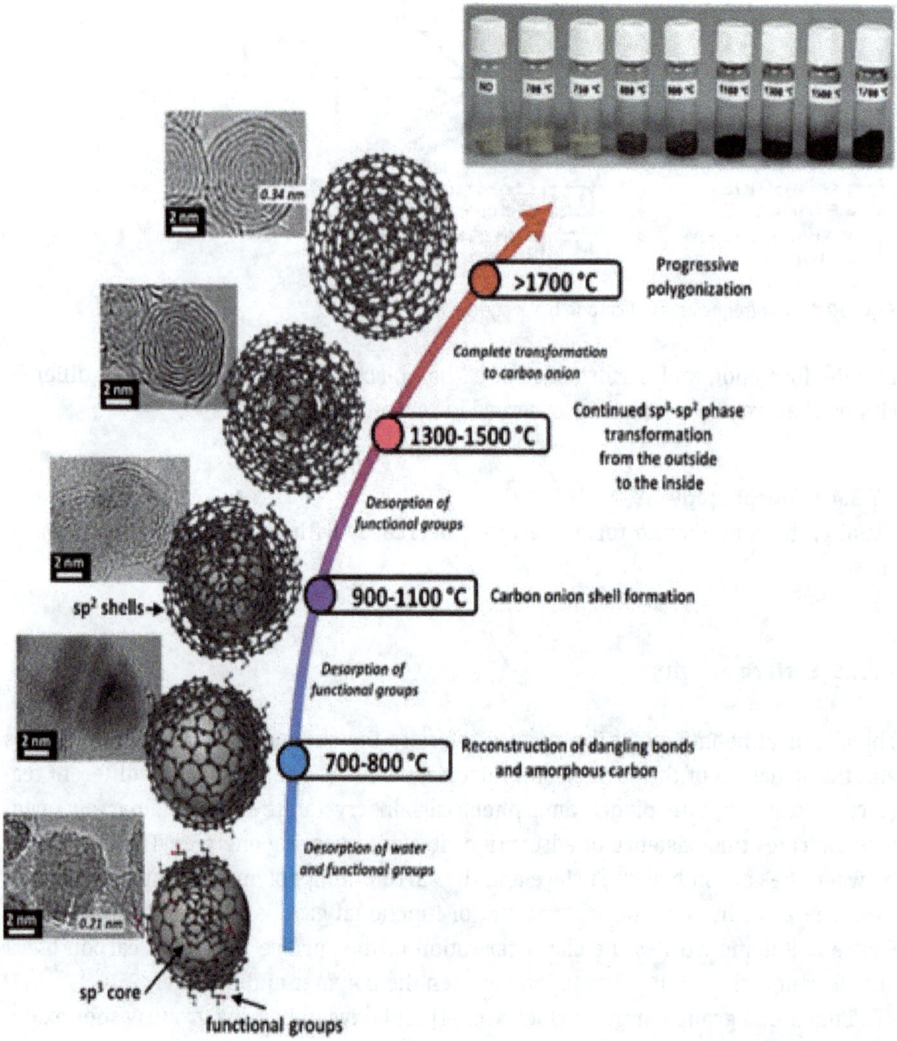


**Figure 2.18:** Characterisation of surface activity of carbon black and its relation to polymer–filler interaction.

## 2.8 Structure of carbon black and rubber properties

The arrangement of carbon black has a significant impact on the behaviour of rubber products during various stages such as mixing, calendering, extrusion, shaping, and reinforcement.

This influence can be observed through measurements like 100%, 200%, and 300% modulus, as well as the cross-linking density of pure test specimens. The 300%

modulus of SBR compounds increases significantly when carbon black with a higher OAN number but the same specific surface area is used. This can be explained by the fact that carbon black aggregates contain a certain amount of rubber within their structure, referred to as 'occluded rubber' or 'bound rubber'. Consequently, these aggregates along with the enclosed rubber act as a single reinforcing entity. The size of these entities increases with higher carbon black structure.

Incorporating carbon black into a polymer matrix creates an interface between a rigid solid phase and a soft solid phase. The characteristics of these interfaces depend on several factors such as the nonporous nature of the rubber grade carbon black, the volume fraction of carbon black in the compound, the specific surface area, and the available surface groups in the black. A higher percentage of bound rubber indicates a greater level of interaction activity, which correlates with a higher degree of reinforcement.

However, the cross-linking density of carbon gel is only a fraction of that found in a vulcanised state. This is evident from the significant swelling observed in different solvents (e.g., toluene, xylene, isooctane, and toluene) for the rubber in the gel, compared to vulcanised rubber. Thermal treatment of carbon black can reduce the formation of bound rubber, and in the case of graphitised carbon black shown in Figure 2.19, the bound rubber is completely eliminated due to the deactivated surface of the carbon black. The rheological properties of rubber filled with carbon black are described in Figure 2.20.

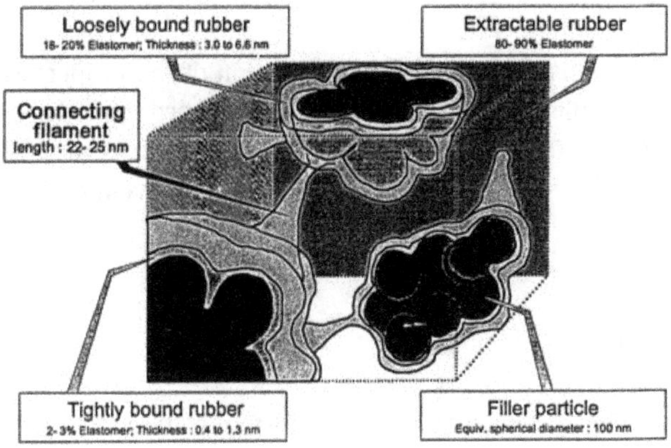


**Figure 2.19:** Graphitisation of carbon black.

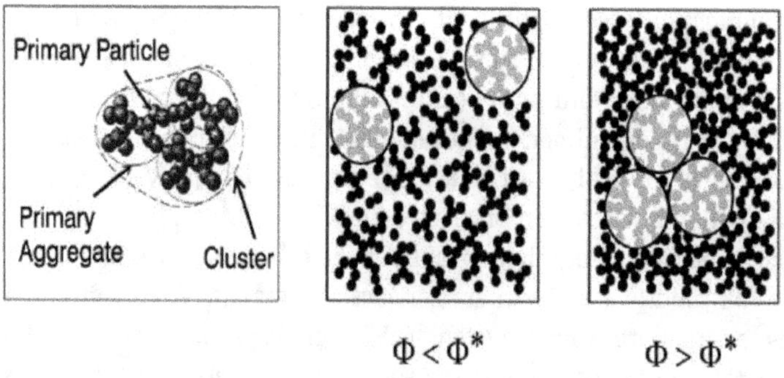


**Figure 2.20:** Rheological property of filled rubber.

## 2.9 Reinforcement

The rubber–filler reinforcement subject is still wide open due to the proliferation of numerous types of polymers and various grades of carbon black. Many ideas have been put forward by many scientists and researchers in chemistry and physics of polymer-filler composite, yet complete conclusive theory with experimental evidence has not emerged. Starting from $sp^2$–$sp^3$ hybridisation during mixing and vulcanisation for the formation of covalent bond, ion-radical species to fullerene concept has been forwarded.

Attempts to provide a comprehensive quantitative description of the properties of reinforced rubber mixtures and vulcanised rubber have only achieved limited success in practical applications so far. However, there are a few noteworthy practical correlations that offer valuable insights into the fundamental mechanisms of carbon black reinforcement.

### 2.9.1 Hydrodynamic effect

For a suspension spherical particle in a Newtonian fluid, Einstein (reference) originally derived a relationship from the change of viscosity due to presence of particles, which was later modified by Guth and Gold:

$$n = n_0\left(1 + 2.5\Phi + 14.1\Phi^2\right);\ \Phi = V_F/V_F + V_R \tag{2.1}$$

where $n_0$ is the viscosity of solvent or dispersing medium; $n$ is the viscosity of mixture; $V_F$ is the volume of filler; $V_R$ is the volume of rubber; $\Phi$ is the volume fraction of filler in the mixture [8].

When the value of $\Phi$ is very small, the third term in the equation becomes negligible. However, this equation does not account for the particle size and structure of the filler. The N900 series black, which has nearly spherical dimensions, conforms well to the equation. However, carbon black with larger specific surface areas and more secondary structure deviates from the theoretical prediction in a positive manner. This can be attributed to the following reasons [9]:

- The carbon black aggregates are not spherical in shape.
- The carbon black aggregates are not free to move within the rubber medium.
- The carbon black aggregates become more entangled with each other, leading to mechanical reinforcement.
- The rubber molecules occupying the empty void spaces within the filler aggregates form rubber–filler aggregates, and are restricted in their movement.
- The rubber molecules occupying the void spaces within the aggregates do not actively participate in the deformation caused by external forces, resulting in an increased effective concentration of the filler.
- Rubber molecules with higher molecular weights are absorbed by the filler molecules, causing the remaining rubber to have lower molecular weight and viscosity.

Therefore, neither $n_0$ nor $\Phi$ accurately represents the real situation in practical terms.

Smallwood showed that for elastic media filled with spherical particles, an equation similar to above equation can be applied. Smallwood's calculation showed that the only parameter that needs to be changed, that is, from viscosity ($n$) should be replaced by elastic modulus $\epsilon$:

$$E = E_0\left(1 + 2.5\Phi + 14.1\Phi^2\right) \tag{2.2}$$

Similar to eq. (2.1), this equation only holds for spherical filler particles and N900 series of blacks only for small strain. However, this relation again does not hold for high-specific surface area and high aggregate structure black. Filler aggregate configuration in rubbers above the percolation point is illustrated in Figure 2.21.

Batchelor and Green and Chen and Acrivos obtained an equation similar to eq. (2.2) by considering liner elastic solids containing spherical particles with two particle interactions:

$$E = E_0\left(1 + 2.5\Phi + 5\Phi^2\right) \tag{2.3}$$

Higher filler loading, typically up to volume fraction of $\Phi \sim 0.35$, a Padê of the series expansion in eq. (2.3) up to the second order in volume fraction leads to another suitable expression:

$$E = E_0\left(1 + 2.5\Phi + 5\Phi^2 + \cdots\right) \sim E_0(1 + 2.5\Phi/1 - 2\Phi) \tag{2.4}$$

A more recent approach to the reinforcement effect of filler in rubber makes use of fractal description of percolating network, as depicted in Figure 2.22.

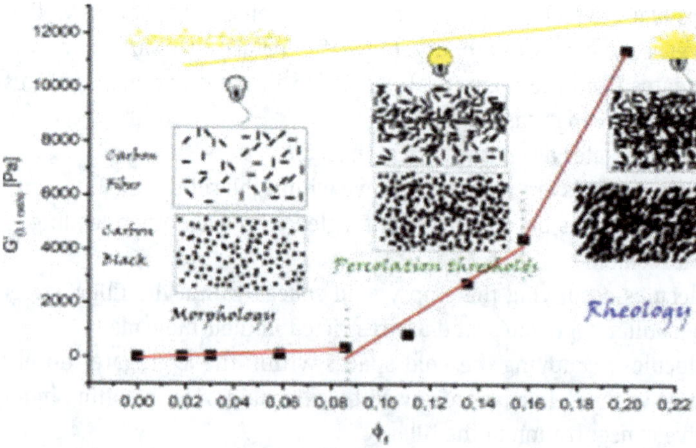


**Figure 2.21:** Filler aggregate configuration in rubbers above the percolation point.

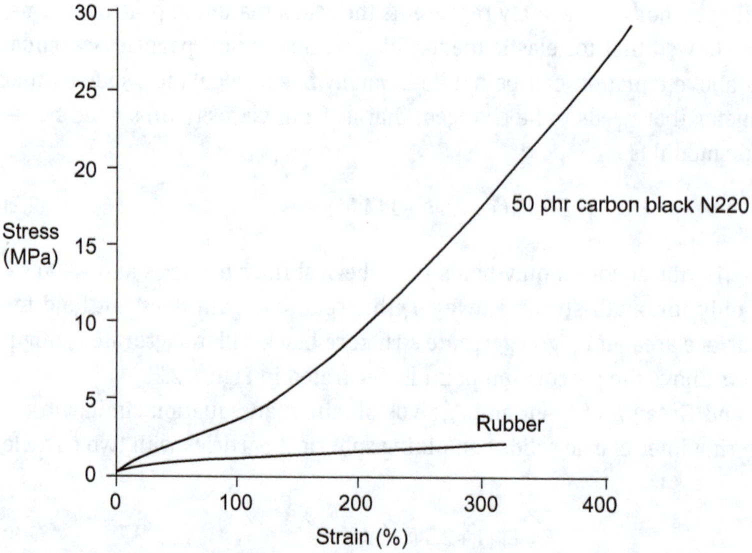


**Figure 2.22:** Percolation curve.

The percolation point is defined as the volume fraction of the filler $\Phi^*$, at which carbon black aggregate come in close contact to form an electrically conductive pathway through the rubber compound.

## 2.9.2 Fractal theory

The principle underlying this concept is that when the length scale of carbon black and the polymer exceeds a certain threshold, the mixing appears to be uniform, resulting in a higher polymer–filler network. Below this length scale, a specific morphological structure is observed, repeating itself at smaller scales, revealing filler–filler networks and leading to increased conductivity [20]. Therefore, it can be stated that the volume fraction of the reinforcing filler should exceed a percolation point:

$$E = \Phi^{3.5} \ldots \text{for } \Phi > \Phi^* \tag{2.5}$$

At these volume fractions, the size of the primary particles or the specific surface area of the carbon black filler does not significantly affect the behaviour.

Although eqs. (2.1) and (2.2) have limited applicability, they provide an initial estimate of the impact of reinforcing fillers on the viscosity and elastic modulus of rubber compounds and vulcanisates. Several refinements have been proposed in the past, with the 'stress concentration' variant being the most well-known. When reforming a reinforced rubber compound, there is a difference between local microscopic and overall macroscopic deformation because the filler particles generally do not deform [10]. In vulcanised rubber, the filler particles bear the load of the tyre. If there is no filler deformation, the deformation will be highest and the load-bearing ability will be minimal [19].

The rubber matrix reinforced with filler can withstand much higher stress than a nonfilled rubber matrix. The stress or force experienced by individual rubber molecules in reinforced rubber is significantly higher than in non-reinforced rubber. This observation, demonstrating the increase in tensile strength of reinforced rubber compared to non-reinforced rubber, is visualised in Figure 2.23.

The low tensile strength of a cross-linked, amorphous rubber is attributed to the uneven distribution of stress within the rubber matrix. Randomly distributed cross-links result in local variations in chain length between the cross-link points. During deformation, stress concentrations primarily occur in the shorter chain segments, which become weak spots, and rupture first. Moreover, these shorter chain segments tend to absorb higher energy due to stress concentration, leading to increased heat generation.

Nevertheless, if the most stressed chain segments have the opportunity to slide along the reinforcing particle, the point of failure shifts to higher stress levels. This enables a more homogeneous distribution of stress among a greater number of effectively loaded chain segments, resulting in higher tensile strength.

In Figure 2.24, the nonextended state, represented by AA', depicts the shortest and the most stressed chain segment. When straining the material and increasing the stress on AA', the stress can only be reduced by allowing the chain segment to slide along the surface of the filler. Simultaneously, chain segment BB' is stretched. With further straining, the stress on BB' is divided through the sliding of rubber molecules adsorbed on the filler particles. All the molecular chains between the filler particles

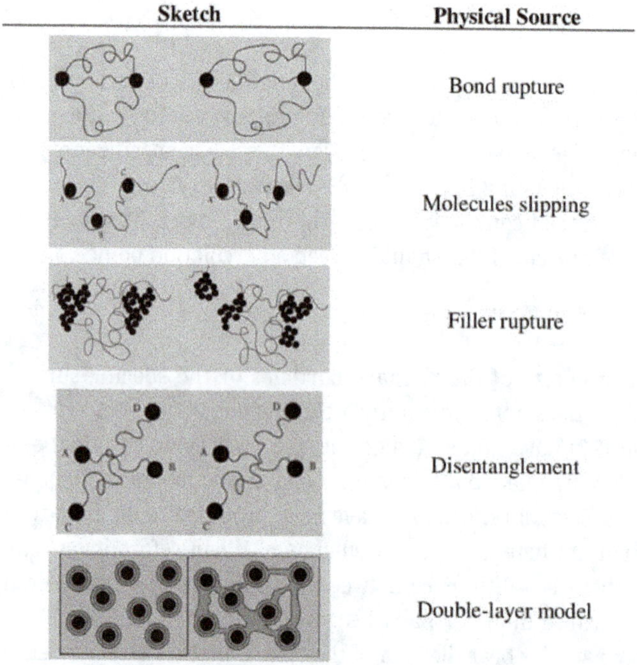


**Figure 2.23:** Tensile strength of filled rubber versus un-filled rubber.

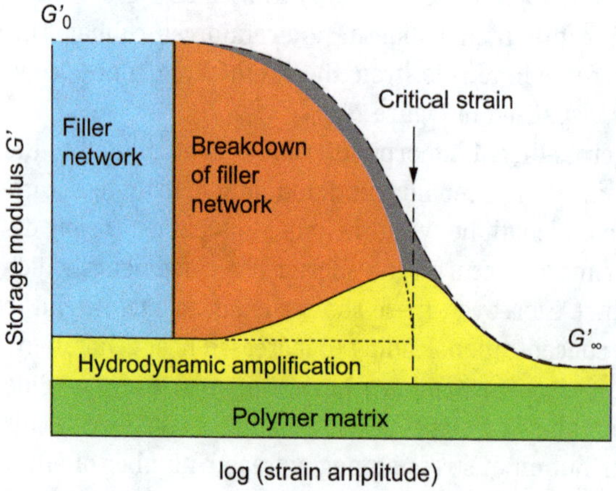


**Figure 2.24:** Physical mechanisms behind the Mullins effect.

experience more or less the same stress. However, upon releasing the stress, the rubber chains between the filler particles do not return to their original length, resulting in different conditions at points W and Z.

This model of reinforcement also explains a phenomenon known as Mullin's effect. This effect refers to the softening of carbon black-reinforced rubber under cyclic deformation.

### 2.9.3 Payne effect

Another model can be described by considering a continuous network of filler embedded within the cross-linked rubber network. When exposed to dynamic mechanical deformation with sinusoidal stress, the behaviour of this filler network can be easily described while the rubber network remains intact. This can be observed in the microscopic properties of vulcanised reinforced rubber, referred to as the Payne effect, shown in Figure 2.26. The complex modulus $E^* = E' + iE''$ in elongation mode or $G^* = G' + iG''$ in stress mode depends on the magnitude (amplitude) of this dynamic deformation.

As the dynamic deformation/strain increases, the storage modulus corresponding to it decreases, while the loss modulus reaches its maximum value. It is important to note that the Payne effect should not be mistaken for the Mullin effect. Figure 2.25 portrays a schematic representation of the Payne effect, highlighting its distinct contributing factors.

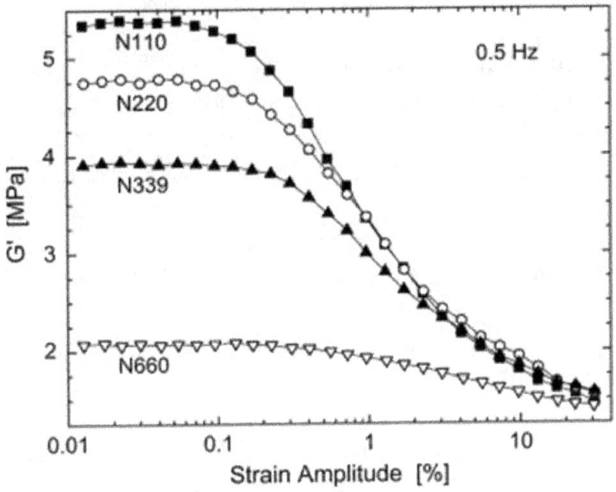


**Figure 2.25:** Schematic representation of Payne effect, resulting from different contributions, as indicated.

The explanation for the combined effect is as follows:

- The filler network contributes to the modulus in a manner that depends on the amplitude, which can be expressed using eq. (2.1).

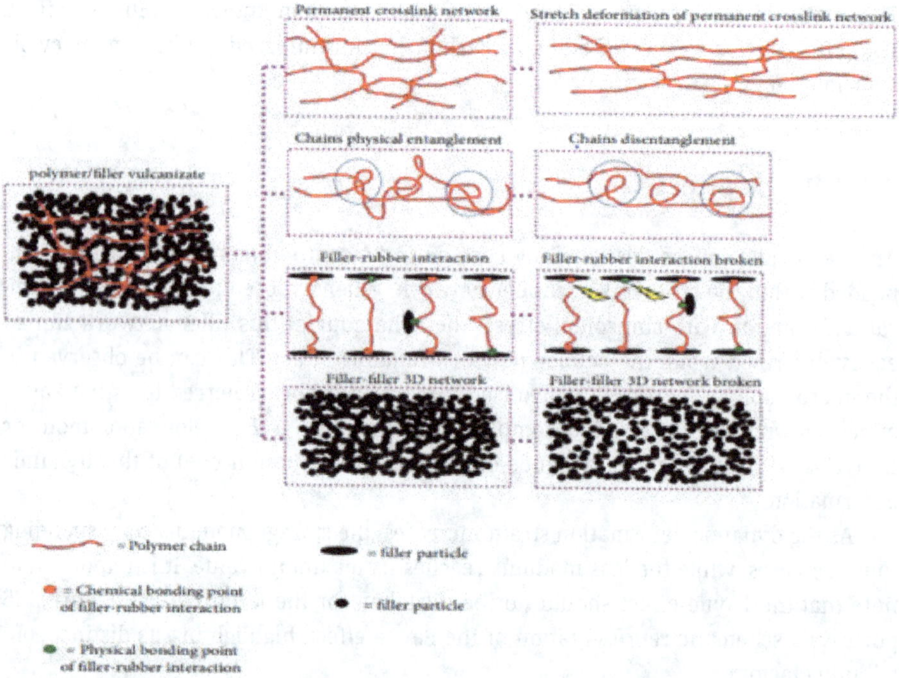


**Figure 2.26:** Graphical representation of the Payne effect for different grades of carbon black.

- The vulcanised polymer network also contributes to the modulus in an amplitude-dependent manner, proportional to the cross-linking density.
- If the polymer on the filler surface undergoes chemical coupling, it provides an additional amplitude-dependent contribution, as if the filler particles form a multifunctional network.
- The most significant contribution to the elasticity modulus of the filler network is the amplitude-dependent or state-dependent breakdown of the filler network at high strains.
- These four contributions determine the elastic modulus of vulcanised rubber with the filler.

The modulus of a carbon black filler network has been known for a long time. While empirical models can explain many observations, a precise quantitative description of carbon black filler reinforcement in rubber is currently impractical. As a result, alternative models have been proposed.

One explanation of the Payne effect involves the destruction and reformation of the interconnected filler network during rubber deformation. A recent model suggests that the interaction between the rubber matrix and carbon black can be modeled similar to the adsorption of gas molecules, using what is known as the Langmuir isotherm. This model assumes a dynamic equilibrium between the adsorption and desorption of gas molecules on a solid substrate.

In this equilibrium, the ratio of occupied to vacant adsorption sites is determined by the vapour pressure of the gas molecules. A higher partial vapour pressure leads to a higher fraction of occupied adsorption sites.

When rubber molecules adsorb onto carbon black particles, the entire rubber chain does not attach to the particle. Instead, the chain molecules wrap around the carbon black particle, forming bonds intermittently along the chain. Due to the thermal Brownian motion, these bonds occasionally dissociate in one location while forming new bonds elsewhere along the chain. This dynamic process establishes equilibrium [11]. When the reinforced rubber is deformed, this equilibrium is disturbed.

As a result, more contact points are released and formed, leading to a decrease in mechanical strength. Conversely, when the force is removed, the equilibrium is reestablished. This model qualitatively describes the Payne effect. A schematic representation of the Mullin effect is shown in Figure 2.27.

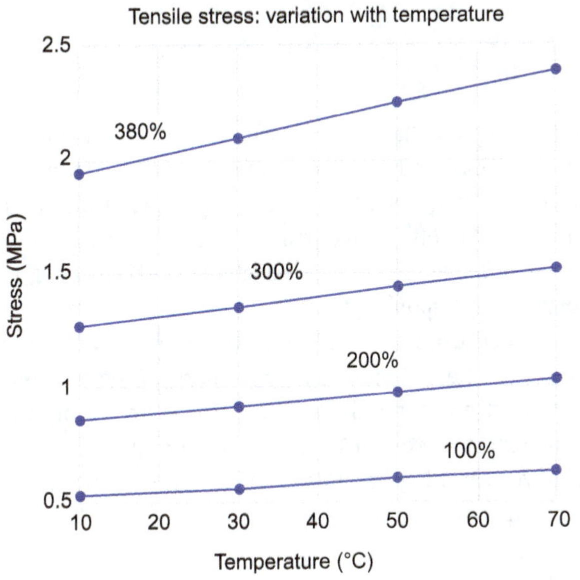


**Figure 2.27:** Schematic representation of the Mullin effect.

Due to the presence of Payne's and Mullin's effects, the mechanical characteristics of reinforced rubber are determined solely by the magnitude of deformation and the time interval between successive large deformations. However, predicting the specific dependence of deformations and the time for individual rubber compounds is challenging. Therefore, when developing rubber products, it is possible to estimate the mechanical properties within a certain range of values, but pinpointing the exact value is extremely difficult.

## 2.10 Thermodynamics of rubber elasticity

Thermodynamics in rubber is not a much studied topic in rubber technology, especially in the case of carbon black application. As we all know that molecules are always in motion and due this movement they can create many different types of energies, rubber elasticity is mainly based on some fundaments of thermodynamics like entropy, enthalpy and Gibbs free energy.

We can also understand the phenomenon of mixing more properly with the understanding of polymer thermodynamics. Thermodynamics of rubber can be divided into two segments – one is the thermodynamics of polymer solutions and the other is thermodynamics in rubber elasticity.

### 2.10.1 Thermodynamics

Temperature plays a significant role in the behaviour of elastomers. When an elastomer is stretched while being heated, it absorbs heat, and upon release of the strain, it contracts with a slight release of heat. Conversely, cooling can cause expansion. This phenomenon can be observed in everyday objects like rubber bands, both with and without reinforcing fillers. When a small rubber product, such as a rubber band, is stretched, it absorbs heat and releases heat upon contraction.

This phenomenon can be explained using the concept of Gibbs free energy. According to the equation $\Delta G = \Delta H - T\Delta S$, where $G$ represents free energy, $H$ denotes enthalpy, and $S$ represents entropy, we find that $T\Delta S = \Delta H - \Delta G$. Since stretching is a nonspontaneous process requiring external work, $T\Delta S$ is generally negative. As temperature ($T$) is always positive (never reaching absolute zero), $\Delta S$ must be negative, indicating that natural rubber chains are normally entangled, but become less entangled when stretched [16].

When the strain is released, the retraction reaction continues, leading to a negative free energy ($\Delta G$) and a cooling effect. This results in a positive $\Delta H$, indicating the release of heat energy.

Therefore, when the tension is released, the reaction becomes spontaneous, causing $\Delta G$ to be negative. As a consequence, the cooling effect leads to a positive $\Delta H$, resulting in positive $\Delta S$.

The above explanation reveals that an elastomer behaves similar to an ideal monatomic gas. Elastic polymers do not store potential energy in a stretched or compressed state, but rather the work is done by the molecules during stretching. When all the work done in the rubber is released (not stored), it appears as thermal energy in the polymer material. Similarly, when other materials lose thermal energy to the surroundings, the elastomers cooling effect occurs.

This phenomenon indicates that the ability of an elastomer to perform work depends solely on changes in entropy and not on any stored energy within the polymer structure [17]. The energy required to perform work is obtained in the form of thermal energy. Positive entropy represents the loss of certain thermal energy as work is done, indicating the unavailability of energy.

## 2.10.2 Experiments

### 2.10.2.1 Variation of tensile stress with temperature

The equilibrium of temperature in a molecular system depends on its energy level. The energy level can be modified by mechanical actions, leading to alterations in entropy. The correlation between changes in energy and changes in entropy is directly related to absolute temperature, as long as the energy remains within the system's thermal state. However, if a stretched rubber material exceeds its limit and enters a nonthermal state, such as the distortion of chemical bonds, this principle no longer holds true.

In situations involving mild to moderate deformation, the principle suggests that the stretching force primarily arises from entropy changes within the network [12]. If this principle is accurate, then the force required for stretching should be directly proportional to the temperature of the sample. In our experiment, we maintained a constant strain on the stretched rubber while adjusting the temperature within the range of 20–50 °C.

These experiments provide evidence of entropy changes that underlie the fundamental mechanism of rubber elasticity. The positive linear relationship between stress and temperature often leads to the misconception that rubber has a negative coefficient of thermal expansion. However, like all other materials, the coefficient of thermal expansion for rubber is always positive [18].

Variation of tensile stress with temperature as strain is held constant at four values (100%, 200%, 300%, and 380%) is shown in Figure 2.28.

## 2.10.3 Snap back velocity

When a piece of rubber, such as a rubber band or strip, is stretched, it undergoes uniform lengthwise deformation, depending on the speed of stretching. At lower strains, every element along the length experiences the same extension throughout the sample. However, at higher speeds, this behaviour deviates. If we release one end, the

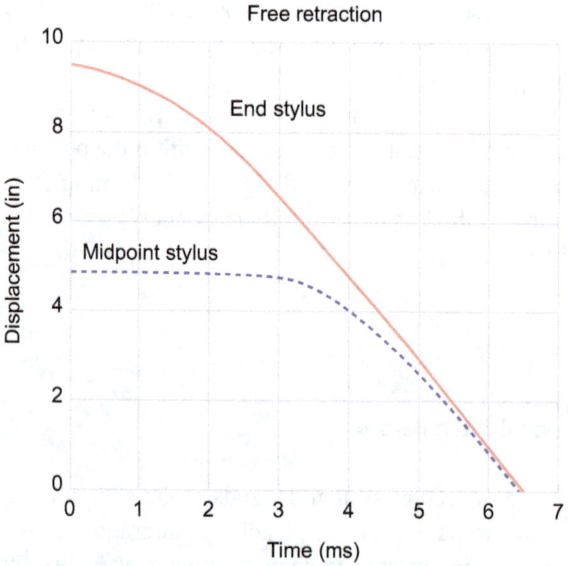


**Figure 2.28:** Variation of tensile stress with temperature as strain is held constant at four values (100%).

sample at lower strain quickly snaps back to its original length, which may not be easily detectable by the naked eye. However, experimental observations by Mrowca et al. reveal a surprising behaviour [13].

They conducted an experiment using a simple apparatus consisting of a rapidly rotating glass cylinder coated with lamp black. Styli were attached to the midpoint and the free end of the rubber sample, maintaining contact with the glass cylinder [14].

After the snap back, the marks left on the paper indicated that at high extension, the energy stored in the stretched rubber network chains is associated with a change in entropy. However, most of the energy is stored in bound conditions, which could involve an entropy change. If the stored potential energy is converted into kinetic energy, the retraction velocity can be calculated using the conservation of energy equation, $E = ½\, mv^2$. Numerical simulations based on molecular kinks can predict velocities [15].

The displacement of the end and the midpoint of the rubber sample over time as it snaps back from high extension is illustrated in Figure 2.29.

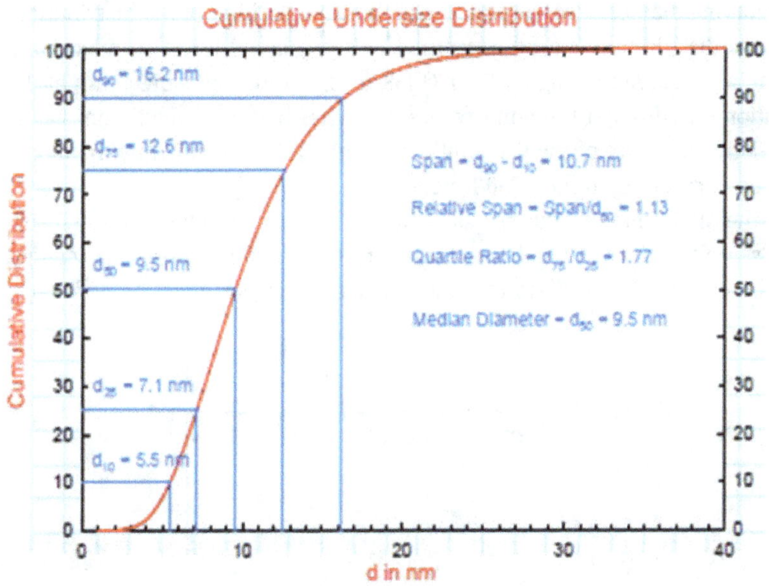


**Figure 2.29:** Displacement of the end and midpoint of a rubber sample versus time as it snaps back from high extension.

## 2.11 Aggregate size distribution and its significance

Carbon black is composed of primary particles that are approximately spherical and have nanometer-scale dimensions. These particles come together to form larger aggregates, which serve as the smallest dispersible unit of carbon black. The overall structure of carbon black is influenced by the shape and size of these aggregates. The aggregate size distribution (ASD) of carbon black is unique to each composition and can be considered a distinctive characteristic. It provides information about the distribution of aggregate sizes within the composition. This distribution has been identified as a significant factor in the reinforcing properties of rubber [22].

If all other features of carbon black are kept constant, smaller aggregates contribute stronger to wear performance. Bigger aggregates can be dispersed easily, but do not contribute as good as small aggregates to the wear performance.

It is common knowledge that the presence of larger aggregates, even in small proportions, can have a significant impact on wear performance. When describing carbon black compositions and differentiating one composition from another, at least two important values come into play.

The ASD provides information about the relative mass distribution at each size. Typically, the ASD exhibits a single peak (unimodal) but is not uniformly sized (monodisperse). The mode, also known as the modal diameter (Dmode), represents the most frequently occurring diameter, corresponding to the peak position in the distribution. The width of the distribution can be quantified by the full width at half maximum, also referred to as ΔD50. According to DIN ISO 15825, ΔD50 represents the width of the mass distribution measured at the point where it reaches half of its maximum value. To obtain a relative measure of the distribution's width, ΔD50 is divided by the modal diameter (Dmode), resulting in the ΔD50/Dmode value.

Although the ΔD50/Dmode value provides insight into the broadness of the distribution curve's peak, it does not indicate the quantity of small or large aggregates. Therefore, another value is necessary to complement the analysis.

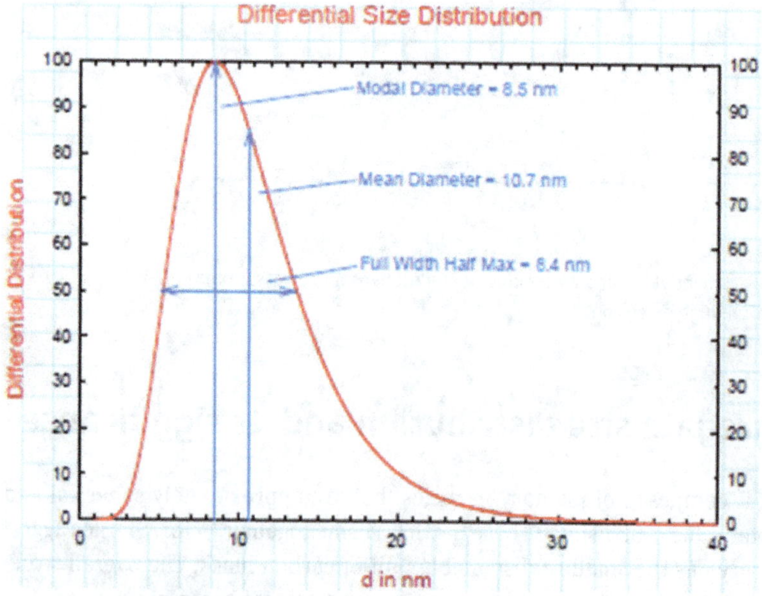


**Figure 2.30:** Cumulative distribution.

Normal distribution is differential distribution, as in Figure 2.31. If one integrates the ASD curve, the cumulative distribution is obtained as shown in Figure 2.30. It reflects the relative amount by mass at or below a particular size.

There are multiple ways to quantify the width of a distribution, based on the cumulative curve. One commonly used measure is the span, which is calculated as the difference between the 90th percentile (D90) and the 10th percentile (D10).

A dimensionless measure of width is the relative span (RS), defined as the ratio of the span to the median diameter (D50), expressed as (D90-D10)/D50. The volume fractions at 10%, 50%, and 90% on the cumulative curve are represented by D10, D50, and D90, respectively. A narrower distribution will result in a smaller absolute measure of width (ΔD50) and a decrease in RS, approaching zero.

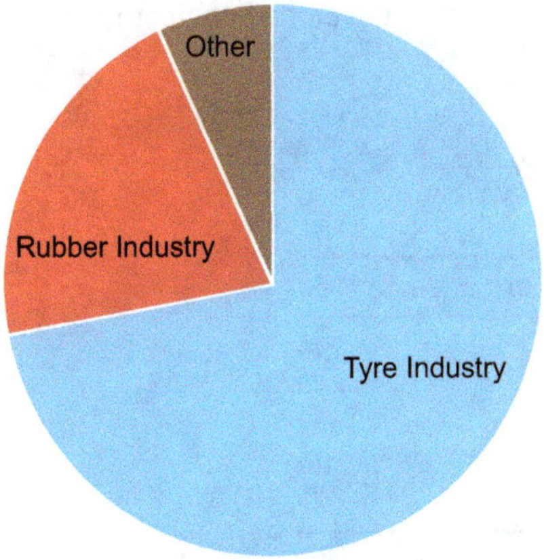


**Figure 2.31:** Differential distribution.

The quartile ratio (QR), expressed as D75/D25, is an additional parameter that describes the mass relationship between smaller and larger particles. According to DIN ISO 15825, the QR value represents the ratio of the upper quartile to the lower quartile of the area under the cumulative curve [23]. In contrast to ΔD50 and RS, a narrower distribution will cause this relative measure of width to approach unity.

Finally, the ratio of the mass-weighted average particle size (Dw) to the number-weighted average particle size (Dn) serves as a measure of the overall particle size homogeneity and is known as the Polydispersity Index or Dw/Dn. Larger values indicate an inhomogeneous distribution, while smaller values indicate a more homogeneous particle size distribution.

## 2.12 Significance of ASD on product property

1. The variation in the ASD of carbon black has a significant impact on its optical properties. This includes a decrease in tinting strength and an increase in the dissymmetry of light scattering.
2. In terms of vulcanisate properties, the effects resulting from differences in ASD are generally minor. However, one notable effect is a reduction in hysteresis for carbon black with a broader distribution.
3. Under conditions of moderately severe service, variations in the breadth of the ASD do not have a significant impact on tread wear, even if they lead to considerable changes in tinting strength.
4. It is important to note that tinting strength alone cannot reliably indicate the quality of tread wear unless carbon black with similar structures and breadth of ASD is compared.
5. Furthermore, these findings are consistent with previous research on carbon black blends, which have demonstrated significant decreases in tread wear resistance due to the presence of highly broadened nodule (particle) sizes and, coincidentally, ASDs.

Thumb rule for the estimation of Shore A Durometer in a developed compound is described in Table 2.4.

**Table 2.4:** Thumb rule for estimation of Shore A Durometer in developed compound.

| For 100 parts of polymer | Base durometer (Shore A) db |
|---|---|
| Polycholoroprene and nitrile rubber | 44 |
| Natural, SBR1502 and synthetic polyisoprene | 40 |
| SSBR and SBR1500 | 37 |
| Butyl rubber | 35 |
| SBR 1730 (oil extended SBR) | 25 |

Increase and decrease of hardness value with the change in filler & oil PHR is shown in Table 2.5.

**Table 2.5:** Increase and decrease of hardness values with change in filler and oil PHR.

| Filler and Oil | Durometer change |
|---|---|
| **N330, N550, N660** | +1/2 of part loading |
| **N220** | +1/2 of part loading + 2 |
| **N110** | +1/2 of part loading + 3 |
| **N774/N772** | +1/3 of part loading |
| **N990** | +1/4 of part loading |
| **Mineral oil** | −1/2 of part loading |

In table 2.6 the equal hardness approaches are illustrated using different grades of carbon black.

**Table 2.6:** The equal hardness approach.

| Replacement of carbon black | Hardness conversion factor for grade to grade replacement | | | | |
|---|---|---|---|---|---|
| | N990 | N774 | N660 | N650/N550 | N330 |
| **N774** | 0.665 | 1.000 | 1.100 | 1.220 | 1.500 |
| **N772** | 0.665 | 1.000 | 1.100 | 1.220 | 1.500 |
| **N660** | 0.600 | 0.900 | 1.000 | 1.110 | 1.360 |
| **N650** | 0.545 | 0.820 | 0.900 | 1.000 | 1.230 |
| **N550** | 0.545 | 0.820 | 0.900 | 1.000 | 1.230 |
| **N330** | 0.445 | 0.670 | 0.740 | 0.820 | 1.000 |

Required carbon black in PHR for 10° rise in compound hardness is described in Table 2.7.

**Table 2.7:** Carbon black required in PHR for 10° rise in compound hardness.

| ASTM no. | NR | SBR | IIR | CR | PBR | NBR | EPDM |
|---|---|---|---|---|---|---|---|
| **N110** | 15 | 18 | 13 | 12 | 22 | 17 | 24 |
| **N220** | 17 | 20 | 16 | 13 | 25 | 19 | 27 |
| **N326** | 21 | 25 | 19 | 17 | 32 | 24 | 34 |
| **N330** | 19 | 23 | 17 | 15 | 28 | 21 | 30 |
| **N339** | 17 | 21 | 16 | 14 | 26 | 20 | 28 |
| **N550** | 23 | 27 | 21 | 18 | 34 | 26 | 37 |
| **N660** | 25 | 26 | 23 | 20 | 38 | 29 | 41 |
| **N774** | 28 | 28 | 25 | 22 | 42 | 32 | 45 |

## 2.13 Carbon black in different applications

The automotive industry represents the largest market for carbon black, accounting for approximately 70% of its usage as a reinforcing agent. Carbon black plays a crucial role in improving wear resistance and ensuring uniform heat distribution in the tread and belt areas of tyres, thereby enhancing their longevity. Furthermore, it helps minimise thermal degradation, extending the service life of tyres. Approximately 20% of global carbon black production is dedicated to mechanical rubber goods, including automobile components, belts, hoses, and other non-tyre rubber products.

Around 8–9% of carbon black production is utilised in pigmented applications such as inks, coatings, paints, fibres, conductive polymers, and plastic masterbatches. It is also employed as an ultraviolet absorbent to mitigate photochemical degradation.

Conductive carbon black finds applications in electrostatic dissipaters, conductive cables, and various conductive coatings. It is also utilised in laser printers, toners, inks, and paints. The high tinting strength and stability of carbon black make it suitable for colouring resins and films.

In the cement industry, carbon black functions as an adhesive, imparting strength after curing. Additionally, it serves as an antistatic additive agent in automobile fuel caps and pipes.

Carbon black is used as a mild insecticide and fertiliser. Vegetable-derived carbon black is employed as a food colouring additive (known as additive E153 in Europe). It has been approved for use as additive 153 (carbon black or vegetable carbon) in colour pigments. Carbon black has been widely used for many years in food and beverage packaging, including in multilayer UHT milk bottles in the United States, parts of Europe and Asia, South Africa, as well as in items such as microwavable meal trays and meat trays in New Zealand.

In 2011, the Canadian government conducted an extensive review of carbon black and concluded that it should continue to be used in various products, including food packaging, as it is typically bound in a matrix and is unavailable for exposure. The review determined that carbon black does not pose a danger to human life or health in Canada.

Carbon black is extensively utilised as a colour pigment in packaging applications worldwide, where compliance with EU or US FDA regulations is necessary. The diverse applications of carbon black are outlined in Figure 2.32.

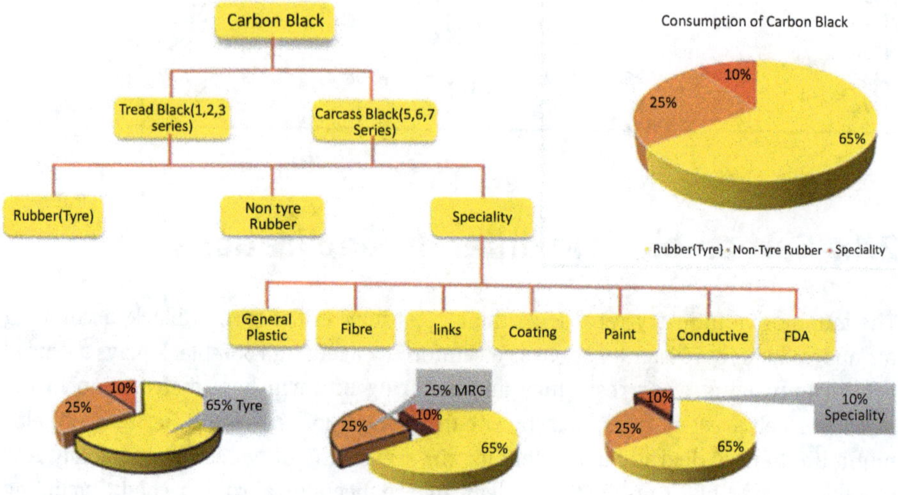


Figure 2.32: Different applications of carbon black.

The detailed applications of different grades of carbon black are given in Table 2.8.

Table 2.8: Applications of different grades of carbon black.

| ASTM grade rubber carbon blacks | Type | Effect on compound properties | Application area |
|---|---|---|---|
| N115 | It is a high fineness grade with medium structure and surface activity under super abrasion black. | It gives good tear strength and cut and chip properties. | It is good for tread compounds in off-road applications and truck tyres. |
| N121 | It is a high fineness, high structure grade with good surface activity under super abrasion black. | It gives excellent tread wear, traction, and durability. | It is used in passenger or truck and bus tread compounds. |

**Table 2.8** (continued)

| ASTM grade rubber carbon blacks | Type | Effect on compound properties | Application area |
|---|---|---|---|
| N134 | It is a high fineness, high structure with high surface activity grade under super abrasion black. | It gives excellent tread wear and traction. | It is mainly used in premium tread compound applications, including high-performance passenger tyres or truck and bus tyres, with optimum tread wear and durability. |
| N220 | It is a medium surface area, medium structure grade and an intermediate super abrasion black. | It gives lower hysteresis with good tear and cut and chip properties are required. | It is good for tread applications, primarily in truck tread compounds. |
| N234 | It is a medium surface area and high surface area grade and a very good general-purpose tread grade. | It gives a good balance of tread wear, traction, and rolling resistance, | It is used in many applications, including in premium radial tyre tread compound applications. |
| N326 | It is a low surface area and low structure carbon black. | It gives low compound viscosity, good fatigue life and crack growth properties. | It is primarily used in steel wire-coat compounds for radial tyres. |
| N330 | It is a general-purpose high abrasion carbon black. | It gives a good balance of properties such as abrasion resistance, hysteresis, and viscosity. | It is used for many body compound applications in tyres and rubber goods |
| N339 | It is a medium surface area and high structure grade under high abrasion black. | It gives a good balance between tread wear and hysteresis. | It is good for general-purpose tread compounds for passenger and truck tyres and body compounds for truck tyres. |
| N347 | It is a medium surface area, high structure grade (i.e., its structure is higher than N-339) under high abrasion black. | It gives a good balance of tread wear and hysteresis. | It is used for general-purpose tread compounds mainly for passenger and truck tyres and body compounds for truck tyres. |
| N375 | It is a medium surface area, low structure grade (i.e., its structure is lower than N339) under high abrasion black. | It gives a good balance of tread wear and hysteresis. | It is used in general-purpose tread compounds for passenger and truck tyres. |

**Table 2.8** (continued)

| ASTM grade rubber carbon blacks | Type | Effect on compound properties | Application area |
|---|---|---|---|
| N550 | It is a medium surface area and structure and a general-purpose fast extruding black. | It gives low heat buildup, long fatigue life, stiffness and tensile strength. | It is used for body compounds in passenger tyres and numerous mechanical rubber goods, especially in profiles. |
| N660 | It is a lower surface area (compared to N550) and medium-low structure and a general-purpose high modulus black. | Compared to N-550, it gives even lower heat buildup, long fatigue life and good processability. | It is used for body compounds in passenger tyres, including inner liners, as well as numerous rubber goods, for example, injection moulding. |
| N772 | It is a semi-reinforcing black with a low surface area and structure. | It is used in applications requiring minimal hysteresis and reinforcement that provides excellent processability. | It is mainly used in inner liners of tyres and in numerous industrial rubber goods. |

Current worldwide production is approximately 13 MMT. Approximately 91% is used in rubber application (tyre and non-tyre), 8% as pigment, plastics, fibres, papers, ink, and 0.8–1% in varied applications for example, cements, fertilisers and other applications as illustrated in Figure 2.33.

## 2.14 Carbon black in tyres

The biggest consumption of carbon black is in tyre application – near about 62% of total capacity of carbon black goes into the tyre application.

Pneumatic tyre is a complicated composite, generally comprising of polymers, fillers, tension members, and bead cord. Different components are used to build a tyre; it varies from 19 to 73 components.

Pneumatic tyre is a complete engineering item, which also integrates its trade design and aspect ratio with the compound of each component, ultimately giving the ability to perform major demands of the tyre manufacturing process, in terms of the following process parameters:

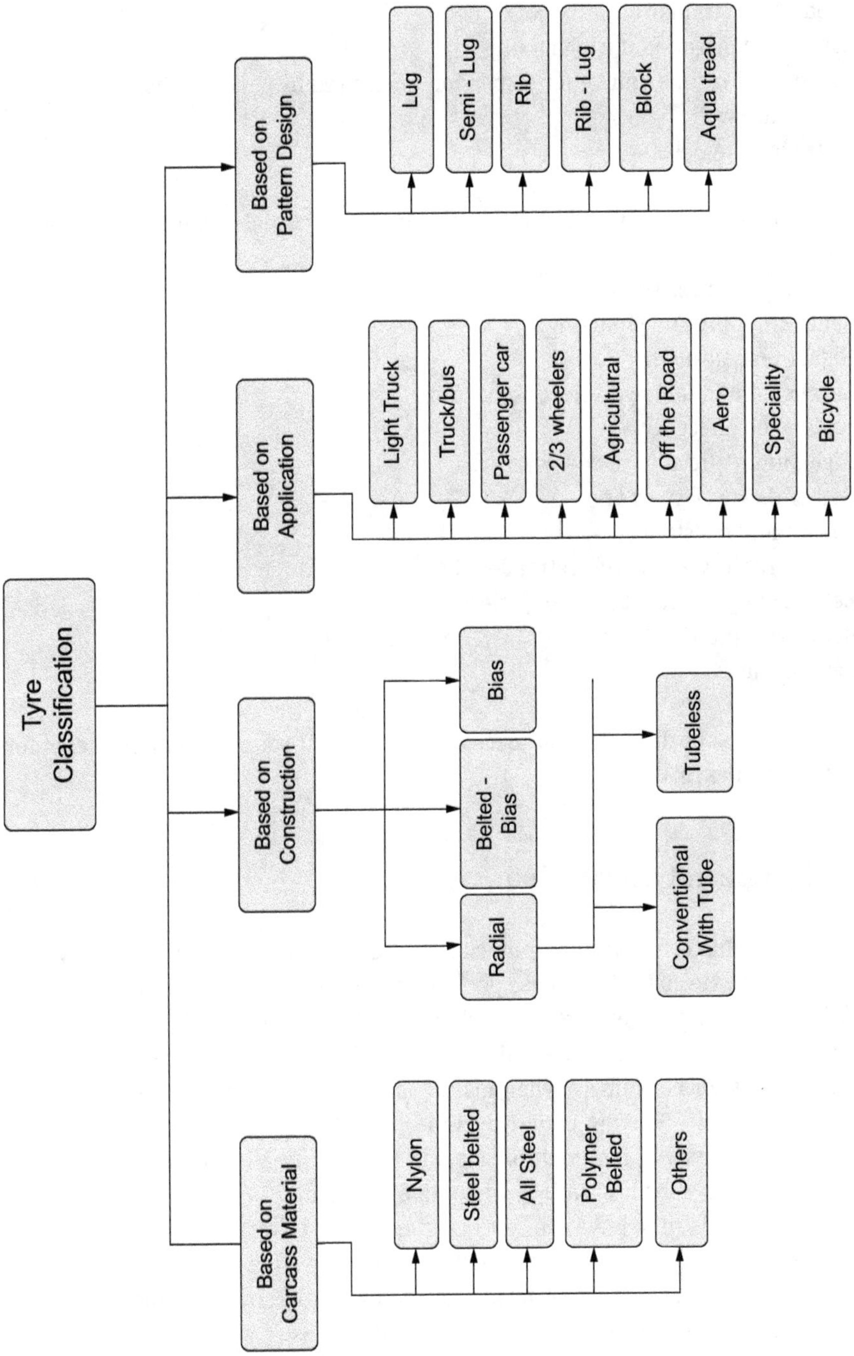


**Figure 2.33:** Carbon black application in different segments.

- Good green strength
- Low heat generation during mixing
- Lower mixing time and better dispersion (measured by conductivity method)
- Lower die swell
- Better building tack

Tyre users look for the following properties in a tyre, which is an integral part of their vehicle.

1. Good abrasion resistance
2. Better chipping and chunking
3. Higher tear strength
4. Low heat build up/low hysteresis/low tanΔ
5. Low Payne effect
6. High Mullin effect
7. Higher die-swell resistance
8. Low rolling resistance
9. High durability and better retread-ability
10. Better riding comfort/Cornering effect/handling
11. Better wet and ride traction
12. Reasonable cost

Regulatory bodies of different countries define different statutory requirements for tyres plying on the road.

## 2.14.1 European tyre labeling

Regulations mandate that tyre manufacturers must disclose the fuel efficiency, wet grip rating, and external rolling noise performance for C1, C2, and C3 tyres, which are primarily used on passenger cars, light vehicles, and heavy-duty vehicles.

Currently, the lack of reliable and comparable information on tyre performance poses challenges for consumers when making purchasing decisions, especially when replacing used tyres. Starting from November 1, 2012, tyre performance data is required to be displayed at the point of sale and included in all technical promotional materials such as catalogues, leaflets, and online marketing.

The objective is to drive a market transformation towards more fuel-efficient, safe, and low-noise tyres, surpassing existing standards. This initiative also encourages competition, based on tyre performance and price, thereby fostering investments in research and development of materials and tyre designs.

The EU Tyre labeling system is implemented by the following three criteria.

## 2.14.2 Fuel efficiency

Fuel efficiency is evaluated based on the rolling resistance (RR) of tyres.

Rolling resistance refers to the resistance experienced when a round object, like a ball or a tyre, moves in a straight line at a constant velocity on a flat surface. This resistance primarily arises from the deformation of the object itself, the deformation of the surface it rolls on, or a combination of both.

Several factors influence rolling resistance, including the radius of the wheel, the speed at which it moves, the adhesion of the surface, and the extent of microsliding between the contacting surfaces. The material composition of the wheel or tyre and the type of ground it interacts with also play significant roles in determining the rolling resistance.

There are seven classes from G (least efficient) to A (most efficient) as described in Table 2.9.

**Table 2.9:** Seven classes from G (least efficient) to A (most efficient).

| Class | PCR (C1) | LTR (C2) | TBR (C3) |
|---|---|---|---|
| A | RRC ≤ 6.5 | RRC ≤ 5.5 | RRC ≤ 4.0 |
| B | 6.6 ≤ RRC ≤ 7.7 | 5.6 ≤ RRC ≤ 6.7 | 4.1 ≤ RRC ≤ 5.0 |
| C | 7.8 ≤ RRC ≤ 9.0 | 6.8 ≤ RRC ≤ 8.0 | 5.1 ≤ RRC ≤ 6.0 |
| D | Empty | Empty | 6.1 ≤ RRC ≤ 7.0 |
| E | 9.1 ≤ RRC ≤ 10.5 | 8.1 ≤ RRC ≤ 9.2 | 7.1 ≤ RRC ≤ 8.0 |
| F | 10.6 ≤ RRC ≤ 12.0 | 9.3 ≤ RRC ≤ 10.5 | 8.1 ≤ RRC |
| G | 12.1 ≤ RRC | 10.6 ≤ RRC | Empty |

Effects may vary among vehicles and driving conditions, but the difference between a G and an A class for a complete set of tyres could reduce fuel consumption by up to 7.5% and even more in the case of trucks.

## 2.14.3 Wet grip

Wet grip refers to the ability of tyres to brake effectively on wet road surfaces, and it directly impacts the safety performance of vehicles. While tyres with low rolling resistance offer higher fuel efficiency, they can pose safety concerns. This is due to the fact that tyres with low rolling resistance tend to have reduced adherence to wet road surfaces. As a result, the European Council mandates that tyre companies disclose information about the adherence (or grip) of their tyres when braking on wet roads. There are seven classes from G (longest braking distances) to A (shortest braking distances) as shown in Table 2.10.

**Table 2.10:** Seven classes from G (longest braking distances) to A (shortest braking distances).

| Class | PCR (C1) | LTR (C2) | TBR (C3) |
|---|---|---|---|
| A | $1.55 \le G$ | $1.40 \le G$ | $1.25 \le G$ |
| B | $1.40 \le G \le 1.54$ | $1.25 \le G \le 1.39$ | $1.10 \le G \le 1.24$ |
| C | $1.25 \le G \le 1.39$ | $1.10 \le G \le 1.24$ | $0.95 \le G \le 1.09$ |
| D | Empty | Empty | $0.80 \le G \le 0.94$ |
| E | $1.10 \le G \le 1.24$ | $0.95 \le G \le 1.09$ | $0.65 \le G \le 0.79$ |
| F | $G \le 1.09$ | $G \le 0.94$ | $G \le 0.64$ |
| G | Empty | Empty | Empty |

The impact of these effects can differ, depending on the vehicle and driving circumstances. However, in the scenario of full braking, switching from a G-class to an A-class set of four identical tyres could potentially result in a braking distance that is up to 30% shorter. For example, for a typical passenger car traveling at a speed of 80 km/h, this reduction in braking distance could amount to up to 18 m.

### 2.14.4 Noise level

Outdoor sound levels are assessed using decibels (dB) and represented by three categories (indicated by vertical lines on the left). The greater the number of vertical lines, the higher the amount of noise produced by the tyres on the road. Along with the decibel measurement (dB(A)) for noise levels, a symbol indicates whether the tyre's external rolling noise exceeds the upcoming mandatory limit set by Europe (three vertical lines indicate a noisier tyre), falls within 3 dB below the future limit (two vertical lines indicate an average noise level), or exceeds 3 dB below the future limit (one vertical line indicates a low-noise tyre).

## 2.15 Types of tyres

This is depicted in Figure 2.34.

One of the major areas of tyre performance depends heavily on the polymer–filler interaction because these two components account for near about 60% of material used in tyres. A correct dose and morphology of filler used will ensure an optimum performance with a suitable construction and tyre design.

Filler morphology generally determines today a correlation with primary structure (STSA) and secondary structure (OAN). The values of STSA on OAN generally give an idea of the reinforcement and processing behaviour of the internal mixture, calender, and extruder.

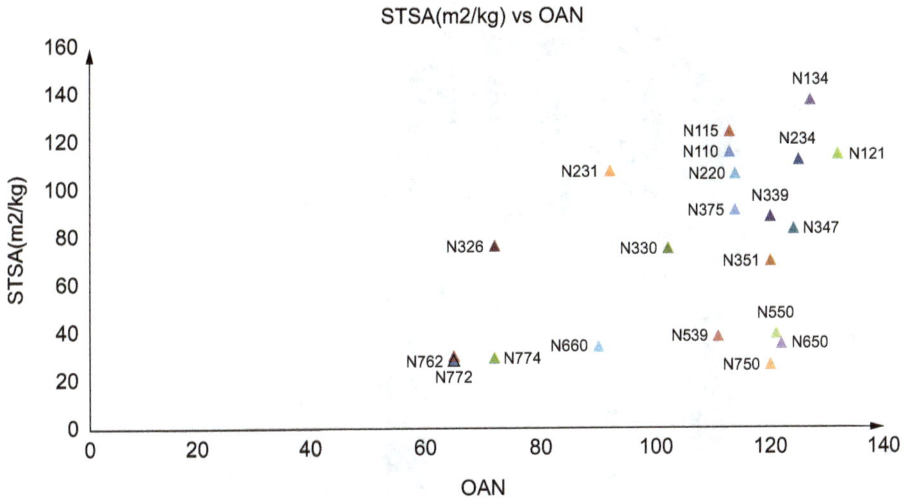


**Figure 2.34:** Tyre classification.

Graph 2.1 shows the position of different grades based on STSA (which is an abstract measurement of reinforcement) against the secondary structure (which is also a major factor for processing behaviours).

## 2.15.1 Tyre construction and use of carbon black in different segments

Different types of tyres are illustrated in Figure 2.35.

Use of carbon black in different components of a tyre is specified in Table 2.11.

The above use patterns are generally in vogue in the tyre industries but in many places, they use highly dispersed silica in tread and carcass compound in a combination of the abovementioned carbon black. Partial replacement of silica gives low rolling resistance in certain types of tyres, but with a compensation of inferior abrasion resistance. There is also a tendency to use dual black in tread and sub tread. The combinations are either few tread (hard black) or one hard and one soft black (carcass black), which generally improve the tear strength and sometimes reduce the heat generation.

## 2.15.2 Special black for tyre

In today's high-speed era, vehicle users want tyres that last longer. Consumers want high mileage passenger car tyres for durability and value (on per kilometer/mile basis). Truck operators also require high durability of their tyres to eliminate in-road failures;

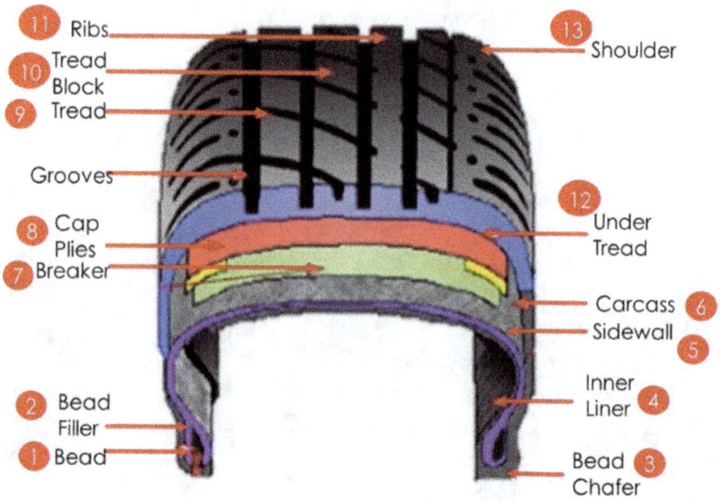


**Figure 2.35:** Different parts of a tyre.

**Table 2.11:** Carbon black application in different components of a tyre.

| S. no. | Component | Grades |
|---|---|---|
| 1 | Bead | 700 series/comb of high filler loading |
| 2 | Bead filler | N660/N650/N550 |
| 3 | Bead chafer | N660/N550 |
| 4 | Inner liner | N660/N650/N550/N539 |
| 5 | Sidewall | N326/comb N326 and N330/N330 and N339/N339 + N234 |
| 6 | Carcass | N650/N660/N550/N330/N326/N351 |
| 7 | Breaker | N330/N220/N339 |
| 8 | Cap plies | N330/N220 |
| 9 | Tread | N234/N134/special black for truck tyre, for synthetic tread N120/N121/tyre gold/tyre gold plus/RTH 300 |
| 11 | Ribs | N134/N234/N220 |
| 12 | Under tread | N220/N330/N339 |
| 13 | Shoulder | N330/N234/N220 |
| 14 | Bead apex | N660/N550 combination/N550 less |
| 15 | Passenger and light truck tyre | N220/N234/N339/N351/N121 |

**Table 2.11** (continued)

| S. no. | Component | Grades |
|---|---|---|
| 16 | Truck and OTR tread | N134/N110/N115/N231/N220/N339/N330 |
| 17 | Belt toping component | N330/N326/N550/N650 |

so the operations of the tyre are maximum. To meet these demands, a tyre manufacturer needs much higher reinforcement black to deliver a tyre with longer life. Different black manufacturers across the globe introduced ultra-reinforced carbon black, which pushed the boundaries of performance, improving tread wear resistance. This black generally has higher surface area per unit mass, with much improvement in its structure.

These categories of black, if chosen properly, will reduce rolling resistance and improve fuel efficiency. We generally know the hysteresis, tanΔ, and the Payne effect when used in a tread compound.

This black can also be combined with conventional ASTM carbon black and deliver lower rolling resistance tyre tread without the need for reactive coupling.

A tyre undergoes tremendous strain due to the different climatic conditions, as revealed in Figure 2.36.

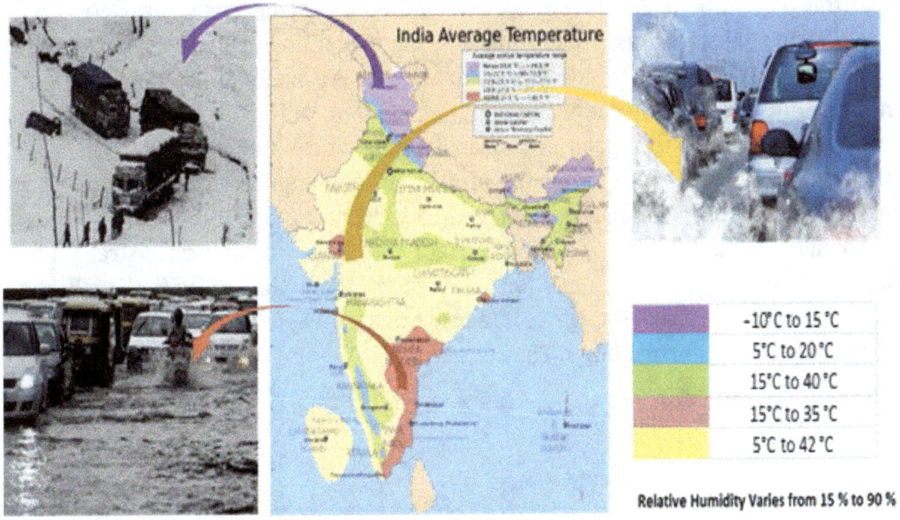


**Figure 2.36:** Different climatic conditions in India.

Performance conscious tyre byers are not only putting pressure on tyre manufacturers; in many countries, their government is implementing new programmes such as tyre labeling to promote the use of different kinds of integrated vehicle technologies that can reduce the green house gases emission. New programs of different countries as for example the US Environmental Protection Agencies' and the European Union Tyre Labeling System seek to educate consumers about the impact of tyre performance on fuel economy and drive the adoption of low RR (rolling resistance) tyres. In fact, a judicial choice from these blacks, either alone or in combination with ASTM blacks, can reduce RR to similar or better than white reinforcing fillers.

## 2.16 Carbon black in non-tyre rubber application (IRP)

This section comprises rubber goods used for industrial applications other than for automobile tyres. This group of products is generally called industrial rubber products (IRP). This section caters to more than 700 products but below is the classification of the major products detailed in Figure 2.37.

Carbon black application in different MRG applications are described in Table 2.12.

### 2.16.1 Carbon black in MRG application

Over the years, the demand from the automobile industries is getting increasingly stringent due to its versatile application coupled with aesthetic looks. Clean and ultraclean blacks are getting introduced in the market. These blacks are of particular value in the manufacture of mechanical rubber goods (MRGs) like extruded profiles and seals. The blacks impart a good processability and contribute to a smooth surface of the final compound. It also eases the production of complicated profiles and parts having different hardness or metal inserts in a particular profile.

Demand for an absolutely smooth surface with higher compressive strength and higher tear energy are the genesis for these blacks. These blacks also find application in the inner liners of tubeless tyres, imparting much less air permeability than if it is compounded with normal black. Figure 2.38 details the use of carbon black in different MRG applications.

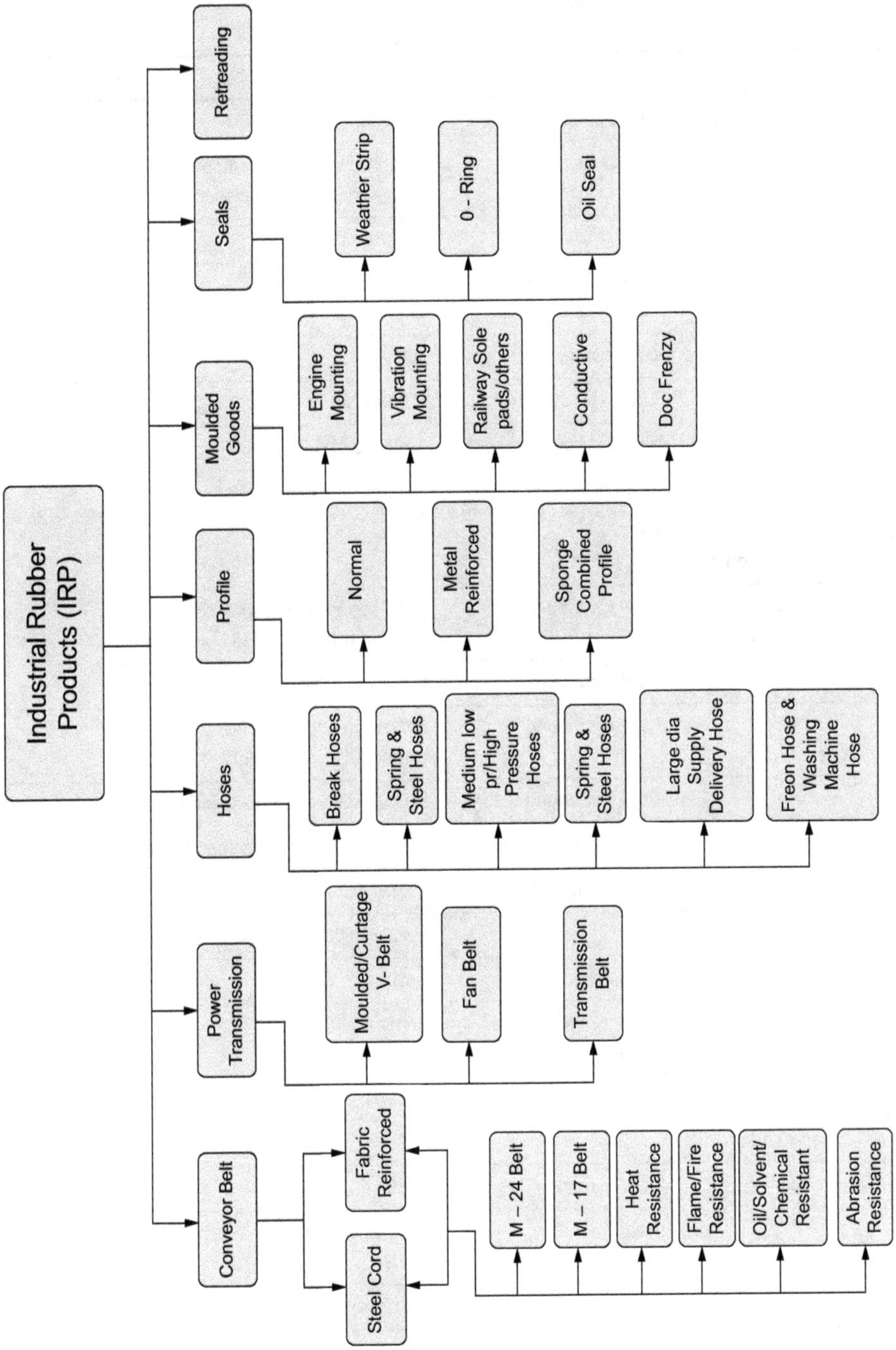


**Figure 2.37:** Classification of industrial rubber products.

**Table 2.12:** Use of carbon black in different MRG applications.

| S. no. | Component | Part | Carbon black |
|---|---|---|---|
| 1 | Conveyor belt | M – 24 Belt | N330/N375 |
| | | M – 17 Belt | Any hard black, depending upon the requirements of installation |
| | | Heat resistance | N220/N339 |
| | | Superior abrasion resistance belt | N234/N134 |
| | | Oil resistance | N339 |
| | | Chemical resistance | N339/N220 |
| | | Flame resistance | N220/N234 |
| | | Carcass all belt | N660/N330/N550/N650 |
| 2 | Power transmission belt | N660 | |
| | V-belt | Base compound | N700 series/N500 series, sometimes in combination with N990 |
| | | Cord compound | N330/N550 series, N660 series |
| | Fan belt | Base compound | N700 series/N500 series, sometimes in combination with N990 |
| | | Cord compound | N330/N550 series, N660 series |
| 3 | Hose | Landing compound | N500/N700 series and special blacks |
| | | Cover compound | N550/N650, sometimes N330 and N220 |
| 4 | Profiles | | Mainly N500, N700 and special blacks |
| 5 | Moulded goods | Cheap product | N330 |
| | | Metal rubber bonded parts | Generally 700 series |
| 6 | Seal | | N550/N220/N330/N660/N650/N700 series |
| 7 | Dock fender | | N330 |
| 8 | Retreading | Bonder gum | N660 |
| | | Tread | N330/N339/N220/N375/N220/N234/N134/special grades |
| 9 | Conductive | | Special grade |

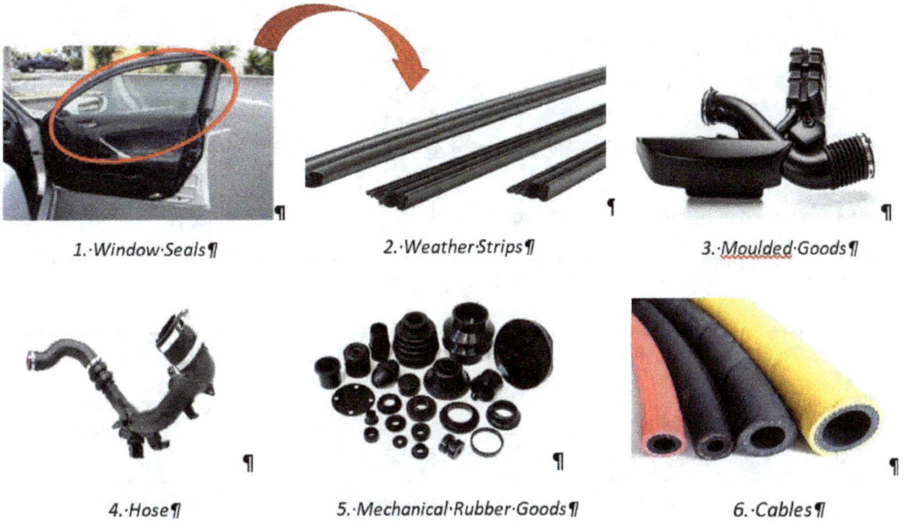


**Figure 2.38:** Use of carbon black in different MRG applications.

## 2.17 Testing and its significance

There are many methods for carbon black testing, which define the properties of carbon black. These properties also help to understand and explain carbon black properly.

Different properties of carbon black and their significance:

- Surface area
  - $I_2$ adsorption no.
  - $N_2$ adsorption no.
  - STSA
  - CTAB (cetyl trimethyl ammonium bromide) adsorption no.
- Structure
  - DBP (di-butyl phthalate) adsorption no.
  - CDBP (compressed DBP) adsorption no.
- Tinting strength – uniformity of particle distribution within the black.
- Toluene discolouration (TD) -
- Contamination
  - Ash
  - Grit
- Pellet quality
  - Pour density
  - Pellet hardness
  - Pellet size distribution

- Fines
- Attrition
- Heat loss (moisture)

Testing of carbon black and its significance are described in Table 2.13.

**Table 2.13:** Testing of carbon black and its significances.

| S no. | Standard | Year of standard | ASTM no. | Unit | Significance |
|---|---|---|---|---|---|
| 1. | $I_2$ adsorption no. | 2016 | D-1510 | mg/g | The $I_2$ adsorption no. is useful in characterising CB. It is related to the surface area of the CB. |
| 2. | $N_2$ adsorption no. | 2016 | D-6556 | $m^2/g$ | It is useful in determining the surface area/particle size of CB. |
| 3. | STSA | 2016 | D-6556 | $m^2/g$ | Measures the actual available surface for reinforcement. |
| 4. | CTAB adsorption | 2004 | D-3765 | $m^2/g$ | It is useful in determining the surface area/particle size of CB. |
| 5. | OAN | 2016 | D-2414 | cc/100 g | The OAN of CB is a measurement of the secondary structure and is related to the processing and vulcanisate properties of rubber compound containing the CB. |
| 6. | COAN | 2016 | D-3493 | cc/100 g | The COAN of CB gives an idea of the primary structure of CB. The difference between OAN and COAN gives an idea of the porosity and the void volume. |
| 7. | Tinting strength | 2015 | D-3265 | %R | Tint generally gives an idea of the jetness of CB. For a broad range of commercial rubber grade CB, tint strength is highly dependent upon particle size.<br>Tint strength is also dependent on the structure and aggregate size distribution. Tint strength is a measurement of uniformity of particle size and its distribution within the grade. |
| 8. | Toluene discolouration (TD) | 2011 | D-1618 | %T | It is used in measure the residual/unburned oil present in CB. |
| 9. | pH | 2015 | D-1512 | | The pH of the carbon black is often used in this industry to indicate the relative acidity or alkalinity of carbon black and will be used in the remainder of these test methods to describe this property. The pH level of a carbon black is known to affect the vulcanisation of some rubber compounds. |

**Table 2.13** (continued)

| S no. | Standard | Year of standard | ASTM no. | Unit | Significance |
|---|---|---|---|---|---|
| 10. | Ash content | 2015 | D-1506 | % | This test indicates the level of inorganic impurities (ionics) coming primarily from the water used for quenching, material of construction of lines, and the additive used for controlling the properties of CB, and also from the feedstock. |
| 11. | Grit/sieve residue | 2015 | D-1514 | ppm | The grit (sieve residue) test provides information on particulate impurities, which may contain coke particle (carbonaceous material), metal or ceramic particles, originating from the production unit or coke particles formed during the production process. |
| 12. | Pour density | D-1513 | 2017 | | Pour density is a function of the degree of compaction during pelletisation. It is strongly influenced by and is inversely proportional to the structure (OAN). Pour density of carbon black is useful for estimating the weight-to-volume relationship for certain applications, such as automatic batch loading systems and for estimating the weights of bulk shipments. |
| 13. | Fines content | 2012 | D-1508 | % | The fines content of carbon black is related to the bulk flow ability, dustiness, and, in some instances, the level of dispersion due to the many other variables that influence dispersion and handling. |
| 14. | Attrition fines | 2012 | D-1508 | % | By comparing the percent fines and attrition, an indication can be obtained of the pellet strength, stability, size and the amount of fines that may be created by pellet degradation in conveying, handling, or transit. |
| 15. | Pellet hardness | 2014 | D-5230 | gm-force | This test method covers a procedure for measuring individual pellet hardness of carbon black by the automated pellet hardness tester. This property is important for dispersion and handling. |
| 16. | Heat loss | 2015 | D-1509 | % | It is used to measure the moisture content of carbon black. |

**Table 2.13** (continued)

| S no. | Standard | Year of standard | ASTM no. | Unit | Significance |
|---|---|---|---|---|---|
| 17. | Pellet size distribution | 2017 | D-1511 | ppm | The variation in the size of the pellets may relate to the level of dispersion and to the ease of handling. Due to the many other variables that influence dispersion and handling, the significance of pellet size must be determined by the user. It is a parameter that affects the flow characteristic of pelletised CBs. A uniform pellet size means a lower bulk density, hence ensuring optimum flow behaviour. |
| 18. | PAH | | | | It measures the presence of polynuclear hydrocarbon in carbon black surface for its use in human contact application. |

### 2.17.1 Common understanding of the major testing

#### 2.17.1.1 Grit or sieve residue

This testing procedure involves measuring the residue left after water washing regular untreated carbon black. However, it may not be suitable for oil-treated carbon blacks as the presence of oil would hinder the effective wetting of the carbon black particles by water.

## 2.18 ASTM standard for testing: ASTM D-1512

**Significance and uses:** The presence of residue on the sieve is of significance in certain moulded or extruded products since it can affect their surface appearance. The maximum allowable residue for each application is typically determined through mutual agreement between the user and the manufacturer.

Grit is mainly the impurities that occur while manufacturing carbon black. There are mainly three types of grits: inorganic grit, organic grit, metallic grit. There are several reasons for grit formation. Inorganic grits mainly occur due to refractory or uniformed flame distribution or improper distribution of nozzle spray, or due to irregular spray pattern.

Organic grits mainly occurs due to coke formation, impure oil quality, nozzle corkage, and so on. Metallic grits are generally from the metallic lining of different equipment.

### 2.18.1 Effect of high grits in product quality

- Frequent corkage/blocking of strainer/wire mesh during straining of butyl or natural rubber-based inner tube profile compounds, resulting in loss of production and probable defects.
- Nonuniform dispersion of carbon black in compounds result in irregular mechanical properties.
- Frequent die cuts during extrusion of automotive tyre tread and sidewall compounds result in loss of extrudate production.
- Due to poor dispersion of carbon black, mainly tensile, tear, and flexing properties are found to be poor.
- Non–mixed carbon black grit in finished rubber compound acts as a flaw or nucleus to initiate crack or failure.
- Due to poor dispersion of carbon black during straining of butyl/natural rubber inner tubes compounds, excessive heat generation takes place, leading to scorching of compounds (loss of costlier compounds).
- Metals act as pro-oxidant of rubber that enhances the oxidation of rubber compound, leading to poorer weather properties.
- Damage to the process equipment's rubber goods.

### 2.18.2 Iodine adsorption number

Carbon black is a colloidal carbon substance that is created through the incomplete combustion of dense petroleum products. Typically, it exists as oligomers, and the size of their aggregates is smaller than 1,000 nm. The iodine adsorption number, measured in mg/g of carbon, indicates the capacity of a specific mass of carbon black to adsorb iodine on its surface.

The iodine adsorption number is influenced by the irregularities on the surface of the carbon black and is therefore a valuable parameter for characterising its surface area. This measurement is conducted following the guidelines outlined in ASTM D 1510-16. The test method described therein is utilised to determine the surface area of carbon black.

## 2.19 Standard of testing: ASTM D-1510 (year 2016)

**Significance and uses:** The iodine adsorption number plays a significant role in characterising carbon black. It provides valuable information about their surface area and is typically correlated with the NSA. Factors such as the presence of volatile components, surface porosity, or extractable substances can impact the iodine adsorption number. Additionally, the iodine number can be influenced by the ageing process of carbon black.

Effect of high iodine number on rubber process properties:

- Scorch time of compound decreases, leading to poor processing safety.
- Incorporation time of carbon black and mill bagging increases.
- Loading capacity and dispersibility of carbon black decreases.
- Viscosity and extrusion smoothness increases.

Effect of high iodine number on rubber vulcanisate properties:

- Compression set is low due to difficulty in dispersion.
- Cut growth resistance, flex resistance, and electrical conductivity increases.
- Tensile strength, modulus, hardness, abrasion resistance, and tear resistance increases.
- Resilience decreases and heat buildup increases.

### 2.19.1 Oil absorption number (OAN)

This test method covers the determination of the OAN of carbon black.

## 2.20 Standard for testing: ASTM D-2414

**Significance and uses:** The OAN of a carbon black is related to the processing and vulcanisate properties of rubber compounds containing the carbon black.

**Effect of high OAN on rubber processing properties:**

- Extrusion shrinkage decreases and extrusion smoothness increases.
- Loading capacity of carbon black decreases and mill bagging increases.
- Incorporation time and dispensability of carbon black increases.
- Viscosity increases and as a result, processing safety of compounds become low.

**Effect of high OAN on rubber vulcanisate properties:**

- Impact on rate of cure, tear resistance, resilience, heat buildup, compression set, and electrical conductivity is not very significant.
- Modulus, hardness, abrasion resistance, and dimensional stability increases.
- Tensile strength, elongation, cut growth resistance, and flex resistance decreases.

**COAN:** This testing procedure involves mechanically compressing a sample of carbon black and determining the OAN of the compressed sample.

## 2.20.1 Standard for testing: ASTM D-3493

### 2.20.1.1 Significance and uses

The OAN of a carbon black is related to the processing and vulcanisate properties of rubber compounds containing the carbon black. It actually indicates the void volume, which contributes to the dispersion and reinforcement properties of CB.

### 2.20.1.2 Tint strength

This test method covers the determination of the tint strength of carbon black, relative to an industry tint reference black (ITRB #9).

## 2.20.2 Standard for testing: ASTM D-3265

### 2.20.2.1 Significance and uses

- Higher the tint, higher will be blackness of compounds. Carbon black having high tint value is widely used in the paint industry.
- Higher tint strength relates to particle size and uniformity.
- Higher value of tint indicates the reinforcing characteristics of carbon black. Generally, higher the value of tint, higher will be the modulus, tensile strength, hardness, abrasion resistance, and tear and flexing resistance.
- In tread black, higher tint generally indicates higher millage.

### 2.20.2.2 Toluene discolouration

This testing procedure focuses on measuring the level of TD caused by extractable substances from carbon black. It is a valuable tool for monitoring and regulating the reaction processes involved in carbon black production. However, this particular test method may not be suitable for carbon black with high levels of extractable substances.

#### 2.20.2.2.1 Effect of TD value

- High TD%: High value of TD is always desirable due to the absence of unreacted oil in carbon black; it improves reinforcing characteristics of rubber compounds.
- Low TD%: Low value of TD causes a lot of adverse impact due to the presence of unreacted conversion oil on the carbon black surface. It causes migration from the

product and is not suitable for human contact application and medicinal usage. It does not provide adequate reinforcement to rubber compounds. Carbon black having specified iodine number but low processing TD value normally gives products with high glossy surface. It sometimes aids to the extrusion process.

#### 2.20.2.3 Fines content

This test method covers the determination of the fines and the attrition of pelleted carbon black.

### 2.20.3 Standard for testing: ASTM D-1508

#### 2.20.3.1 Significance and uses

The presence of fine particles in carbon black is associated with characteristics such as bulk flowability, dustiness, and, in certain cases, the level of dispersion. However, it should be noted that various other factors also influence dispersion and handling.

By examining the percentage of fines and assessing attrition, valuable insights can be gained regarding pellet stability and the potential generation of fine particles resulting from pellet degradation during conveying, handling, or transportation.

#### 2.20.3.2 Effect of fines content on product quality

**High fines:**

- Carbon black having higher values of fines is not suitable for rubber compound mixing. It causes more incorporation time of carbon black in rubber, in rubber compound mixing. The degree of carbon black dispersion is also affected by the presence of high fines content.
- Flowability of carbon black in handling during mixing (carbon black silos to Banbury) is very much affected by the higher fines content of carbon black.

**Low fines:** Low value of fines content is highly desirable, since it gives improved/uniform carbon black dispersion. As a result technical and service properties of rubber compound are very good.

#### 2.20.3.3 Pour density

This test method covers the determination of the pour density of pelleted carbon black. This is a very useful parameter for handling and transportation of carbon black.

### 2.20.4 Standard: ASTM D-1513 (2017)

#### 2.20.4.1 Significance and uses

The pour density of carbon black is determined by the level of compaction during the pelletisation process. It is notably affected by the structure of the carbon black (measured by OAN) and follows an inverse relationship. The pour density provides valuable information for estimating the weight-to-volume ratio in specific applications, including automatic batch loading systems, as well as for approximating weights in bulk shipments.

#### 2.20.4.2 Effect of variation of pour density on carbon black property

Pour density value directly contributes to the better understanding of the packing behaviour of carbon black pellet as well as in rubber mixing. It also contributes to the flow pattern of carbon black from silo to Banbury.

#### 2.20.4.3 Pellet size distribution (PSD)

This testing procedure involves measuring the distribution of pellet sizes in carbon black samples.

### 2.20.5 Test standard: ASTM D-1511 (2017)

#### 2.20.5.1 Significance and uses

The differences in pellet sizes can impact both the level of dispersion and the ease of handling. Hence, it is important to note that there are numerous other factors that also influence dispersion and handling.

#### 2.20.5.2 pH

The pH of carbon black is commonly employed in this industry to determine its relative acidity or alkalinity. Throughout the remaining test methods, the pH will be utilised to describe this characteristic.

### 2.20.6 Standard method of testing: ASTM D-1512 (2015)

#### 2.20.6.1 Significance and uses

The pH level of carbon black can impact the vulcanisation process of certain rubber compounds.

#### 2.20.6.2 Effect of pH on property

- High pH: For a specified dosage of sulphur and accelerators, the rate of cure of rubber compounds become faster, leading to less processing safety (chances of more processing scrap).
- Low pH: For a specified dosage of sulphur and accelerators, the rate of cure of rubber compound become slower resulting in less curing of rubber compounds (chances of more vulcanisation scrap generation).

The optimum pH of carbon black for better processing and vulcanising properties is generally recommended to be 7–9.

#### 2.20.6.3 Pellet hardness

This testing procedure entails using an automated pellet hardness tester to measure the hardness of the individual pellets of carbon black.

### 2.20.7 ASTM standard for testing: ASTM D-5230 (2016)

#### 2.20.7.1 Significance and uses

- Pellet hardness generally has an optimum value for a particular application; higher or lower values lead to mal-dispersion.

#### 2.20.7.2 Nitrogen surface area (N2SA)

This testing procedure involves determining the total surface area of a substance using the Brunauer, Emmett, and Teller (BET) theory, which relies on multilayer gas adsorption behaviour and utilises multipoint determinations. Additionally, the external surface area is determined using the STSA method.

### 2.20.8 ASTM standard for testing: ASTM D-6556

#### 2.20.8.1 Significance and uses

This test method is used to measure the total and external surface area of carbon black, based on multipoint nitrogen adsorption. The NSA measurement is based on the BET theory and it includes the total surface area, inclusive of microspores – pore diameters less than 2 nm (20 Å). The external surface area, based on STSA, is defined as the specific surface area that is accessible to rubber.

#### 2.20.8.2 Effect of variation of N2SA on product quality

- It is generally affected by porosity, surface impurity, and surface oxidation.
- It is a much reliable and accurate method than iodine adsorption no. for the determination of surface area of carbon black.

#### 2.20.8.3 Statistical thickness surface area

The external surface area, based on STSA, is defined as the specific surface area that is accessible to rubber.

STSA can be calculated from the empirical BET equation:

$$\frac{P}{V_a(P_0 - P)} = \frac{1}{V_m C} + \frac{C-1}{V_m C} \times \frac{P}{P_0} \tag{2.6}$$

where $P$ is the manometer pressure in kPa, $P_0$ is the saturation vapour pressure of nitrogen in kPa, $V_m$ is the volume of nitrogen per gram that covers one monomolecular layer in standard $cm^3/g$, $C$ is the BET constant and its numerical value depends on the heat of adsorption of the monolayer.

The statistical surface area is determined by using a plot of volume of nitrogen gas adsorbed per gram of sample at STP ($V_a$) versus the statistical layer thickness ($t$), where $t$ is the statistical layer thickness of carbon black = $0.088(P/P_0)^2 + 0.645\ (P/P_0) + 0.298$.

STSA to the nearest 0.1 $m^2/g$ is computed as follows:

$$STSA = M \times 15.47 \tag{2.7}$$

where $M$ is the slope of the $V_a - t$ plot; 15.47 is a constant for the conversion of nitrogen gas to liquid volume, and conversion of units to $m^2/g$.

STSA gives a very close correlation of reinforcement either through mechanical method or the swelling method.

#### 2.20.8.4 Ash content

These test methods cover the determination of the ash content of carbon black.

### 2.20.9 ASTM standard for testing: ASTM D-1506

#### 2.20.9.1 Significance and uses

The ash content of carbon black refers to the quantity of non-carbon components remaining after combustion. The main contributors to ash are water from the manufacturing process and the catalyst in the feedstock. Higher ash content corresponds to lower strength in rubber articles.

#### 2.20.9.2 Heat loss

These testing methods pertain to measuring the heating loss of carbon black at a temperature of 125 °C. This loss primarily includes moisture although other volatile substances may also evaporate. So, these testing methods are not suitable for treated carbon black that contains added volatile materials if the objective is to measure moisture loss.

These testing methods can also be applied to determine the heating loss of recovered carbon fillers (rCF/rCB) at 125 °C. It is important to note that these materials were not included in precision studies, and therefore, the precision statements provided in this standard may not be applicable to them.

## 2.20.10 ASTM standard of testing: ASTM D-1509

### 2.20.10.1 Significance and uses

Besides assessing the heating loss, primarily attributed to moisture content, these drying conditions are also employed to prepare carbon black samples before conducting various other tests.

Carbon black possesses hygroscopic properties, meaning it can absorb moisture. The quantity of moisture absorbed depends on factors such as the surface area of the black, relative humidity, ambient temperature, and duration of exposure to the environment.

### 2.20.10.2 Effect of heat loss on product quality:

- Higher the moisture content, scorcher is the product.
- Higher moisture leads to the high porosity surface defect in rubber goods.
- Irregular behaviour leads to surface undulation and surface blisters may occur.

Effects of carbon black particle size and structure on vulcanisate properties are described in Table 2.14.

**Table 2.14:** Effects of carbon black particle size and structure on vulcanisate properties.

| Vulcanisate properties | Decreasing particle size | Increasing structure |
|---|---|---|
| Rate of cure | Decrease | Little effect |
| Tensile strength | Increases | Decreases |
| Modulus | Increases to maximum and then decreases | Increases |
| Hardness | Increases | Increases |
| Elongation | Decreases to minimum and then increases | Decreases |
| Abrasion resistance | Increases | Increases |
| Tear resistance | Increases | Little effect |
| Cut growth resistance | Increases | Decreases |
| Flex resistance | Increases | Decreases |
| Resilience | Decreases | Little effect |
| Heat buildup | Increases | Increases slightly |
| Compression set | Little effect | Little effect |
| Electrical conductivity | Increases | Little effect |

Different types of carbon black properties and their effects on plastic applications are shown in Table 2.15.

**Table 2.15:** Different types of carbon black properties and their effects on plastic applications.

| Masstone | Darker |
|---|---|
| Tint | Stronger |
| Undertone | Browner |
| Surface area | Higher |
| Porosity | Greater |
| DBP absorption | Higher |
| pH | Lower |
| Moisture pickup | Higher |
| Volatilities | Higher |
| Dispersibility | Higher |
| Reinforcement | Better |
| UV protection | Better |
| Electrical conductivity | Higher |
| Thermal conductivity | Lower |

# References


[1] Medalia, I. A. In: Sichel, E. K. (ed) Carbon Black – Polymer Composites. New York: Marcel Dekker, Inc (1982).

[2] Patterson, W. J., Wang, M.-J., Mahmud, K., Tire technology, International '98, p. 33 (1998).

[3] Spahr, M. E., Rothon, R. Carbon Black as Polymer Filler. Berlin, Heidelberg: Springer (23rd April 2016).

[4] Medalia, A. I., Rivin, D., Sanders, D. R. A comparison of carbon black with soot. Science of the Total Environment 31(1):1–22 (Oct 1983). doi: 10.1016/0048-9697(83)90053-0. PMID: 6197752 – Elsevier.

[5] Watson, A. Y., Valberg, P. A. Carbon black and soot: Two different substances. pubmed.gov.

[6] Donnet, J.-B., Bansal, R. C., Wang, M.-J. (eds) Carbon Black, Science and Technology. New York: Marcel Dekker, Inc. (1993). Chapt. 2.Rubber technology and manufacture, 2nd edition, Blow, C. M., Hepburn, C. (eds). London: Buttenvorths (1982).

[7] Wang, M.-J., Wolff, S., Freund, B. Filler-Elastomer interactions. Part XI. Investigation of the carbon-black surface by scanning tunneling microscopy. Rubber Chemistry and Technology 67:27 (1994).

[8] Wang, M.-J., Wolff, S., Donnet, J.-B. Fifty years of research and progress on carbon black, Filler—Elastomer interactions. Part III. Carbon-black-surface energies and interactions with elastomer analogs. Rubber Chemistry and Technology 64:714 (1991).

[9] Wang, M.-J., Wolff, S. Filler-Elastomer interactions. Part VI. Characterization of carbon blacks by inverse gas chromatography at finite concentration. Rubber Chemistry and Technology 65:890 (1992).

[10] Faraday, M. Products of combustion. In: Crooker, W. (ed) The Chemical History of a Candle. New York: Viking Press, 60–69 (1960).

[11] U.S. Pat. 2,564,700, J. C. Krejci (to Phillips Petroleum Co.) (1951).

[12] U. S. Pat. 3,490,869, G. L. Heller (to Columbia Carbon Co.) (1970).

[13] U. S. Pat. 5,554,739, J. A. Belmont (to Cabot Corp.) (1996).

[14] U.S. Pat. 4,366,138, Eisenmenger, E., Engel, R., Kuehner, G., Reck, R., Schaefer, H., Voll, M. (to Degussa AG) (1982).

[15] U.S. Pat. 2,439,442, Amon, F. H., Thornhill, F. S. (to Cabot Corp.) (1948).
[16] U.S. Pat. 5,159,009, Wolff, S., Gorl, U. (to Degussa AG) (1992).
[17] U.S. Pat. 5,708,055, Joyce, G. A., Little, E. L. (to Columbian Chemical Co.) (1998).
[18] Mahmud, K., Wang, M.-J. Francis, R. A., Elastomeric compound incorporating silicon-treated carbon black. U.S. Pats. 5,830,930 (1998) and 5,877,238, K. Mahmud, M.-J. Wang, and R. A. Francis (1999).
[19] Francis (both to Cabot Corp.) U.S. Pats. 5,904,762 (1999) and 6,211,279 K. Mahmud (2001).
[20] U.S. Pat. 6,364,944, Mahmud, K., Wang, M.-J., Kutsovsky, Y. to Cabot Corp. (2002).
[21] Medalia, A. I. Morphology of aggregates: VII. Comparison chart method for electron microscopic determination of carbon black aggregate morphology. Journal of Colloid and Interface Science 32:115 (1970).
[22] Wang, M.-J. Effect of polymer-filler and filler-filler interactions on dynamic properties of filled vulcanizates. Rubber Chemistry and Technology 71:520 (1998).
[23] Nikula, K. J., Thomassen, D. G. HEI Research Report Number 68. Cambridge, Mass.: Health Effects Institute (1994).
[24] Nikula, K. J., Snipes, M. B., Barr, E. B., Griffith, W. C., Henderson, R. F., Mauderly, J. L. In: Mohr, U., Dungworth, D. L., Mauderly, L. L., Oberdorster, G. (eds) Toxic and Carcinogenic Effects of Solid Particles in the Respiratory Tract. Washington: ILSI Press 1337–1346 (1997).

# Dedicated to Mr. Anurag Choudhury, Managing Director of Himadri Speciality Chemical Ltd

The author expresses his deep gratitude to his wife, Mrs. Krishna Chakraborty, for her constant encouragement.

The author also wishes to express his deepest gratitude to Ms. Suravi Bhaskar, who has given constant help in writing and also interpreted the most scientific facts. Without her help, the present edition would not have reached completion.

Dipak Kumar Setua

# Chapter 3 Elastomer blend and compatibility: a flow visualisation study

**Abstract:** An experimental study on the influence of several potential compatibilising agents on mixing of binary blends of (i) nitrile rubber (NBR) and ethylene–propylene copolymer (EPM), (ii) polyisoprene (IR) and EPM, and (iii) epoxidised natural rubber (ENR) and EPM have been made. The compatibilising agents used were different types of elastomers, for example, (a) chlorinated polyethylene (CM), (b) chlorosulfonated polyethylene (CSM), (c) chlorinated isobutylene isoprene copolymer (CIIR), (d) *trans*-polyoctenylene (TOR), and (e) polybutadiene (BR). These compatibilisers have varied levels of relative polarity, solubility parameter, and chemical compositions. Addition of a suitable compatibiliser, in a small concentration, in different binary elastomer blends was found to increase the rate of mixing with a corresponding reduction of the blend homogenisation time. Further, the addition of an effective compatibiliser generates finer phase morphology with respect to dispersion and distribution as well as a substantial reduction of size of final dispersed particles in the matrix phase.

**Keywords:** Elastomer blends, flow visualisation in internal mixer, compatibility, optical and scanning electron microscopy on phase morphology

## 3.1 Introduction

Elastomer blends have potential commercial interest, where the component phases are either mixed physically or by doing some chemical modifications of one or more phases prior to the commencement of blending in order to achieve compatibility. The published literature on elastomer blends constitutes studies on their mixing and processing characteristics in a two-roll mixing mill, an extruder, or an internal mixer; evaluation of phase morphology and microstructure; and also on determination of mechanical and functional properties of the vulcanisates from the application point of view. The rubber–rubber blends are broadly classified as two types: miscible and immiscible types. However, in case of thermoplastic elastomers (TPEs), which are based on melt mixed rubbers with thermoplastics, they are generally immiscible. But these immiscible systems possess much wider applications and, therefore, attract attention of both researchers as well as industries for generation of technical compatibilisation of these systems. Examples of a few miscible binary rubber blends are styrene–butadiene rubber (SBR) with isobutylene-co-isoprene (butyl) rubber (IIR), acrylonitrile-co-butadiene (nitrile) rubber (NBR) with other NBRs having different acrylonitrile contents, and natural rubber (NR) with vinyl

https://doi.org/10.1515/9783110668537-003

butyl rubber. In the case of TPEs, a blend of NBR with polyvinyl chloride (PVC) is a unique example of miscible rubber–plastic blend. However, other commercial blends of rubbers, for example, NR/polybutadiene rubber (BR), NR/ethylene propylene diene monomer rubber (EPDM), NBR/polypropylene (PP), EPDM/PP and NBR/EPDM, etc. are immiscible.

The term 'blending' describes the process generally conducted in the melt in mixing or extrusion equipment, in which two polymers are compatibilised to form a polyblend. The term 'alloy' is used to indicate, in some cases, a chemical reaction between the functional groups on the polymers, which resulted in covalent bond formation. However, the effectiveness of hydrogen and ionic bonding in promoting the action of a compatibilising agent in a polymer blend indicates that this is certainly a distinction without a difference. Various methods have been adopted to improve the miscibility or generation of compatibility in rubber–rubber and rubber–plastic blends. For example, the mechanical type of compatibilisation is achieved via mixing of micronised rubber powders, rubber latexes, or by addition of a third component externally during the processing and the prepared compounds do not show any phase ripening, that is, abstain from formation of any large-scale coalescence of the dispersed particles, which segregate on prolonged storage. In the case of physical compatibilisation, the process recourses to different routes, viz., (a) modification of the microstructure to enhance miscibility through specific physical interactions between the phase components, (b) controlled crystallisation of one or more phases in order to develop a lock-in type of morphology of the macromolecular structure, or (c) an addition of a block or graft copolymer as a compatibiliser. However, in case of a chemical compatibilisation process, a chemical reaction is pertinent for establishing cross-linking between the phase components by: (a) formation of an interpenetrating or a semi-interpenetrating network, (b) formation of reversible or permanent cross-linking of the participating phases, or (c) an exchange reaction through in situ grafting/polymerisation/ester exchange (reactive blending), etc.

Compatibility refers to a polymeric mixture of two or more polymeric materials to form a homogeneous composition, which has useful properties and does not separate into component parts during the expected lifetime of the product. Technological compatibilisation is a result of the technique for improving ultimate properties by making polymers in a blend, which is less incompatible. It is, therefore, not the application of a technique that induces 'thermodynamic compatibility', where the polymers exist in a single molecularly blended homogeneous phase. The compatibilising agent acts as a polymeric surfactant, lowering the surface tension and promoting the interfacial adhesion between the dispersed phase/s and the matrix. They are added to polymers undergoing mixing in an internal mixer and as a result, control the dispersed phase size like in the case of a low molecular weight surfactant in an oil and water emulsion. An effective compatibilising agent is effective in small concentrations, for example, 0.1 to less than 5 wt% of total parts of the polymers. The minimum molecular weight of a compatibilising agent may vary from a single reactive functionality to a higher molecular weight polymer, but a copolymer having any segment with molecular weight above 100,000 is a poor compatibilising agent.

## 3.2 Blends of thermoplastics

It has long been known that the introduction of a small quantity of certain additives to blends of thermoplastics can lead to major changes in the phase morphology and also significant improvement in mechanical properties [1–7]. A compatibilising agent permits blending of otherwise incompatible polymers to yield polyblends or alloys with unique properties, generally not attainable from individual components. Block and graft copolymers with segmental structural identity or solubility parameter difference less than 1.0 of the competing polymers are effective as compatibilising agents. They may either be preformed or added to the polymers undergoing blending or may be generated in situ. The nature of bonding between segments of the compatibilising agent and polymers are either covalent, ionic, hydrogen bonding, or a result of donor-acceptor interaction.

## 3.3 Elastomer blends for cost reduction

There are manifold reasons for development of elastomers blends for cost reduction due to synthesis of new polymers to meet high performance as well as functional properties of the end products. Cutting time overrun for newer synthesis and getting their legislation/QA/QC/FDA approval are also significant issues. Blends of NBRs with other rubbers can improve their ozone, weather, thermal, oil, and fuel resistances. Manufacture of TPEs and TPE vulcanisates (TPVs) based on NBRs have also become commonplace. Examples of such commercially available blends are NBR/PVC, where the components are miscible in all proportions, as mentioned in Tables 3.1 and 3.2. The final products have very good weatherability, oil/fuel resistances, better low temperature flexibility for use as seals, O-ring gaskets, especially in strategic defence equipment. TPVs prepared by melt-mixing of NBRs with varieties of olefinic thermoplastics have thermal or mechanical reversibility, besides recyclability. Blends of cross-linked NBRs with polypropylene (PP) are commercially available as Zeolast® manufactured by Advanced Elastomer Systems, USA. Similarly the blends comprised PP and EPDM are available in the trade name Santroprene®, as introduced by Du Pont, USA. Examples of some NBR- and EPDM-based blends and common products are shown in Table 3.2. Development of high performance TPEs by blending of NBRs with Nylon 6 in an internal mixer and also by a reactive mixing process in a twin screw extruder intended for advanced defense and aerospace applications has been reported on by Setua [8].

**Table 3.1:** Typical applications of NBR.

| | |
|---|---|
| Hoses for fire, irrigation, gasoline, hydraulic and marine sectors | Expanded insulation for air-conditioning |
| Wire and cable insulation: mining, oil well, and welding cables | Textiles, cots, and aprons |
| Footwear: military, safety, and industrial boots | Microcellular sheets |
| Weather seals in building such as door and window | Conveyor belts – mining and food processing |
| Automotive, air-conditioning hose insulations | Gaskets, seals, diaphragm, etc. |
| Flooring, roofing, and chemical tanks | Pond liners and geomembranes |

**Table 3.2:** Some applications of NBR and PVC and NBR and EPDM blends.

| NBR/PVC blend | NBR/EPDM blend |
|---|---|
| Automotive moulded products, e.g., bellows, grommets, and seals | Automotive moulded products, e.g., bellows, grommets, and seals |
| Automotive extruded products, e.g., window and door seals | Automotive extruded products, e.g., window and door seals |
| Microcellular sheets for footwear | Outdoor weather ducts, expansion joints |
| LPG tubing, etc. | Bridge-bearing pads, etc. |

## 3.4 EPDM rubbers

EPDM rubbers, with varied ethylene and propylene content and composed of different diene types, possess excellent mechanical properties and resistance towards ozone, heat, weather, and so on due to the presence of saturated carbon–hydrogen backbone structure as well as very good low-temperature flexibility. Thus, they become more attractive than NR or SBR in a variety of rubber-related components where an extended lifetime of the goods is extremely important. The synthetic version of NR is *cis*-1,4-polyisoprene (IR), and IR and EPDM blends play an important role in the tie industry associated with the development of sidewall compounds of passenger, aero-tyres and in the retreading of tyres for life extension and reuse [9]. However, there have been only few basic studies of the nature of mixing and assessment of phase morphology for this system [10–14]. These studies, made so far, have used phase contrast optical microscopy (OM), scanning electron microscopy (SEM), or transmission electron microscopy (TEM) for determination of dispersed phase size in samples prepared by either a solution or by dry blending processes. Preferential carbon black distributions as well as anisotropic swelling of the different phase components have also been used to study the blend compatibility [15–18]. Incidentally, chlorobutyl rubber (CIIR) has been added to IR and EPDM blends to minimise the covulcanisation problem for preparation of their vulcanisates [19]. Lohmar [20] has reported on the phase morphology in blends of NBR and EPDM containing various proportions of trans- polyoctenylene rubber (TOR) using SEM and TEM. His observation states that a relatively small amount of TOR (10–20% of the total weight of the polymers) favours mechanical dispersion of these incompatible rubbers due to formation of a thin 'TOR-skins' separating the NBR and EPDM domains, thereby improving the distribution of EPDM particles inside NBR with reduced dimensions. The average dispersed phase size was reported to be of the order of 1 μm in a blend containing 40% NBR (Perbunan N 2810, Bayer AG), 40% EPDM (Buna AP 341, Buna Werke Hüls), and 20% TOR (Vestenamer 8012, Hüls). When NR is chemically modified to make ENRs, its properties such as air permeability, oil resistance, and wet grip are better than NR. ENRs also have inherent strain-induced crystallisation property like NR, and thus offer very good adhesion and green strength. Besides, their reactivity is enhanced due to the presence of epoxy group, which leads to opening of its ring structure and forming carboxylic acid group in situ during blending with other rubbers or plastics. The oil and fuel resistance of ENR and maleic acid-grafted EPDM (MA-g-EPDM) blends were found to be excellent at room temperature and fairly good even up to 70 °C. Use of these blends for high temperature and oil-resistant rubber seal, O-ring gaskets suitable for continuous use in defense and aerospace systems have been reported by Setua [21]. In general, Setua and coworkers [22–26] and others [27–30] have reported on the use of sophisticated analytical techniques to characterise the compatibility in a variety of elastomer blends and TPEs.

## 3.5 Flow visualisation studies

Flow visualisation studies in a model internal mixer were initially used by Freakley and Wan Idris [31] and Asai et al. [32]. Min and White [33–37] studied a modified process by installation of front and transverse glass windows on the mixer. This enabled them to investigate visually the flow behaviours of pure polymers and polymer blends during the blending operation. Depending on temperature and the nature of polymers, they categorise the mixing into four different regimes. The effect of various rotor designs, that is, Banbury-type two-wing to four-wing rotors and intermeshing type of rotors were used in these studies to create different flow fields, which were found to result in better control of distributive mixing and goodness of the phase morphologies. The flow visualisation through glass windows also demonstrated shearing, stretching, and tearing motions of the elastomers and plastics. Flow visualisation of a scaled-down rubber compounding process in an internal mixer and the addition of carbon black and oil were studied by Morikawa et al. [38–40]. The times required for rubber bale 'pickup' by the rotors, bale homogenisation time of various elastomers, and the times required to incorporate black and oil were measured. Subsequently the apparatus has successfully been used to study the transient phase morphology development during various stages of mixing of the immiscible elastomers by Setua and White [41–44]. It is, therefore, worthwhile to investigate the relative efficiencies of various potential compatibilising agents on the phase emulsification characteristics of some important elastomer blends having potential commercial applications. This will allow creating a rich tapestry of behaviour of different compatibilising agents in blends with varying levels of polarity. The present chapter gives a detailed description of blending of copolymers of ethylene propylene (EPM) and NBRs and IR and EPM and influence of compatibilising agents on the rate of mixing and development of final phase morphology. Different types of compatibilisers, viz., CM, CSM, CIIR, TOR, and BR have been tried to find out their efficacy as compatibilising agents. To better understand the mixing characteristics and morphology of IR and EPM blend, a contrast has been made by studies on ENR (25) and EPM, and ENR (50) and EPM blends. The choice of EPM in place of EPDM is primarily because of the greater ability of EPDM to extract it from vulcanisates through solvent etching from the IR and EPM, and ENR and EPM blends. The ENR, IR, and EPM, due to their carbon–hydrogen backbone cause poorer phase contrast and cause difficulty in distinguishing them in the SEM observations. The average size of the dispersed EPM particles could easily be measured from the size of holes left out on the cross sectional surfaces after solvent etching. The blends ratio was, therefore, adjusted as 80:20 with respect to either ENR or IR and EPM, but 50:50 in case of NBR and EPM blends, where etching was not necessary.

## 3.6 Experimental studies

### 3.6.1 Materials

The characteristics and source of different elastomers, for example, NBRs of different acrylonitrile concentrations, EPM, IR, ENR (25), ENR (50), and varieties of the compatibilising agents, for example, CIIR, BR, CR, CM, CSM, and TOR are summarised in Table 3.3. A list of staining and vulcanising additives used is given in Table 3.4.

## 3.7 Flow visualisation apparatus and procedures

The internal mixer used for the blending studies was a modified Haake-Buchler Rheocord 750 laboratory internal mixer specially designed with separate heaters for uniform temperature control and a front glass window through which videotapes were made, as shown in Figure 3.1. The mixing chamber has a capacity of 70 cm$^3$ and double-winged non-intermeshing Banbury-type rotors with a speed ratio of 7:6 (Figure 3.2).

**Table 3.3:** Characteristics and source of elastomers and compatibilisers.

| S. no. | Elastomer type | Designation | Commercial grade | Manufacturer |
|---|---|---|---|---|
| 1. | Ethylene–propylene copolymer (% ethylene = 68) | EPM | EPM 306 | Polysar (Bayer AG) |
| 2. | Acrylonitrile–butadiene copolymer | NBR | | |
| | (a) 28% acrylonitrile | | Chemigum N 715 B | Goodyear |
| | (b) 38% acrylonitrile | | Chemigum N 336 B | Goodyear |
| | (c) 45% acrylonitrile | | Chemigum N 206 | Goodyear |
| 3. | *Cis*-1,4-Polyisoprene | IR | Natsyn 2200 | Goodyear |
| 4. | Epoxidised natural rubber | ENR | | |
| | (a) 25 mol% epoxidation | | ENR (25) | Guthrie Latex |
| | (b) 50 mol% epoxidation | | ENR (50) | Guthrie Latex |
| 5. | Chlorinated isobutylene isoprene copolymer (% chlorine = ~2) | CIIR | CIIR 1068 | Exxon |
| 6. | Chlorinated polyethylene (% chlorine = 36) | CM | Tyrin (R) CM 0136 | Dow Chemical |
| 7. | Chlorosulfonated polyethylene (% chlorine = 29 ± 1.5) | CSM | Hypalon 20 | Du Pont |
| 8. | *Trans*-Polyoctenylene | TOR | Vestenamer 8012 | Hüls AG |
| 9. | Polybutadiene | BR | Taktene | Polysar (Bayer AG) |

**Table 3.4:** Source and structure of staining and vulcanising agents.

| S. no. | Material/trade name | Chemical composition | Supplier/manufacturer of the agents |
|---|---|---|---|
| 1. | Ruthenium chloride | $RuCl_3, 3H_2O$ | Sigma Chemical |
| 2. | Durax | *N*-Cyclohexyl-2-benzothiazole sulfenamide | Vanderblit, USA |
| 3. | Methyl tuads | Tetramethyl thiuram disulfide | Vanderblit, USA |
| 4. | Sulphur | – | Sigma Aldrich |

Our procedure was as follows: The elastomers were first compounded with less than 1 wt% of red pigment (added to EPM rubber) and yellow (in cases of NBRs, ENR, and IR) at 80 °C in the internal mixer at 40 rpm for 2 min to produce samples having good colour contrast as well as development of combination colours, which could be easily detected through video-recording.

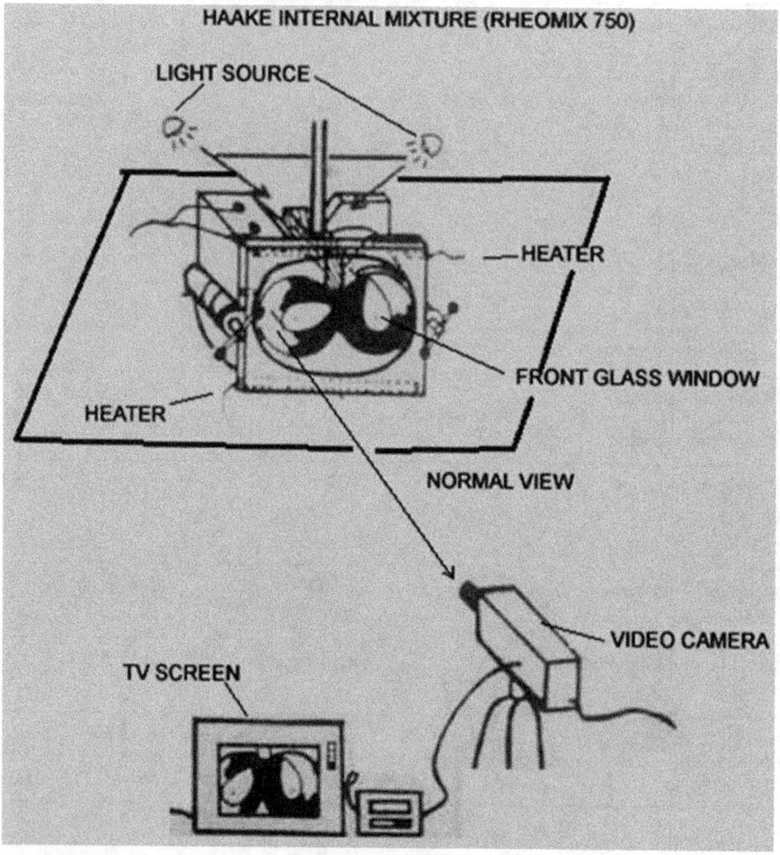


**Figure 3.1:** Schematic diagram of the apparatus of flow visualisation.

Compatibilising agents (BR, CIIR, TOR, CSM, and CM, wherever used, were compounded with the EPM phase (at concentration of 10 wt% with respect to EPM along with 1 wt% of red pigment) also at 80 °C and 40 rpm for 2 min. The mixture compounds were then left in a mould (12 mm thickness under compression) in an electrically heated hydraulic press for 10 min at 120 °C. Bale shaped samples (of dimension 38 × 19 × 12 mm) were cut from the compression moulded sheets.

In order to reproduce a simulated compounding process, which might actually take place in a commercial internal mixer (e.g., Banbury), bale-shaped rubber samples were fed into the internal mixer at 80 °C and 40 rpm. The specially designed apparatus allows us to make quantitative monitoring of the blending process. Yellow (NBRs, ENR, or IR) and red (either EPM alone or EPM mixed with a compatibilising agent) pigmented bales

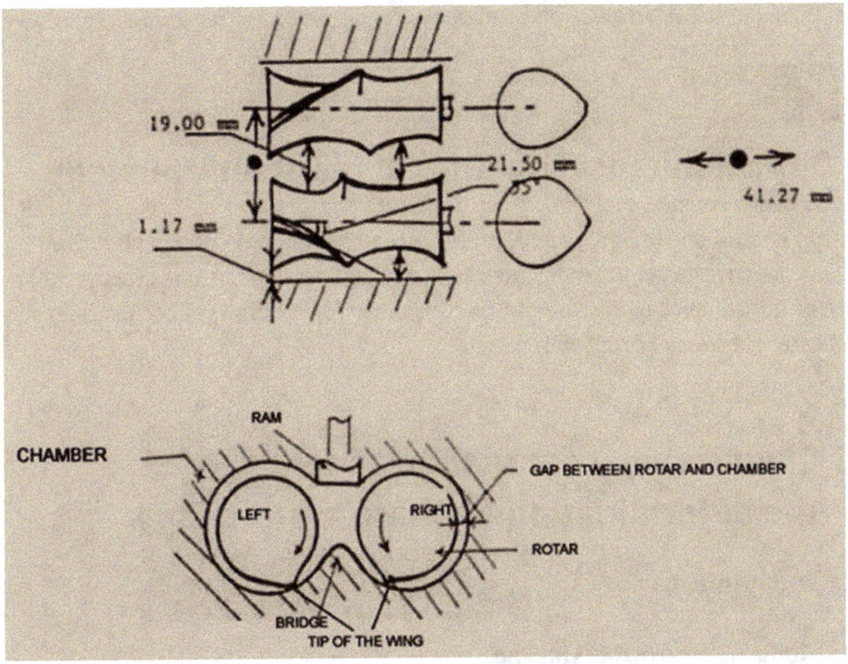


**Figure 3.2:** Schematic diagram of rotor designs, mixer, and terminologies used in the study.

were alternately added into the internal mixer in either 50:50 (NBRs and EPM) or 80:20 (IR and EPM or ENRs and EPM) weight ratio, until the mixing chamber reached a fill factor of 0.7, that is, 70% of the volume capacity of the mixing chamber. Commencement of the mastication of rubbers with the process of mixing demonstrated the yellow- and the red-coloured bales were broken down into small pieces and subsequently started mixing with each other. They were homogenised to an orange colour after a certain period of time. In this manner a 'blend homogenisation time' was determined from the videotapes as the duration from the introduction of the first bale to the development of a uniform orange colour of the blends.

## 3.8 Scanning electron microscopy (SEM) on phase morphology

A high-resolution scanning electron microscope (ISI, S × 40) was used to investigate the phase morphology of the blends. In the case of morphological investigations using the SEM, the sample surfaces were first carefully cut without toughing the surface and were sputter-coated with gold. These gold-coated surfaces were then observed under the SEM and the photomicrographs were recorded in a Polaroid camera. Details of the sample preparation and sample storage techniques for SEM studies have been described earlier [45].

## 3.9 Optical microscopy (OM) of phase morphology

An optical microscope (model Leitz HM POL light polarising microscope) fitted with a travelling stage was used to record the phase morphology of the same cross-sectional surface of the samples as in case of SEM but the gold coating was not necessary at magnification between 50 and 500×.

## 3.10 Chemical methods followed for the determination of phase morphology

### 3.10.1 Ruthenium

#### 3.10.1.1 Ruthenium tetroxide staining

A ruthenium tetroxide staining process for unsaturated elastomers (e.g., IR and ENRs) was used [46, 47]. The staining solution was prepared by dissolving 02 g of $RuCl_3{\cdot}3H_2O$ in 10 mL of 5.25% aqueous sodium hypochlorite. The fresh mixture was dark red. Samples roughly of 02 $cm^3$ sizes were soaked in this solution for 20 min and were blotted dry. Subsequently these samples were vitrified and fractured in liquid nitrogen and the fractured surfaces were examined under the SEM.

## 3.11 Cross-linking of unsaturated elastomers (IR and ENRs) to the ebonite stage by dipping in a molten sulphur bath followed by solvent etching

The blends were first fractured in liquid nitrogen and the fractured surfaces were dipped in a molten sulphur bath at 130 °C for 10 h. Subsequently these surfaces were solvent etched in boiling cyclohexane for 08 h to remove uncross-linked EPM phase. They were then dried in a hot-air oven at 70 °C for 24 h, gold-coated, and examined under the SEM.

### 3.11.1 Cross-linking of the unsaturated elastomers (IR and ENRs) to the ebonite stage via addition of a curative (sulphur and accelerators) package followed by solvent etching

A total of 25 phr (parts per hundred parts of rubber) of sulphur, 3 phr of Durax, and 1.5 phr of methyl tuads were added to the blend while mixing in the internal mixer. Curing of the unsaturated phases to the ebonite stage was carried out in an electrically heated hydraulic press for 1 h at 150 °C under compression. The samples were then fractured in liquid nitrogen and the cross section of these samples was solvent-etched in boiling cyclohexane for 8 h to remove the saturated uncross-linked EPM phase. They were then dried in a hot-air oven at 70 °C for 24 h, gold-coated, and examined under the SEM.

## 3.12 Results and discussions

### 3.12.1 NBR/EPM blends

#### 3.12.1.1 Flow visualisation

NBRs of different acrylonitrile (ACN) concentrations varying from 28–45% were used in this study (Table 3.3). Bales of EPM (pigmented red) and NBRs (all pigmented yellow) were fed alternately into the internal mixer at 80 °C and 40 rpm in 50:50 weight ratio until the mixing chamber reached a fill factor 0.7. In the beginning of the mixing process, both NBRs and EPM bales were pulled out by the tip of the wing of the faster moving rotor. The NBRs always showed a higher 'bale pickup time' than EPM. This is due to higher breaking strength and stretching characteristics of NBRs compared to EPM. The EPM bales thus picked up were broken into chunks at the junction between the tip of the wing and the bridge and subsequently subdivided into finer particles. These smaller particles then started circulating inside the chamber around the rotor until they were pumped into the inter-rotor region. Here a stagnant segregated mass of the crumbled rubber was developed. The incoming bales also got accumulated. Actual homogenisation commenced when the NBR bales started breaking up into smaller chunks and initiated incorporation of the already broken EPM particles. At a later stage the slower moving rotor started pulling out the segregated mass and further tearing of the chunks occurred. They would again then be pumped into the stagnant region. With the progress of mixing, the NBR chunks, which adhered to the EPM particles, were further subdivided into smaller fragments by shearing between the chamber wall and the rotors and finally the bales were homogenised into a uniform orange mass, which is the 'blend homogenisation time'. The individual blend homogenisation time was determined for each repair of the blends from their respective videotapes and is given in

Table 3.5. Figure 3.3(a–e) shows a sequence of still photos recorded by the camera during the homogenisation process of NBR (45% ACN)–EPM blends without a compatibiliser. The general mixing pattern of all the blends is similar excepting variation of the time and rate of homogenisation.

**Table 3.5:** Homogenisation time of NBRs and EPM (with and without compatibiliser) in 50:50 blend ratio.

| Compatibiliser type | Blend homogenisation time of NBRs with different ACN content, (sec) | | |
|---|---|---|---|
| | 45% | 38% | 28% |
| Nil | 270 | 235 | 180 |
| BR | 280 | 240 | 150 |
| CIIR | 270 | 220 | 140 |
| TOR | 250 | 200 | 90 |
| CSM | 240 | 175 | 90 |
| CM | 150 | 120 | 75 |

## 3.13 Phase morphology development

SEM studies on the nature of dispersion, distribution, and dispersed NBR particles as well as their average sizes are given in Table 3.6. On an average, 20 particles were considered for calculation of the dispersed phase size. Both optical and SEM were performed on the cross-sectional area of the freeze-fractured surface of NBR and EPM blends. Figure 3.4(a) gives the SEM photomicrograph of phase morphology of NBR (45% ACN)–EPM blend without a compatibiliser, while Figure 3.4(b) shows the optical photomicrograph of the same surface.

## 3.14 Effect of compatibilising agents on rate of blend homogenisation

In these cases, the EPM bales were mixed with 10 wt% of one of the compatibilising additives, viz., CIIR, TOR, CSM, or CM along with 1 wt% of the red pigment. Blends of EPM and NBRs were prepared in 50:50 weight ratios, as described in the previous section. 'Blend homogenisation times' were determined in all cases and are summerised in Table 3.7. Sharp differences in the mixing behaviour of NBR and EPM were observed when an effective compatibiliser was added to the system. In these cases, broken NBR chunks were adhered to the EPM particles at an earlier stage of mixing due

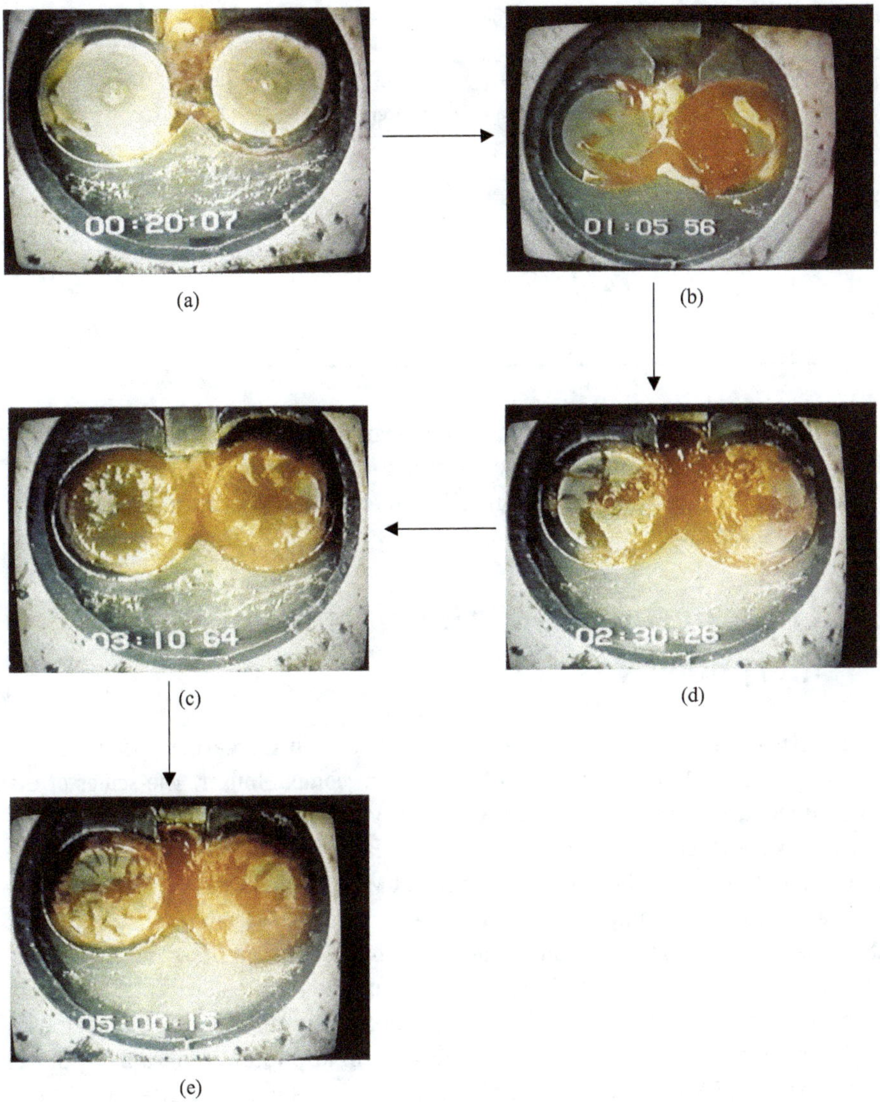


**Figure 3.3:** (a–e) Sequence of still photos showing the homogenisation of NBR (45% ACN)–EPM blends without a compatibiliser.

to enhancement of miscibility between the component phases. The compatibilisers were found to form a skin on the surface of EPM particles and help comparatively larger NBR chunks to adhere to them. They were further subdivided together into smaller fragments until the onset of homogenisation. Significant reduction in the blend homogenisation times was thus observed in these cases. Figure 3.5(a–g) shows the sequence of still photos on the homogenisation of NBR (45% ACN)–EPM blends

**Table 3.6:** Phase morphology of NBR and EPM blends.

| Compatibiliser type | Average dispersed phase size of NBRs with different ACN content (µm) | | |
|---|---|---|---|
| | 45% | 38% | 28% |
| Nil | 3.5 | 6.5 | 10.5 |
| BR | 3.5 | 6.5 | 4.5 |
| CIIR | 3.5 | 6.5 | 4.5 |
| TOR | 3.5 | 3.5 | 2.5 |
| CSM | 2.0 | 2.0 | 2.5 |
| CM | 0.7 | 1.5 | 2.0 |

with CM as a compatibiliser. However, an ineffective compatibiliser did not vary the mixing behaviour and the homogenisation times remained more or less the same as applicable to the blends, for a particular ACN content, without a compatibiliser.

## 3.15 Effect of compatibilising agents on phase morphology

Average sizes of the dispersed phase were reduced with increasing miscibility between EPM and NBR in the presence of an efficient compatibiliser. The scales of the dispersed globules have been measured and are given in Table 3.8. The variation of the acrylonitrile level in NBRs influences their polarity, solubility parameter, and the effectiveness of a compatibilising agent as well. CM has been found to be the most effective amongst all the compatibilising additives investigated. The presence of CM not only caused a significant reduction in the blend homogenisation times but also generated the lowest dispersed phase dimensions compared to other competitors. Especially in the case when the ACN concentration in NBR is 45%, only the CM behaved as a suitable compatibiliser. Figure 3.6(a) shows the SEM photomicrograph of phase morphology of NBR (45% ACN)–EPM blend with CM as a compatibiliser, while Figure 3.6(b) shows the optical photomicrograph of the same. This is also reflected in the results in Tables 3.3 and 3.4. For the blends containing 28% NBR, both CSM and TOR showed similar blend homogenisation times and dispersed phase sizes. However, at ACN concentration of 38% or 45% in NBRs, the TOR gave lower levels of compatibilisation than either CSM or CM. Higher ACN content in NBRs reduced the level of mixability of NBR and EPM to such an extent that only the higher chlorine-containing additives such as CSM or CM could become effective compatibilisers. The addition of either CIIR or BR could result in moderate compatibilisation only for the blend containing 28% NBR, but they are still less effective than TOR in this case.

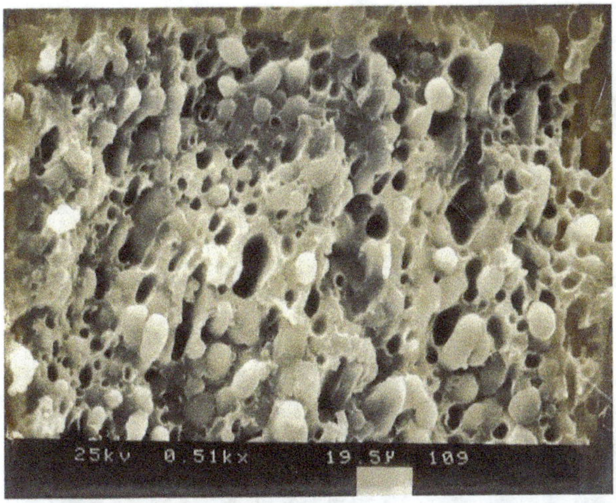


(a)

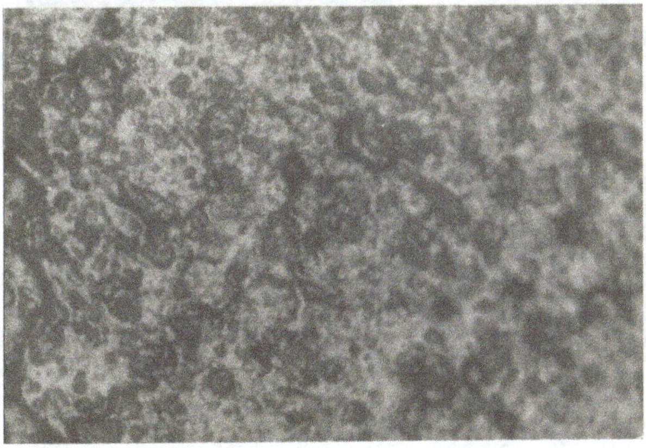

(b)

**Figure 3.4:** (a) SEM photomicrograph of phase morphology of NBR (45% ACN)–EPM blend without a compatibiliser and (b) optical photomicrograph of phase morphology of NBR (45% ACN)–EPM blend without a compatibiliser (50×).

# 3.16 IR and EPM blends

## 3.16.1 Flow visualisation

Yellow (IR) and red (EPM alone) bales were fed into the internal mixer in 80:20 blend ratio at 80 °C and 40 rpm until the mixing chamber reached a fill factor of 0.7. The EPM bales were broken up into chunks homogeneously from the beginning of the mastication process. The IR bales were stretched at the junction between the rotor and the bridge

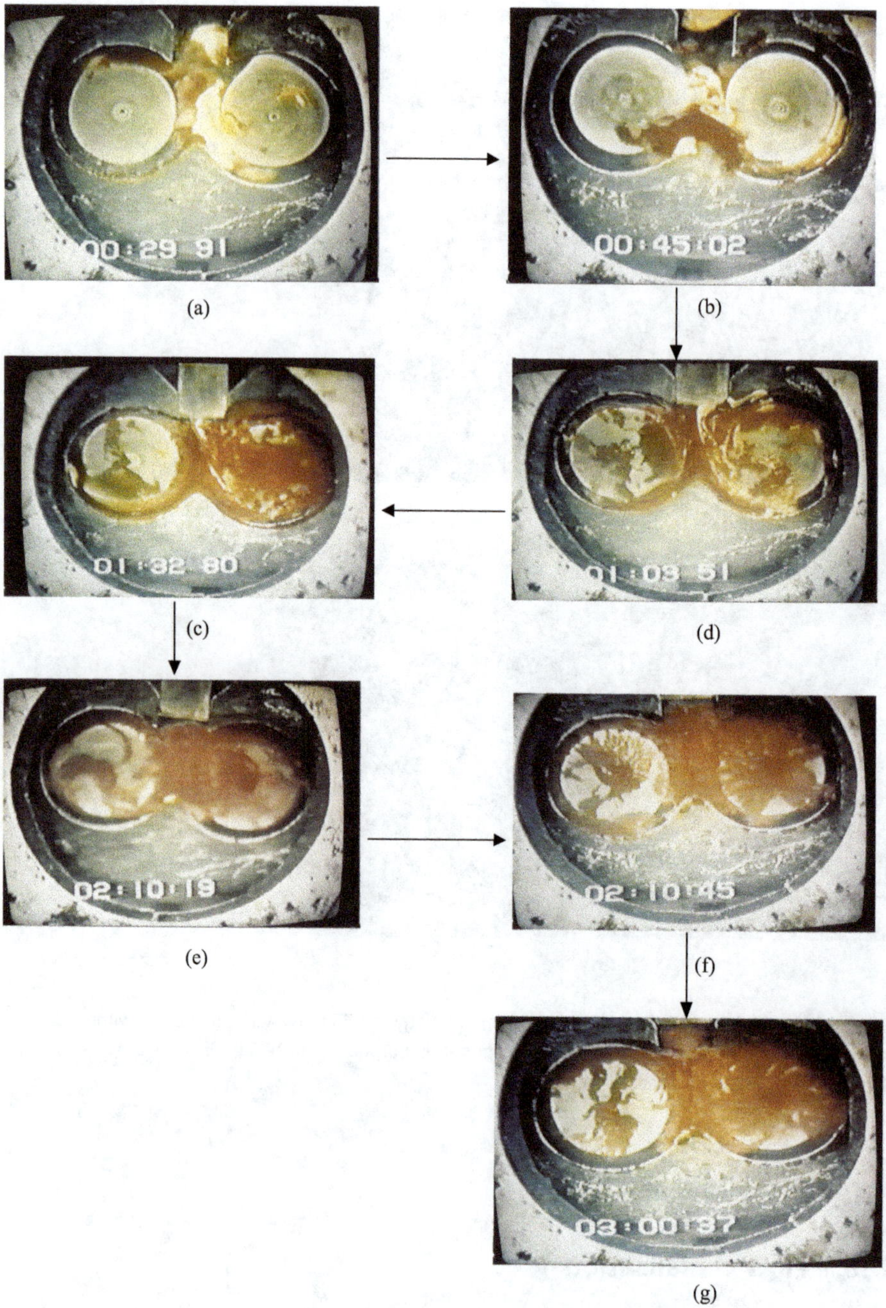


**Figure 3.5:** (a–g) Sequence of still photos showing the homogenisation of NBR (45% ACN)–EPM blends with chlorinated polyethylene (CM) as a compatibiliser.

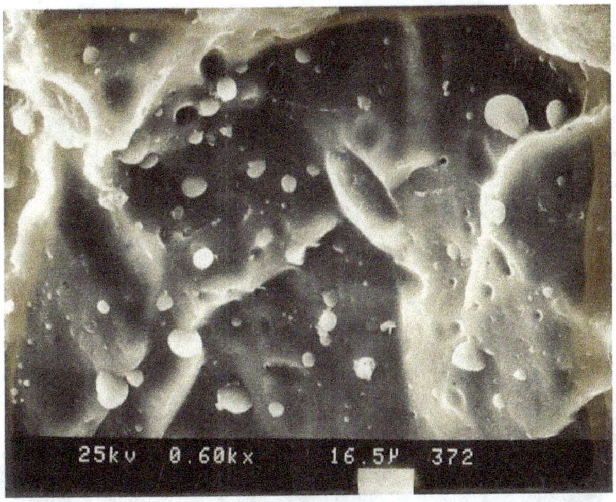


(a)

(b)

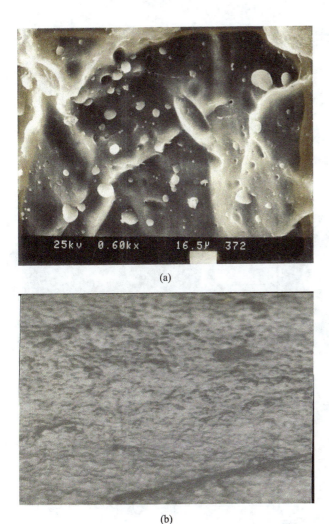

**Figure 3.6:** (a) SEM photomicrograph of phase morphology of NBR (45% ACN)–EPM blend with CM as a compatibiliser and (b) optical photomicrograph of phase morphology of NBR (45% ACN)–EPM blend with CM as a compatibiliser (50x).

and remained as large chunks (distorted and merged bales) for a longer time. After some time the IR chunks formed a tight band around the rotors and the smaller broken pieces of EPM were circulating on top of the IR bands. Partial incorporation of the EPM particles into IR occurred at this stage and the unhomogenised mass was pumped into the inter-rotor region. Repeated withdrawal of the accumulated mass from this region by the rotors and extended shearing inside the chamber wall finally caused homogenisation of the blend. Figure 3.7(a–e) shows a sequence of still photos recorded on the homogenisa-

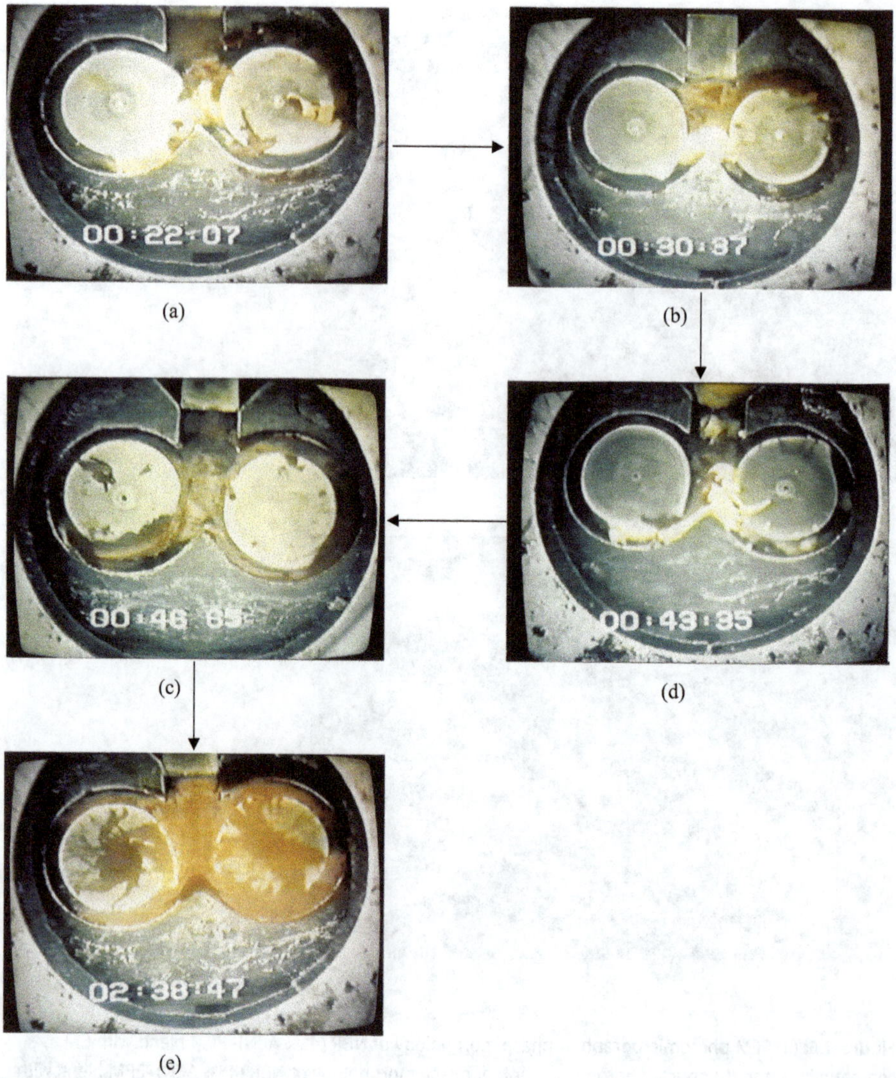


**Figure 3.7:** (a–e) Sequence of still photos showing the homogenisation of IR–EPM blends without any compatibiliser.

tion of IR–EPM blends without a compatibiliser. The blend homogenisation time was also noted and given in Table 3.7.

**Table 3.7:** Homogenisation time of IR / EPM Blends.

| Compatibiliser type | Blend homogenisation time (s) |
|---|---|
| Nil | 145 |
| TOR | 95 |
| CIIR | 110 |
| CSM | 190 |
| CM | 195 |

## 3.16.2 Phase morphology development

Different experimental methods were followed to investigate the phase morphology of this blend as stated in Section 3.13. Similar electron densities due to carbon hydrogen structures of IR and EPM result in poorer phase contrast and therefore, the two phases could not be distinguished clearly. The $RuO_4$-treated cross-sectional area showed some improvement of phase contrast, but it was still far too difficult to determine accurately the dispersed phase geometry and dimension. Sharp differences were obtained when the molten sulphur curative method (Section 3.11) or the addition of curative package method (Section 3.11.1) were followed by solvent extraction of the EPM phase. Dispersed EPM domains formed holes in the cross-sectional area as they were leached out during solvent extraction process. However, the curative package addition method was chosen for the shorter time required obtaining optimum results in determining the distribution, shape, and size of the dispersed EPM phase. The surface was studied under the SEM and OM. Figure 3.8(a) shows the SEM photomicrograph of phase morphology of cross section of IR–EPM blend (without any compatibiliser) via cross-linking to ebonite stage and extraction of EPM in cyclohexane, and Figure 3.8(b) shows the optical photomicrograph of the surface. At least 20 holes were considered for calculation of the scale of the dispersed EPM phase. The average size of EPM inside the continuous IR matrix was found to be of the order of 0.8 μm as depicted in Table 3.8.

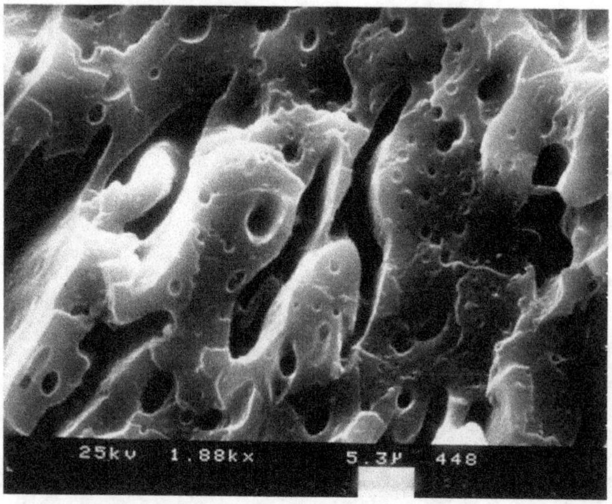


(a)

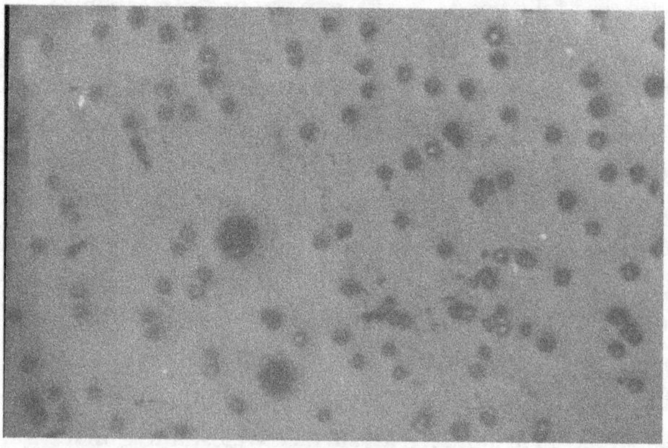

(b)

**Figure 3.8:** (a) SEM photomicrograph of phase morphology of cross section of IR–EPM blend without any compatibiliser and via cross-linking to ebonite stage and extraction of EPM in cyclohexane and (b) optical photomicrograph of phase morphology of IR–EPM blend without a compatibiliser (500x).

**Table 3.8:** Phase morphology of IR/EPM blends.

| Compatibiliser type | Average dispersed phase size, (μm) |
|---|---|
| Nil | 0.8 |
| CIIR | 0.7 |
| TOR | 0.7 |
| CSM | 0.9 |
| CM | 1.0 |

## 3.17 Effect of compatibilising agents on blend homogenisation

EPM was previously mixed with either CIIR, TOR, CSM, or CM along with 1 wt% of the red pigment and the blends were prepared in the internal mixer with yellow (IR) and red (EPM mixed with a compatibiliser) in 80:20 weight ratio. Homogenisation times were noted from the video tapes and are illustrated in Table 3.9. Although the general mixing pattern as observed in the case of IR/EPM blends without any compatibiliser remained the same, the addition of CIIR and TOR resulted in faster homogenisation. But the presence of either CSM or CM caused an extended homogenisation period. The efficiency of TOR in the blend homogenisation is more than that of CIIR.

### 3.17.1 Effect of compatibilising agents on phase morphology

Neither any significant change in the phase morphology nor any substantial variation in the scale of the dispersed phase was found to occur due to addition of any of the compatibilising agents (Table 3.8). Figure 3.9 is the SEM photomicrograph of phase morphology of the IR–EPM blend with TOR as a compatibiliser via extraction of EPM in cyclohexane. Average sizes of the dispersed EPM particles left out as a hole on the cross-sectional etched surface of the sample were observed to be of the same order, that is, between 0.8 and 1.0 μm in all cases.

IR exhibited obviously striking stress-induced crystallisation unlike EPM. This was reflected in the flow visualisation characteristics of the blending of these two elastomers. Stretching of the IR bales by the rotors caused distortion and elongations of the bales, which resulted in stress-induced crystallisation and thereby, formation of a tight band around the rotors for better gum strength. The lower homogenisation time of these blends is due to higher level of miscibility between the IR and EPM compared to the case of NBR and EPM blends, which are immiscible. CIIR and TOR were found to have some compatibilising effects, while CSM and CM were not due to chlorine content in these two polymers giving rise to polarity and making them highly incompatible with both IR and EPM.

## 3.18 ENR and EPM blends

### 3.18.1 Flow visualisation

Different types of ENRs were used for this study: (a) ENR with 25 mol% epoxidation and (b) ENR with 50 mol% epoxidation. Both the ENRs were pigmented yellow and were fed to the internal mixer at 80 °C and 40 rpm along with the red-pigmented EPM

**Figure 3.9:** SEM photomicrograph of phase morphology of cross section of IR–EPM blend with TOR as a compatibiliser and via cross-linking to ebonite stage and extraction of EPM in cyclohexane.

bales in 80:20 weight ratios and at a fill factor of 0.7. Blend homogenisation times were determined and are shown in Table 3.9.

At a 25% epoxidation level, ENR (25) and EPM blends showed almost similar mixing characteristics to those of IR and EPM blends. Figure 3.10(a–e) shows the sequence of still photos on the homogenisation cycle of ENR (25)–EPM blend without a compatibiliser. The only difference was that ENR formed bands around the rotors at a slower rate compared to that observed in the case of IR (Figure 3.7(a–e)). As expected, the blend homogenisation time for ENR (25) and EPM blends was found to be slightly higher due to a more polar nature of ENR (25) than IR, which caused less miscibility with EPM (compare Tables 3.6 and 3.8).

**Table 3.9:** Homogenisation time of ENR (25)/EPM and ENR (50)/EPM blends.

| Compatibiliser type | Blend homogenisation time of ENRs with different levels of epoxidation, (s) | |
|---|---|---|
| | **25%** | **50%** |
| Nil | 175 | 220 |
| CIIR | 160 | 210 |
| TOR | 145 | 185 |
| CSM | 145 | 170 |
| CM | 140 | 160 |

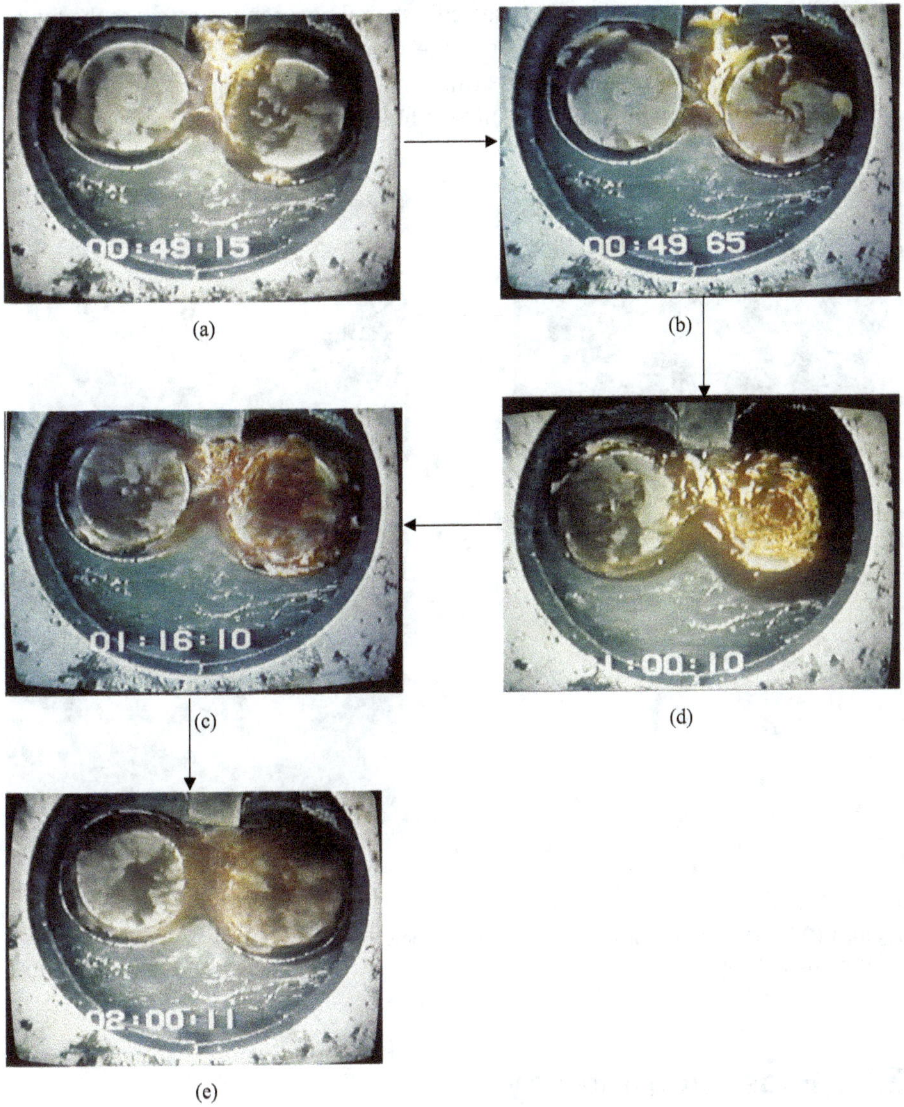


**Figure 3.10:** (a–e) Sequence of still photos showing the homogenisation of ENR (25)–EPM blend without a compatibiliser.

However, ENR (50) and EPM blend showed a large difference in the mixing characteristics. ENR (50) bales were broken into chunks and remained as distorted and elongated chunks for a longer time compared to ENR (25) before they formed a tight band around the rotors. While the EPM bales, like earlier, were broken down to smaller particles. The extent of pumping of the material into the inter-rotor region was found to be greatly reduced and this leads to the formation of a starved region below the ram. Unhomogenised material was repeatedly shared between the rotors at the top of the bridge and finally shearing between the moving rotors and the chamber wall caused blend homogenisation. Figure 3.11(a–c) illustrates the sequence of still photos of course of homogenisation of ENR (50)–EPM blend without a compatibiliser.

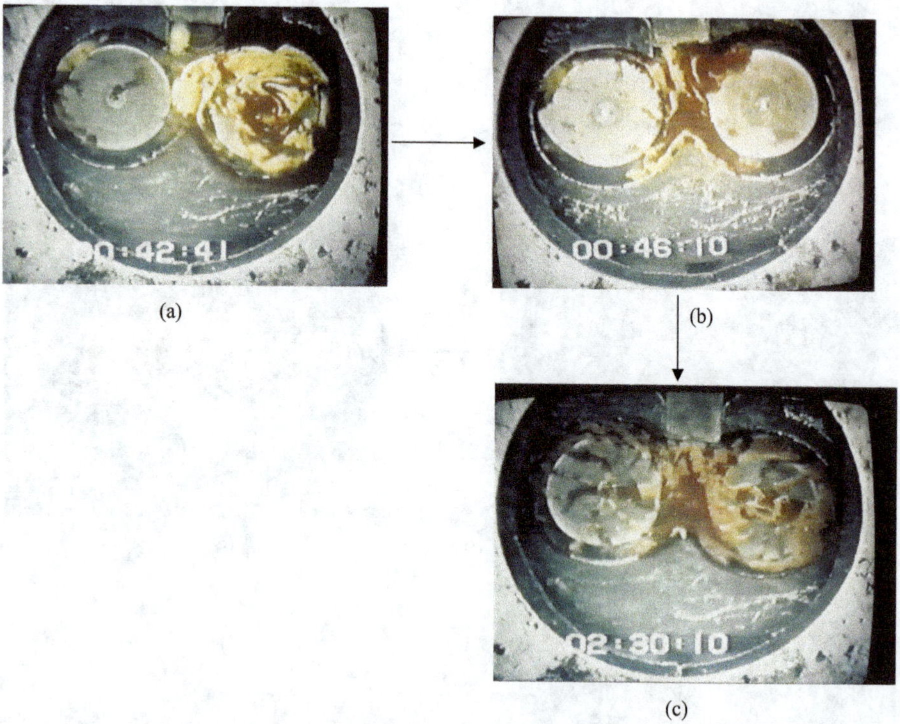


**Figure 3.11:** (a–c) Sequence of still photos showing the homogenisation of ENR (50)–EPM blend without a compatibiliser.

## 3.19 Phase morphology

While the ENR (25) and EPM blend showed similar phase morphology to that of IR and EPM system, extensive phase growth was found to occur in the case of ENR (50) and EPM blend. An average dispersed phase size of 1.5 μm for ENR (25) and EPM blend compared to that of 0.8 μm for IR and EPM system was obtained. However, in the case of ENR (50) and EPM system, the dispersed phase grows abruptly to 4.1 μm (Table 3.10, and Figures 3.12 and 3.13).

**Table 3.10:** Phase morphology of ENR (25)/EPM and ENR (50)/EPM blends.

| Compatibiliser type | Average dispersed size of ENRs with different levels of epoxidation (μm) | |
|---|---|---|
| | 25% | 50% |
| Nil | 1.5 | 4.1 |
| CIIR | 1.0 | 3.5 |
| TOR | 0.8 | 3.0 |
| CSM | 0.7 | 2.5 |
| CM | 0.7 | 2.0 |

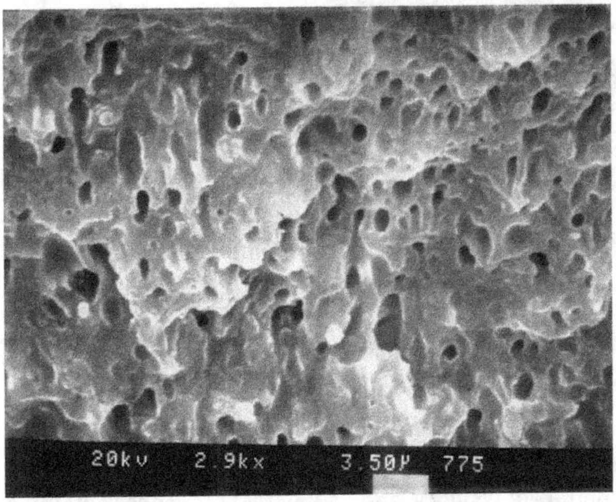


**Figure 3.12:** SEM photomicrograph of phase morphology of cross section of ENR (25)–EPM blend without a compatibiliser and via cross-linking to ebonite stage and extraction of EPM in cyclohexane.

## 3.20 Effect of compatibilising agents on blend homogenisation

Both the ENRs were blended with EPM containing CIIR, TOR, CSM, or CM (10 wt% with respect to EPM) in 80:20 weight ratios in the internal mixer as described earlier. Blend homogenisation times were recorded and are given in Table 3.9. In the case of 25 mol% epoxidised ENR, the presence of any compatibilising agent in EPM reduced the blend homogenisation time compared to that of the ENR (25) and EPM blend without a compatibiliser. However, TOR, CSM, and CM showed almost equal efficiency in

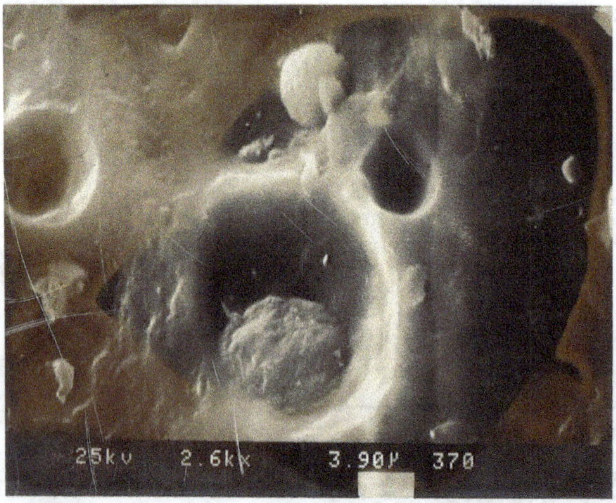


**Figure 3.13:** SEM photomicrograph of phase morphology of cross section of ENR (50)–EPM blend without a compatibiliser and via cross-linking to ebonite stage and extraction of EPM in cyclohexane.

the compatibilisation process compared to CIIR (Table 3.9). The presence of either CM or CSM in ENR (50) and EPM blends markedly lowers the blend homogenisation time followed by TOR and CIIR; they showed poorer emulsification characteristics and caused only a marginal reduction in the blend homogenisation times (Table 3.9).

### 3.20.1 Effect of compatibilising agents on phase morphology

The average size of the dispersed EPM phase calculated from the SEM photomicrographs of the cross section of both ENR (25)/EPM and ENR (50)/EPM blends containing different types of compatibilising agents are mentioned in Table 3.9. The addition of compatibilising agents leads to a lower dimension of the dispersed domains inside the continuous ENR (25) matrix. The average dispersed phase size was observed to be of the order of ENR (25)/(EPM + CM) ≈ ENR (25) and (EPM + CSM) ≈ (ENR (25)/(EPM + TOR) < ENR (25) and (EPM + CIIR) < ENR (25)/EPM. In the case of ENR (50) and EPM blends, addition of the compatibilisers causes a reduction of the scale of the dispersed phase and the extent of phase size reduction closely resembles their efficacies of phase emulsification (Table 3.10). Figure 3.14 shows the SEM photomicrograph of phase morphology of cross section of ENR (50)–EPM blend with CM as a compatibiliser obtained via cross-linking of ENR to ebonite stage and subsequent extraction of EPM in cyclohexane.

The mixing behaviours of IR or ENR with EPM in the internal mixer varied. While IR rapidly formed a tight band on the rotors, ENRs showed a decreased tendency to form so, as the level of epoxidation increased. This is certainly associated with a decreasing tendency to strain-induced crystallisation. Epoxidation of NR generates polarity in ENR and makes it more hostile to nonpolar EPM. The dispersed phase size has larger values in ENR and EPM blends compared to those of IR and EPM blends. At 25 mol% epoxidation level, the efficiency of TOR as a compatibilising agent is more effective than CIIR in reducing both the blend homogenisation time as well as the dispersed phase size, while its presence in the IR and EPM system resulted in a faster homogenisation of the blend but not the dispersed phase size as compared to CIIR. When the epoxidation level is raised to 50 mol%, the ENR (50) and EPM blends closely resemble the mixing behaviours and the morphological features of the NBR and EPM blends both in the absence and presence of a compatibiliser.

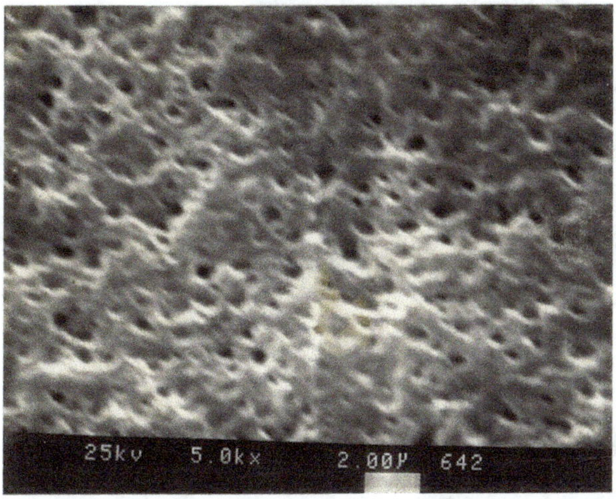


**Figure 3.14:** SEM photomicrograph of phase morphology of cross section of ENR (50)–EPM blend with CM as a compatibiliser and via cross-linking to ebonite stage and extraction of EPM in cyclohexane.

## 3.21 Conclusions

Flow visualisation studies in the internal mixer have been successfully used as a tool to investigate the effect of miscibility between the pairs of different elastomers from the rate as well as the nature of mixing. Effect of several potential compatibilising agents on rate of mixing and phase morphology of immiscible blends of EPM and NBR (varying nitrile content) have been studied. Chlorinated polyethylene (CM) was found to be most effective in increasing mixing rate and reducing scales of morphology. However, for NBRs with low acrylonitrile concentration, chlorosulfonated polyethylene (CSM), polychloroprene (CR), hydrogenated nitrile (HNBR), and polybutadiene (BR) were also found as suitable compatibilising agents. The behaviour of IR and ENR in the internal mixer chamber varied from that of NBR. The IR rapidly formed a tight band on the rotors. However, the ENR showed a decreased tendency to form tight bands as the level of epoxidation increased. This is certainly associated with a decreasing tendency to strain-induced crystallisation with increasing epoxidation level. The EPM on the other hand rapidly tears into crumbs as observed in case of NBR but less rapidly. Addition of CIIR into IR and EPM blends caused faster mixing whereas that due to CM resulted in a reduction in the rate of mixing. However, the addition of either CIIR or CM did not cause any significant change in the final phase morphology. CIIR, therefore, has some compatibilising efficiency with IR while CM does not. Epoxidation of NR produces ENR, which generates polarity and thereby lowers its miscibility with EPM. Dispersed phase size, as a consequence, showed larger values in ENR and EPM blends compared to that of IR and EPM blends.

At 25 mol% epoxidation level, both CM and CIIR act as efficient compatibilisers in the ENR (25) and EPM blends, which caused lower blend homogenisation time as well as reduced scale of the dispersed phase compared to those of the IR and EPM blends. When the epoxidation level is raised to 50 mol% in ENR (50), its blends with EPM closely resemble the mixing behaviours and morphological features of NBR and EPM blends both in the presence of CIIR or CM and without any compatibiliser.

## References


[1] Heikens, D., Barentsen, W. M. Particle dimensions in polystyrene/polyethylene blends as a function of their melt viscosity and of the concentration of added graft copolymer. Polymer 18:69 (1977).

[2] Heikens, D., Hoen, N., Barentsen, W. M., Piet, P., Ladan, H. Tensile properties and morphology of copolymer modified blends of polystyrene and polyethylene. Journal of Polymer Science, Polymer Symposia 62:309 (1978).

[3] Barentsen, W. M., Heikens, D., Piet, P. Phase equilibria in a polyethylene–polystyrene system. Polymer 15:119 (1974).

[4] Ide, F., Hasegawa,. Polymer-polymer interface in polypropylene/polyamide blends by reactive processing. Journal of Applied Polymer Science 18:963 (1974).

[5] Chen, C. C., Fontan, E., Min, K., White, J. L. Compatibilizing agents in polymer blends: Interfacial tension, phase morphology, and mechanical properties. Polymer Engineering and Science 28:69 (1988).

[6] Endo, S., Min, K., White, J. L., Kyu, T. Polyethylene – polycarbonate blends; Interface modification, phase morphology, and measurement of orientation development during cast film and tubular extrusion by infrared dichroism. Polymer Engineering and Science 26:45 (1986).

[7] Yoshida, M., Ma, J. J., Min, K., White, J. L., Quirk, R. P. Po;yester – polystyrene block copolymers and their influence on phase morphology, and mechanical properties in polymer blends. Polymer Engineering and Science 30:30 (1990).

[8] Setua, D. K. Development of thermoplastic elastomers by blending of nitrile rubbers with Nylon 6 in an internal mixer and by reactive mixing in a twin screw extruder. Journal of Biosciences 10:1 (2016).

[9] Setua, D. K. Advances in Tyre retreading and scope of run-flat tyres and elastomer nanocomposites. In: Banerjee, B., (ed) Tyre Retreading, 2nd edition. Berlin: De Gruyter, Walter de Gruyter, GmbH, Ch. 5 109–138 (2019).

[10] Walters, M. H., Keyte, D. N. Transactions Institution Rubber Industry 36:T40 (1962).

[11] Cardiner, J. B. Evaluation of physical and electrical properties of chloroprene rubber and natural rubber blends. Rubber Chemistry and Technology 43:370 (1970).
[12] Smith, R. W., Andries, J. C. Study on rubber blends: Influence of epoxidation on phase morphology and inter phase. Rubber Chemistry and Technology 47:64 (1974).
[13] Corish, P. J., Powell, B. D. Miscibility studies in blends of bromobutyl rubber and natural rubber. Rubber Chemistry and Technology 47:481 (1974).
[14] 26. Kresge, E. N. Blends of isotactic polypropylene and nitrile rubber: Morphology, mechanical properties and compatibilization. Journal of Applied Polymer Science 29:37 (1984).
[15] Marsh, P. A., Mullens, T. J., Price, L. D. Elastomers and rubber technology. Rubber Chemistry and Technology 43:400 (1970).
[16] Marsh, P. A., Voet, A., Price, L. D., Mullens, T. J. Curing studies of elastomer blends With special reference to NR/ SBR and NR/BR blends. Rubber Chemistry and Technology 41:344 (1968).
[17] Avgeropoulos, G. N., Weissert, F. G., Biddison, P. H., Bohm, G. G. A. Characterization of polymer blends miscibility, morphology and interfaces. Rubber Chemistry and Technology 49:93 (1976).
[18] 31. Callan, J. E., Hess, W. M., Scott, C. E. Evaluation of physical and electrical properties of chloroprene rubber and natural rubber blends. Rubber Chemistry and Technology 44:814 (1971).
[19] Joseph, R., George, K. E., Francis, D. J., Thomas, K. T. Polymer grafting and crosslinking. International Journal of Polymeric Materials 12(1):29 (1987).
[20] Lohmar, J. Electronmicroscopic investigations of the phase morphology in rubber blends containing poly-trans-octenylene. Kautschuk Und Gummi, Kunststoffe 39:1065 (1986).
[21] Setua, D. K., Nando, G. B. High-performance oil/fuel-resistant blends of ethylene propylene diene monomer (EPDM) and epoxidized natural rubber (ENR). In: Visakh, P. M., Semkin, A. O., (eds) High Performance Polymers and Their Nanocomposites. USA: Scrivener Publishing LLC, Ch. 9 315–346 (2019).
[22] Setua, D. K., Pandey, K. N., Saxena, A. K., Mathur, G. N. Characterization of elastomer blend and compatibility. Journal of Applied Polymer Science 74:480 (1999).
[23] Srivastava, R., Awasthi, R., Setua, D. K. Image processing of elastomer blend morphology. Kautschuk Und Gummi, Kunststoffe 53(5):268 (2000).
[24] Setua, D. K., Soman, C., Bhowmick, A. K., Mathur, G. N. Oil resistant thermoplastic elastomers of nitrile rubber and high density polyethylene blends. Polymer Engineering and Science 42:10 (2002).
[25] Pandey, K. N., Setua, D. K., Mathur, G. N. Determination of the compatibility of NBR-EPDM blends by an ultrasonic technique, modulated DSC, dynamic mechanical analysis, and atomic force microscopy. Polymer Engineering and Science 45:1265 (2005).
[26] Setua, D. K., Gupta, Y. N. On the use of micro thermal analysis to characterize compatibility of nitrile rubber blends. Thermochimica Acta 462:32 (2007).
[27] Setua, D. K. Indian Thermal Analysis Society (ITAS. Bulletin 4(2):1 (2011).
[28] Agarwal, K., Prasad, M., Chakraborty, A., Vishwakarma, C. B., Sharma, R. B. Studies on phase morphology and thermo-physical properties of nitrile rubber blends. Journal of Thermal Analysis and Calorimetry 104(3):1125 (2011).
[29] Soares, B. G., Sirqueira, A., Oliveira, M. G., Almeida, M. S. M. Rice husk fibers and their extracted silica as promising bio-based fillers for EPDM/NBR rubber blend vulcanizates. Kautschuk Und Gummi, Kunststoffe 55(9):454 (2002).
[30] Pandey, K. N., Setua, D. K., Mathur, G. N. Characterization of liquid carboxy terminated copolymer of butadiene acrylonitrile modified epoxy resin. Polymer Engineering and Science 45(9):1187 (2005).
[31] Freakley, P. K., Wan Idris, W. Y. Numerical investigations of the effect of fill factor In an internal mixer for tyre manufacturing process. Rubber Chemistry and Technology 52:134 (1979).
[32] Asai, T., Fukui, T., Inoue, K., Kariyama, M. International Rubber Conference Paris paper 111–114 (1982).

[33] Min, K., White, J. L. Compounding polymer blends. Rubber Chemistry and Technology 58:1024 (1985).
[34] Min, K., White, J. L. Flow visualization of the rubber compounding cycle in an internal mixer based on elastomer blends. Rubber Chemistry and Technology 60:361 (1987).
[35] Min, K. Modelling of the processing of incompatible polymer blends. International Polymer Processing 1:179 (1987).
[36] Min, K. Development of dispersion in the mixing of calcium carbonate into polymer blends in an internal mixer. Advances in Polymer Technology 7:243 (1987).
[37] Min, K. Flow visualization in internal mixers. In: White, J. L. (ed) Mixing and Compounding of Polymers Theory and Practice. KG: Carl Hanser Verlag GmbH & Co., Ch. 10 337–361 (2009).
[38] Morikawa, A., Min, K., White, J. L. Comparative studies on the properties of SAN/SBS blend prepared by batch and continuous process. Advances in Polymer Technology 8:383 (1983).
[39] Morikawa, A., White, J. L., Min, K. An experimental and theoretical study of starvation effects on flow and mixing of elastomers in an internal mixer Kautschuk Und Gummi, Kunststoffe 41:1226 (1988).
[40] Morikawa, A., White, J. L., Min, K. Flow visualization of the rubber compounding cycle in an internal mixer based on elastomer blends International Polymer Processing 4:23 (1989).
[41] Setua, D. K., White, J. L. Flow visualization and phase morphology development in blends of nitrile and ethylene propylene rubber in an internal mixer. Kautschuk und Gummi, Kunststoffe 44:137 (1991).
[42] Setua, D. K., White, J. L. Influence of compatibilizing agents on rate of mixing and phase morphology development in blends of nitrile and ethylene propylene rubber in an internal mixer : a flow visualization study. Kautschuk Und Gummi, Kunststoffe 44:542 (1991).
[43] Setua, D. K., White, J. L. Flow visualization of the influence of compatibilizing agents on the mixing of elastomer blends and the effect on phase morphology. Polymer Engineering and Science 31:1742 (1991).
[44] Setua, D. K., White, J. L. Flow visualization of the mixing and blend phase morphology of binary and ternary blends of polychloroprene, acrylonitrile--butadiene copolymer and ethylene propylene diene monomer rubber. Kautschuk Gummi Kunststoffe 44:821 (1991).
[45] Setua, D. K., Chakaraborty, S. K., De, S. K., Dhindaw, B. K. Scanning electron microscopy studies on the mechanism of rubber tear. Scanning Electron Microscopy Part III:973 (1982).
[46] Vitali, R., Montani, E. Ruthenium tetroxide as a staining agent for unsaturated and saturated polymers. Polymer 21:1220 (1980).
[47] Trent, J. S., Scheinbeim, J. I., Couchman, P. R. Ruthenium tetraoxide staining of polymers for electron microscopy. Macromolecules 16:539 (1983).

Mrinmoy Debnath and Abhijit Adhikary

# Chapter 4
# Use of graphene in rubber nanocomposite, its processability, and commercial advantages

## 4.1 Introduction

High-performance elastomeric composites are important materials in the new product development strategy for divergent applications that require apparently contradictory performance requirements. Particulate fillers like carbon black, silica, and other nanofillers are widely used with natural and synthetic rubbers in appropriate doses to enhance various processing and performance characteristics. Not only do these fillers significantly improve the reinforcing, dynamic mechanical, and other end use properties of the product, they are also used to dilute the elastomeric matrix for material cost reduction. However, with the incorporation of fillers in higher doses, the friction between the filler particles increases, causing higher energy loses in dynamic applications, resulting in higher heat generation, thereby limiting its use. Secondly, the higher specific gravity of fillers, as compared to rubbers, increase the specific gravity of the composite with increased filler loading, which acts against the strategy of developing low-weight rubber components.

In 2004, Professor Andre Geim and Professor Konstantin Novoselov of The University of Manchester discovered graphene [1]. Since its discovery, as a potential filler, graphene attracted attention of the scientific fraternity throughout the world owing to its astonishing properties. In 2010, due to their outstanding contribution in science, both professors were awarded the Nobel Prize in Physics [1]. Graphene is made up of a single-layer hexagonal honeycomb lattice structure. It possesses very high elastic modulus (~1 TPa) and extraordinary intrinsic strength (~130 GPa). Graphene shows high specific surface area (~2,630 $m^2/g$) and its gas barrier property is better than that of clay [2–4]. Thermal conductivity (~5,000 W/mK) and electrical conductivity (~6,000 $S/cm^2$) [5] of graphene are also quite high. Such properties of graphene largely depend on its different manufacturing processes [6]. Few common methods that are commonly practiced are chemical vapour deposition (CVD), thermal exfoliation, and mechanical exfoliation in dispersed condition [7–9]. Out of these methods, the thermal exfoliation process is capable of producing too many layers of graphene, which could show interesting properties like single-layer graphene.

Due to the surface characteristics of graphene, it tends to aggregate, and faces challenges in dispersing properly and interact effectively with rubber, thus affecting graphene-rubber reinforcement. To attain an even and stable dispersion of any filler in

https://doi.org/10.1515/9783110668537-004

rubber matrix, aggregation of filler particles needs to be reduced by increasing rubber-filler interfacial interaction. Such aggregation between graphene sheets can be minimised by their appropriate functionalisation with the incorporation of suitable hydrophilic/hydrophobic groups or by their bulky size [10]. This helps to improve the graphene-rubber interaction and to achieve a consistent dispersion of graphene in the rubber matrix. Thus, the process ultimately facilitates improvement of stress transfer between rubber and filler, which consequently improves the rubber reinforcement. Mixing methods such as melt mixing by using mechanical shearing in molten state of polymer, solution intercalation and/or latex blending, and in situ polymerisation have been employed for effective dispersion of graphene in the rubber nanocomposite, taking into account the cost and time involved in the manufacturing process [11–14]. In general, the melt mixing method is preferred in the industry as it combines the short span of mixing and the low-cost of operation. However, this method cannot result in a proper dispersion of graphene in the rubber phase, leading to inadequate development of its characteristic properties. Lower affinity of graphene with rubber, van der Waals force, and the π–π interaction between graphene sheets make their dispersion difficult in the rubber matrix by this process. Therefore, it is required to exfoliate the graphene sheet for a better interfacial interaction. Suitable exfoliation of graphene sheets is achieved by using special techniques such as sonication and high shear mixing. Solution intercalation/latex phase blending is carried out to achieve proper dispersion of the exfoliated graphene sheets in the rubber matrix. In the solution blending method, the rubber remains in the solvent phase, while latex blending is like solution-blending process, where the elastomer is in the latex form. Removal of the solvents used during the solution blending process and the cost of solvent are of important concern for solvent-based mixing, whereas latex blending method is environment friendly. In situ polymerisations are carried out by mixing fillers with monomers, followed by in situ polymerisation.

Replacement of conventional fillers by a small amount of graphene, functionalised graphene (FG), graphene oxide (GO), reduced graphene oxide (RGO), and thermally exfoliated graphene oxide (thermally RGO (TRGO)) in rubber nanocomposite resulted in substantial improvement in the mechanical property, thermal stability, electrical properties, gas impermeable properties, and so on [6]. Thus, a small dose of graphene or its precursor can replace a large quantity of conventional fillers like carbon black and silica with significant reduction in the filler–filler friction. For tyre components, it results in significant reduction in rolling resistance, heat generation, and improvement in wear resistance.

The success of producing the right quality of graphene–rubber nanocomposite depends on many parameters such as the type and quantity of functional groups present on the modified graphene surface, the type of polymer, the type of equipment used for exfoliation and dispersion, the quantity of graphene used as a replacement of filler, etc. A combination of a suitable type and quality of graphene or its derivative and use of an appropriate manufacturing technique could improve the performance of graphene substantially as a filler in the rubber nanocomposite. Suitable adoption of technology and selection of the right graphene material would find its use in applications like tyre components, automobile parts, gasket and seals, hoses, footwear, various types of clothing, and thermal and electrically conductive applications. Of late, few bicycle tyre manufacturers have implemented graphene or its derivatives in tread in selective product lines. Experimental passenger tyres have been shown to significantly improve the rolling resistance (fuel economy) and mileage, keeping the traction performance comparable to its silica counterpart. Apart from the lightweight and strong polymeric composite, graphene has been proposed and is under development in the field of electronics, biological engineering, filtration, photovoltaic, and other energy storage devices.

Graphene is toxic in nature. It depends on several factors such as size, shape, functional groups, dispersion stage, synthesis methods, quantity, exposure time, etc. Research has shown that 10 μm 'few layered' graphene flakes can enter the cell membrane in solution. The biological effects of this toxic effect are vague and is comparatively still a subject under proper investigation. As of now, graphene is costlier than the conventional fillers, but with its wider commercial use, its manufacturing cost is expected to come down substantially. The potential of graphene is too big to be ignored; it is where our future lies.

## 4.2 Graphene

Graphene is a two-dimensional, single layer, hexagonal honeycomb lattice of graphite [15]. A typical carbon atom has a diameter of about 0.33 nm. Van der Waals force of attraction keeps such hundreds of thousands of graphene layers bonded together and essentially make graphite where, nearly 3 million of graphene layers form 1 mm of graphite. The structural makeup of graphite and graphene are slightly different. Interlayer spacing of graphite is ~0.34 nm and the bond distance between the carbon atoms is ~0.14 nm as described in Figure 4.1 [16, 17]. As carbon atom of graphite is $sp^2$ hybridised; the fourth loosely bound π-electron of carbon is highly mobile and mainly triggers electrical properties. It is very soft and can be easily broken on application of low pressure.

Graphene is an expensive material and relatively difficult to produce. Much effort has been spent to find out effective and inexpensive ways to make defect-free, single to few layers, low-cost graphene. Properties like physical, mechanical, electrical etc. depend mainly on the methods of synthesis of graphene. Out of the various methods, two main graphene fabrication approaches are the bottom-up process and the top-down process.

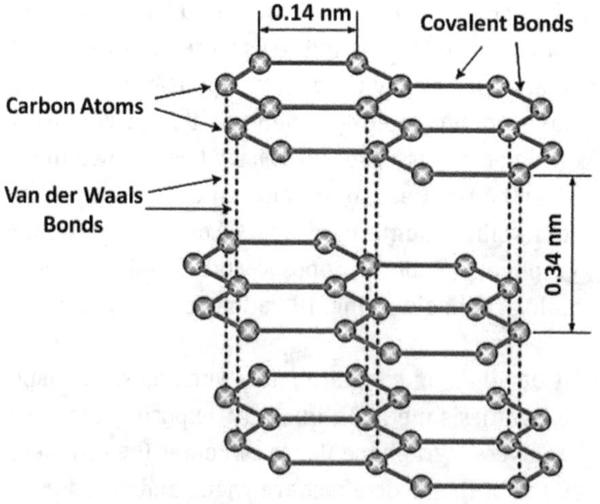


**Figure 4.1:** Schematic representation of distances in three-dimensional structure of graphite.

# 4.3 Bottom-up process

By these methods, carbonaceous materials are converted to make graphene. Some methods like CVD [15, 18–22], epitaxial growth on SiC [23–28], unzipping of carbon nanotubes (CNTs) [29, 30], arc discharge [31, 32], chemical conversion, and self-assembly of surfactants [33–35] could be chosen. Out of these methods, CVD on transition metal has been widely used in the process to manufacture graphene. A high aspect ratio graphene can be produced by this method. CVD method was first established in 2008 to synthesise graphene [36]. Several transition metals as substrates also were experimented in the CVD method [37–40] and copper has been found to be the better one. The scanning electron microscope (scanning electron microscopy (SEM)) images of growth of graphene by the CVD method is shown in Figure 4.2 [21]. Although this method could make very good quality graphene, few problems are accompanied with the method. First, it is the separation of graphene from the metal substrate, which could potentially destroy the structure and changes the properties of graphene. Second, it is an expensive process as it requires quite a high energy. Third, it is required to revive the metal substrates on every cycle. Addition of a roll-to-roll synthesis step could scale up this process, where Cu-coated rolls are used as substrate to grow graphene on it and then transfer to polymer films [40–42].

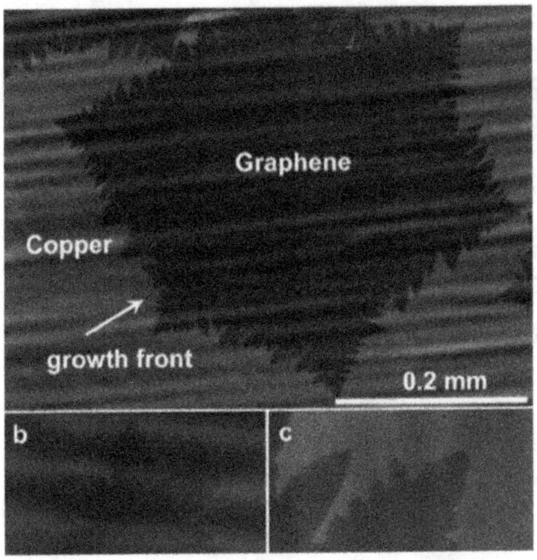


**Figure 4.2:** SEM images of graphene growth on the surface of copper by the CVD method. (a) Growth of graphene at 1,035 °C and ~6 μm/min on average. (b) Formation of graphene nuclei at the early stage of growing. (c) The arrow in (a) shows the growth front of high-surface-energy graphene.

## 4.4 Top-down method

Due to the small interlayer spacing of graphitic layers, polymer chains find difficulties to go inside the gallery spacing of graphite, resulting in an improper dispersion of graphite flakes [43] in the polymer matrix. Therefore, numerous physical and chemical methods have been worked out to increase the spacing between the graphite planes. Incorporation of many oxygen-containing functional groups by oxidation is one such noteworthy process. This process makes graphite more hydrophilic, and its layers become easily detachable. At this stage, the materials are intercalated or exfoliated.

Many covalent and ionic types of chemicals could be reacted with graphite to form graphite intercalated compound (GIC) through the intercalation process, which increases the interlayer spacing of graphite layers [44–46] and they are classified as covalent intercalation compound and ionic intercalation compound, respectively. Covalently intercalated compounds are only restricted to graphite oxide and graphite fluoride and have limited applications as they are electrically insulated. Hence, the GICs are mainly referred to ionic types such as graphite-acid type compounds, which give ionic or polar type characteristic to GICs. The well-known graphite–acid salt is graphite bisulfate (graphite–$H_2SO_4$) and could be produced by the chemical oxidation reaction of graphite with conc. $H_2SO_4$ or by electrolysis [47–53]. Several oxidising agents could be used for the intercalation of graphite. The surface area of the exfoliated graphite depends on the extent of intercalate. In the exfoliation process, when graphite bisulfate is heated at nearly 1,050 °C, the graphite enlarges irrevocably by nearly hundreds of its volume and forms expanded graphite (EG) [54]. A schematic design is illustrated in Figure 4.3, and forms EG [54].

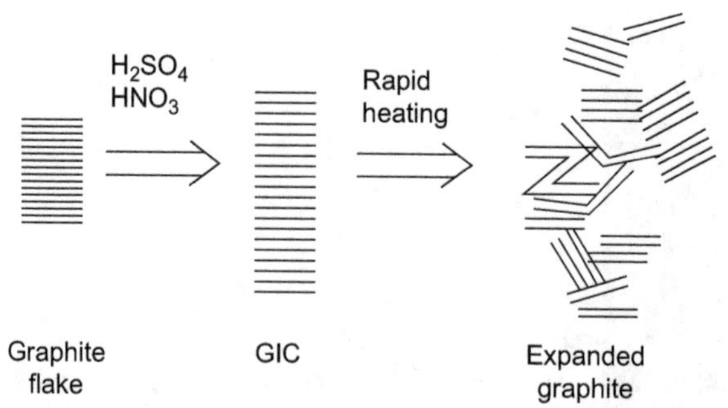


**Figure 4.3:** Schematic design of producing expanded graphite (EG) by the nitric acid and sulphuric acid route.

Oxidation of graphite could produce single atomic layered material called GO. Nearly 150 years ago, it was synthesised by the Brodie's method [55] where graphite was oxidised with $KClO_3$ and fuming $HNO_3$. Later, many scientists such as Staudenmaier [56] and Hummers and Offeman [57] modified this method. On oxidation of graphite using conc. $H_2SO_4$, $NaNO_3$, $KMnO_4$, and so on, scientists found an oxygenated functional group in the GO [58, 59]. As GO is hydrophilic in nature, it can be exfoliated in water with the help of ultrasonication. This process helps to make single to few layers of FG. GO can be easily dispersed finely in water as well as in other polar solvents due to its different functional groups. It can also be simply mixed with different rubbers and other materials, which enhance many properties of the composite. Like these merits, it possesses some shortcomings like poor conductivity and inferior thermal and electrical properties. These shortfalls of GO can be regenerated to a certain amount by chemical or thermal reduction process in the presence of an inert gas atmosphere. Many reducing chemicals such as $NaBH_4$ [60], hydrazine monohydrate [61], L-ascorbic acid [62], $LiALH_4$ [63], $NH_2OH$ [64], and HI [65] could be used to partially reduce oxygen-containing functional groups on GO and to regain few of its original double bonds [66].

Thermal reduction of GO is performed in an inert atmosphere at a very high and rapid heating condition (1,000 °C) [67] to make RGOs. Chemically RGO (CRGO) is produced by the chemical method whereas TRGO [57, 68] is obtained by the thermal route. The RGOs still contain oxygen, which makes it easier to use [69], but this graphene is not in its pure or pristine form and has many defects. The oxidation and reduction of the material leads to a partial degradation of the graphitic structure [70].

Graphene in its original, pure, nonoxidised form is called pristine graphene, which enjoys superior properties when compared to its oxidised counterpart. Large-scale production of single-layered pristine graphene by nondestructive exfoliation from graphite is still a tough challenge [71–73].

Generally, the exfoliation of graphite is carried out by oxidation, followed by exfoliation and reduction process. This process induces many structural defects on graphene and significantly deteriorates its basic properties. Hence, many attempts were made to make pristine graphene by direct exfoliation so that the $sp^2$ carbon lattice of layers are not damaged [74].

Another method, termed as thermal exfoliation, is used to produce near completely exploited single to few layers of graphene by employing high instantaneous thermal shock. This is a very rapid and solvent-free method in comparison to mechanical exfoliation and takes place within seconds at a high temperature. A fast microwave heating can be applied to the acid intercalated graphite, which rapidly expands its layers by instantaneous vapourisation of the acid inside the graphite layers [75].

This process could produce approximately 10 nm thick graphite nanoplates. During the quick heat treatment, degradation of oxygenated functional groups of graphitic planes produces gases, which generate pressure between the adjoining graphitic planes.

If this pressure exceeds the Van der Waals force of attraction between the graphitic layers, it causes exfoliation of the layers. Generally, graphite oxide, EG, GICs, etc. rather than pure graphite are used in this process due to the presence of functionalities in them. Normally, near about 1,050 °C temperature, approximately 2,000 °C/min heating rate, and 700 $m^2/g$ pressures are crucial parameters for attainment of complete exfoliation of graphite [76, 77].

Graphene could be produced in bulk by this method, which still shows some interesting properties of single-layer graphene. Due to such features, thermal exfoliation process is most popular currently.

Mechanical exfoliation was first attempted by Novoselov et al. [78] in 2004 by using scotch tape to make single-layer graphene. The process involves repeated peeling of graphene layers out of graphite; hence, the number of layers and their thickness could not be controlled. Although this process can make graphene with astonishing quality and property, it could not be scaled up due to difficulties and constraints associated with this process [1, 78].

In liquid phase exfoliation, mechanical force like high-speed stirring, shaking, or ultrasonication are applied in the presence of a selective solvent, which facilitates the exfoliation of graphite by suitable interaction between graphite and the solvent, thereby reducing the energy requirement for the process and lowers the re-aggregation of the exfoliated sheets. Re-aggregation of the exfoliated sheets occurs due to the Van der Waals force of interactions between the graphene sheets. Many intercalants, surfactants, and a functionalisation process are adopted in this method to keep these sheets under dispersion in the solution. Organic solvents like N-methyl-pyrrolidone [79], dimethylformamide [80], dimethylsulfoxide, tetrahydrofuran [81], and per fluorinated aromatics [89] are used in this process [90].

This is relatively an efficient method because interaction between graphene and the solvent generates extra energy, which also smoothens the exfoliation process [81–86]. Hence, the selection of a suitable solvent is one of the most important criterion in this method. Surface tension of the solvent is the determining factor for solvent selection. A good solvent should have surface tension higher than graphene-to-graphene interaction energy.

Usually, organic solvents work effectively but water and surfactant-based suspensions are also confirmed to be effective [91]. There is also a recent report published on the exfoliation of graphite in inorganic salt solution of $CuCl_2$ and NaCl [92]. The main benefit of this method is the availability of an array of solvents that could be used to produce graphene with good quality and has huge viable applications. The yield of graphene produced by this method is very less. Solution intercalation of graphene and blending with elastomer solution is quite easy and can be directly used to prepare graphene–elastomer nanocomposites in the same solvent. Nevertheless, this process is not environment friendly due to the using of a solvent and is expensive as well.

## 4.5 Graphene–rubber nanocomposite

### 4.5.1 Characterisation

The worth of graphene depends on its quality but interestingly, all its applications do not require its best quality. For example, GO functionalised with oxygen and hydrogen is inexpensive but could be used for many potential applications. The intended improvement of properties by the addition of graphene, such as mechanical, chemical, electrical, thermal, gas barrier, abrasion, dynamic mechanical, and abrasion, depend on many factors. So, after completion of synthesis of the nanocomposite, it is necessary to characterise the materials to understand the interactions between the graphene and the elastomer, which play a critical role on the physicochemical properties of the composite. The relationships between the dispersion, orientation and distributions of fillers, their interaction with the matrix, cure behaviours, etc. are the property characterisation techniques adopted to assess the concentrations, dispersion, and distributions of nanoparticles as well as the nature of the interface between the polymer and the nanoparticles.

Dispersion of the graphene sheet in the rubber matrix depends on its quantity, dimension, and its functionality. The orientation of the graphene sheet also regulates many properties of the composites. Chemical modification of fillers and its interactions with the polymer chains affect the processing properties of the composites; hence, an analysis of many other properties is required to determine the true potential of the nanocomposites in the application. For this reason, several analytical characterisation techniques are applied to understand the surface and physicochemical properties of graphene–elastomer nanocomposites. The techniques include Fourier-transform infrared spectroscopy (FTIR), X-ray diffraction (XRD), X-ray photoelectron spectroscopy, SEM, transmission electron microscopy (TEM), Raman spectroscopy, cure behaviour, and contact angle measurement [87–117].

## 4.6 Fourier-transform infrared spectroscopy (FTIR)

FTIR technique is a versatile analytical procedure to characterise and identify organic, polymeric, and inorganic materials. It can also be employed to understand the interaction of functional groups of elastomers with fillers. S. Yaragalla and coworkers [114] used TRGOGO, prepared by the thermal reduction technique, with natural rubber (NR) to prepare their nanocomposite. FTIR analysis of the nanocomposite is depicted in Figure 4.4. It was observed that the virgin NR, which is made up of *cis*-1,4-polyisoprene units, displays distinctive IR adsorption due to stretching vibrations of =CH, C=C bonds at 2,916 $cm^{-1}$ and 1,596 $cm^{-1}$, respectively. Due to the incorporation of 3 wt% of GO to NR, the intensity as well as peak associated with =CH stretching are loosened and moves slightly to the lower frequency side, whereas the C=C peak completely diminished, compared to neat NR. It is an indication of a strong noncovalent type of interaction due to the π-electrons of GO and carbon–carbon double bonds of the isoprene unit of NR.

The FTIR data shows prominent chemical interactions in the case of 3 wt% TRG, in comparison to 0.5 and 2 wt% TRG. On the contrary, Habib et al. [88] observed a characteristic stretching peak (969 $cm^{-1}$), measured by FTIR analysis of butadiene unit of GO and nitrile rubber (GO/NBR) composite; it was more sharp and intense. Interaction of double bond of butadiene unit with π-electrons of graphene might cause such phenomena. Kang et al. [89] fabricated GO/XNBR nanocomposite by the latex route and observed a shift in the FTIR peak of the carbonyl group of XNBR from 1,729 $cm^{-1}$ to 1,735 $cm^{-1}$ in GO/XNBR on addition of 1.9% of GO with XNBR. This might have happened on account of hydrogen bond formation between the carboxylic and hydroxyl groups of GO and the carboxylic groups of XNBR rubber.

Such interactions in the composite also restrict the flexibility of the XNBR rubber chains, which is evident by the shift in the glass transition temperature from −29.9 to −26.7 °C.

### 4.6.1 X-ray diffraction (XRD)

XRD is one of the most significant methods used to characterise nanoparticles. This method could be extensively utilised to determine the amount of exfoliated and intercalated graphene layers in the elastomeric phase. Wu et al. [90] incorporated GO in NR latex to prepare NR/GO nanocomposites. XRD experiment as such did not show visible characteristic diffraction in the nanocomposite.

It is an indication of the homogeneous dispersion and exfoliation of GO in nanocomposites. Broad diffraction peak around 20° is observed due to the noncrystalline structure of NR, where the diffraction peaks between 30–50° were noticed due to the presence of zinc oxide particles in the vulcanisates [91]. N. Habib et al. [88] prepared GO/NBR nanocomposites and carried out XRD analysis to characterise the dispersion of

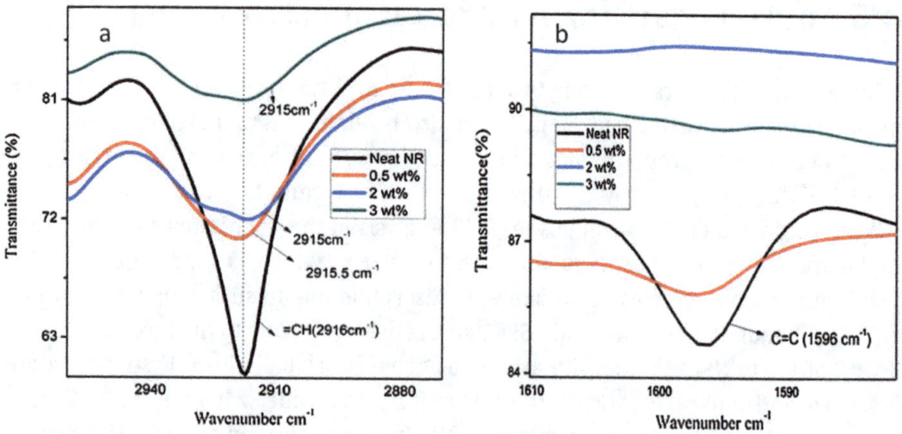


**Figure 4.4:** FTIR spectrum of TRG-NR nanocomposites at a selective normalised frequency range.

GO sheet in the nanocomposite. XRD analysis of GO displayed a distinctive diffraction peak at $2\theta = 11.39°$whereas unfilled NBR rubber as well as the nanocomposite filled with 0.2, 0.4, and 0.8 wt% doses of GO exhibited a characteristic wide XRD peak at $2\theta = 20°$, which are indications of complete dispersion, disappearance of crystalline periodic structure, and complete exfoliation of GO sheets in the matrix rubber. GO/acrylonitrile–butadiene rubber (GO/NBR) nanocomposites were prepared by Mensah et al. [92] and XRD analysis showed diffraction peak approximately at 6.4° (1.53 nm) whereas XRD analysis of GO exhibited diffraction peak at $2\theta = 10.7°$(0.92 nm), which signify a stabbed and interpenetrated intercalation of GO within the rubber compound. Similar type of observation was reported by Bai et al. [93] in the exfoliated GO and hydrogenated carboxylated nitrile–butadiene rubber (HXNBR) composites and by Xiong et al. [94] in their ionic-liquid-modified GO and bromo butyl rubber nanocomposites.

Kang et al. [89] observed a comparable diffraction pattern of XNBR and GO/XNBR nanocomposites by XRD; this is another indication of very good exfoliation and dispersion of GO in the XNBR phase. The difference in storage modulus ($G'$) between the lower and higher oscillatory shear strain is considered a significant parameter to evaluate the filler-filler network as well as the filler-polymer interaction in the graphene-polymer nanocomposites [95–99]. This is known as the Payne effect. Figure 4.5 (d) shows a strain-dependent storage modulus ($G'$) of GO and XNBR nanocomposite [89].

The increase in storage modulus with increasing GO doses in the composite are indications of easy π–π interactions of GO sheets with the polar groups of nitrile butadiene rubber (NBR) segments. Observations of noticeable extended plateau zone of storage modulus with the addition of the said doses of GO with rubber indicate a very high distribution of GO sheets in the XNBR matrix, which causes its strong filler networks within the rubber matrix. These improve the dispersion of the oxide sheets in the XNBR matrix and form sturdy filler–polymer interactions.

## 4.6.2 Electron microscopy

The morphological analysis of graphitic fillers and their dispersibility in graphitic nanocomposites could be characterised appropriately by electron microscopy techniques [100]. Internal composition of the nanocomposites is preferably measured by high-magnification TEM [101] and filler dispersion at the fracture sides of the nanocomposites is well characterised by SEM [102]. Habib et al. [88] observed improved dispersion with minimal appearance of agglomeration of GO sheets by SEM in the NBR rubber matrix, which was achieved by solution blending followed by mixing in an internal mixer. Liu et al. [103] characterised the nanocomposites made of GO and a thermoplastic polyurethane (TPU) by analytical techniques and observed uniform dispersion of GO in the rubber matrix. SEM micrographic analysis additionally confirms adhesion between GO and the rubber matrix.

It has been observed that graphene sheets convert to highly crumpled form after modification by chemical process as well as by the application of thermal treatment. Such typical property of graphene makes their characterisation very difficult. In spite of these problems, scientists have reported their attempts to measure the dispersion level of graphene in elastomer matrix by TEM image analysis. Zhan et al. [104] prepared NR latex/graphene nanocomposites by in situ reduction of GO during ultrasonic latex mixing process. Fully exfoliated graphene layers were observed at 20 times the dilution level of latex by TEM image analysis. Yang et al. [105] successfully quantificated the dispersion of 5 phr EG in styrene butadiene rubber (SBR) and NR by TEM. Wu et al. [90] captured TEM and SEM images shown in Figure 4.6 from the fractured surfaces of a tensile sample specimen of NR/GO nanocomposites, prepared by latex blending method with 2 phr of GO having different lateral sizes, designated as G1, G2, and G3. SEM images in Figure 4.6 analysis show a good dispersion and no obvious aggregations of GO in the NR matrix.

The tensile fractured surfaces are rough and no cavities are found between the GO sheets and NR. All these characterisations indicate strong interactions of GO with the NR matrix. TEM image analysis shows few dark spots in the images, which may have been originated due to the use of higher doses of antioxidants compared to GO. TEM image analysis further confirms lower GO sheets length in the nanocomposites compared to the original, which might have happened due to the application of dynamic shear force during the mill mixing stage. Mao et al. [107] fabricated GO/SBR nanocomposite by the latex blending process. In the cross section of the nanocomposite, few wrinkled GO sheets were observed while measuring the dispersion of GO sheets by SEM techniques.

High-resolution TEM images of nanocomposite with different GO doses did not show any indications of multilayer stacking of GO sheets. Kang et al. [89] studied the dispersion of GO sheets in the XNBR compound by high-resolution TEM (HRTEM) depicted in Figure 4.5(a), where they observed very less agglomerated GO sheets, which may be due to the unexfoliated GO during the preparation of the GO suspension. Red arrows in the HRTEM image in Figure 4.5(b) are used to emphasise the dispersion of GO sheets, with each layer having a thickness of nearly 1–2 nm in the matrix rubber. Several experimental studies prove that GO sheets exfoliate well and disperse uniformly in the rubber phase. Microscopy techniques are also proven to measure the dispersion of these nanofillers in the nanocomposites.

### 4.6.3 Raman spectroscopy

Raman spectroscopy is very sensitive to geometric structure and the bonding of molecules and is widely used to study graphene elastomer nanocomposites. Graphene shows strong effective resonance on Raman scattering and displays three different distinctive bands at significantly different wavelength, which are D band at ~1,330 $cm^{-1}$, G band at ~1,580 $cm^{-1}$, and the 2D at ~2,650 $cm^{-1}$ [107] with significantly different relative intensities. The position of G band gives very accurate information regarding the quantity of graphene layers whereas both position and intensity of the G band provide measurable information regarding layer thickness of graphene [108, 109]. Graphene always shows a strong 2D band.

Its position and shape are used to measure the layer thickness of graphene. It broadens and up shifts with increasing number of graphene layers. The ratio of 2D and G band provides indicative information regarding the number of graphene layers. If the ratio is nearly 2.0, then it is considered a monolayer graphene whereas a lower value indicates few too many layers of graphene [106]. The D band is associated with defects or disorders of monolayer graphene, which is generally very weak for prime quality graphene and absent in defect-free monolayer graphene. Its intensity proportionately increases with the level of defects in graphene.

The ratio of the intensity of the D and G bands signifies the extent of defects generated during the production of graphene and the damages occurred during its exfoliation and edge formation [111].

The most significant use of Raman spectroscopy in graphene characterisation is to get accurate information about the number of graphene layers. Malardand and his coworkers [110] have found out that Raman spectroscopy is indeed a useful tool to characterise monolayer graphene and to differentiate monolayer from few layers of graphene. Young et al. [109] characterised graphene layers by Raman spectroscopy (illustrated in Figure 4.7), and observed many valuable information. D band is not present in top spectrum of Figure 4.7, which indicates the presence of a defect-free monolayer. Bottom spectrums show widening of the 2D and G band ratio from monolayers to multilayers of graphene. Zheng-Tian Xie et al. [113] studied the interfacial

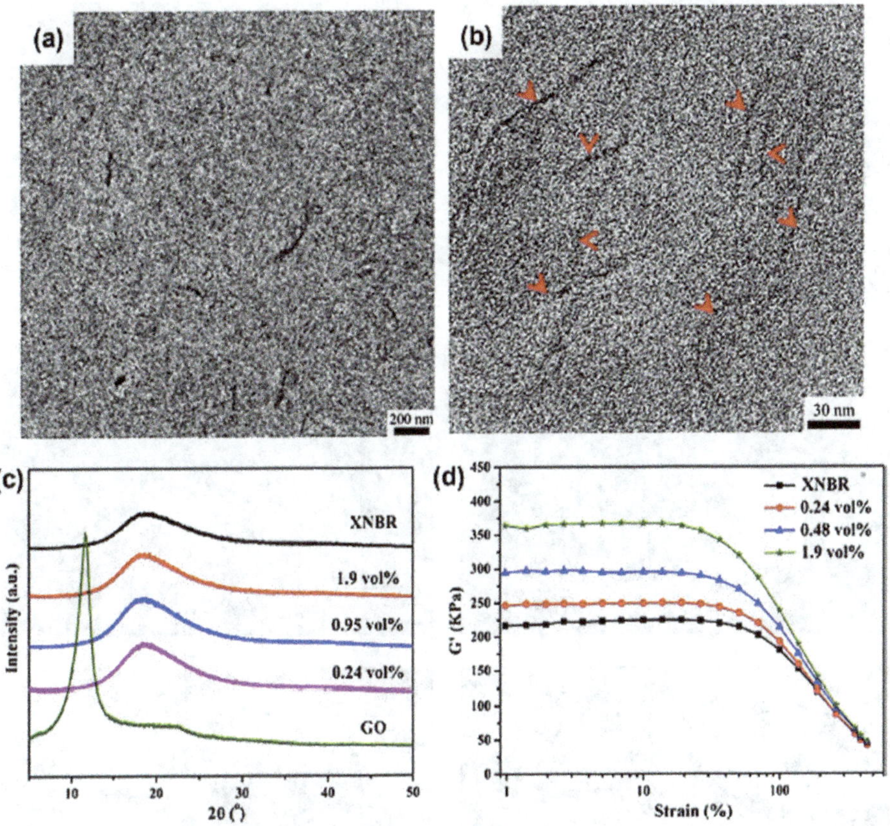


**Figure 4.5:** HRTEM images of GO/XNBR nanocomposite at different magnification levels are shown in picture (a) and picture (b) containing 0.48 vol% GO; (c) XRD patterns of GO, XNBR, and GO/XNBR composites with various GO content; (d) Payne effect diagram for GO/XNBR nanocomposites at different GO content levels.

interaction between graphene, isoprene rubber, and organosilane-grafted (Si-69) GO and isoprene rubber.

The graphene and isoprene rubber forms unmodified graphene–isoprene (IR) nanocomposite. The organosilane-grafted GO was reduced in the solution phase to prepare surface-modified graphene–isoprene rubber nanocomposite (SGE/IR). Raman characterisation has demonstrated the increase of thicker bound rubber on SGE than that on unmodified graphene, which is an indication of comparatively higher interfacial interaction of SGE with IR. As an outcome, IR/SGE showed slower chain dynamics and lower strain-induced crystallisation than IR/graphene. Yaragalla et al. [114] also carried out Raman spectroscopic analysis of NR/TRG nanocomposite to understand the probable chemical interactions of TRG in NR matrix. On adding TRG with NR, the G

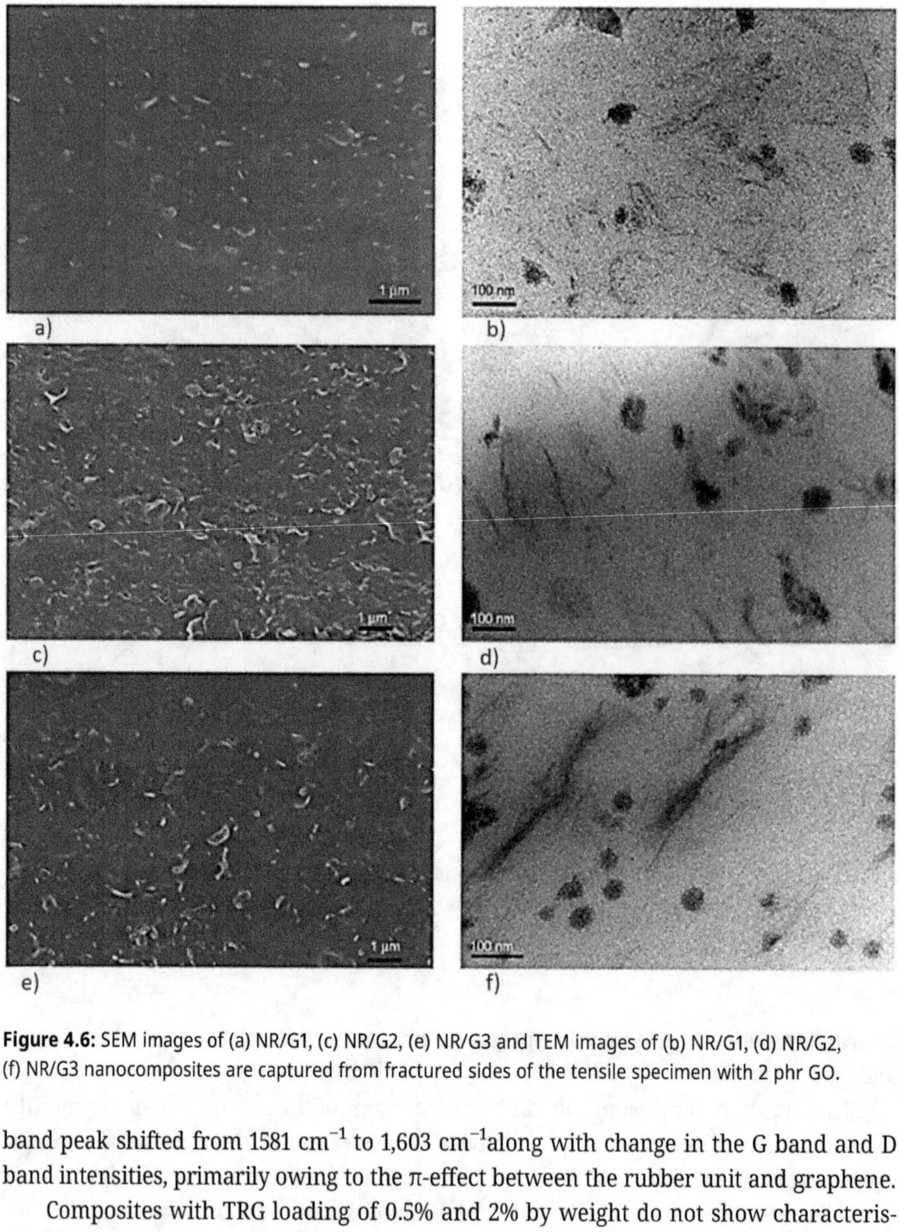


**Figure 4.6:** SEM images of (a) NR/G1, (c) NR/G2, (e) NR/G3 and TEM images of (b) NR/G1, (d) NR/G2, (f) NR/G3 nanocomposites are captured from fractured sides of the tensile specimen with 2 phr GO.

band peak shifted from 1581 $cm^{-1}$ to 1,603 $cm^{-1}$along with change in the G band and D band intensities, primarily owing to the π-effect between the rubber unit and graphene.

Composites with TRG loading of 0.5% and 2% by weight do not show characteristic differences in intensities of D and G band whereas complete elimination of the D band and the shift of G band are observed at 3% w/w of TRG in the NR composite, which indicates strong interaction between TRG and NR.

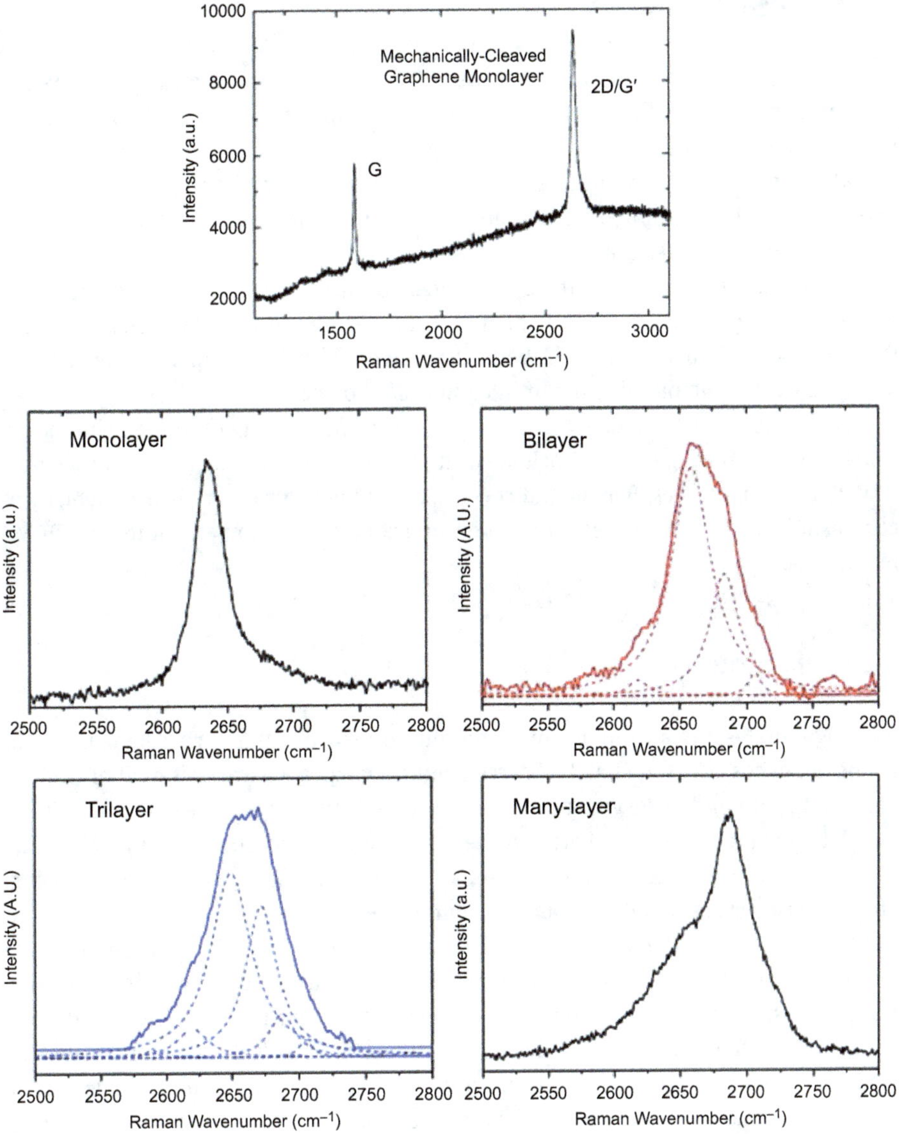


**Figure 4.7:** Complete Raman spectrum of monolayer graphene (top-left) and bilayer graphene with the G and 2D/*G′* bands (top-right). Raman spectra with 2D/*G′* bands for mono-, bi-, tri-, and many-layer graphene (bottom) are shown in detail.

### 4.6.4 Contact angle

Contact angle of nanocomposites with specific liquids are used to measure the extent of interaction of nanofillers present in the nanocomposites with the liquid. A lower contact angle with the liquid signifies a higher surface-free energy of the liquid in contact with the nanocomposite and vice versa, based on hydrophilicity and hydrophobicity of the nanocomposite and delivers useful information regarding filler dispersion in the elastomer matrix.

Chen et al. [115] measured the contact angle of the nanocomposite made up of GO and ethylene–propylene–diene rubber (EPDM)/petroleum resin (PR) blend, where they observed slight variations in both dispersive and polar components of the surface-free energy compared to the unfilled EPDM/PR blends.

This indicates a significance of homogeneous dispersion of GO in the blend of the rubber matrix. During contact angle measurements of graphene-polydimethylsiloxane (PDMS) nanocomposites, Berean and coworkers [116] observed more hydrophobicity of composite rubber membranes on addition of graphene, in comparison to neat PDMS membrane.

### 4.6.5 Rheology

Rheology analysis is an important technique to measure the curing behaviours of nanocomposites. Mensah et al. [92] carried out rheological characterisation of GO/NBR nanocomposite and observed an increase in scorch time ($t_{s2}$), cure rate index (CRI), and delta torque (difference between the maximum and minimum torque, $M_H - M_L$) on addition of GO, which are indications of processing safety, accelerated curing, and increased cross-link density of the composites.

On the other hand, Das et al. [117] observed reduced scorch time, slower cure rate index, and higher maximum torque on adding ionic liquid-integrated graphene nanoplates (GnPs) in bromobutyl rubber (BIIR) matrix, in comparison with unfilled rubber, which signify the effect of ionic liquid in cure characteristic of nanocomposite. Wu et al. [118] observed accelerated vulcanisation with lower scorch safety ($t_{10}$), lower optimum cure time ($t_{c90}$), and higher delta torque ($M_H - M_L$), on addition of graphene in the NR matrix. Quick decrease of $t_{10}$ was observed with less than 1 phr graphene dose, followed by a slow decrease of $t_{10}$, which signifies reduction of the vulcanisation induction period.

Decrease of $t_{90}$ is also strongly dependent on graphene loading of lower than 1 phr. A higher delta torque is assumed to be proportionately related to the cross-linking density, which increases with the addition of small quantity of graphene.

## 4.7 Use of graphene in rubber nanocomposite

Rubber is used as an indispensable material in our day-to-day life, and we cannot even imagine our world without rubber products. There are never ending approaches to modify rubber products to enhance their quality and service life. Thus, to improve various performances of rubber products, many developmental activities are going on worldwide. Graphene, as a nanomaterial, has shown many exceptional properties in many domains. It is considered to improve numerous properties in nanocomposites made with graphene and rubber.

Therefore, such nanocomposites are extensively explored by academics and industries worldwide to enhance the performance of graphene-induced rubber nanocomposites and to widen the application of graphene in the rubber field. Like several other nanomaterials, achieving a homogeneous and thermodynamically stable dispersion of graphene in the rubber matrix is the key challenge, and anticipated properties are possible when its proper dispersion is achieved in the matrix rubber. Several significant new and modified techniques are applied during the nanocomposite processing stages to improve the properties of rubbers goods.

The section is a review of the use of graphene in rubber nanocomposites, highlighting the preparation of nanocomposites by several processing techniques. The mechanical, dynamical-mechanical, electrical, gas barrier and thermal properties, and applications of graphene–rubber composites will also be discussed. This section also talks about few commercial examples of use of graphene in the graphene–rubber nanocomposites.

### 4.7.1 Mechanical properties

Reinforcing fillers improve mechanical properties like tensile strength, modulus of elasticity, tear, and abrasion resistance of the elastomeric composite [119]. As traditional reinforcing fillers, carbon black and precipitated silica with silane coupler are being extensively used for rubber reinforcement for the past many years. Homogeneous dispersion of fillers in the rubber matrix improves the polymer–filler interaction, which is the basic requirement of rubber reinforcement to get a compound with a low hysteresis property.

However, apart from the dispersion of fillers, the improvement of many other properties of the compounds depends on their size, shape, orientation, aspect ratio, and interfacial interaction of fillers with the rubber macromolecules [120]. If fillers form aggregates, bundles or stacking in the composite, it will act as defects in the compound and will cause an early failure of the product. In general, better reinforcement is caused due to the lower filler–filler [121–123] and higher filler–polymer interactions [124].

A nanomaterial with average particle size in the range of 1–100 nm is useful for rubber reinforcement. Carbon black, silica [125, 126], and CNTs/nanofibre are in two- and three-dimension in the nanometer range, respectively [127, 128], whereas layered silicates, expanded graphene, graphene sheets, and so on contain one dimension in the nanometer range [129–132]. Several nanofillers like CNTs [133–135], clays [136], and others [137] have been used in polymer composites but the defect-free single-layered graphene can add many unique properties to the nanocomposites.

Many research works are going on for past several years on the use of graphene and its precursors in rubber compounds. As discussed earlier, graphene possesses very high Young's modulus of elasticity and tensile strength [1–4, 138, 139]. If the dispersion of graphene in the elastomer is properly achieved, it could positively improve the Young's modulus and tensile strength of the composites. Crack propagation resistance of the composites is also observed to be improved on incorporation of graphene sheet to the rubber matrix owing to the mechanical interaction between the graphene sheets and the rubber matrix [140].

If nanomaterials are properly compounded, in comparison to traditional fillers, nanofillers with very small quantity can achieve uniform dispersion in the elastomer matrix and enhance the composite properties significantly.

### 4.7.2 Tensile properties

Graphene as a nanofiller has been extensively evaluated in natural and synthetic rubber nanocomposite vulcanisates to assess its effects in improving the tensile properties. Potts et al. prepared RGO and NR nanocomposites via 'solution treatment' and 'two-roll mill' mixing methods [141]. They observed good modulus improvement in solution-treated samples at low elongations, and in milled samples at high elongations. Decrease of elongation-at-break on addition of R-GO was much noticeable in samples mixed by the 'two-roll' method, in comparison to solution-treated samples. In another study [142], they attempted to disperse TEGO in NR latex by the 'ultrasonically assisted latex co-coagulation' process and subsequent mixing in the two-roll mill.

Stress–strain properties of the compounds are shown in the Figure 4.8. A drop of 100% and 300% elongation moduli were observed with 2 phr and 3 phr loading of the TEGO in NR compounds, respectively, whereas minor improvement of moduli were noticed with 4 and 5 phr TEGO loading, compared to neat NR. On other hand, the nanocomposites (L-TEGO)/NR prepared with TRGO by latex the 'premixed' process demonstrated nearly 38% improvements in modulus at 100% elongation and better tensile strength in comparison to 5 phr TEGO/NR nanocomposite. Yaragalla et al. [114] mixed TRGO with NR by the melt mixing method using the Haake internal mixer and observed good modulus improvement in the nanocomposite. It has been observed that the improvement in modulus with 3% w/w of TRGO is nearly comparable with the value reported by Potts et al. [142] with 5% w/w of TRGO. This improvement in modulus could be explained due to the π-effect between the rubber macromolecular segment and GO, and originates as tight anchorage points and restricts mobility of the rubber macromolecular chains.

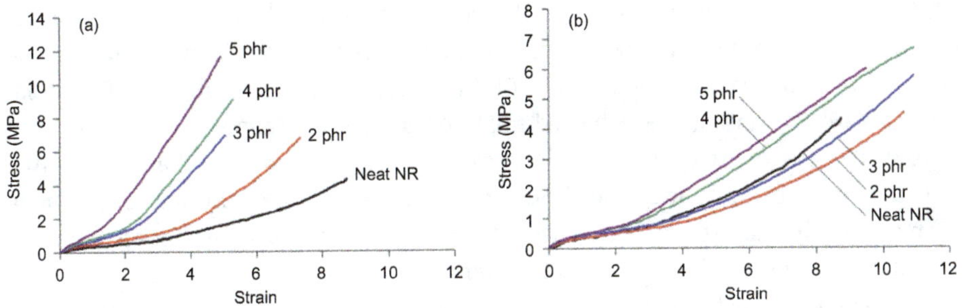


**Figure 4.8:** Illustrative curves of stress-strain analysis of (a) L-TEGO/NR and (b) TEGO/NR nanocomposites at various loading of fillers.

Zhan et al. [104] made nanocomposites (graphene/NR) of graphene and NR by applying ultrasonication during the latex phase of mixing, followed by in situ reduction and coagulation. Compared to raw NR, tensile and tear strength of the nanocomposite with 2 wt% graphene was reported to be increased by 47% and 50%, respectively, which is an indication of the homogeneous dispersion and proper exfoliation of graphene in NR matrix.

Wu et al. [144] observed a higher reinforcing effect in bis(triethoxysilylpropyl) tetrasulfide (Si69)-functionalised GO and NR nanocomposite (SGO/NR), compared to the nonfunctionalised counter part of the nanocomposite (GO/NR) due to the additional coupling interaction between GO and NR macromolecules. Hernandez et al. [145] made nanocomposites with NR and FG sheet (FGS) by the two-roll milling method. With low FGS concentration (0.1 phr), the nanocomposite shows negative tensile strength, which could be due to an alteration of the cross-link density in the nanocomposite. With further addition of FGS content of 0.5 phr, exceptionally high tensile strength and modulus improvement were observed. Tensile strength at 500% elongation with FGS content 1.0 phr was noted to be as high as 135% of that with 0.1 phr FGS content. Ozbas et al. [146] observed that 1 wt% of FGS can deliver a comparable level of modulus improvement on using 16 wt% of carbon black in SBR, NR, and PDMS elastomers.

Xing et al. [147] noticed 260% improvement of tensile strength and 140% improvement of strain-at-break on addition of 0.3 phr graphene in the SBR/graphene nanocomposite, which indicates promising reinforcement of graphene in SBR. It was further observed that the tensile strength of the composite is 11 times higher on addition of 7 phr of graphene and the strain-at-break remains the same as pure SBR. The tensile strength of the nanocomposite was also reported to be between that of 40 phr and 30 phr carbon black or 40 phr of fumed nano-silica-filled SBR compound. Schopp et al. [148] used CRGO, TRGO, and RGO to prepare their nanocomposites with SBR latex.

For comparison purpose, neat SBR compound along with similar compound with different fillers of 25 phr like rubber carbon black, CNTs, multilayer graphene having 350 $m^2/g$ surface area (MLG350) and expanded graphene having 40 $m^2/g$ surface area

(EG40) was used. An improvement in the mechanical properties was observed in accordance with the filler dispersion and its surface area. A maximum of 240% tensile strength was observed in SBR/25 phr TRGO nanocomposite and the highest 260% modulus at 300% elongation was achieved in SBR/25 phr CRGO. Tang et al. [149] observed improvement of the stress-strain properties in both the GO- and graphene-containing SBR nanocomposites. Song et al. [150] observed improvement in modulus, tensile strength, etc. in acid-treated and thermal-treated graphite over natural graphite.

Mao et al. [151] observed a good improvement of the tensile, tear strength, and moduli of GO and styrene-butadiene rubber-SBR latex nanocomposite without compromising the elongation-at-break. Bai et al. [93] observed an improvement in the modulus as well as tensile strength in nanocomposites of hydroxylated carboxylated acrylonitrile butadiene rubber and GO.

Yang et al. [152] prepared carboxylated acrylonitrile butadiene rubber (XNBR) and expanded graphene (EG) nanocomposite by ultrasonically assisted latex blending method and noticed a substantial improvement in tensile strength if graphite doses are in the range of 0–20 phr, whereas the improvement was less in higher graphite loading. Mensah et al. [92] observed an improvement of tensile modulus in nanocomposite made of acrylonitrile–butadiene rubber and GO.

The tensile strength and stress-at-break were observed to be higher with filler content of 0.10 and 2.0 phr. Kang et al. [89] noticed an increase in tensile strength and tensile modulus of GO/XNBR while elongation-at-break decreased. Frasca et al. [153] also observed very high improvement of elastic modulus in a nanocomposite of chlorobutyl rubber and multilayer graphene (nearly 10 layers) by ultrasonic solution, intermix, and two-roll mill mixing, compared to one with only melt mixing process.

Thus, the aim of filler surface modification and adoption of a suitable mixing method is to improve the compatibility and interfacial interaction between the fillers and the matrix elastomer. It has been observed from several experiments that the extent of improvement of tensile properties depends on better filler and polymer interaction. Once interaction between the fillers and the rubber is properly established, fillers will be homogeneously dispersed in the rubber matrix [143].

### 4.7.3 Understanding the reinforcing mechanism of graphene nanofiller

It has already been well observed that a small amount of graphene can dramatically enhance the mechanical properties of elastomer due to the reinforcing effect of graphene. Therefore, the reinforcing mechanism of graphene as a filler in the elastomer matrix also needs to be properly understood. Surface chemistry of any filler decides its interaction and compatibility with the elastomer; likewise, for graphene, the surface characteristics play a pivotal role in determining its thermodynamically stable dispersion and interfacial adhesion with the elastomer.

The high surface area and wrinkled structure of graphene nano-platelets contributes to stronger interfacial interaction with the polymer chains [149]. It has also been observed that strain-induced crystallisation [155] of NR plays a key role in the improvement of mechanical properties in NR composites.

Li et al. [156] carried out stress–strain analysis of gum NR and graphene and GO-filled NR composites. Addition of small quantities of graphene or GO substantial enhances the tensile strength of the respective composites. NR nanocomposite with 1 phr of graphene shows 163% improvement whereas 1 phr of GO shows 128% improvement in tensile strength compared to gum NR compound.

It shows that graphene has a better reinforcing effect than GO on NR. A higher dose of carbon black is required to obtain a similar tensile strength level of graphene or GO. Further study on NR/graphene reinforcements was conducted using Synchrotron WAXD and the theoretical Entanglement-Bound Rubber Tube Model. Enhanced strain-induced crystallisation is understood to be the prime factor behind such improvement in the tensile properties of NR-based graphene/GO nanocomposites.

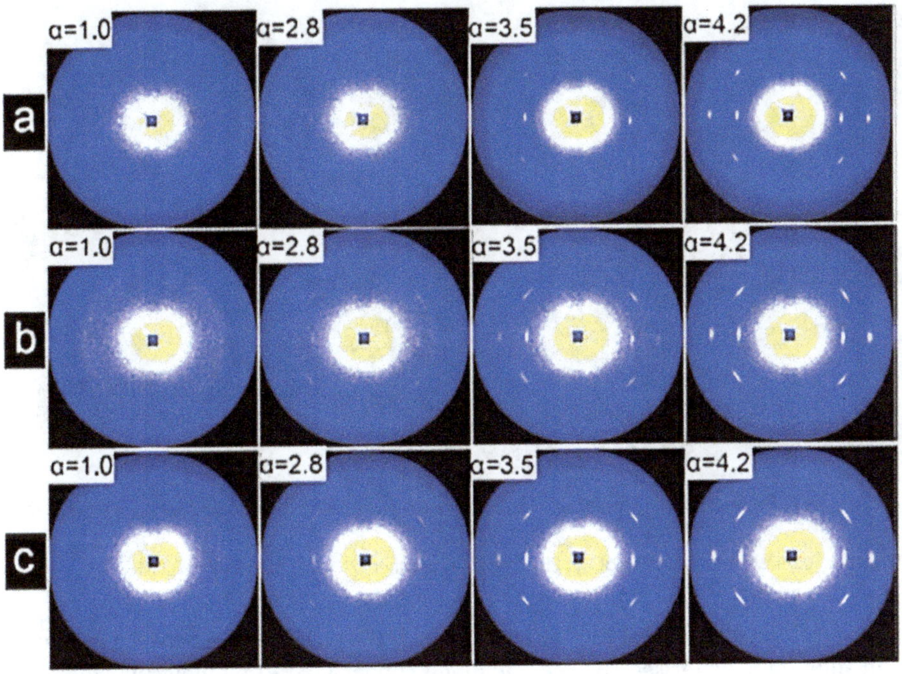


**Figure 4.9:** Successive change of WAXD pattern of (a) NR, (b) 0.7 GO/NR, and (c) 0.7 graphene/NR with strain (mentioned at the left-top patterns).

The typical progresses on elongation of the WAXD patterns of gum NR compound and NR composites with 0.7 phr GO and 0.7 phr graphene are depicted in Figure 4.9. The above observation proves that graphene generates crystallinity and SIC earlier than GO and corresponds with the change in tensile strength. Xie et al. [113] carried out WAXD experiment on unfilled isoprene rubber and surface-modified graphene-isoprene rubber nanocomposite and observed that addition of 0.5 phr on such modified graphene can originate SIC much faster and improve the tensile strength by 37.3% than unfilled isoprene rubber. Ozbas et al. [157] observed lower onset of strain to induce crystallisation in composited containing 1 wt% of graphene compared with 16 wt% of carbon black in NR matrix [154].

To understand the basics of graphene rubber reinforcement phenomena, the entanglement bound rubber tube (EBT) model [156] has been conceptualised. As per the model, the rubber network on the filler surface is graded as 'tightly bound rubber', followed by 'loosely bound rubber', 'transition zone', and 'bulk rubber' [156]. The EBT model assumes the formation of few entanglements between the tightly absorbed bound and bulk rubbers far away from filler surface phase, which is portrayed in Figure 4.10. The filler and rubber interaction are prevalent in the transition zone due to the improved short chin chemical cross-linkages between graphene and rubber, which increases with increase of graphene content.

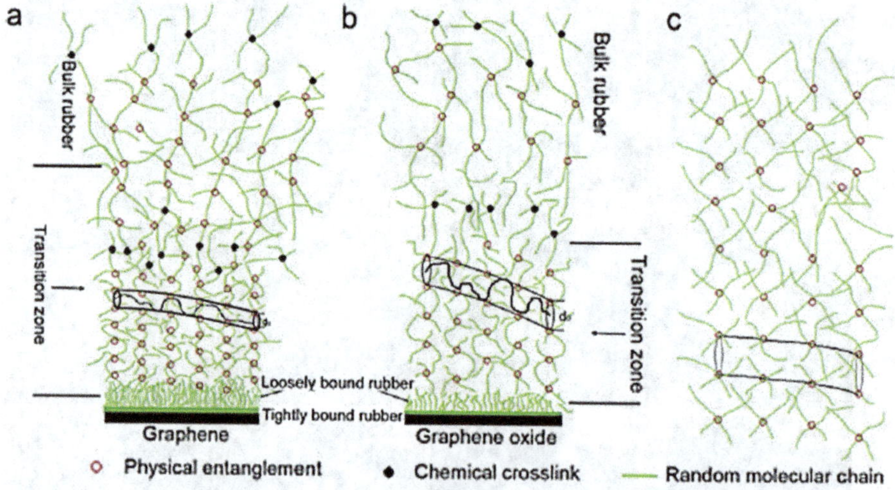


**Figure 4.10:** Illustrative depiction of the EBT model (a) graphene/NR, (b) GO/NR, and (c) NR.

Luo et al. [158] investigated the molecular level interfacial bonding between graphene and solution-polymerised styrene-butadiene rubber (S-SBR) by molecular dynamic simulations approach and observed improvement of the pull-out energy and shear stress in the nanocomposite. It was also noticed that increased vinyl content of the rubber increases the graphene-rubber interfacial interaction and hence the bonding energy.

### 4.7.4 The dynamic mechanical properties

Dynamic mechanical analyser (DMA) is used to measure the viscoelastic properties of rubber composites as a function of temperature, frequency of stress, etc. Viscoelastic properties such as storage modulus (G'), loss modulus (G''), and loss factor or damping coefficient (tan $\delta$) are three important parameters, which are mostly talked about for characterisation and developing new rubber vulcanisates. Loss factor or damping coefficient (tan $\delta$) is a valuable indicator that provides significant information about the performance of composites at its various operating temperatures at the specified frequencies of stress and strain. Higher values of tan $\delta$ at 0 °C signify better wet grip, and higher values of tan $\delta$ at 30 °C indicate better dry grip, whereas lower values of tan $\delta$ at 60–70 °C are indicators of better rolling resistance of the tyre.

Thus, for improved fuel economy (low rolling resistance), tan $\delta$ should be lower at 60–70 °C and for improved wet grip, the compound should show higher tan $\delta$ at 0 °C. The temperature corresponding to the peak of the tan $\delta$ curve signifies the glass transition temperature ($T_g$) of the composite, while lower $T_g$ values of the composite indicate better abrasion resistance and higher $T_g$ composites have better grip.

Fillers such as carbon black and silane-coupled silica have set benchmarks in their respective field although new generation carbon blacks are made, or improved silica fillers are manufactured and used in combination with new generation of rubber to achieve further innovations.

To understand the viscoelastic behaviour of graphene-elastomer nanocomposite, many studies have been performed. Wan et al. [159] modified GO with *p*-phenylenediamine (PPD) and analysed the dynamic mechanical properties of PPD-modified GO in S-SBR rubber composites with varying graphene contents, where GO1 to GO5 represent the graphene content from 1 phr to 5 phr (parts per hundred) of SSBR in the formulation. Table 4.1 gives tan $\delta$ values of the nanocomposites at 0 and 60 °C temperatures.

**Table 4.1:** Formulation of PPD modified graphene oxide nanocomposites.

| Sample | SSBR | ZnO | Stearic Acid | Sulphur | TBBS | Carbon Black | PPD-GO |
|---|---|---|---|---|---|---|---|
| SSBR | 100 | 2.18 | 0.73 | 1.27 | 1.00 | 50.00 | 0.00 |
| PPD-GO1 | 100 | 2.18 | 0.73 | 1.27 | 1.00 | 50.00 | 1.00 |
| PPD-GO2 | 100 | 2.18 | 0.73 | 1.27 | 1.00 | 50.00 | 2.00 |
| PPD-GO3 | 100 | 2.18 | 0.73 | 1.27 | 1.00 | 50.00 | 3.00 |
| PPD-GO4 | 100 | 2.18 | 0.73 | 1.27 | 1.00 | 50.00 | 4.00 |
| PPD-GO5 | 100 | 2.18 | 0.73 | 1.27 | 1.00 | 50.00 | 5.00 |

Higher storage modulus (E') and higher peak height of composites were observed in Figures 4.11 (a) **and** (b), which indicate higher cross-link density due to the sturdy physical interaction between the modified GO and the rubber matrix. Table 4.2 shows the tan

**Table 4.2:** tanδ of the nanocomposites at 0 °C and 60 °C temperature.

| Sample | SSBR | PPD-GO1 | PPD-GO2 | PPD-GO3 | PPD-GO4 | PPD-GO5 |
|---|---|---|---|---|---|---|
| 0 °C | 0.145 +/-0.0002 | 0.153 +/-0.0001 | 0.145 +/-0.0002 | 0.146 +/-0.0001 | 0.148 +/-0.0001 | 0.147 +/-0.0002 |
| 60 °C | 0.123 +/-0.0002 | 0.119 +/-0.00023 | 0.116 +/-0.0002 | 0.120 +/-0.0003 | 0.115 +/-0.0001 | 0.112 +/-0.0001 |

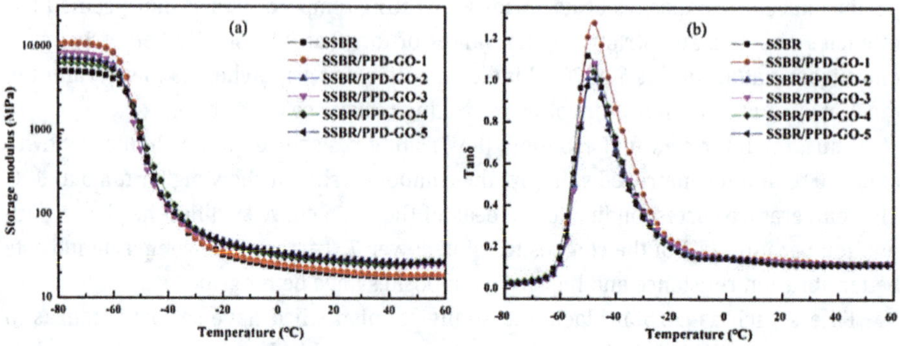


**Figure 4.11:** DMA curves of change of (a) storage modulus with temperature and (b) tan $\delta$, with temperatures of the nanocomposites.

$\delta$ values of the compounds. The value of tan $\delta$ at 0 °C of the nanocomposite with GO1 was observed to be 0.153, an increase of 5.5%, whereas for a nanocomposite with GO5, tan $\delta$ was observed to be lowest, lower by 8.9% in comparison to unfilled SSBR compound.

Wet traction and rolling resistance indicators of the nanocomposite with GO4 were found to be 2.1% and 6.5% higher than that of gum SSBR compound. Kang et al. [160] evaluated the DMA properties of graphene/NR nanocomposites and observed reduction of the tan $\delta$ value at 60 °C with increasing amounts of graphene – specified in Figure 4.12 (b). Lower value of loss factor (tan $\delta$) at temperature of 60 °C indicates better rolling resistance of the tyre. To achieve this performance, better filler and better rubber adhesion and/or interaction are required for tyre tread compounds.

Tang et al. [161] applied the dynamic mechanical analysis of graphene and styrene butadiene rubber nanocomposite to validate the incorporation of graphene, which acts as an oxygen scavenger, delays the oxidative induction time, and maintains well the thermo-oxidative resistance at low temperatures.

Yaragalla et al. [114] carried out a DMA experiment and observed a 25% increase of storage modulus on addition of TRGO into neat NR. The tan $\delta$ versus temperature curve of TRGO vs. neat NR shows additional relaxation due to the strong interfacial interaction of TRGO with the rubber matrix.

A 15 °C shift of $T_g$ was observed, which may be due to the strong $\pi$-effect between the $\pi$-bonds of TRGO and NR. Similar findings in the improvement of storage modulus

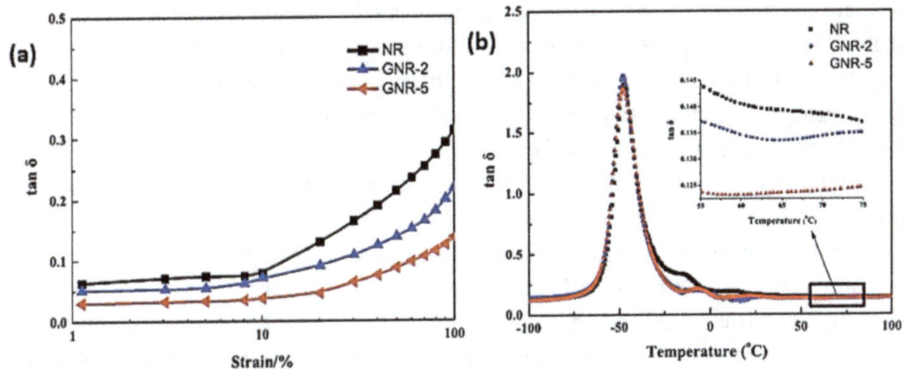


**Figure 4.12:** DMA curve of (a) tan $\delta$ versus strain (b) tan $\delta$ versus temperature of NR and graphene/NR nanocomposites with different loadings of graphene.

were also reported by Potts et al. [141] and Zhan et al. [104] but the shift of $T_g$ value was insignificant. Araby et al. [162] made a nanocomposite by solution phase mixing of ~3 nm thick graphene plates (GnPs) in an SBR matrix and observed that the $T_g$ of the nanocomposite increased by 4 °C and an increase of the storage modulus at the glassy region and at the rubbery region by three-factor and seven-factor, respectively, on addition of 10.4 vol% of GnPs.

The shifting of $T_g$ and the improvement of storage modulus indicate improved interaction between the rubber macromolecules and the nanofillers. X. Bai et al. [93] observed increase in the dynamic $T_g$ value by 1.6 °C in GO/HXNBR nanocomposites on addition of 1.3 vol% GO, which is on account of its good dispersion and interaction due to the hydrogen bond formation between the several oxygenic functional moieties of the GO surface and the -COOH group of HXNBR. Potts et al. [141] prepared a RGO NR composite (NR) by latex, solution phase, and mill mixing methods.

The nanocomposite prepared by the latex phase mixing method shows significant change in the peak height and shape of the tan $\delta$ curve, whereas the solution phase and mill-mixed compounds displayed decrease in both peak height and width of the tan $\delta$ curve. The drop of the peak width is more distinct compared to mill-mixed samples. These are indications of sufficiently strong filler and rubber interactions in the rubber-filler interface.

### 4.7.5 Thermal properties

A polymeric compound must be stable at its required service temperature and time. Thus, thermal stability of the polymeric compound is an important requirement for its performance at the service temperature. A polymeric composite should be stable at the service temperature and the compound properties should not deteriorate due to thermo-oxidative ageing. Thermogravimetric analysis is utilised to find out the controlled degradation patterns of polymers and their composites, either in inert or in oxygen or in their suitable combination. The process has also been widely used in polymeric nanocomposites to understand their thermal stability on addition of the nano fillers in the polymeric matrix.

Many research works have been reported on the thermal stability of the graphene–elastomer nanocomposites [92, 94, 104, 148, 163–170]. It has been observed that uniformly dispersed graphene improves the degradation temperature and eases heat transfer in the nanocomposite of SBR, BR, and their blend [35].

Kim et al. [166] observed an improvement in the thermal stability of multilayered graphene and styrene butadiene rubber nanocomposites due to the π-effect between graphene and phenyl groups of matrix rubber. Wang et al. [171] observed an improvement of thermal stability from 318 to 350 °C on addition of 1 phr of graphene in a composite of XNBR rubber containing 40 phr of carbon black.

However, additional use of graphene does not improve the degradation temperature further. On the other hand, Schopp et al. [148] noticed no changes in the degradation temperature as well as $T_g$ in a nanocomposite containing 25 phr TRGO in a styrene butadiene rubber matrix. Li et al. [142] noticed a shift of degradation temperature of a nanocomposite made of silicone rubber and 1 wt% of TRGO from 356 to 417 °C while with the further addition of TRGO, the degradation temperature of the nanocomposite further shifts to 489 °C. Tang et al. [161] carried out an extensive study on thermal ageing at 90 °C for 13 days with unfilled SBR and 7 phr of graphene-SBR nanocomposite (SBR/GE-7).

Experimental results show that the addition of graphene improves the flexibility of the nanocomposite, the deterioration rate of strain-at-break, thermo-oxidative ageing, tensile strength of elastomer materials, and cross-linking density on ageing. Overall, graphene showed more antioxidative effect than carbon black.

Graphene has an excellent theoretical thermal conductivity (5,300 W/m/K) [172–175] and is measured to be much higher than that of CNTs, graphite, diamond, etc.; it is found to be the best filler to enhance this property in polymeric nanocomposites. Thermal conductivity of rubber composites is required for many applications. Zhan et al. [104] found improvement of thermal conductivity by nearly 12% with 5 wt% of graphene in graphene-NR nanocomposites. Song et al. [176] noticed improvement in the thermal conductivity, approximately 20%, with the addition of 5 wt% of graphene in SBR-graphene nanocomposites. Potts et al. [141] observed improved of thermal conductivity in GE–NR composite prepared by the latex stage mixing process rather than by the mill-mixing method. Zhang et al. [177] noticed substantial improvement of thermal conductivity in various silane-coupled graphene nanoplates with 2 phr of graphene content in silicon rubber composite in the presence of a surfactant. Xue et al. [205] also found significant improvement of such property with the addition of 3 phr GOs in SBR and XNBR rubber matrix blend.

## 4.7.6 Electrical properties

### 4.7.6.1 Electrical conductivity

Rubbers are insulating in nature. A common procedure of incorporating electrical conductivity to rubbers is the incorporation of conductive filler particles at a certain concentration and state of dispersion, called percolation threshold which forms a three-dimensional network structure in the rubber matrix. The composite under such a situation is electrically conductive and follows the power law relationship [178, 179].

Electrically conductive fillers can efficiently make nonconductive rubbers to electrically semiconductive materials. This transformation depends on the purity and loading of the conductive filler. Graphene is such a perfect nanofiller used to prepare electrically conductive nanocomposites due to its intrinsic electric conductivity. Improved filler dispersion facilitates the formation of an interconnecting filler network due to the proximity of the filler particles, and is the basic requirements to obtain electric conductivity in rubber nanocomposites.

This happens due to the rapid movement of electrons from the interconnecting filler surface through the rubber layers and it is called the tunneling mechanism [182, 193]. Percolation threshold critically depends on the orientation of fillers. It was observed that parallel orientation of fillers shows improved percolation threshold compared to a random arrangement [156, 167, 169]. Suitable functionalisation of graphene can improve its interactions with the rubber matrix and further enhance the electrical conductivity of the composite.

Several works have been reported on the electric conductivity of graphene-rubber nanocomposite. Kim et al. [166] observed the rise of electric conductivity of multilayered graphene sheet and SBR nanocomposite to $4.56 \times 10^{-7}$ S/cm from $4.52 \times 10^{-13}$ S/cm on increasing the multilayered graphene sheet from 0.10 to 5.0 wt%. Potts et al. [141] observed improved electric conductivity in GE–NR nanocomposites made by the latex blending method compared to the melt mixing method.

Outstanding improvement of such property was observed by He et al. [181] in graphene-epoxidized NR nanocomposite with 0.23 vol% of graphene. Zhan et al. [182] observed approximately 5 times higher electrical conductivity with the addition of 1.78 vol% graphene in NR-based nanocomposite processed by the latex blending method compared to the regular process. Matos et al. [183] observed 4 times drop in electrical resistivity in NR-graphene nanocomposite with a 2 wt% graphene content, compared to gum NR compound – exemplified in Figure 4.13.

Xing et al. [147] observed improvement of electrical conductivity by 11 factors in graphene- and SBR-based nanocomposites with 7 phr graphene loading compared to unfilled SBR vulcanisate. Kim et al. [166] noticed remarkable change in the electrical property of graphene-SBR rubber nanocomposite with a small addition of graphene. Ozbas et al. [146] experimented the dependency of electrical conductivity with the surface area of graphene in poly dimethylsiloxane base nanocomposites. Post vulcanisation

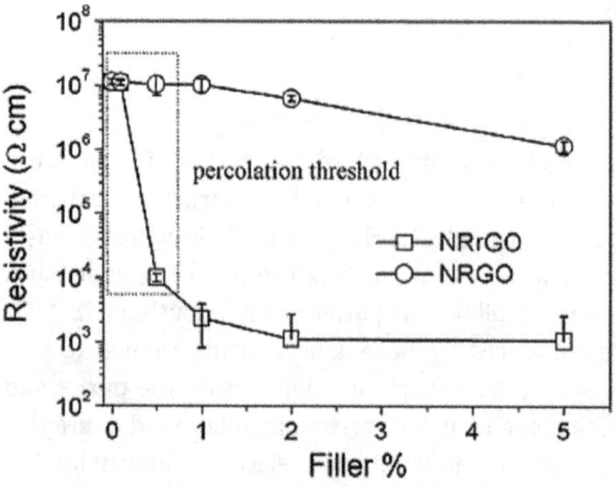


**Figure 4.13:** Variation of resistivity of rGO/NR and GO/NR nanocomposites with loading of a filler.

improvement in volume resistivity by 7 times in EPDM and graphene nanocomposite was observed by Araby et al. [189].

#### 4.7.6.2 Dielectric properties

Dielectric property is a fundamental property of a material. Dielectric materials are poor conductor of electricity and an ideal dielectric material does not show electrical conductivity if an electric field is applied. Dielectric materials show dielectric polarisation on application of an electric field and almost all such materials show little conductivity, usually which increases with increasing temperature and applied electric field.

Nanocomposites, on exposure to an electric field, show ionic, interfacial, and dipole polarisation. Similarly, nanocomposites made of graphene and elastomer show dielectric properties. Parameters like dispersion of graphene in the elastomer matrix, reinforcement and thickness of the composite together affect the dielectric properties of the graphene-elastomer nanocomposites. Mensah et al. [92] observed that the real part in the permittivity value of insulation-blended GO/NBR nanocomposite was about five times higher than that of the gum compound. Singh et al. [184] observed an increase in both permittivity parts with the addition of RGO in solution-blended RGO/NBR nanocomposites.

### 4.7.7 Gas barrier properties

Elastomers, except butyl and halo butyl rubbers, usually show poor barrier properties. Improvement of gas and liquid barrier property of elastomer nanocomposites is one of the most important requirements. The barrier property could be improved by mixing lamellar fillers with rubber. Few examples are clay tactoids, and graphene and its precursors. These types of fillers mostly have very high aspect ratio. Due to this reason, they can alter the straight diffusion route through the matrix rubber. Graphene as a reinforcing filler in rubber nanocomposites, when properly dispersed, can create a percolating type of network. This network can create tortuous path and inhibit diffusion through the matrix [186]. These factors play very significant roles in lowering the gas permeability through the rubber composite.

It is quite difficult to produce a rubber nanocomposite without the wrinkling nature of the graphene sheet. Wu et al. [144] observed higher gas barrier property in functionalised GO/NR composite than in TRGO/NR composite on account of the wrinkling of TRGO sheets during the mixing process of the compound. Scherillo et al. [187] noticed significantly improved permeability in the natural latex phase mixing of rubber-graphene nanocomposite with nonsegregated GO morphology, which is due to better interaction with the rubber matrix compared to its segregated counterpart. Y. Li et Al. [112] also found the advantages of the formation of a segregated network of functionalised GO on increasing the gas barrier properties of bromobutyl rubber composite.

Yaragalla et al. [114] measured the oxygen permeability of TRG/NR nanocomposites – depicted in Figure 4.14(a and b). Significant improvement in gas permeability was observed in the nanocomposite on adding TRG to the NR matrix. This could be due to better-quality dispersion and π-effect of TRG in the NR phase. The dimension of fillers also contributes to the gas permeability property. It was also observed that TRG shows comparatively low gas permeability when compared to graphite- and GO-based NR nanocomposites.

Xing et al. [147] measured the oxygen permeability of graphene/SBR nanocomposite at 25 °C. The oxygen permeability of the nanocomposite was drastically lowered with the addition of graphene in the SBR matrix. It was observed that the property decreased by 67.2% and 87.8% on the addition of 3 and 7 phr graphene, respectively, in the SBR matrix when compared to the unfilled SBR vulcanisate. Wu et al. [144] similarly observed a drastic drop of the air permeability value, by 52%, on the addition of only 0.3 wt% SGO in NR-based composite. It was further demonstrated that the same amount of SGO shows comparable air permeability property with 16.7 wt% of pristine clay in NR compound [188]. Das et al. [117] noticed 64.2% reduction in the gas transmission rate in GnPs-BIIR with 5 phr of GnPs, compared to unfilled BIIR.

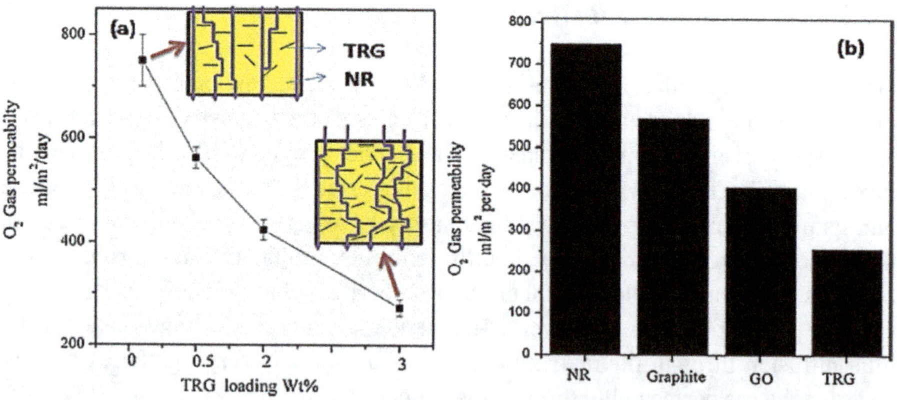


**Figure 4.14:** (a) Gas permeability of TRG/NR nanocomposites with different loadings of TRG (a) TRG alone. (b) Comparison of gas permeability between graphite, GO, and TRG.

## 4.8 Processability of graphene–elastomer nanocomposite

Processing is the most critical step in elastomer compounding and is the way to control the ultimate properties of the composite. The major aim of proper processing is to achieve thermodynamically stable and homogeneous dispersion of filler in the rubber phase. This can be achieved by improving the polymer and filler interaction and suppressing filler agglomeration. Thus, a strong polymer-filler interaction is necessary for better adhesion of the polymer and filler, and for better mechanical interactions between the filler and rubber. To attain good dispersion and to achieve good polymer–filler interaction for nanofillers, various methods like melt mixing, solution/latex blending, in situ polymerisation, etc. have been adopted. These are extensively used and described in literature to prepare nanocomposites. All the methods are different, have their own merits, and add different properties to the derived nanocomposites.

### 4.8.1 Melt mixing

In this method, fillers are directly mixed with rubber by applying very high shear force, either in a two-roll mill or in an internal mixer. The melt mixing method has some advantages and is preferred by the industry due to its low operating cost and high productivity [144, 189]. The prime advantage of this method is that it does not require any organic solvent to carry out the mixing operation. Due to this the method, it is favoured and preferred by the industry over in situ polymerisation and solution intercalation mixing process. Chemical modifications of different ingredients like fillers are also possible by this process. To improve the polymer-filler interaction, this method also allows the use of compatibiliser and is utilised to significantly mix both the polar and nonpolar elastomers [190]. Although the method has many advantages, it is associated with a few disadvantages. Mixing of fillers in the rubber matrix requires higher mixing temperature, which may make the rubber matrix prone to degradation. High filler fraction and high viscosity of rubber can hinder proper filler dispersion.

The high share of mixing utilised to lower the viscosity of the rubber matrix can break the graphene sheets [104]. Thus, the melt mixing process is not suitable and is not preferred for preparing graphene-elastomer nanocomposites due to the improper dispersion of fillers in matrix elastomers, compared to the other two methods [191, 192]. In spite of the associated problems, the melt-mixing process has been extensible and very precisely used in academic research works to prepare graphene-elastomer nanocomposites and outstanding properties are also obtained. Malas et al. [193] had prepared different graphene-containing composites with SBR and BR rubber blend by the melt-blending method in a laboratory-scale two-roll mill.

Araby et al. [189] prepared graphene-EPDM nanocomposites by this process and obtained good dispersion of the graphene EPDM matrix. They also observed good electrical and stress-strain properties. Song et al. [176] observed improved thermal as well as stress–strain properties of graphene and silicone rubber nanocomposites prepared by this process.

### 4.8.2 Solution intercalation/latex blending

A solution intercalation technique is frequently used in academic experimentations till date to produce elastomer nanocomposites [128, 139, 163, 182, 185, 192]. This method is solvent-based and is dependent on a common solvent-based system. The elastomer needs to be properly solubilised in its common solvent. Initially, graphene is dispersed homogeneously in the same solvent, employing either high shear mixer or ultrasonication or mechanical stirring, followed by mixing the elastomer solution and graphene dispersion. By the process, elastomeric chains become intercalated within the interlayer spacing of graphene once the solvent is taken out by the subsequent applicable procedures. Important criteria of this method are the type and polarity of the solvent used [194, 195]. This process does not require any noteworthy additional steps to incorporate graphene suspensions in the elastomer phase. It has reported that this process ensured suitable dispersion and exfoliation of sheet-like inorganic nanofillers in the elastomer matrix and elastomer reinforcement without chemical modification of the filler [196].

In the latex blending method, the elastomer remains in the latex form. On account of environmental concerns, this method shows advantages over solution blending method as toxic organic solvents are not used in this method. Several steps have been inducted in this process to attain strong interactions and good dispersion of nanofillers the in the elastomer matrix to achieve reinforcement. Initially, a stable dispersion of graphene is made in water either by high shear mixer or ultrasonication by using a suitable surfactant. The graphene dispersion is subsequently mixed with latex by mechanical stirring, followed by co-coagulation. These steps can improve the dispersion of the fillers in the rubber phase by preventing kinetical aggregation of the fillers and by accelerating the coagulation process [11, 89, 104, 141, 182, 185, 187].

Although solution mixing has several advantages, effective removal of solvent is a big concern in this process. High cost and disposal of solvent restricts its adaptability and scaleup by industries. The quality of the nanocomposites prepared by this process depends on many parameters. Solvent quality and quantity preliminarily decide the exfoliation of the graphene sheets on application of high shear. A suitable sonication is also used for the exfoliation of graphene. A proper mixing time also determines its extent of exfoliation. Wu et al. [144] prepared TESPT silane-grafted GO dispersion and mixed it with NR by the solution intercalation process. It was observed that the successful grafting of TESPT silane on the GO surface results in its good dispersion in NR solution. Xing et al. [147] prepared dispersion of graphene by ultrasonication and in the presence of a surfactant. The graphene/SBR nanocomposite was made by latex blending of graphene dispersion and SBR latex. Good dispersion of graphene was observed in the composite on account of the high interfacial interaction with the matrix rubber. Matos et al. [183] observed the effect of cetyl trimethyl ammonium bromide and used it as a surfactant to improve the interaction of graphene and GO to prepare a nanocomposite with NR latex via the latex blending process. Substantial enhancements in the dielectric, chemical, and stress-strain properties were observed in the nanocomposites due to the improved polymer-filler interaction [180]. J.R. Potts [142] observed substantial improvement in the properties of the TEGO-NR nanocomposite prepared by the latex blending method using the two-roll mill mixing method. Xue et al. [198] followed the latex blending method to prepare SBR/XNBR/GO nanocomposites with a small dose of XNBR. Kang et al. [89] used the latex blending process to prepare GO/XNBR nanocomposites [197].

### 4.8.3 In situ polymerisation

In this polymerisation process, nanofillers are initially dispersed properly in the monomer of the polymer or in a solution of the monomer, followed by its polymerisation. This process has been significantly used to prepare elastomeric nanocomposites with layered-silicate fillers and has demonstrated effective incorporation of exfoliated silicate sheets in elastomeric macromolecules [129, 199–202]. The Figure 4.15 described various kinds of structures of nanocomposites which were produced based on the nature of the filler and the preparation method of the nanocomposite [192]. As graphene possesses a layered structure similar to layered-silicates, this process has also been adopted to make graphene-elastomer nanocomposites [192, 203, 204]. This method is not as popular as it requires low viscosity of elastomers. Elastomer macromolecular chains may become effectively incorporated to the exfoliated graphene sheets. This improves the interaction between graphene and the elastomer and causes better stress transfer between rubber and the filler. Subsequently, better mechanical reinforcement properties are obtained by this method.

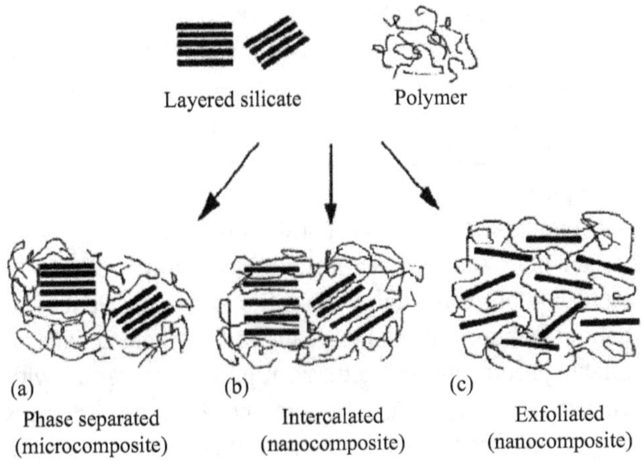


**Figure 4.15:** Schematic representation of different types of nanocomposites obtained due to interaction of layered silicates with the polymers: (a) phase-separated; (b) intercalated, and (c) exfoliated.

## 4.9 Commercial advantages

Although the graphene industry is till now in its early stage, is growing rapidly and finding application in many sectors due to its miracle properties. Numerous experiments exhibited superior reinforcement efficiencies of graphitic filler over conventional carbon black and precipitated silica. Substitution of common fillers by small amounts of graphene in different rubbers adds more mechanical strength and reduces the overall weight of the composite by decreasing its density. Along with improvement of mechanical properties, graphene also improves electrical conductivity, thermal conductivity, and gas retention capability of the rubber composites. Graphene can be prepared from naturally available inexpensive graphite. Replacement of carbon black by graphene enables significant reduction of oil dependency in the rubber industry. There are many successful examples of commercialisation of graphene in rubber products and the application of graphene has also begun in the elastomer sector due to its own merits.

Vittoria, an Italian tyre maker has become well known for adding the wonder material to its mountain bicycle tyre for several years. At present, Vittoria is using graphene to achieve high speed, improved wet traction, better mileage, and puncture-proof resistance in their tyres. They use he new generation graphene 2.0 (G 2.0) to significantly reduce rolling resistance at high speed and abrasion resistance to improve mileage. Goodyear has developed a proprietary compound enhanced with graphene and 'next-generation amorphous (noncrystalline) spherical silica'.

This is expected to provide low rolling resistance, improved grip in the dry and wet road conditions, and long-term durability. In 2015, Chengdu Trust Well Company developed a process to prepare graphene-NR masterbatch and its composites in larger scale, which enhanced the air retention and electric conductive properties. In 2016, Double Star Company from China developed a graphene-based tyre. Vorbeck Materials Corp introduced Vor-flex™50, a graphene-enhanced engineered nanocomposite of HNBR, which shows very high modulus at low elongation and the composite can even resist temperature of nearly 93 °C. Vor-flex's high temperature stability finds its use in automotive and petrochemical applications.

It can be used to make selective gaskets. Cabot Corporation has been in the business of producing carbon black for more than 135 years and they are developing technologies for the production and use of graphene by using its wide experiences in carbon materials. With the help of in-house developed team and external collaborations, they have developed scalable production routes that can produce materials with unique morphologies and properties as per specific application needs of the customers and increasing their production capabilities to enable commercial adoptions. Gratomic Inc. made graphene-enhanced tyres ('Gratomic Tyres') with surface-enhanced graphene and it exhibited improved abrasion resistance, fuel economy, and ice and dry traction.

First Graphene and its partner, New Gen Group, has developed graphene-enhanced rubber wear liners for mining shovel buckets at a major iron ore mine in Australia. The initial results indicated improvement of at least 100% reduction in wear rates, which result in major cost savings for mine operators and can be achieved with just a 1% blend of graphene by volume to the rubber compound.

In 2016, Inov-8, a UK-based company, in collaboration with the National Graphene Institute has developed shoe sole with 50% more elasticity and harder wearing property. In May 2021, The Sixth Element (Changzhou) Materials Technology Co. Ltd. revealed the development of a range of latex-based graphene rubber masterbatches along with Qingdao Lanwan ENE Carbon Material, developing tyres with enhanced performance using a graphene NR masterbatch in the tyre tread formulation of passenger car tyres and heavy truck tyres. Performance evaluation by road testing confirmed substantial improvement in rolling resistance as well as wear resistance in PCR and TBR tyres.

Graphene manufacturing processes are improving continuously to produce high quality and economical graphene. Thus, with higher volume availability and consistent quality, the costs of graphene will gradually fall, which will subsequently draw the attention of end users and will ensure continuous commercial growth of graphene. Although graphene is a high-performing, high potential material, being nanofiller, it is a matter of concern from handling, environmental, health and safety aspects, which need to be investigated suitably and stringent norms to be defined for storage, handling, and use like similar other nano materials.

# References


[1] Gerstner, E. Nobel Prize 2010: Andre Geim& Konstantin Novoselov. Nature Phys 6:836 (2010). https://doi.org/10.1038/nphys1836

[2] Yang, G., Li, L., Lee, W. B., Man Cheung, N. Structure of graphene and its disorders: A review. Science and Technology of Advanced Materials 19(1):613–648 (2018). 10.1080/14686996.2018.1494493.

[3] Gadipelli, S., Xiao Guo, Z. Graphene-based materials: Synthesis and gas Sorption, storage and separation. Progress in Materials Science 69:1–60 (2015). ISSN 0079-6425 https://doi.org/10.1016/j.pmatsci.2014.10.004

[4] Cui, Y., Kundalwal, S. I., Kumar, S. Gas barrier performance of graphene/polymer nanocomposites. Carbon 98:313–333 (2016). ISSN 0008-6223 https://doi.org/10.1016/j.carbon.2015.11.018

[5] Balandin, A. Thermal properties of graphene and nanostructured carbon materials. Nature Mater 10:569–581 (2011). https://doi.org/10.1038/nmat3064

[6] Papageorgiou, D. G., Kinloch, I. A., Young, R. J. Mechanical properties of graphene and graphene-based nanocomposites. Progress in Materials Science 90:75–127 (2017). ISSN 0079-6425 https://doi.org/10.1016/j.pmatsci.2017.07.004

[7] Bhuyan, M. S. A., Uddin, M. N., Islam, M. M., et al. Synthesis of graphene. Int Nano Lett 6:65–83 (2016). https://doi.org/10.1007/s40089-015-0176-1

[8] McAllister, M. J., Li, J.-L., Adamson, D. H., Schniepp, H. C., Abdala, A. A., Liu, J., et al. Single sheet functionalized graphene by oxidation and thermal expansion of graphite. Chem, Mater 19(18):4396–4404 (2007). https://doi.org/10.1021/cm0630800

[9] Min, Y., Shen, Z. A review on mechanical exfoliation for the scalable production of graphene. Journal of Materials Chemistry A 3(22):11700–11715 (2015).

[10] Georgakilas, V., Otyepka, M., Bourlinos, A. B., Chandra, V., Kim, N., Kemp, K. C., Hobza, P., Zboril, R., Kim, K. S. Functionalization of Graphene: Covalent and Non-Covalent Approaches. Derivatives and Applications, Chemical Reviews 112(11):6156–6214 (2012). doi 10.1021/cr3000412.

[11] Luo, Y., Zhao, P., Yang, Q., Dongning, H., Kong, L., Peng, Z. Fabrication of conductive elastic nanocomposites via framing intact interconnected graphene networks. Composites Science & Technology 100:143–151 (2014). ISSN 0266-3538 https://doi.org/10.1016/j.compscitech.2014.05.037

[12] Song, S. H., Jeong, H. K., Kang, Y. G., *et al.* Physical and thermal properties of acid-graphite/styrene-butadiene-rubber nanocomposites. Korean J. Chem. Eng. 27:1296–1300 (2010). https://doi.org/10.1007/s11814-010-0178-7

[13] Zhang, H., Xing, W., Hengyi, L., ZhengtianXie, G. H., Jinrong, W. Fundamental researches on graphene/rubber nanocomposites. Advanced Industrial and Engineering Polymer Research 2 (1):32–41 (2019). ISSN 2542-5048 https://doi.org/10.1016/j.aiepr.2019.01.001

[14] Li, H., Wu, S., Wu, J., *et al.* A facile approach to the fabrication of graphene-based nanocomposites by latex mixing and in situ reduction. Colloid Polym Sci 291:2279–2287 (2013). https://doi.org/10.1007/s00396-013-2959-0

[15] Kim, H., Abdala, A. A., Macosko, C. W. Graphene/Polymer Nanocomposites. Macromolecules 43(16):6515–6530 (2010). 10.1021/ma100572e.

[16] Kharisov, B. I., Kharissova, O. V. Conventional Carbon Allotropes. Carbon Allotropes: Metal-Complex Chemistry, Properties and Applications. Cham: Springer (2019), 9–33. https://doi.org/10.1007/978-3-030-03505-1_2

[17] Heyrovska, R. The Coulombic Nature of the van der Waals Bond Connecting Conducting Graphene Layers in Graphite. Graphene 5:35–38 (2016). 10.4236/graphene.2016.52004.

[18] Wang, X., You, H., Liu, F., Li, M., Wan, L., Li, S., Li, Q., Xu, Y., Tian, R., Yu, Z., Xiang, D., Cheng, J. Large-Scale Synthesis of Few-Layered Graphene using CVD. Chem. Vap. Deposition 15:53–56 (2009). https://doi.org/10.1002/cvde.200806737

[19] Kim, K., Zhao, Y., Jang, H., et al. Large-scale pattern growth of graphene films for stretchable transparent electrodes. Nature 457:706–710 (2009). https://doi.org/10.1038/nature07719

[20] Dervishi, E., Li, Z., Watanabe, F., Biswas, A., Xu, Y., Biris, A. R., Saini, V., Biris, A. S. Large-scale graphene production by RF-cCVD method. Chemical Communication 4061 (2009). 10.1039/B906323D.

[21] Xuesong, L., Cai, W., Jinho, A., Kim, S., Nah, J., Yang, D., Piner, R., Aruna Velamakanni, I. J., Tutuc, E., Banerjee, S. K., Colombo, L., Ruoff, R. S. Large-Area Synthesis of High-Quality and Uniform Graphene Films on Copper. Science 324(5932):1312–1314 (2009). June 10.1126/science.1171245.

[22] Di, C.-A., Wei, D., Yu, G., Liu, Y., Guo, Y., Zhu, D. Patterned Graphene as Source/Drain Electrodes for Bottom-Contact Organic Field-Effect Transistors. Advances in Materials 20:3289–3293 (2008). https://doi.org/10.1002/adma.200800150

[23] Rollings, E., Gweon, G.-H., Zhou, S. Y., Mun, B. S., McChesney, J. L., Hussain, B. S., Fedorov, A. V., First, P. N., de Heer, W. A., Lanzara, A. Synthesis and characterization of atomically thin graphite films on a silicon carbide substrate. Journal of Physics and Chemistry of Solids 67(9–10):2172–2177 (2006). https://doi.org/10.1016/j.jpcs.2006.05.010

[24] De Heer, W. A., Berger, C., Xiaosong, W., First, P. N., Conrad, E. H., Xuebin, L., Tianbo, L., Sprinkle, M., Hass, J., Sadowski, M. L., Potemski, M., Martinez, G. Epitaxial graphene. Solid State Communications 143(1–2):92–100 (2007). https://doi.org/10.1016/j.ssc.2007.04.023

[25] Ni, Z. H., Chen, W., Fan, X. F., Kuo, J. L., Yu, T., Wee, A. T. S., Shen, Z. X. Raman spectroscopy of epitaxial graphene on a SiC substrate. Phys. Rev. B 77:115416 (2008). https://doi.org/10.1103/PhysRevB.77.115416

[26] Camara, N., Rius, G., Huntzinger, J. R., Tiberj, A., Mestres, N., Godignon, P., Camassel, J. Selective epitaxial growth of graphene on SiC. Applied Physics Letters 93(12):123503 (2008).

[27] Seyller, T., Bostwick, A., Emtsev, K. V., Horn, K., Ley, L., McChesney, J. L., Ohta, T., Riley, J. D., Rotenberg, E., Speck, F. Epitaxial graphene: A new material. Phys. Stat. Sol. (B) 245:1436–1446 (2008). https://doi.org/10.1002/pssb.200844143

[28] Sprinkle, M., Soukiassian, P., de Heer, W. A., Berger, C., Conrad, E. H. Epitaxial graphene: The material for graphene electronics. Phys. Stat. Sol. (RRL) 3:A91–A94 (2009). https://doi.org/10.1002/pssr.200903180

[29] Janowska, I., Ersen, O., Jacob, T., Vennégues, P., Bégin, D., Ledoux, M.-J. Cuong Pham-Huu, Catalytic unzipping of carbon nanotubes to few-layer graphene sheets under microwaves irradiation. Applied Catalysis A: General 371(1–2):22–30 (2009). https://doi.org/10.1016/j.apcata.2009.09.013

[30] Dai, W., Wang, D. Cutting Methods and Perspectives of Carbon Nanotubes. The Journal of Physical Chemistry C 125(18):9593–9617 (2021). 10.1021/acs.jpcc.1c01756.

[31] Xiaohui, W., Liu, Y., Yang, H., Shi, Z. Large-scale synthesis of high-quality graphene sheets by an improved alternating current arc discharge method. RSC Advancesances 6:93119–93124 (2016). 10.1039/c6ra22273k.

[32] Subrahmanyam, K. S., Panchakarla, L. S., Govindaraj, A., Rao, C. N. R. Simple Method of Preparing Graphene Flakes by an Arc-Discharge Method. The Journal of Physical Chemistry C 113(11):4257–4259 (2009). 10.1021/jp900791y.

[33] Yuan, Z. K., Xiao, X. F., Li, J., Zhao, Z., Yu, D. S., Li, Q. Self-Assembled Graphene-Based Architectures and Their Applications. Adv. Sci. 5:1700626 (2018). https://doi.org/10.1002/advs.462

[34] Striolo, A., Patrick Grady, B. Surfactant Assemblies on Selected Nanostructured Surfaces: Evidence, Driving Forces, and Applications. Langmuir 33:8099–8113 (2017). 10.1021/acs.langmuir.7b00756.

[35] Wei, W., Bai, F., Fan, H. Surfactant-Assisted Cooperative Self-Assembly of Nanoparticles into Active Nanostructures. iScience 11:272–293 (2019). https://doi.org/10.1016/j.isci.2018.12.025

[36] Yu, Q., Lian, J., Siriponglert, S., Li, H., Chen, Y. P., Pei, S. S. Graphene segregated on Ni surfaces and transferred to insulators. Appl Phys Lett 93(11):113103 (2008).

[37] Sutter, P., Flege, J. I., Sutter, E. Epitaxial graphene on ruthenium. Nature Mater 7:406–411 (2008). https://doi.org/10.1038/nmat2166

[38] Coraux, J., N'Diaye, A. T., Busse, C., Michely, T. Structural Coherency of Graphene on Ir(111). Nano, Letters 8(2):565–570 (2008). 10.1021/nl0728874.

[39] Sutter, P., Sadowski, J. T., Sutter, E. Graphene on Pt(111): Growth and substrate interaction. Physical Review B 80:245411 (2009). https://doi.org/10.1103/PhysRevB.80.245411

[40] GedengRuan, Z. S., Peng, Z., Tour, J. M. Growth of Graphene from Food, Insects, and Waste. ACS Nano 5(9):7601–7607 (2011). 10.1021/nn202625c.

[41] Li, X., Colombo, L., Ruoff, R. S. Synthesis of Graphene Films on Copper Foils by Chemical Vapor Deposition. Adv. Mater 28:6247–6252 (2016). https://doi.org/10.1002/adma.201504760

[42] Xuesong, L., Magnuson, C. W., Venugopal, A., Tromp, R. M., Hannon, J. B., Vogel, E. M., Colombo, L., Ruoff, R. S. Large-Area Graphene Single Crystals Grown by Low-Pressure Chemical Vapor Deposition of Methane on Copper. Journal of the American Chemical Society 133(9):2816–2819 (2011). 10.1021/ja109793s.

[43] Malas, A., Hatui, G., Pal, P., Das, C. K. Synergistic effect of expanded graphite/carbon black on the physical and thermo-mechanical properties of ethylene propylene diene terpolymer. PolymPlastTechnol Eng 53(7):716–724 (2014). https://doi.org/10.1080/03602559.2013.877928

[44] Chapter 10 – Graphite Intercalation Compounds (GIC). In: Brandt, N. B., Chudinov, S. M., Ponomarev, Y. G. ((Eds.)) Modern Problems in Condensed Matter Sciences. North Holland: Elsevier, Vols. 20, 1 (1988). 197–321 ISSN 0167-7837, ISBN 9780444870490 https://doi.org/10.1016/B978-0-444-87049-0.50016-0

[45] Akuzawa, N. Chapter 6 – Intercalation Compounds. In: Yasuda, E.-I., Inagaki, M., Kaneko, K., ENDO, M., Asao, O. Y. A., Tanabe, Y. (eds.) CarbonAlloys. Netherlands / London, England: Elsevier Science (2003). 99–108 ISBN 9780080441634 https://doi.org/10.1016/B978-008044163-4/50006-1

[46] Shioyama, H. The interactions of two chemical species in the interlayer spacing of graphite. Synthetic Metals 114(1):1–15 (2000). https://doi.org/10.1016/S0379-6779(00)00222-8

[47] Rüdorff, W., Hofmann, U. ÜberGraphitsalze. Z Anorg Allg Chem 238:1–50 (1938). https://doi.org/10.1002/zaac.19382380102

[48] Hennig, G. R. Interstitial Compounds of Graphite. In: Cotton, F. A. (ed.) Progress in Inorganic Chemistry. New York: Inter science Publishers Inc. (1959), 125–205. https://doi.org/10.1002/9780470166024.ch2

[49] Thiele, H. Über die Quellung von Graphit. Z. Anorg. Allg. Chem 206:407–415 (1932). https://doi.org/10.1002/zaac.19322060409

[50] Berlouis, L. E. A., Schiffrin, D. J. The electrochemical formation of graphite bisulphate intercalation compounds. Journal of Applied Electrochemistry 13(2):147–155 (1983).

[51] Beck, F., Junge, H., Krohn, H. Graphite intercalation compounds as positive electrodes in galvanic cells. Electrochimica Acta 26(7):799–809 (1981). ISSN 0013-4686 https://doi.org/10.1016/0013-4686(81)85038-4

[52] Chuan, X.-Y., Chen, D., Zhou, X. Intercalation of CuCl2 into expanded graphite. Carbon 35:311–313 (1997).

[53] Fukuda, K., KazuhikoKikuya, K., Yoshio, M. Foliated natural graphite as the anode material for rechargeable lithium-ion cells. Journal of Power Sources 69(1–2):165–168 (1997). ISSN 0378-7753 https://doi.org/10.1016/S0378-7753(97)02568-8

[54] Chen, G.-H., Wu, D.-J., Weng, W.-G., He, B., Yan, W.-L. Preparation of polystyrene–graphite conducting nanocomposites via intercalation polymerization. Polym. Int 50:980–985 (2001). https://doi.org/10.1002/pi.729

[55] Brodie, B. C. Sur le poidsatomique du graphite. Ann. Chim. Phys 59:466–472 (1860).

[56] Staudenmaier, L. VerfahrenzurDarstellung der Graphitsaure. Ber Deut Chem Ges 31:1481–1499 (1898).

[57] Hummers, W. S., Offeman, R. E. Preparation of Graphitic Oxide. Joural Am Chem Soc 80(6):1339 (1958).

[58] Dreyer, D. R., Park, S., Bielawski, C. W., Ruoff, R. S. The chemistry of graphene oxide. Chemical Society Reviews 39(1):228–240 (2010).

[59] Zhu, Y., Murali, S., Cai, W., Li, X., Suk, J. W., Potts, J. R., Ruoff, R. S. Graphene and Graphene Oxide: Synthesis, Properties, and Applications. Adv. Mater 22:3906–3924 (2010). https://doi.org/10.1002/adma.201001068

[60] Shin, H.-J., Kim, K. K., Benayad, A., Yoon, S.-M., Park, H. K., Jung, I.-S., Jin, M. H., Jeong, H.-K., Kim, J. M., Choi, J.-Y., Lee, Y. H. Efficient Reduction of Graphite Oxide by Sodium Borohydride and Its Effect on Electrical Conductance. Advanced Functional Materials 19:1987–1992 (2009). https://doi.org/10.1002/adfm.200900167

[61] Stankovich, S., Dikin, D., Dommett, G., *et al.* Graphene-based composite materials. Nature 442:282–286 (2006). https://doi.org/10.1038/nature04969

[62] Zhang, J., Yang, H., Shen, G., Cheng, P., Zhang, J., Guo, S. Reduction of graphene oxide via L-ascorbic acid. Chemical Communications 46(7):1112–1114 (2010). https://doi.org/10.1039/B917705A

[63] Ambrosi, A., Chua, C. K., Bonanni, A., Pumera, M. Lithium aluminum hydride as reducing agent for chemically reduced graphene oxides. Chem, Mater 24(12):2292–2298 (2012). https://doi.org/10.1021/cm300382b

[64] Zhou, X., Zhang, J., Wu, H., Yang, H., Zhang, J., Guo, S. Reducing graphene oxide via hydroxylamine: A simple and efficient route to graphene. Journal of Physical Chemistry C 115(24):11957–11961 (2011). https://doi.org/10.1021/jp202575j

[65] Cataldo, F., Ursini, O., Angelini, G. Graphite oxide and graphene nanoribbons reduction with hydrogen iodide. Fuller Nanotub Carbon Nanostruct 19(5):461–468 (2011). https://doi.org/10.1080/1536383X.2010.481064

[66] Robinson, J. T., Keith Perkins, F., Snow, E. S., Wei, Z., Sheehan, P. E. Reduced graphene oxide molecular sensors. Nano Letters 8(10):3137–3140 (2008). https://doi.org/10.1021/nl8013007

[67] Song, N. J., Chen, C. M., Lu, C., Liu, Z., Kong, Q. Q., Cai, R. Thermally reduced graphene oxide films as flexible lateral heat spreaders. Journal of Materials Chemistry A 2(39):16563–16568 (2014). https://doi.org/10.1039/C4TA02693D

[68] Stankovich, S., Dikin, D. A., Piner, R. D., Kohlhaas, K. A., Kleinhammes, A., Jia, Y., Wu, Y., Nguyen, S. B. T., Ruoff, R. S. Synthesis of graphene-based nanosheets via chemical reduction of exfoliated graphite oxide. Carbon 45(7):1558–1565 (2007). https://doi.org/10.1016/j.carbon.2007.02.034

[69] Liu, W., Speranza, G. Tuning the Oxygen Content of Reduced Graphene Oxide and Effects on Its Properties. ACS Omega 6(9):6195–6205 (2021). Published 2021 Mar 1 10.1021/acsomega.0c05578.

[70] Tiwari, S. K., Sahoo, S., Wang, N., Huczko, A. Graphene research and their outputs: Status and prospect. Journal of Science: Advanced Materials and Devices 5(1):10–29 (2020). ISSN 2468-2179 https://doi.org/10.1016/j.jsamd.2020.01.006

[71] Lin, L., Peng, H., Liu, Z. Synthesis Challenges for Graphene Industry. Nat. Mater. 18:520–524 (2019). 10.1038/s41563-019-0341-4.

[72] Zhu, Y. W., Ji, H. X., Cheng, H. M., Ruoff, R. S. Mass Production and Industrial Applications of Graphene Materials. Natl. Sci. Rev. 5:90–101 (2018). 10.1093/nsr/nwx055.

[73] Wang, X., Narita, Y., Mullen, A., Precision, K. Synthesis Versus Bulk-Scale Fabrication of Graphenes. Nat. Rev. Chem. 0100 (2018). 10.1038/s41570-017-0100.

[74] Wencheng, D., HongboGeng, Y. Y., Zhang, Y., Rui, X., Li, C. C. Pristine graphene for advanced electrochemical energy applications. Journal of Power Sources 437:226899 (2019). ISSN 0378-7753 https://doi.org/10.1016/j.jpowsour.2019.226899

[75] Xiang, J., Drzal, L. T. Thermal conductivity of exfoliated graphite nanoplatelet paper. Carbon 49(3):773–778 (2011). https://doi.org/10.1016/j.carbon.2010.10.003

[76] Chen, W., Yan, L., Bangal, P. R. Preparation of graphene by the rapid and mild thermal reduction of graphene oxide induced by microwaves. Carbon 48(4):1146–1152 (2010).

[77] Zhang, H.-B., Wang, J.-W., Yan, Q., Zheng, W.-G., Chen, C., Zhong-Zhen, Y. Vacuum-assisted synthesis of graphene from thermal exfoliation and reduction of graphite oxide. J. Mater. Chem 21:5392–5397 (2011). 10.1039/C1JM10099H.

[78] Novoselov, K. S., Geim, A. K., Morozov, S. V., Jiang, D., Zhang, Y., Dubonos, S. V., Grigorieva, I. V., Firsov, A. A. Electric field effect in atomically thin carbon films. Science 306:666–669 (2004). https://doi.org/10.1126/science.1102896

[79] Hernandez, Y., Nicolosi, V., Lotya, M., *et al.* High-yield production of graphene by liquid-phase exfoliation of graphite. Nature Nanotech 3:563–568 (2008). https://doi.org/10.1038/nnano.2008.215

[80] Blake, P., Brimicombe, P. D., Nair, R. R., Booth, T. J., Jiang, D., Schedin, F., Ponomarenko, L. A., Morozov, S. V., Gleeson, H. F., Hill, E. W., Geim, A. K., Novoselov, K. S. Graphene-based liquid crystal device. Nano Lett 8(6):1704–1708 (2008 Jun). 10.1021/nl080649i.

[81] Park, S., An, J., Jung, I., Piner, R. D., An, S. J., Li, X., Velamakanni, A., Ruoff, R. S. Colloidal suspensions of highly reduced graphene oxide in a wide variety of organic solvents. Nano Lett 9(4):1593–1597 (2009 Apr). 10.1021/nl803798y.

[82] Paredes, J. I., Villar-Rodil, S., Martínez-Alonso, A., Tascón, J. M. Graphene oxide dispersions in organic solvents. Langmuir Oct 7 24(19):10560–10564 (2008). 10.1021/la801744a.

[83] Bourlinos, A. B., Georgakilas, V., Zboril, R., Steriotis, T. A., Stubos, A. K. Liquid-Phase Exfoliation of Graphite Towards Solubilized Graphenes. Small 5:1841–1845 (2009). https://doi.org/10.1002/smll.200900242

[84] Ciesielski, A., Samori, P. Graphene via sonication assisted liquid-phase exfoliation. Chemical Society Reviews 43(1):381–398 (2014). 10.1039/C3CS60217F.

[85] Lotya, M., Hernandez, Y., King, P. J., Smith, R. J., Nicolosi, V., Karlsson, L. S., Blighe, F. M., De, S., Wang, Z., McGovern, I. T., Duesberg, G. S., Coleman, J. N. Liquid phase production of graphene by exfoliation of graphite in surfactant/water solutions. Journal of the American Chemical Society Mar 18 131(10):3611–3620 (2009). 10.1021/ja807449u.

[86] Niu, L., Li, M., Tao, X., Xie, Z., Zhou, X., Raju, A. P., et al. Salt-assisted direct exfoliation of graphite into high-quality, large-size, few-layer graphene sheets. Nanoscale 5(16):7202–7208 (2013).

[87] Berki, P., László, K., Tung, N. T., Karger-Kocsis, J. Natural rubber/graphene oxide nanocomposites via melt and latex compounding: Comparison at very low graphene oxide content. Journal of Reinforced Plastics and Composites 36(11):808–817 (2017). https://doi.org/10.1177/0731684417690929

[88] Abdullah Habib, N., BuongWoeiChieng, N. M., Rashid, U., RobiahYunus, S. A. R. Elastomeric Nanocomposite Based on Exfoliated Graphene Oxide and Its Characteristics without Vulcanization. Journal of Nanomaterials 2017:11 (2017). Article ID 8543137 https://doi.org/10.1155/2017/8543137

[89] Kang, H., Zuo, K., Wang, Z., Zhang, L., Liu, L., Guo, B. Using a green method to develop graphene oxide/elastomers nanocomposites with combination of high barrier and mechanical performance. Composites Science and Technology 92:1–8 (2014). 10.1016/j.compscitech.2013.12.004.

[90] Wu, X., Lin, T. F., Tang, Z. H., Guo, B. C., Huang, G. S. Natural rubber/graphene oxide composites: Effect of sheet size on mechanical properties and strain-induced crystallization behavior. Express Polym Lett 9(8):672–685 (2015). 10.3144/expresspolymlett.2015.63.

[91] Premanathan, M., Karthikeyan, K., Jeyasubramanian, K., Manivannan, G. Selective toxicity of ZnO nanoparticles toward Gram-positive bacteria and cancer cells by apoptosis through lipid peroxidation. Nanomedicine: Nanotechnology, Biology and Medicine 7:184–192 (2011). 10.1016/j.nano.2010.10.001.

[92] Mensah, B., Kim, S., Arepalli, S., Nah, C. A study of graphene oxide-reinforced rubber nanocomposite. *J. Appl*. Polym. Sci 131:40640 (2014). 10.1002/app.40640.

[93] Bai, X., Wan, C., Zhang, Y., Zhai, Y. Reinforcement of hydrogenated carboxylated nitrile–butadiene rubber with exfoliated graphene oxide. Carbon 49(5):1608–1613 (2011). https://doi.org/10.1016/j.carbon.2010.12.043

[94] Xiong, X., Wang, J., Jia, H., Fang, E., Ding, L. Structure, thermal conductivity, and thermal stability of bromobutyl rubber nanocomposites with ionic liquid modified graphene oxide. Polymer Degradation and Stability 98(11):2208–2214 (2013). https://doi.org/10.1016/j.polymdegradstab.2013.08.022

[95] Payne, A. R. The dynamic properties of carbon black-loaded natural rubber vulcanizates. Part I. J. Appl. Polym. Sci 6:57–3 (1962). https://doi.org/10.1002/app.1962.070061906

[96] Allegra, G., Raos, G., Vacatello, M. Theories and simulations of polymer-based nanocomposites: From chain statistics to reinforcement. Progress in Polymer Science 33(7):683–731 (2008). https://doi.org/10.1016/j.progpolymsci.2008.02.003

[97] Fröhlich, J., Niedermeier, W., Luginsland, H.-D. The effect of filler–filler and filler–elastomer interaction on rubber reinforcement. Composites Part A: Appl. Sci. Manuf. 36(4):449–460 (2005). https://doi.org/10.1016/j.compositesa.2004.10.004

[98] Edwards, D. C. Polymer-filler interactions in rubber reinforcement. J Mater Sci 25:4175–4185 (1990). https://doi.org/10.1007/BF00581070

[99] Bokobza, L. The Reinforcement of Elastomeric Networks by Fillers. Macromolecul Mater Engineer 289:607–621 (2004). https://doi.org/10.1002/mame.200400034

[100] Malas, A., Kumar Das, C. Development of modified expanded graphite-filled solution polymerized styrene butadiene rubber vulcanizates in the presence and absence of carbon black. Polymer Engineering and Science 54:33–41 (2014).

[101] Smith, D. J. Chapter 1: Characterization of Nanomaterials Using Transmission Electron Microscopy. Nanocharacterisation, Angus I Kirkland, Sarah J Haigh (eds.), United Kingdom: The Royal Society of Chemistry, (2). 1–29 (2015). eISBN: 978-1-78262-186-7 10.1039/9781782621867-00001.

[102] Karak, N. Chapter 1 – Fundamentals of Nanomaterials and Polymer Nanocomposites, Nanomaterials and Polymer Nanocomposites. Netherlands: Elsevier, 1–45 (2019). ISBN 9780128146156 https://doi.org/10.1016/B978-0-12-814615-6.00001-1

[103] Liu, S., Tian, M., Yan, B., Yao, Y., Zhang, L., Nishi, T. Nanying Ning, High performance dielectric elastomers by partially reduced graphene oxide and disruption of hydrogen bonding of polyurethanes. Polymer 56:375–384 (2015). https://doi.org/10.1016/j.polymer.2014.11.012

[104] Zhan, Y., Wu, J., Xia, H., Yan, N., Fei, G., Yuan, G. Dispersion and Exfoliation of Graphene in Rubber by an Ultrasonically-Assisted Latex Mixing and In situ Reduction Process. Macromol. Mater. Eng 296:590–602 (2011). https://doi.org/10.1002/mame.201000358

[105] Yang, J., Tian, M., Jia, Q.-X., Shi, J.-H., Zhang, L.-Q., Lim, S.-H., Zhong-Zhen, Y., Mai, Y.-W. Improved mechanical and functional properties of elastomer/graphite nanocomposites prepared by latex compounding. Acta Materialia 55(18):6372–6382 (2007). https://doi.org/10.1016/j.actamat.2007.07.043

[106] Escobar-Alarcón, L., Espinosa-Pesqueira, M. E., Solis-Casados, D. A., *et al.* Two-dimensional carbon nanostructures obtained by laser ablation in liquid: Effect of an ultrasonic field. Appl. Phys. A 124:141 (2018). https://doi.org/10.1007/s00339-018-1559-8

[107] Ferrari, A. C., Meyer, J. C., Scardaci, V., Casiraghi, C., Lazzeri, M., Mauri, F., Piscanec, S., Jiang, D., Novoselov, K. S., Roth, S., Geim, A. K. Raman Spectrum of Graphene and Graphene Layers. Physical Review Letters 97(18):187401 (2006). https://doi.org/10.1103/PhysRevLett.97.187401

[108] Tang, B., Guoxin, H., Gao, H. Raman Spectroscopic Characterization of Graphene. Applied Spectroscopy Reviews 45(5):369–407 (2010). 10.1080/05704928.2010.483886.

[109] Young, R. J., Kinloch, I. A. Graphene and Graphene-based Nanocomposites,in. Nano Science: Volume 1: Nanostructure through Chemistry. United Kingdom: The Royal Society of Chemistry, 145–179 (2013).

[110] Malard, L. M., Pimenta, M. A., Dresselhaus, G., Dresselhaus, M. S. Raman spectroscopy in graphene. Physics Reports 473(5–6):51–87 (2009). https://doi.org/10.1016/j.physrep.2009.02.003

[111] Zhang, W., Zhang, Y., Tian, Y., Yang, Z., Xiao, Q., Guo, X., Jing, L., et al. Insight into the capacitive properties of reduced graphene oxide. ACS Applied Materials & Interfaces 6(4):2248–2254 (2014).

[112] Yingjun, L., Qin, H., Zhang, H., Liu, A., Zhiyu, H., Yintao, L., Zhou, Y. Functionalised graphene oxide-bromobutyl rubber composites with segregated structure for enhanced gas barrier properties. Plastics, Rubber and Composites, (2021). https://doi.org/10.1080/14658011.2021.2008702

[113] Xie, Z.-T., Xuan, F., Wei, L.-Y., Luo, M.-C., Liu, Y.-H., Ling, F.-W., Huang, C., Huang, G., Wu, J. New evidence disclosed for the engineered strong interfacial interaction of graphene/rubber nanocomposites. Polymer 118:30–39 (2017). ISSN 0032-3861 https://doi.org/10.1016/j.polymer.2017.04.056

[114] Srinivasarao Yaragalla, M. A. P., Kalarikkal, N., Thomas, S. Chemistry associated with natural rubber–graphene nanocomposites and its effect on physical and structural properties. Industrial Crops and Products 74:792–802 (2015). ISSN 0926-6690 https://doi.org/10.1016/j.indcrop.2015.05.079

[115] Bokobza, I., Bruneel, J.-L., Couzi, M. Raman spectroscopy as a tool for the analysis of carbon-based materials (highly oriented pyrolitic graphite, multilayer graphene and multiwall carbon nanotubes) and of some of their elastomeric composites. Vibrational Spectroscopy 74:57–63 (2014). https://doi.org/10.1016/j.vibspec.2014.07.009

[116] Berean, K. J., Ou, J. Z., Nour, M., Field, M. R., Alsaif, M. M., Wang, Y., et al. Enhanced gas permeation through graphene nanocomposites. J Phys Chem C 119(24):13700–13712 (2015). https://doi.org/10.1021/acs.jpcc.5b02995

[117] Das, A., Leuteritz, A., Nagar, P. K., et al. Improved Gas Barrier Properties of Composites Based on Ionic Liquid Integrated Graphene Nanoplatelets and Bromobutyl Rubber. *International*. Polymer Science and Technology 43(6):1–8 (2016). 10.1177/0307174X1604300601.

[118] Wu, J., et al. Vulcanization kinetics of graphene/natural rubber nanocomposites. Polymer (2013). http://dx.doi.org/10.1016/j.polymer.2013.04.044

[119] Pramanik, M., Srivastava, S. K., Samantaray, B. K., Bhowmick, A. K. Synthesis and characterization of organosoluble, thermoplastic elastomer/clay nanocomposites. J Polym Sci B: Polym Phys 40:2065–2072 (2002). 10.1002/polb.10266.

[120] Bokobza, L. Multiwall carbon nanotube elastomeric composites: A review. Polymer 48 (17):4907–4920 (2007). ISSN 0032-3861 https://doi.org/10.1016/j.polymer.2007.06.046

[121] Payne, A. R., Whittaker, R. E. Low Strain Dynamic Properties of Filled Rubbers. Rubber Chemistry and Technology 44(2):440–478 (1971). 1 May https://doi.org/10.5254/1.3547375

[122] Waddell, W. H., Beauregard, P. A., Evans, L. R. Use of nonblack fillers in tire Compounds, Vol.69, Issue 3, 1996, Rubber Chemistry and Technology, Rubber Division, American Chemical Society. Tire Technol. Int. 24 (1995).

[123] Wang, M.-J. The Role of Filler Networking in Dynamic Properties of Filled Rubber. Rubber Chemistry and Technology 72(2):430–448 (1999). 1 May https://doi.org/10.5254/1.3538812

[124] Pliskin, I., Tokita, N. Bound rubber in elastomers: Analysis of elastomer-filler interaction and its effect on viscosity and modulus of composite systems. J. Appl. Polym. Sci 16:473–492 (1972). 10.1002/app.1972.070160217.

[125] Jana, S. C., Jain, S. Dispersion of nanofillers in high performance polymers using reactive solvents as processing aids. Polymer Jul 1 42(16):6897–6905 (2001).

[126] Saujanya, C., Radhakrishnan, S. Structure development and crystallization behaviour of PP/nanoparticulate composite. Polymer 42(16):6723–6731 (2001).

[127] Endo, M., Strano, M. S., Ajayan, P. M. Potential Applications of Carbon Nanotubes. In: Jorio, A., Dresselhaus, G., Dresselhaus, M. S. ((Eds.)) Carbon Nanotubes. Topics in Applied Physics. Berlin, Heidelberg: Springer, Vol. 111, 13–62 (2007). https://doi.org/10.1007/978-3-540-72865-8_2

[128] Coleman, J. N., Khan, U., Blau, W. J., Gun'ko, Y. K. Small but strong: A review of the mechanical properties of carbon nanotube–polymer composites. Carbon 44(9):1624–1652 (2006). https://doi.org/10.1016/j.carbon.2006.02.038

[129] Alexandre, M., Dubois, P. Polymer-layered silicate nanocomposites: Preparation, properties and uses of a new class of materials. Mater Sci Eng R Reports 28(1–2):1–63 (2000). https://doi.org/10.1016/S0927-796X(00)00012-7

[130] Theng, B. K. G. The Chemistry of Clay-Organic Reactions, UK. London: Taylor & Francis 343 (1974).

[131] Chen, G.-H., Wu, D.-J., Weng, W.-G., Yan, W.-L. Preparation of polymer/graphite conducting nanocomposite by intercalation polymerization. J. Appl. Polym. Sci 82:2506–2513 (2001). https://doi.org/10.1002/app.2101

[132] Pan, Y.-X., Yu, -Z.-Z., Ou, Y.-C., Hu, G.-H. A new process of fabricating electrically conducting nylon 6/graphite nanocomposites via intercalation polymerization. J. Polym. Sci. B Polym. Phys 38:1626–1633 (2000). https://doi.org/10.1002/(SICI)1099-0488(20000615)38:12

[133] Frogley, M. D., Ravich, D., Wagner, H. D. Mechanical properties of carbon nanoparticle-reinforced elastomers. Composites Science and Technology 63(11):1647–1654 (2003). https://doi.org/10.1016/S0266-3538(03)00066-6

[134] Ponnamma, D., Sadasivuni, K. K., Grohens, Y., Guo, Q., Thomas, S. Carbon nanotube based elastomer composites – An approach towards multifunctional materials. Journal of Materials Chemistry C 2:8446–8485 (2014).

[135] Mensah, B., Gil Kim, H., Lee, J.-H., Arepalli, S., Nah, C. Carbon nanotube-reinforced elastomeric nanocomposites: A review. International Journal of Smart and Nano Materials 6(4):211–238 (2015). 10.1080/19475411.2015.1121632.

[136] Lan, T., Pinnavaia, T. J. Clay-reinforced epoxy nanocomposites. Chem. Mater 6(12):2216–2219 (1994).

[137] Favier, V., Chanzy, H., Cavaille, J. Polymer nanocomposites reinforced by cellulose whiskers. Macromolecules 28(18):6365–6367 (1995).

[138] Lee, C., Wei, X., Kysar, J. W., Hone, J. Measurement of the elastic properties and intrinsic strength of monolayer graphene. Science 321(5887):385–388 (2008)

[139] SherifAraby, L. Z., Kuan, H.-C., Dai, J.-B., Majewski, P., Jun, M. A novel approach to electrically and thermally conductive elastomers using graphene. Polymer 54(14):3663–3670 (2013). ISSN 0032-3861 https://doi.org/10.1016/j.polymer.2013.05.014

[140] Wakabayashi, K., Pierre, C., Dikin, D. A., Ruoff, R. S., Ramanathan, T., Brinson, L. C., et al. Polymer-graphite nanocomposites: Effective dispersion and major property enhancement via solid-state shear pulverization. Macromolecules 41(6):1905–1908 (2008).

[141] Potts, J. R., Shankar, O., Du, L., Ruoff, R. S. Processing morphology property relationships and composite theory analysis of reduced graphene oxide/natural rubber nanocomposites. Macromolecules 45(15):6045–6055 (2012).

[142] Potts, J. R., Shankar, O., Murali, S., Du, L., Ruoff, R. S. Latex and two-roll mill processing of thermally exfoliated graphite oxide/natural rubber nanocomposites. Composites Science & Technology 74:166–172 (2013).

[143] Wang, M.-J. Effect of Polymer-Filler and Filler-Filler Interactions on Dynamic Properties of Filled Vulcanizates. Rubber Chemistry and Technology 71(3):520–589 (1998). 1 July https://doi.org/10.5254/1.3538492
[144] Wu, J., Huang, G., Li, H., Wu, S., Liu, Y., Zheng, J. Enhanced mechanical and gas barrier properties of rubber nanocomposites with surface functionalized graphene oxide at low content. Polymer 54(7):1930–1937 (2013).
[145] Hernández, M., Del Mar Bernal, M., Verdejo, R., Ezquerra, T. A., López-Manchado, M. A. Overall performance of natural rubber/graphene nanocomposites. Composites Science and Technology 73:40–46 (2012). ISSN 0266-3538 https://doi.org/10.1016/j.compscitech.2012.08.012
[146] Ozbas, B., O'Neill, C. D., Register, R. A., Aksay, I. A., Prud'homme, R. K., Adamson, D. H. Multifunctional elastomer nanocomposites with functionalized graphene single sheets. Journal of Polymer Science: Polymer Letters Edition, Part B, Polymer Physics 50:910–916 (2012). http://dx.doi.org/10.1002/polb.23080
[147] Xing, W., Tang, M., Jinrong, W., Huang, G., Hui, L., Lei, Z., Xuan, F., Hengyi, L. Multifunctional properties of graphene/rubber nanocomposites fabricated by a modified latex compounding method. Composites Science and Technology 99:67–74 (2014). https://doi.org/10.1016/j.compscitech.2014.05.011
[148] Schopp, S., Thomann, R., Ratzsch, K.-F., Kerling, S., Altstädt, V., Mülhaupt, R. Functionalized Graphene and Carbon Materials as Components of Styrene-Butadiene Rubber Nanocomposites Prepared by Aqueous Dispersion Blending. Macromol. Mater. Eng 299:319–329 (2014). https://doi.org/10.1002/mame.201300127
[149] Tang, Z., Zhang, L., Feng, W., Guo, B., Liu, F., Jia, D. Rational design of graphene surface chemistry for high-performance rubber/graphene composites. Macromolecules 47(24):8663–8673 (2014). https://doi.org/10.1021/ma502201e
[150] Song, S. H., Jeong, H. K., Kang, Y. G. Preparation and characterization of exfoliated graphite and its styrene butadiene rubber nanocomposites. Journal of Industrial and Engineering Chemistry 16(6):1059–1065 (2010). https://doi.org/10.1016/j.jiec.2010.07.004
[151] Mao, Y., Wen, S., Chen, Y., *et al.* High Performance Graphene Oxide Based Rubber Composites. Sci Rep 3:2508 (2013). https://doi.org/10.1038/srep02508
[152] Yang, J., Zhang, L. Q., Shi, J. H., Quan, Y. N., Wang, L. L., Tian, M. Mechanical and functional properties of composites based on graphite and carboxylated acrylonitrile butadiene rubber. Journal of Applied Polymer Science Jun 5 116(5):2706–2713 (2010).
[153] Frasca, D., Schulze, D., Wachtendorf, V., Morys, M., Schartel, B. Multilayer graphene/chlorine-isobutene-isoprene rubber nanocomposites: The effect of dispersion. Polym Adv Technol (2016). Available from http://dx.doi.org/10.1002/pat.3740
[154] Huneau, B. STRAIN-INDUCED CRYSTALLIZATION OF NATURAL RUBBER: A REVIEW OF X-RAY DIFFRACTION INVESTIGATIONS. Rubber Chemistry and Technology 84(3):425–452 (2011). 1 September https://doi.org/10.5254/1.3601131
[155] Nie, Y., Gu, Z., Wei, Y., *et al.* Features of strain-induced crystallization of natural rubber revealed by experiments and simulations. Polym J 49:309–317 (2017). https://doi.org/10.1038/pj.2016.114
[156] Fayong, L., Ning, Y., Yanhu, Z., Guoxia, F., Hesheng, X. 7, Probing the reinforcing mechanism of graphene and graphene oxide in natural rubber, January (2013). https://doi.org/10.1002/app.38958
[157] Ozbas, B., Toki, S., Hsiao, B. S., Chu, B., Register, R. A., Aksay, I. A., Prud'homme, R. K., Adamson, D. H. Strain-induced crystallization and mechanical properties of functionalized graphene sheet-filled natural rubber. J. Polym. Sci. B Polym. Phys 50:718–723 (2012). https://doi.org/10.1002/polb.23060
[158] Luo, Y., Wang, R., Zhao, S., Chen, Y., Su, H., Zhang, L., Chan, T. W., Wu, S. Experimental study and molecular dynamics simulation of dynamic properties and interfacial bonding characteristics of

graphene/solution-polymerized styrene-butadiene rubber composites. RSC Adv 6:58077 (2016). 10.1039/C6RA08417F.
[159] Wan, S., Lu, X., Zhao, H., Chen, S., Cai, S., He, X., Zhang, R. Effect of Graphene Oxide Modified with Organic Amine on the Aging Resistance, Rolling Loss and Wet-skid Resistance of Solution Polymerized Styrene-Butadiene Rubber. Materials (Basel) Feb 25 13(5):E1025 (2020). 10.3390/ma13051025.
[160] Kang, H., Tang, Y., Yao, L., Yang, F., Fang, Q., Hui, D. Fabrication of graphene/natural rubber nanocomposites with high dynamic properties through convenient mechanical mixing. Composites Part B: Engineering 112:1–7 (2017). https://doi.org/10.1016/j.compositesb.2016.12.035
[161] Tang, M., Xing, W., Wu, J., Huang, G., Xiang, K., Guo, L., Lia, G. Graphene as a prominent antioxidant for diolefin elastomers. Journal of Materials Chemistry A 3:5942–5948 (2015). https://doi.org/10.1039/C4TA06991A
[162] SherifAraby, Q. M., Zhang, L., Kang, H., Majewski, P., Tang, Y., Jun, M. Electrically and thermally conductive elastomer/graphene nanocomposites by solution mixing. Polymer 55(1):201–210 (2014). https://doi.org/10.1016/j.polymer.2013.11.032
[163] Chen, B., Ma, N., Bai, X., Zhang, H., Zhang, Y. Effects of graphene oxide on surface energy, mechanical, damping and thermal properties of ethylenepropylene-diene rubber/petroleum resin blends. RSC Adv 2(11):4683–4689 (2012). 10.1039/C2RA01212J.
[164] Yan, D., Zhang, H. B., Jia, Y., Hu, J., Qi, X. Y., Zhang, Z., et al. Improved electrical conductivity of polyamide 12/graphene nanocomposites with maleated polyethylene-octene rubber prepared by melt compounding. ACS Applied Materials Interfaces 4(9):4740–4745 (2012). https://doi.org/10.1021/am301119b
[165] Valentini, L., Bolognini, A., Alvino, A., Bittolo Bon, S., Martin-Gallego, M., Lopez-Manchado, M. A. Pyroshock testing on graphene based EPDM nanocomposites. Composites Part B: Engineering 60:479–484 (2014). https://doi.org/10.1016/j.compositesb.2013.12.022
[166] Sil Kim, J., Ho Yun, J., Kim, I., Eun Shim, S. Electrical properties of graphene/SBR nanocomposite prepared by latex heterocoagulation process at room temperature. Journal of Industrial and Engineering Chemistry 17(2):325–330 (2011). https://doi.org/10.1016/j.jiec.2011.02.034
[167] Lian, H., Li, S., Liu, K., Xu, L., Wang, K., Guo, W. Study on modified graphene/butyl rubber nanocomposites. I. Preparation and characterization. Polym. Eng. Sci. 51(11):2254–2260 (2011).
[168] Gan, L., Shang, S., Wah Marcus Yuen, C., Jiang, S.-X., Mei Luo, N. Facile preparation of graphene nanoribbon filled silicone rubber nanocomposite with improved thermal and mechanical properties. Composites Part B: Engineering 69:237–242 (2015). https://doi.org/10.1016/j.compositesb.2014.10.019
[169] Xing, W., Wu, J., Huang, G., Li, H., Tang, M., Fu, X. Enhanced mechanical properties of graphene/natural rubber nanocomposites at low content. Polym. Int 63:1674–1681 (2014). https://doi.org/10.1002/pi.4689
[170] Li, C., Feng, C., Peng, Z., Gong, W., Kong, L. Ammonium-assisted green fabrication of graphene/natural rubber latex composite. Polym Composite 34:88–95 (2013). https://doi.org/10.1002/pc.22380
[171] Wang, J., Jia, H., Tang, Y., *et al.* Enhancements of the mechanical properties and thermal conductivity of carboxylated acrylonitrile butadiene rubber with the addition of graphene oxide. J Mater Sci 48:1571–1577 (2013). https://doi.org/10.1007/s10853-012-6913-1
[172] Huxtable, S. T., Cahill, D. G., Shenogin, S., Xue, L., Ozisik, R., Barone, P., Usrey, M., Strano, M. S., Siddons, G., Shim, M., Keblinski, P. Interfacial heat flow in carbon nanotube suspensions. Nature Mater Nov 2(11):731–734 (2003). 10.1038/nmat996. Epub 2003 Oct 12. PMID: 14556001
[173] Kashiwagi, T., Grulke, E., Hilding, J., Groth, K., Harris, R., Butler, K., Shields, J., Kharchenko, S., Douglas, J. Thermal and flammability properties of polypropylene/carbon nanotube nanocomposites. Polymer 45:4227–4239 (2004). 10.1016/j.polymer.2004.03.088.

[174] Ma, W., Li, J., Deng, B., *et al.* Properties of functionalized graphene/room temperature vulcanized silicone rubber composites prepared by an *In-situ* reduction method. J. Wuhan Univ. Technol.-Mat. Sci. Edit 28:127–131 (2013). https://doi.org/10.1007/s11595-013-0653-1

[175] Alofi, A., Srivastava, G. P. Thermal conductivity of graphene and graphite. Phys. Rev. B 87(11):115421 (2013) https://doi.org/10.1103/PhysRevB.87.115421

[176] Song, Y., Jinhong, Y., Lianghao, Y., Alam, F. E., Dai, W., Chaoyang, L., Jiang, N. Enhancing the thermal, electrical, and mechanical properties of silicone rubber by addition of graphene nanoplatelets. Materials & Design 88:950–957 (2015). https://doi.org/10.1016/j.matdes.2015.09.064

[177] Zhang, G., Wang, F., Dai, J., Huang, Z. Effect of Functionalization of Graphene Nanoplatelets on the Mechanical and Thermal Properties of Silicone Rubber Composites. Materials (Basel) 9(2):92 (2016). Published 2016 Feb 2 10.3390/ma9020092.

[178] Marsden, A. J., Papageorgiou, D. G., Cristina Valles, A. L., Palermo, V., Bissett, M. A., Young, R. J., Kinloch, I. A. Electrical percolation in graphene–polymer composites. 2D Materials 5(3):032003 (2018).

[179] Lu, C. T., Weerasinghe, A., Maroudas, D., *et al.* A Comparison of the Elastic Properties of Graphene- and Fullerene-Reinforced Polymer Composites: The Role of Filler Morphology and Size. Sci Rep 6:31735 (2016). https://doi.org/10.1038/srep31735

[180] Noël, A., Faucheu, J., Chenal, J.-M., Viricelle, J.-P. Elodie Bourgeat-Lami, Electrical and mechanical percolation in graphene-latex nanocomposites. Polymer 55(20):5140–5145 (2014). https://doi.org/10.1016/j.polymer.2014.08.025

[181] He, C., She, X., Peng, Z., Zhong, J., Liao, S., Gong, W., Liao, J., Kong, L. Graphene networks and their influence on free-volume properties of graphene–epoxidized natural rubber composites with a segregated structure: Rheological and positron annihilation studies. Physical Chemistry Chemical Physics 17(18):12175–12184 (2015).

[182] Zhan, Y., Lavorgna, M., Buonocore, G., Xia, H. Enhancing electrical conductivity of rubber composites by constructing interconnected network of self-assembled graphene with latex mixing. J. Mater. Chem 22(21):10464–10468 (2012). 10.1039/C2JM31293J.

[183] Matos, C. F., Galembeck, F., Zarbin, A. J. G. Multifunctional and environmentally friendly nanocomposites between natural rubber and graphene or graphene oxide. Carbon 78:469–479 (2014). 10.1016/j.carbon.2014.07.028.

[184] Singh, V. K., Shukla, A., Patra, M. K., Saini, L., Jani, R. K., Vadera, S. R., Kumar, N. Microwave absorbing properties of a thermally reduced graphene oxide/nitrile butadiene rubber composite. Carbon 50(6):2202–2208 (2012). https://doi.org/10.1016/j.carbon.2012.01.033

[185] Wu, S., Tang, Z., Guo, B., Zhang, L., Jia, D. Effects of interfacial interaction on chain dynamics of rubber/graphene oxide hybrids: A dielectric relaxation spectroscopy study. RSC Adv 3(34):14549–14559 (2013).

[186] Paul, D. R., Robeson, L. M. Polymer nanotechnology: Nanocomposites. Polymer 49(15):3187–3204 (2008). https://doi.org/10.1016/j.polymer.2008.04.017

[187] Scherillo, G., Lavorgna, M., Buonocore, G. G., Zhan, Y. H., Xia, H. S., Mensitieri, G., Ambrosio, L. Tailoring assembly of reduced graphene oxide nanosheets to control gas barrier properties of natural rubber nanocomposites. ACS Appl. Mater. Interfaces 6(4):2230–2234 (2014). https://doi.org/10.1021/am405768m

[188] Wu, Y. P., Wang, Y. Q., Zhang, H. F., Wang, Y. Z., Yu, D. S., Zhang, L. Q., et al. Rubber-pristine clay nanocomposites prepared by co-coagulating rubber latex and clay aqueous suspension. Compos Sci Technol 65(7):1195–1202 (2005).

[189] Araby, S., Zaman, I., Meng, Q., Kawashima, N., Michelmore, A., Kuan, H. C., Majewski, P., Ma, J., Zhang, L. Melt compounding with graphene to develop functional, high-performance elastomers. Nanotechnology Apr 26 24(16):165601 (2013). 10.1088/0957-4484/24/16/165601.

[190] Kumar Sadasivuni, K., Deepalekshmi Ponnamma, S. T., Grohens, Y. Evolution from graphite to graphene elastomer composites. Progress in Polymer Science 39(4):749–780 (2014). ISSN 0079-6700 https://doi.org/10.1016/j.progpolymsci.2013.08.003

[191] Zhang, H.-B., Zheng, W.-G., Yan, Q., Yang, Y., Wang, J.-W., Zhao-Hui, L., Guo-Ying, J., Zhong-Zhen, Y. Electrically conductive polyethylene terephthalate/graphene nanocomposites prepared by melt compounding. Polymer 51(5):1191–1196 (2010). ISSN 0032-3861 https://doi.org/10.1016/j.polymer.2010.01.027

[192] Kim, H., Miura, Y., Macosko, C. W. Graphene/polyurethane nanocomposites for improved gas barrier and electrical conductivity. Chem. Mater 22(11):3441–3450 (2010). https://doi.org/10.1021/cm100477v

[193] Potts, J. R., Dreyer, D. R., Bielawski, C. W., Ruoff, R. S. Graphene-based polymer nanocomposites. Polymer 52(1):5–25 (2011). https://doi.org/10.1016/j.polymer.2010.11.042

[194] Theng, B. K. G. Formation and Properties of Clay-polymer Complexes. Amsterdam /Oxford/ New York: Elsevier Scientific Publishing Company, 362 (1979).

[195] Ganter, M., Gronski, W., Reichert, P., Mülhaupt, R. Rubber Nanocomposites: Morphology and Mechanical Properties of BR and SBR Vulcanizates Reinforced by Organophilic Layered Silicates. *Rubber*. Chemistry and Technology 1 74(2):221–235 (2001). May doi: https://doi.org/10.5254/1.3544946

[196] Sadhu, S., Bhowmick, A. K. Morphology study of rubber based nanocomposites by transmission electron microscopy and atomic force microscopy. Journal of Materials Science 40(7):1633–1642 (2005).

[197] Vasudeo Rane, A., Krishnan Kanny, V. K. A., Thomas, S. Chapter 5 – Methods for Synthesis of Nanoparticles and Fabrication of Nanocomposites. In: Bhagyaraj, S. M., Oluwafemi, O. S., Kalarikkal, N., Thomas, S. (eds) Micro and Nano Technologies, Synthesis of Inorganic Nanomaterials. Sawston, Cambridge: Woodhead Publishing (2018). 121–139 ISBN 9780081019757 https://doi.org/10.1016/B978-0-08-101975-7.00005-1

[198] Zhang, X., XiaodongXue, Q. Y., Jia, H., Wang, J., Qingmin, J., Zhaodong, X. Enhanced compatibility and mechanical properties of carboxylated acrylonitrile butadiene rubber/styrene butadiene rubber by using graphene oxide as reinforcing filler. Composites Part B: Engineering 111:243–250 (2017).

[199] Vassiliou, A. A., Chrissafis, K., Bikiaris, D. N. In situ prepared pet nanocomposites: Effect of organically modified montmorillonite and fumed silica nanoparticles on pet physical properties and thermal degradation kinetics. Thermochim. Acta 500(1–2):21–29 (2010).

[200] Leroux, F., Besse, J.-P. Polymer inter leaved layered double hydroxide: A new emerging class of nanocomposites. Chem. Mater 13(10):3507–3515 (2001). https://doi.org/10.1021/cm0110268

[201] Paul, M.-A., Alexandre, M., Degée, P., Calberg, C., Jérôme, R., Dubois, P. Exfoliated Polylactide/Clay Nanocomposites by In-Situ Coordination–Insertion Polymerization. Macromol Rapid Commun 24:561–566 (2003). https://doi.org/10.1002/marc.200390082

[202] Jun, M., Jian, X., Ren, J.-H., Zhong-Zhen, Y., Mai, Y.-W. A new approach to polymer/montmorillonite nanocomposites. Polymer 44(16):4619–4624 (2003). ISSN 0032-3861 https://doi.org/10.1016/S0032-3861(03)00362-8

[203] Paszkiewicz, S., Szymczyk, A., Zdenkošpitálský, J. M., Kwiatkowski, K., Rosłaniec, Z. "Structure and properties of nanocomposites based on PTT-block-PTMO copolymer and graphene oxide prepared by in situ polymerization". European Polymer Journal 50:69–77 (2014).

[204] Lee, Y. R., Raghu, A. V., Mo Jeong, H., Kyu Kim, B. "Properties of waterborne polyurethane/functionalized graphene sheet nanocomposites prepared by an in situ method". Macromolecular Chemistry and Physics 210(15):1247–1254 (2009).

[205] Xue, X., Yin, Q., Jia, H., Zhang, X., Wen, Y., Qingmin, J., Zhaodong, X. "Enhancing mechanical and thermal properties of styrene-butadiene rubber/carboxylated acrylonitrile butadiene rubber blend by the usage of graphene oxide with diverse oxidation degrees". Applied Surface Science 423:584–591 (2017).

Asit Baran Bhattacharya and Kinsuk Naskar

# Chapter 5
# Transmission rubber V-belt technology

## 5.1 Introduction

This chapter reviews the different types of transmission belts, mainly for automotive and industrial applications. The definition and application areas of the different types of belts are discussed thoroughly. Materials like compounding aspect are also discussed. The different types of manufacturing techniques and failure mode of the belts have been described elaborately. Compound development aspects for a V-belt have been described in detail. The physical properties, heat buildup (HBU), adhesion strength, DMA study, and finally, belt cost optimisation and life is also discussed [1].

## 5.2 Concept of transmission belts

### 5.2.1 Definition of transmission belts

Power transmission by belts is described as the transmission of power from the prime mover to one or more driven machines through a flexible nonmetallic member. That nonmetallic member is generally layers of various rubber compounds that are reinforced with cords and textiles. Transmission belts consist of a combination of fabric, cord, and elastomeric compounds, the whole being bonded together uniformly and shaped by following the best manufacturing practices.

The function of a belt is to transfer the rotation from one powered pulley to one or more driven pulleys as shown in Figure 5.1. The perfect belt design can transfer the power efficiently and reliably [2].

## 5.3 Types of transmission belts

### 5.3.1 Types of wrapped V-belts

A) *Classical V-belt:* A V-belt is a loop of flexible material used to link two or more rotating, most often parallel, pulleys mechanically. Belts may be used as a source of motion to transmit power efficiently. It is made of compounds as a significant component, polymer fibre cords as a reinforcement member, and rubber-coated

https://doi.org/10.1515/9783110668537-005

fabric as a jacketing material. Dimensions of this type of belts are trapezium-type and height of the belt is 60% of the top width.

B) *Wedge belt:* The function, components, and materials all are same as the classical V-belts. The only difference is the dimension. The height of the belt is 80% of the top width.

C) *Banded belt:* A belt made up of two or more wedge belts, or V-belts joined across the top surface with a reinforced band of flexible material. Figure 5.2 shows the structure of different types of V–belts.

**Figure 5.1:** V-belt with pulley.

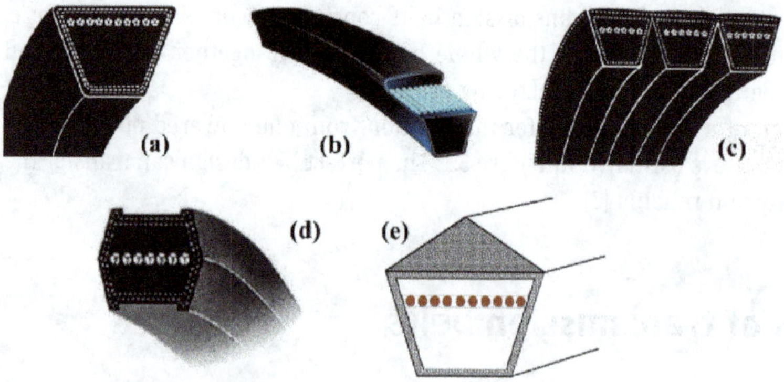


**Figure 5.2:** Cross section of different types of V-belts: (a) classical V-belt, (b) wedge V-belt, (c) banded V-belt, (d) hexagonal V-belt, and (e) crested V-belt.

There are a few more types of V-belts:

D) *Hexagonal V-belts* (application: agriculture machineries, textile machineries, rice mills)

E) *Crested V-belts* (application: tiles or ceramic manufacturing industry)

F) *Profile top V-belts* (application: conveying industry) [1–2], etc.

### 5.3.2 Types of raw edge belts

In raw edge belts, as the name signifies, the edges (sidewalls) are cut from sleeves, and the sidewalls are not covered, i.e., side walls are raw. There is no fabric wrapping on the sides as in standard belts. Further, the bending resistance reduces and also mechanical losses during bending decrease. This means improved power transmission with better heat dissipation than V-belts. Application areas are: automobile applications, agriculture machineries, textile machineries, and general engineering industries as shown in Figure 5.3.

**Figure 5.3:** (a) Cross section of raw edge belt and (b) cross section of poly-V belt.

Raw edge cogged belts, raw edge plain (REP) belts, and raw edge laminated belts are illustrated in Figure 5.4.

### 5.3.3 Types of poly-V belts or multi-V-belts

A poly-V belt or multi-V-belt is a single, continuous belt used to drive multiple peripheral devices in an automotive engine, such as an alternator, power steering pump, water pump, air conditioning compressor, and air pump.

The belt may also be guided by an idler pulley and/or a belt tensioner (which may be spring-loaded, hydraulic, or manual) [1–2].

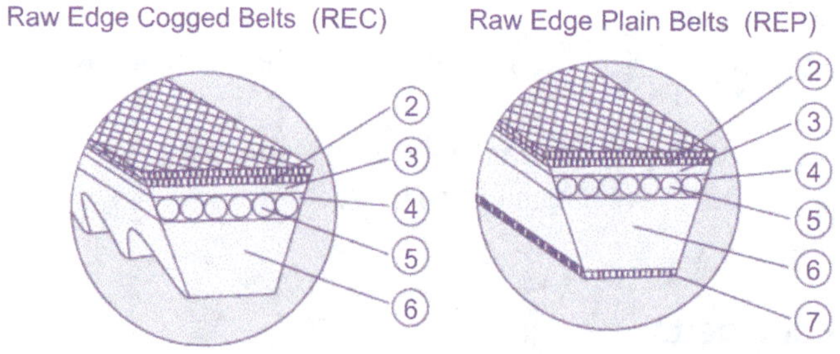


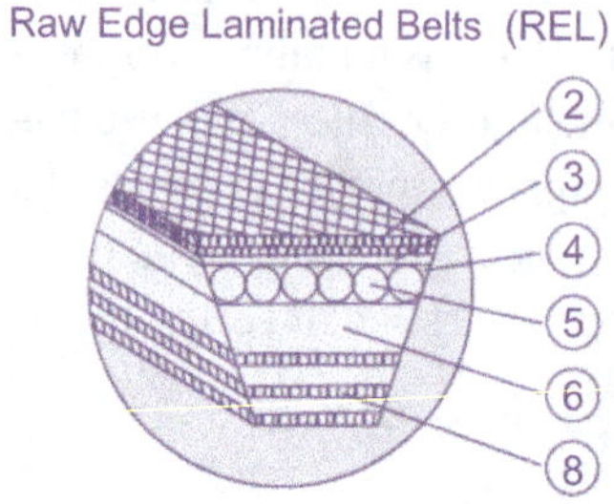


| No. | Name |
|---|---|
| 1 | Cover Fabric |
| 2 | Top Fabric |
| 3 | Top Cushion Rubber |
| 4 | Adhesion Rubber |
| 5 | Tension Member |
| 6 | Compression Rubber |
| 7 | Bottom Fabric |
| 8 | Laminate Fabric |

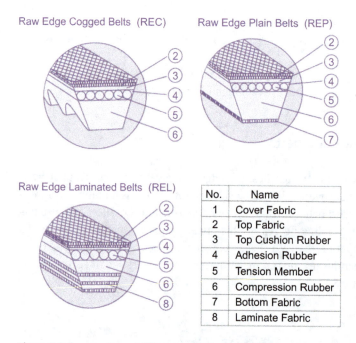

**Figure 5.4:** Cross sections of different types of raw edge belts.

## 5.4 History of transmission belts: pros and cons

Power transmission belts have played an essential role in the industries of the world for more than 250 years. Flat belts are generally made of rubber, leather, and fabrics and piled up layer by layer. Usually, leather is used for flat belts because it has a high coefficient of friction and can transmit more power. To achieve the desired thickness of the flat belt, the number of layers of rubber, fabric, and leather is manipulated. The handling of the flat belt is done between the pulleys. Bending of a flat belt is very easy around the pulleys. But the major disadvantage is the slippage of the belt from the pulleys, which causes machine shutdown and maintenance. Flat belts require more width to transmit more power and more thickness to withstand more load. They are of made of piled-up leather and regular cotton or hemp rope [1].

To overcome the slippage problem of flat belts and the requirement of power transmission capacity for industrial machinery for electric motors, V-belts replaced the flat belts slowly in the market. Different types of V-belts have been developed over the years depending upon application requirements. The type and design of the V-belt that operates through pulleys can be changed based on groove width and operating diameter as depicted in Figure 5.5.

With time, the development of transmission belts also accelerated depending upon requirements. To increase the transmission power efficiency, the V-belt design was con-

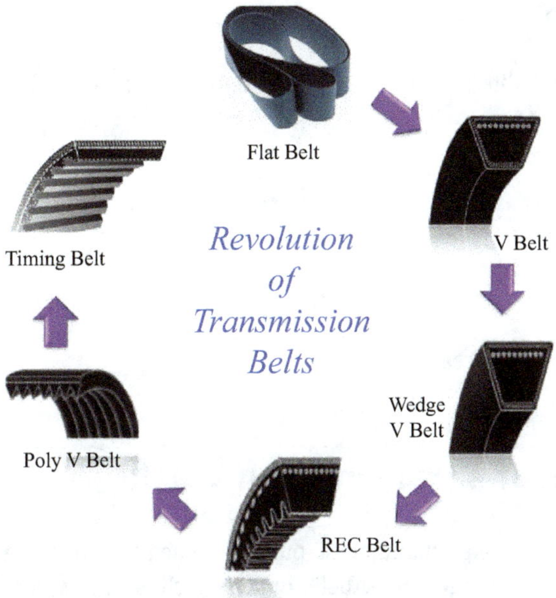


**Figure 5.5:** Development of transmission belts.

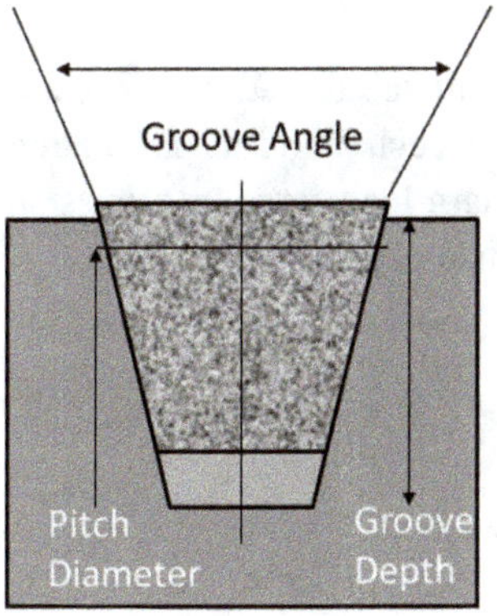


**Figure 5.6:** Cross section of V-belt and pulley groove.

verted into the wedge V-belt and then raw edge belt and raw edge cogged belt to increase the flexibility. Then poly-V or multi-V-belt, moulded poly-V belt, and then timing belt with latest trends in the industrial and automotive sector. The vulcanised rubber V-belts were introduced around the year 1917 and within a few years of introduction, they dominated the belt market due to the various advantages [2]. The cross section of a V-belt is portrayed in Figure 5.6.

## 5.5 Application areas

The wide variety of belt sizes available permits the application of V-belts in a wide range of applications such as [2]:

1) Machine tools
2) Transmission engineering
3) Conveyance technology
4) Textile machines
5) Paper mill
6) Stone crushers
7) High power generation automobile areas, etc.

## 5.6 Drive mechanism of a transmission V-belt

The mechanism of a V-belt employs a V-belt mounted on pulleys or sheaves having an annular V-shaped groove for receipt of the belt. V-belts include inclined faces (less than 40°) which make with the pulley groove. The tension in the belt causes the belt to be wedged into sidewall contact with the pulley groove. They run smoothly between belts and driving pulleys. However, improvements in belt engineering allow the use of V-belts in place of chains or gears.

V-belts have generally made of rubber-like material of reinforced construction riding on metal pulleys. The coefficient of friction between such materials is less and results in good power transmission. In environments where high force transmission is required, rubber belts have limitations of stress, tension, and side pressures.

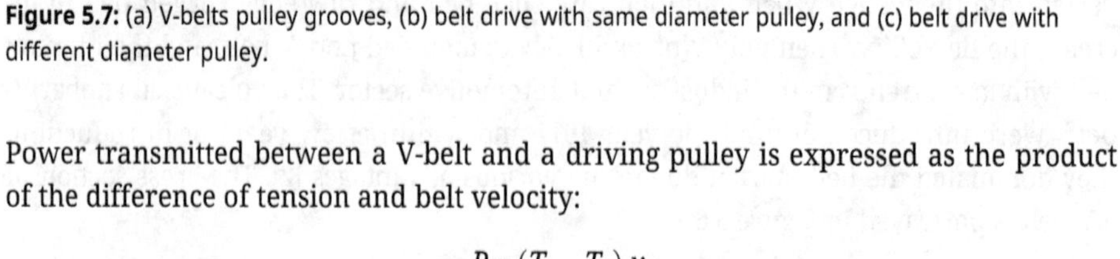

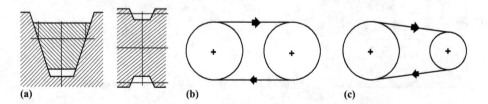


**Figure 5.7:** (a) V-belts pulley grooves, (b) belt drive with same diameter pulley, and (c) belt drive with different diameter pulley.

Power transmitted between a V-belt and a driving pulley is expressed as the product of the difference of tension and belt velocity:

$$P = (T_1 - T_2).v$$

where $T_1$ and $T_2$ are tensions on the tight side and slack side of the belt respectively. They are related as:

$$T_1/T_2 = e.\mu.\alpha$$

where $\mu$ is the coefficient of friction and $\alpha$ is the angle (in radians) subtended by the contact surface at the centre of the pulley as shown in Figure 5.7 [3].

## 5.7 Factors affecting on durability or life

### 5.7.1 Theory of heat resistance

It is not easy to give a quantitative definition of heat resistance in power transmission belts due to the wide variety of factors encountered in various types of belts and application areas. In addition, length of the exposed area of the belt, service life, and the presence of other factors such as raw materials and chemicals, which might accelerate the effects of temperature, all add to the complexity of such a definition. Some of the main factors that have to be considered for V-belts are:
1) Excessive heat generation in rubber products during dynamic applications, which helps further curing of the rubber product and as a result, it becomes hard and brittle. Thus, cracks are easily formed in the product after the dynamic stress–strain cycle.
2) Environment exposure and effect of ozone accelerates for deterioration.
3) Internal heat is created by constant flexing of the components.
4) High ambient temperatures increase both internal and external heat.

If one is able to reduce or remove these factors, it can quickly improve the belt durability life and drive efficiency. Thus, it can result in two changes, either increasing the heat resistance or decreasing the HBU generation in the applications.

The goal is to decrease the HBU of the belts during applications by developing a new compound.

Each type of power transmission belt has its own characteristics of HBU. Belt operating temperature is a function of pulley diameter, the load being transmitted, belt flex rate, belt type and construction, ambient temperature, etc. There are still many parameters like HBU, fatigue life, and the stiffness of the belts that have to be improved to get better results [2].

### 5.7.2 Theory of fatigue resistance

Webster's has defined the term 'fatigue' as follows: Fatigue is the action that takes place in materials, causing deterioration and failures after a repetition of stress. According to Mars et al. there are many factors that affect the fatigue crack nucleation and growth process in rubber. The fatigue life of rubbers depends upon mechanical loading history, environmental conditions, formulation of the rubber compound, and to some extent, the stress–strain constitutive behaviour. Environmental factors like temperature, oxygen, ozone, and static electrical charges, can affect both the short- and long-term fatigue behaviour of rubber.

Compounding variations may affect the fatigue life of a rubber specimen in many different and complicated ways. It affects the number and stability of the cross-links. The dispersion of the compound system will affect the uniformity of the distribution of cross-linking sites. Also compounding variation affects the hysteresis properties of the vulcanised compounds.

In testing for effects on fatigue life, it is necessary to know whether the end use will involve constant energy input or constant amplitude vibration. Constant amplitude tends to penalise high modulus compounds since the energy level will be greater.

The effect of the curing system on fatigue life appears quite clear since most agree that nonelemental sulphur vulcanisates are weaker than conventional sulphur-accelerator vulcanisates. There is some agreement that among these systems, those giving a high proportion of poly-sulfidic to mono-sulfidic cross-links provide improved fatigue resistance [1–3].

## 5.8 Failure modes

There are several types of failure modes of V-belts during the application, as listed below:

1. *Bottom crack*
   i) Slippage of the belt causing HBU and gradual hardening
   ii) Overloading of the belt due to dropping out of belts or undersigned drive
2. *Snapping of belt*
   i) Excessive shock load
   ii) Wearing out of reinforcing member
   iii) Belt falling out of the drive due to misalignment of the pulley
3. *Burning of bottom and side*
   i) Slippage of the belt while starting and stalling load
   ii) Wearing out of pulley grooves
4. *Excessive wearing of the outer cover (enveloping fabric)*
   i) Rubbing of the belt against some obstruction
   ii) Difference between belt and pulley angles
   iii) High-temperature buildup

## 5.9 Construction of transmission belts

A V-belt has a different layer of components, like rubber, cord, and fabric. Here we are only giving the construction of a V-belt [1]. The belt was constructed as per the standard **JIS K 6323** and the details are illustrated in Table 5.1 below for B Section and in Figure 5.8.

**Table 5.1:** Belt construction dimensions.

| Component | Gauge |
|---|---|
| Topping or cushioning layer | 1.50 ± 0.10 mm |
| Reinforcement layer | 01 |
| Base layer | 8.70 ± 0.10 mm |
| Fabric layer | 0.50 ± 0.05 mm |

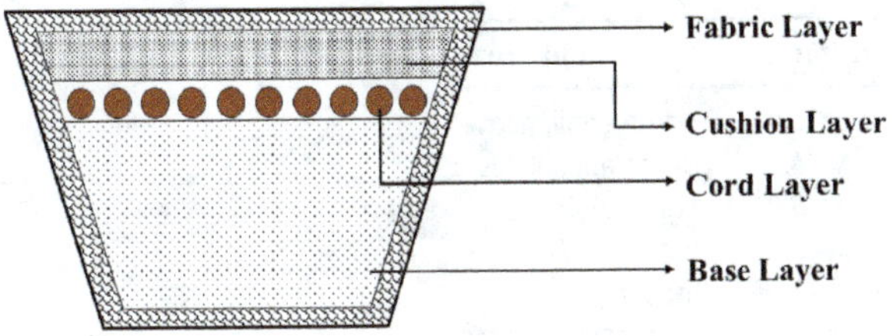


**Figure 5.8:** Cross section of a V-belt showing different construction layers.

## 5.10 Belt nomenclature

Depending upon the dimensions of the V-belts, they are divided into different sections. For classical belts there are Z, A, B, C, D, and E sections and the top width (*W*) is 10 mm, 13 mm, 17 mm, 22 mm, 32 mm, and 38 mm, respectively, and the thickness or height (*T*) is the 60% of the top width (*W*).

For the space saver or wedge belts, a different nomenclature is used. The letters 'SP' are added to the respective section: SPZ, SPA, SPB, etc. The top width (*W*) is the same as in classical belts. Thickness (*T*) is the 80% of the top width (*W*) portrayed in Figure 5.9.

In the case of joined V-belts (banded belts), the letter **'J'** is added to the section: JZ, JA, JB, or JSPZ, JSPA, JSPB, etc.

When it is raw edge V-belt the nomenclature is changed slightly: for cogged type the letter **'X'** term is added to the respective section: ZX, AX, BX, or SPAX, SPBX, etc. The REP types: (a) industrial applications: REPZ, REPA, REPB and (b) automotive applications: AV10, AV13, AV17, AV10X, AV13X, etc.

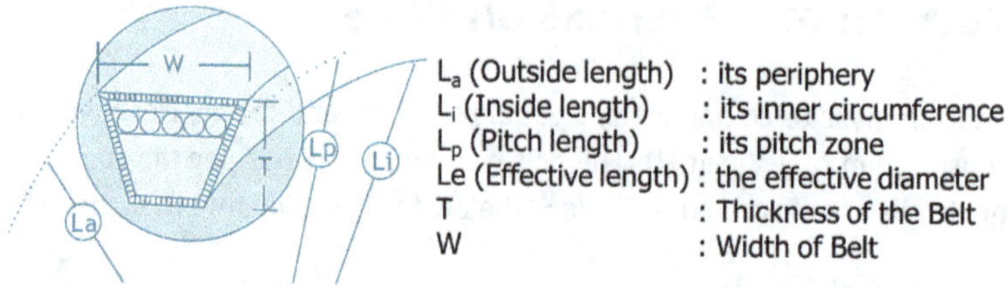


**Figure 5.9:** Cross section of V belt dimension designations.

The nomenclature of the poly-V belts is slightly different from the other two types. Here another term 'pitch' is observed [1]. For different sections, pitch and height are different: PI, PJ, PK, PL, PM, etc. The pitch and height are detailed in Table 5.2.

**Table 5.2:** Different sections, pitch and heights of rubber V belt.

| Belt section | Pitch (mm) | Height (mm) | Application areas |
|---|---|---|---|
| **PI** | 1.60 | 3.0 | Home appliances |
| **PJ** | 2.34 | 4.0 | Home appliances |
| **PK** | 3.56 | 6.0 | Automobile |
| **PL** | 4.70 | 10.0 | Industry |
| **PM** | 9.40 | 17.0 | Industry |

## 5.11 Materials for manufacturing a transmission belts

### 5.11.1 Rubbers

Rubber or elastomer has the ability to undergo large elastic deformations, that is, to stretch and return to their original shape in a reversible way. Rubber is defined by the ASTM standard **D1566** as 'a material that is capable of recovering from large deformations quickly and forcibly and can be, or already is, modified to a state in which it is essentially insoluble (but can swell) in boiling solvent, such as benzene, methyl ethyl ketone, or ethanol-toluene azeotrope' [4].

The essential requirement for a substance to be rubbery is that it should consist of long flexible chain-like molecules. The molecules themselves must, therefore, have a 'backbone' of many noncollinear single valence bonds, about which rapid rotation is possible as a result of thermal agitation. In most applications for rubber products, there are no alternative materials except other rubbers. Rubber has very high deformability. It consists of very high molecular mass molecules that can be crosslinked together to form a network.

Elastomers have very low moduli and for this reason can be easily deformed under the influence of relatively low stresses. Elongations up to 500–1,000% can be achieved, and the original shape is recovered once the stress is removed, i.e., the behaviour is elastic. Flexible polymers with a low glass transition temperature (below the working temperature) and whose chains are capable of being cross-linked may form elastomers. The degree of cross-linking needs to be low as highly cross-linked materials give rigid polymers with low deformability [4].

#### 5.11.1.1 Natural rubber (NR)

**Natural rubber (NR)** can be isolated from more than 200 different species of plant including surprising examples such as dandelions. Only one tree source, *Hevea brasiliensis,* is the commercially important raw material used for the production of these grades are field coagulum mainly tree lace and cup lumps. The latex typically contains 30–40% dry rubber by weight, and 10–20% of the collected latex is concentrated by creaming, or centrifuging, and used in its latex form [5].

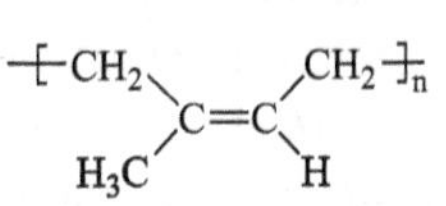


**Figure 5.10:** Polyisoprene monomer unit.

NR has the chemical name polyisoprene. It is also important to note that there is a unique feature of its structure that accounts for its unique properties. This concerns the possible isomers that can occur in a polyisoprene chain as shown in Figure 5.10.

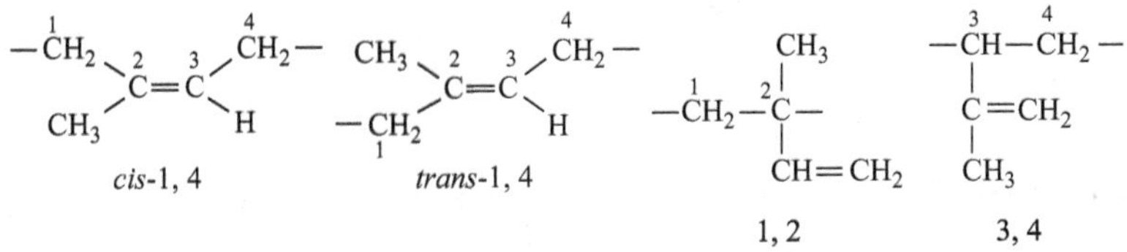


**Figure 5.11:** Polyisoprene microstructures.

It turns out that NR consists of polymer chains all having an almost perfect *cis*-1,4 structure; hence, the actual chemical name for this polymer is *cis*-1,4-polyisoprene. When the chain units in a macromolecule all consist of the same isomer, the polymer is said to be stereoregular. Due to this remarkable regularity, to NR chains can attain an excellent regularity, especially when the rubber is stretched. NR crystallises on stretching, resulting in high gum tensile strength. NR is vulcanised with sulphur compounds that can cross-link the chains because of the presence of the reactive double bonds (unsaturation) [5]. A polyisoprene microstructure chain is depicted in Figure 5.11.

There are many types of NR available in the market, like ISNR, SMR, pale crepe, RSS, TSR, etc. ISNR is the most competitive grade of NR available in the market. NR (ISNR) has good tensile strength – gum and compounded both – excellent crack propagation, resistance property, and so on.

Uses of NR are in:
1) V-belts
2) tyres
3) Conveyor belts
4) Footwear
5) Coated fabrics
6) Adhesives, etc.

#### 5.11.1.2 Styrene–butadiene rubber (SBR)

**Styrene–butadiene rubber (SBR)** describes the family of synthetic rubbers derived from styrene and butadiene. SBR has excellent abrasion resistance, crack initiation resistance, and ageing stability when protected by additives. The styrene:butadiene ratio influences the properties of the polymer: with high styrene content, the rubbers are harder and less rubbery. It is preferred in applications such as conveyor belts, shoe soles, V-belts, and extruded and moulded rubber goods [4, 5].

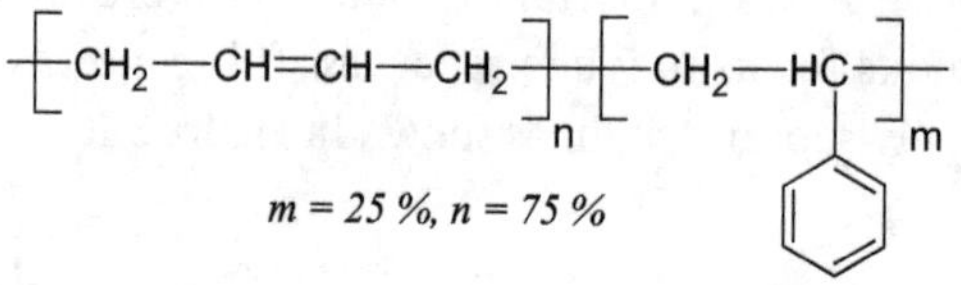


**Figure 5.12:** Styrene–butadiene rubber monomer unit.

There are two processes followed for making SBR: emulsion polymerisation and solution polymerisation. These processes are analogous to those for BR.

**Emulsion SBR:** Emulsion polymerisation produces SBR of high molecular weight. It is a free-radical-initiated process; the composition of the resultant chains is governed by the statistics of polymerisation, with units of styrene and butadiene randomly spaced throughout. Further, there is little stereochemical control of the insertion of the butadiene. SBR production is dominated by the emulsion process [4].

**Solution SBR:** In the 1960s, anionic polymerised S-SBR began to challenge E-SBR in the automotive tyre market. This is now being complemented by solution polymerisation using Ziegler–Natta transition metal catalysts. In S-SBR, use of the anionic or coordination initiators allows control of the *trans/cis*-microstructure and the 1,2 content of the polymerised butadiene. S-SBR is similar to E-SBR in terms of tensile strength, modulus, and elongation, but it has better flex resistance, lower HBU, and higher resilience and lowers rolling resistance in tyres [5].

In Figure 5.12, a monomer unit of SBR is shown.

**Uses of SBR are in:**
1. Passenger car tyres
2. Truck and bus tyres
3. Automotive
4. Mechanical goods
5. V-belts, etc.

#### 5.11.1.3 Polybutadiene rubber (BR)

**Polybutadiene rubber (BR)** is a polymer formed from the polymerisation of the monomer 1, 3-butadiene. The primary applications for BR include tyre treads, V-belts, conveyor belts, sportswear, golf balls, automotive components, conveyor belts, and shoe heels and soles. BR is also well suited for the production of flexible rollers and mechanical goods due to its superior elasticity, resilience, less HBU and abrasion properties [6].

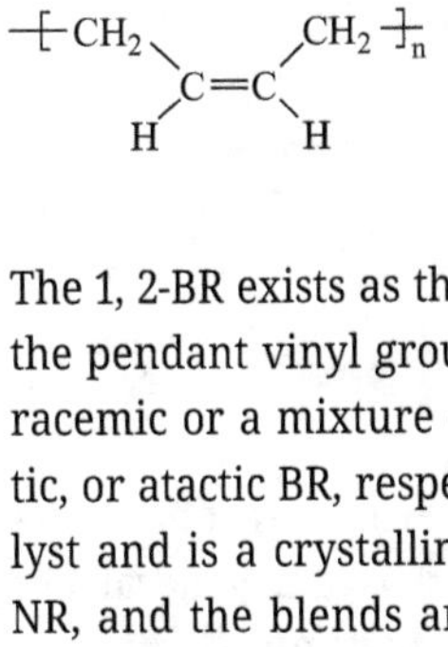

**Figure 5.13:** Polybutadiene rubber monomer unit.

The 1, 2-BR exists as three isomers. The 1,2-insertion forms a chiral carbon attached to the pendant vinyl group. The tacticity of this group can be either exclusively meso or racemic or a mixture of the two. This orientation corresponds to syndiotactic, isotactic, or atactic BR, respectively. Syndiotactic BR is made with cobalt Ziegler–Natta catalyst and is a crystalline thermoplastic melting at about 220 °C. It is compatible with NR, and the blends are excellent thermoplastic elastomers. Intrachain mixtures of syndiotactic and atactic isomers lead to lower melting points and a variety of BRs containing different fractions of syndiotactic sequences are commercially available [5].

A polybutadiene rubber monomer unit is shown in Figure 5.13.

**Uses of BR are in:**
1) Tyre treads
2) Carcass and sidewall
3) Automotive industries
4) V-belts, etc.

#### 5.11.1.4 Polychloroprene rubber (CR)

Polychloroprene rubber is prepared by emulsion polymerisation of chloroprene monomer. This is rubber with the unsaturated backbone that can be cross-linked by sulphur. Due to the presence of the chlorine unit, it is an oil-, fuel-, and flame-resistant rubber. As the structure of the monomer unit is almost similar to the NR backbone, it also exhibits excellent dynamic properties like NR. It is to be noted that the adhesion property of this rubber is excellent – especially *trans*-polychloroprene is very good for use as adhesive. In support of this reason, it is called all-rounder rubber [4–6]. In Figure 5.14, a monomer unit of polychloroprene rubber is depicted.

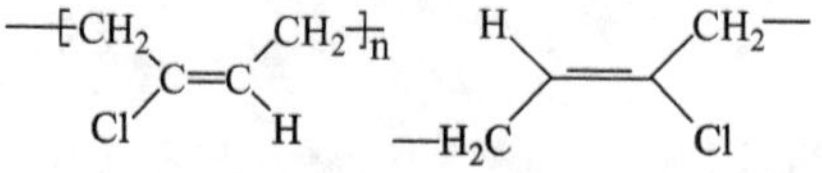

**Figure 5.14:** Polychloroprene rubber monomer unit.

**Uses of CR are in:**
1. Hoses
2. Belts
3. Axle-boots
4. Wiper blades
5. Mounts
6. Air springs
7. Cable jackets, etc.

### 5.11.2 Fillers and additives

Fillers are significant ingredients in rubber compounds for achieving target hardness and as well as reinforcement. Amount of fillers 02–60 PHR can be used in general, but sometimes it can be more than that. These ingredients are used to reinforce physical properties and to impart specific processing characteristics or to reduce cost [4]. Reinforcing filler enhances:
1. Hardness
2. Tensile strength
3. Modulus
4. Tear strength of a compound

An extending filler or diluent is loading and non-reinforcing material. It is typically selected to:

a) reduce cost or to impart specific desirable processing properties;
b) provide uncured firmness (green strength);
c) offer smooth extrusion.

In general, in the rubber compound, we are using inorganic fillers widely, but there are also organic fillers. The inorganic fillers are classified as:

1. Reinforcement types:
   A. Fibrous – glass fibre
   B. Particulate
      a. Black – carbon black
      b. Non-black – silica and activated $CaCO_3$
2. Non-reinforcement types:
   China clay, $TiO_2$, mica, asbestos, and so on [4].

#### 5.11.2.1 Carbon black

When we are trying to describe rubber filler, carbon black is always coming first in front of us. The most important event that was to have the most significant influence on the usage of carbon black occurred at the turn of the century and involved the discovery of the reinforcing effect of carbon blacks when added to NR, a discovery that was destined to become the most significant milestone in the rubber and automotive industry. By using carbon black as reinforcing filler, the service life of rubber products was significantly increased, ultimately making it possible to achieve durability. Today carbon blacks play an essential role, not only as reinforcing filler for tyres and other rubber products but also as a pigment for printing inks, coatings, plastics, and a variety of other applications [7].

Before beginning, there is merit in reviewing some basic definitions in black carbon technology. Although it is not attempted to present a comprehensive list of definitions, several important ones will be given, and the reader is referred to **ASTM D3053** for additional carbon black terminology: carbon black is an engineered material, primarily composed of elemental carbon, obtained from the partial combustion or thermal decomposition of hydrocarbons, existing as aggregates of aciniform morphology, which are composed of primary spheroidal particles and turbostatic layering within the primary particles [7].

The majority of industrial carbon blacks produced today are also based on the process of incomplete combustion of hydrocarbons. However, a second process is also used, namely that of thermal decomposition, during which the carbon black is formed in the absence of oxygen. These two process definitions may serve as a preliminary classification, which subsequently will be subdivided further [8]. The electron micrographs of carbon black are shown in Figure 5.15.

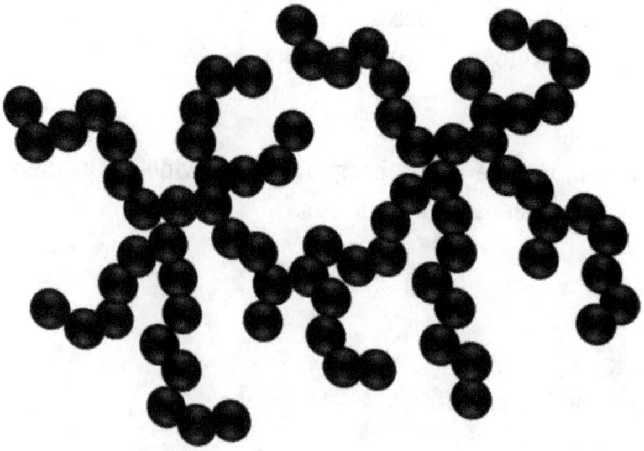

**Figure 5.15:** Electron micrographs of carbon black.

### 5.11.2.2 Silica

Precipitated silica is the highest reinforcing nonblack filler and is closest to carbon black in compound properties. The formation of precipitated silica is a chemical reaction of sodium silicate (water glass), and sulphuric acid. By-products of the reaction are sodium sulfate, which must be washed out, and water. The chemical reaction is an equilibrium reaction, which can be influenced by process parameters such as pH, temperature, and concentration. The essential properties: surface area, structure, and silane group density, are controlled in the precipitation process [8].

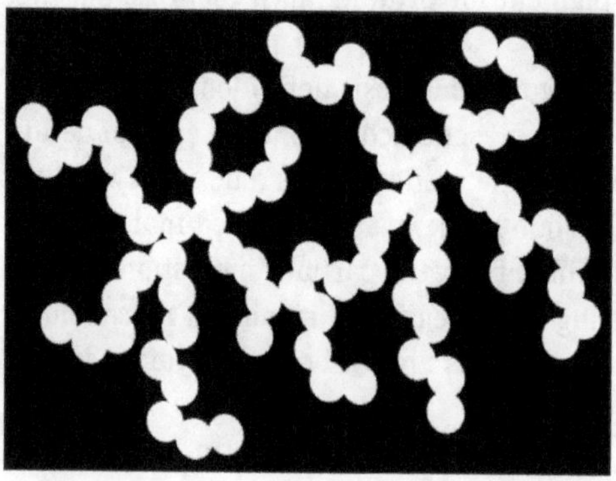

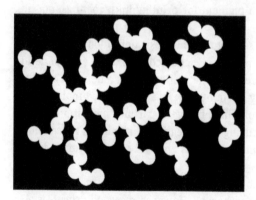

**Figure 5.16:** Electron micrographs of silica.

The above Figure 5.16 is a transmission electron micrograph of precipitated silica. It is surprising, in spite of the significant difference in manufacturing methods, that the morphology of silica is similar to that of carbon black, except that silica fusion tends to be a little more extensive. Recent work indicates that highly dispersible silica can be obtained through optimum precipitation conditions of high pH value with short drying time. Electron micrographs of silica are depicted in Figure 5.16.

## 5.11.3 Reinforcing unit: cords and fabrics

### 5.11.3.1 Cord

During the last decade, there was a remarkable growth in the production of polymer fibres and especially cords made from those fibres, like rayon, nylon, polyester, and aramid. Due to their fantastic performance in different types of rubber products (hose, V-belt, conveyor belt, etc.) as reinforcement member, they are widely used [9]. Polyester is a category of polymers that contains the ester functional group in their main chain. As a specific material, it most commonly refers to a type called polyethylene terephthalate (PET). Polyester is a high-tenacity synthetic fibre with high strength and low elongation. It is UV-resistant, abrasion-resistant, and chemical-resistant. The polyester repeating units are shown in Figure 5.17.

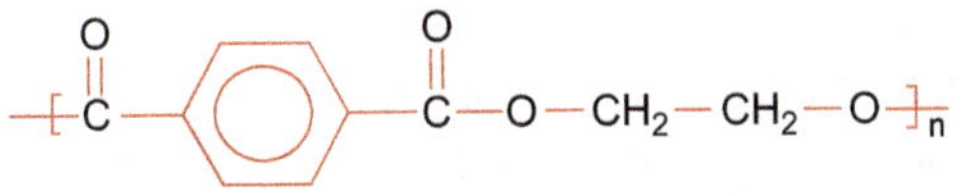

**Figure 5.17:** Polyester repeating units.

**Figure 5.18:** RFL-treated polyester cord.

Depending on the product and reinforcement, either single-end yarns or cords (several twisted yarns twisted together), or in some cases, cord-fabric form, where the cords have been assembled together into a woven cord fabric are used. For attaining better adhesion strength in the rubber products, RFL (resorcinol formaldehyde and rubber latex) dipping is widely used. And according to the application, there are different types of dipping techniques followed by industries [9]. RFL-treated polyester cord is represented in Figure 5.18.

#### 5.11.3.2 Fabrics

Commonly, there are three techniques trailed by fabric formation technology like weaving, knitting, and nonwoven. The conversion of yarn into the woven fabric is accomplished by interlacing warp and weft on a weaving machine or loom. In a weaving machine, the warp yarns are passed from a warp beam to the fabric beam.

**Figure 5.19:** Woven fabrics and knitted fabrics.

Knitting is the second most frequently used method, having various properties such as wrinkle resistance, stretchability, and better fit, mainly in demand due to the rising popularity of sportswear, and casual wears. A knitted fabric may be made with a single yarn, which is formed into interlocking loops with the help of hooked needles [9]. Images of woven and knitted fabric construction are illustrated in Figure 5.19.

##### 5.11.3.2.1 Polyester

Production of polyester is gaining popularity in fibres and fabrics, for technical applications at a lower price compared to PA and viscose fibres. It contains the ester functional group in their main chain and the chemical name is polyethylene terephthalate. Polyester has excellent resistance towards acid and moisture. Due to its excellent moisture resistance, it is used as the reinforcement fabric in water hoses. Polyester is a high-tenacity synthetic fibre with high strength and low elongation [9].

##### 5.11.3.2.2 Cotton

Cotton is 99% cellulose and cellulose is a macromolecule – a polymer made up of a long chain of glucose molecules linked by C-1 to C-4 oxygen bridges with the elimination of water (glycoside bonds). The number of repeat units linked together to form the cellulose polymer is referred to as the 'degree of polymerisation'. Cotton fabric shows excellent durability and utility, is a chemically stable material, and has high water-absorbing

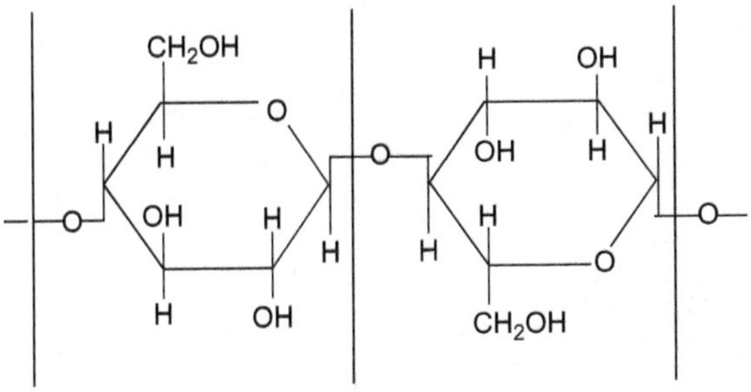


**Figure 5.20:** Cellulose chemical structures.

capacity and deficient elasticity. That is why cotton fabrics are widely used in many applications. The cellulose chemical structure is portrayed in Figure 5.20.

Depending on the product and requirement the cotton and polyester percentages are varied by the suppliers. To complete the requisite specifications, these two fabrics are blended and polycotton blended fabric is manufactured [9].

## 5.12 Manufacturing processes

1) The rubber compounds are mixed in an open mill or internal mixer.
2) The base and cushion rubber compounds are calendered into 1.0–1.5 mm sheets, while the jacketing fabric is skimmed with friction compounding 3-roll calenderer.
3) The cord is pre-stretched or heat-stretched in the dipping process. In the case of rayon cords, an RFL dip is used and in the case of polyester, a dip containing RFL and isocyanate is used.
4) The belts are then assembled in the building. There are two different methods for building V-belts [1]. The manufacturing flow chart is described in Figure 5.21.

### 5.12.1 Drum building process

- At first, the base sheet is plied without air-entrapment over an expandable drum (mandrel).
- The cord which is under tension is wound spirally from one end to the other end and then topping layer is applied over the cord.
- After that, further layers of cushion and filler rubber are built over it.
- The built-up cylinder is slit into rectangular sections of the desired width with circular rotating knives.

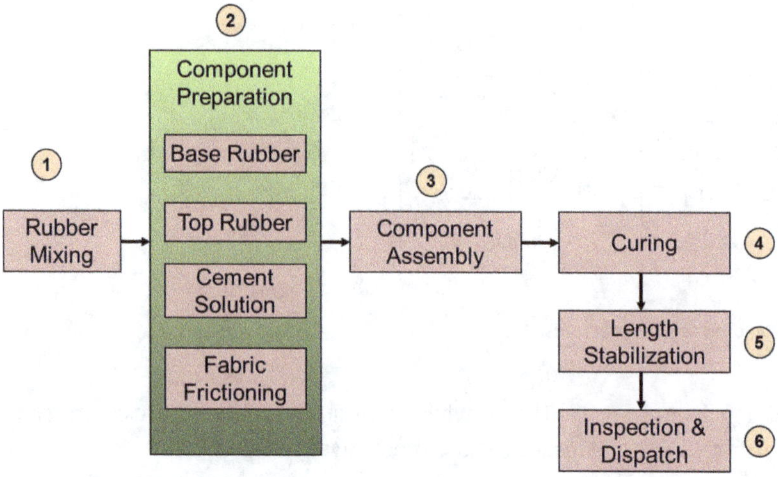


**Figure 5.21:** Manufacturing process flow chart.

- The cut sections are skived to trapezoidal shape with two circulars angularly placed rotating knives.
- The raw belts are jacketed with rubberised fabric in a jacketing machine [1].

The drum building process is portrayed in Figure 5.22.

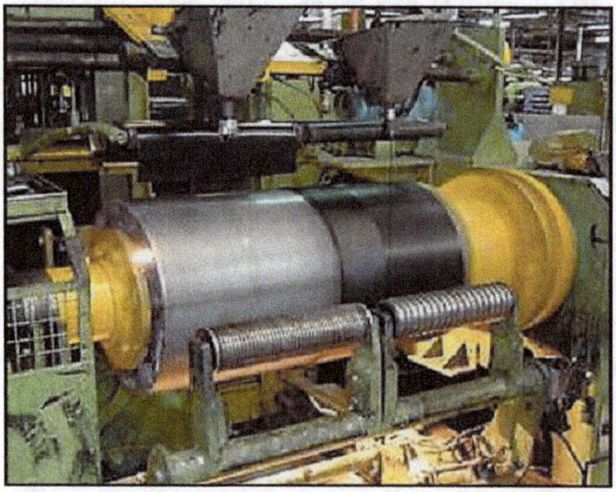

**Figure 5.22:** Drum building process.

The general manufacturing process of V-belt is demonstrated in Figure 5.23.

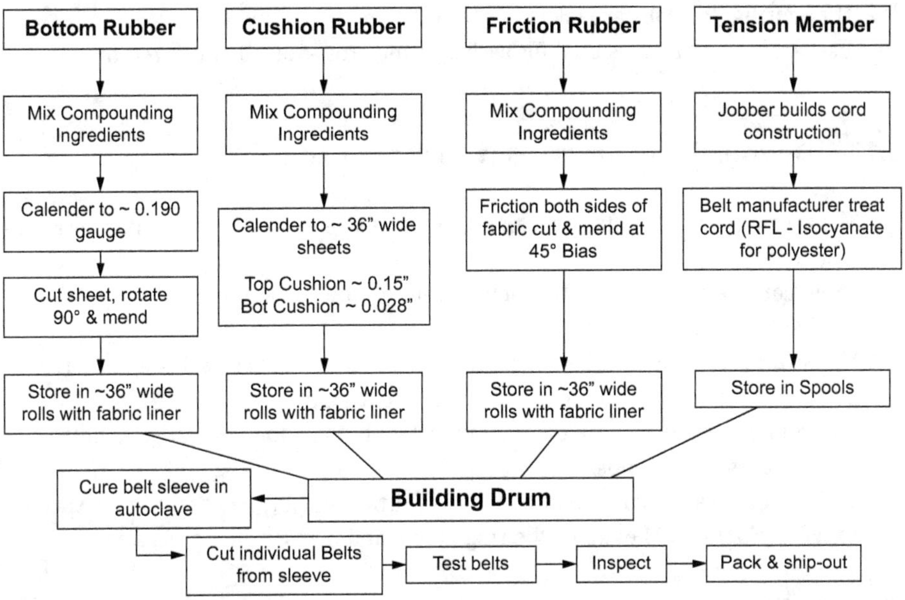


**Figure 5.23:** General manufacturing process of V-belt.

## 5.12.2 Cord band process

- The RFL- dipped and the heat-set polyester cord is encapsulated with rubber in a T-head extruder for providing tack to the cord [1].
- Cord band is formed by spirally winding the cord between adjacent strands over a two-drum winding machine where one of the drums is movable to enable to set the required cord band length.
- According to the belt section, the cord band is cut in many small bands with a certain number of cords.
- The topping is applied on one side, and the base is applied on the other side of the cord band.
- The whole assembly is covered with jacketing fabric.

## 5.12.3 Ring moulding

- Jacketed raw belts are placed between two rings of a multi-cavity circular mould.
- The moulds with the belts are then bolted up axially and wrapped with wet wrapping cloth under tension so that pressure can be exerted on the top of the belt during cure.
- The wrapped mould is then placed in an autoclave for vulcanisation in open steam.

- After curing, the belts are taken out from the mould cavity.
- A separate set of rings is used for each size and cross-section of belt required [1, 3].

### 5.12.4 Double day-light press with stretched rollers

- The green belts are loaded over the platen and stretch rollers and cured for a specific time.
- After completion of cure, the belt is rotated and a non-vulcanised portion is brought into the curing area.
- In this process, while the belt is cured in steps, between two such steps, there will be an overlapping portion that will undergo curing twice.
- To avoid overcuring in the overlapping portion, the cooling arrangement uses to provide on both the ends.
- This process can accommodate curing of different belt lengths up to a specific range [1]. Figure 5.24 exhibits the ring mould and press curing processes.

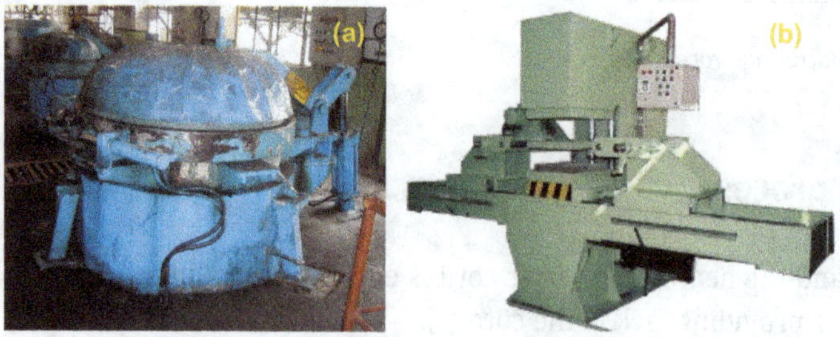


**Figure 5.24:** The v – belt curing processes using (a) Ring Mould system and (b) Flat bed press curing system.

### 5.12.5 Rotocuring

- This process is a continuous vulcanisation method.
- The belts are loaded in between a set of grooved drum and stretch roller.
- The drum is heated by steam and the pressure applied over the belts by a full band, which is electrically heated.
- The drum rotates, and thereby the belt gets vulcanised.

The polyester cord undergoes thermal shrinkage while curing the belts. To arrest this, a slight stretch is applied to the belt by stretch rollers and quenched in chilled water to arrest the residual shrinkage [1].

## 5.13 Manufacturing defects

- Bareness: caused due to lower weight, that is, material content, less pressure, lower temperature, lower cutting width, and so on.
- Pinch: the top position of the belt protrudes out of the edges because of higher weight and material content, higher pressure, free flipping, and so on.
- Jacket gap: Jacket fabric not overlapped at the bottom of the belt.
- Surface defect: This is caused due to entrapment of foreign matter on the belt surface, the formation of the crease.
- Air/spongy: it happens due to improper application of pressure and temperature during curing.
- Seam up: Jacket fabric is overlapped more than required on the driving surface.
- Knuckle: It occurs due to hot coolant.
- Building: Top surface of the belt becomes convex-shaped resulting in the cord position of the belt building upward.
- Twisting: Number of cords comes down towards the base resulting in twisting of cured belts.
- Neckless: The surface of the belt becomes down, as a result, the cord layer comes down [3].

## 5.14 Development of V-belt compound

There are a wide variety of materials required for V-belt compounds. Traditional V-belts were based on general-purpose rubbers like NR and SBR for wrapped type belts. The friction compound usually made from polychloroprene provides some resistance to oils, grease, ozone, and flame. With the development of raw edge belts, all compounds – base, cushion, and friction are based on polychloroprene. From the commercial grades of polychloroprene available, the sulphur-modified G grades are most suitable. The base compound, especially in case of raw edge belts is made hard and firm by the addition of cotton or synthetic fibre flock. The cushioning compound may contain dry bonding agents like HRS (*Hexa Resorcinol Silica*) to get adhesion with polyester cords [10–13]. Importance of the different rubber compound layers in the V-Belt has been tabulated in the Table 5.3 [15].

### 5.14.1 Design criteria for V-belt compound

1. Low elongation for better set properties in service.
2. Lower growth during power transmission.
3. Right wedge action for better power transmission.
4. Low heat generation for higher belt life.

5. Low slip for less heat generation and more power transmission by effective utilisation.
6. High tensile strength for better power rating and life.
7. Cushion compound should have low heat development and better heat dissipation characteristics.
8. The compound should have superior resistance to water, chemicals, and heat ageing [12].

## 5.15 Essential requirement for various components

**Table 5.3:** Importance of the different rubber compound layers in the V-Belt [15].

| Base compound | Gum compound | Jacket fabric | Reinforcing members |
|---|---|---|---|
| High modulus | Good tear resistance | Good wear resistance | High tenacity |
| Low dynamic heat buildup | Good interface adhesion | Good heat ageing property | Good fatigue resistance |
| Resistance to flex cracking | Good heat ageing property | Good oil resistance property | Low elongation |
| Good heat ageing property | Good cured adhesion to the cord | Good cured adhesion with base rubber and gum rubber | Good thermal stability |
| Good hardness | | High coefficient of friction | Good bending resistance |
| Good interface adhesion | | | Low moisture absorption |
| | | | Low creep |
| | | | Better adhesion with rubber compounds |

## 5.16 Compound development

To troubleshoot the failure, some points that play a significant role in the service life of the belts come up. They are:

A. Base rubber compound
B. Fabric type
C. Friction compound

As our target is to increase service life, so there are many approaches for developing the rubber compounds which can improve the service life of a V-belt. The findings can be like: (1) decreasing the HBU of the compounds and belts during applications, (2) increasing the fatigue life of the compound and as well as a belt, (3) increasing the adhesion strength between the fabric to rubber and cord to rubber [11, 14] etc. In support of better understanding, some case studies are discussed below on the above three points.

### Case 1: To decrease the heat buildup

**Table 5.4:** Formulation of a V-belt base rubber compound.

| Ingredients[a] | CB | | | | | | RR | | | | |
|---|---|---|---|---|---|---|---|---|---|---|---|
| | 1 | 2 | 3 | 4 | 5 | 6 | 1 | 2 | 3 | 4 | 5 |
| NR | 100.0 | 0.0 | 55.0 | 55.0 | 55.0 | 55.0 | 45.0 | 40.0 | 35.0 | 35.0 | 30.0 |
| SBR | 0.0 | 100.0 | 45.0 | 45.0 | 45.0 | 45.0 | 55.0 | 60.0 | 60.0 | 55.0 | 55.0 |
| BR | | | 0.0 | | | | 0.0 | 0.0 | 5.0 | 10.0 | 15.0 |
| Silica | | | 5.0 | | | | | | 15.0 | | |
| HAF | 20 | 20 | 25 | 25 | 30 | 30 | | | 25.0 | | |
| FEF | 50 | 50 | 50 | 45 | 40 | 35 | | | 35.0 | | |
| SRF | 15 | 15 | 15 | 20 | 20 | 25 | | | 20.0 | | |
| Wood resin | | | | | | 3.0 | | | | | |
| Sulphur | | | | | | 3.0 | | | | | |
| Accelerator (MBTS) | | | | | | 2.0 | | | | | |
| Accelerator (TMTD) | | | | | | 0.25 | | | | | |

[a]This indicates that the amounts are in PHR.

**Table 5.5:** Physical properties of the CB series compounds.

| Sample name | Condition | Tensile strength in MPa | Retention of TS % | EB % | Modulus in MPa | | | Hardness (IRHD) |
|---|---|---|---|---|---|---|---|---|
| | | | | | 50% | 100% | 200% | |
| CB01 | Initial | 12.4 | 92 | 262 | 2.7 | 4.9 | 9.9 | 80 |
| | Ageing | 11.4 | | 194 | 3.5 | 6.5 | 11.5 | 80 |
| CB02 | Initial | 14.8 | 100 | 242 | 3.1 | 6.0 | 12.9 | 81 |
| | Ageing | 14.8 | | 187 | 4.0 | 7.9 | – | 84 |
| CB03 | Initial | 13.8 | 99 | 254 | 3.1 | 5.7 | 11.8 | 82 |
| | Ageing | 13.6 | | 182 | 4.2 | 8.0 | – | 83 |
| CB04 | Initial | 14.0 | 97 | 273 | 2.6 | 5.1 | 11.0 | 80 |
| | Ageing | 13.6 | | 197 | 3.6 | 7.1 | 8.6 | 81 |
| CB05 | Initial | 13.5 | 93 | 240 | 3.0 | 5.8 | 12.3 | 81 |
| | Ageing | 12.6 | | 156 | 4.2 | 8.3 | – | 84 |
| CB06 | Initial | 13.3 | 96 | 237 | 3.0 | 5.6 | 11.8 | 84 |
| | Ageing | 12.8 | | 169 | 4.1 | 7.8 | – | 83 |

Case study 1 formulation has been tabulated in the Table 5.4 [34]. Table 5.5 demonstrates the mechanical properties of the samples before and after ageing of CB series and similarly Table 5.6 depicts the RR series [34]. If tensile strength is the only concern, then CB02 (100% SBR) would be the best one. But 100% SBR compound will be costlier than regular compound; that is why it cannot be considered for the next step. And, according to our application, elongation at break, modulus, retention percentage after ageing, hardness, and abrasion resistance are highly considered, and in that case, CB04 is the best compound [15–18].

Both CB02 and CB04 show excellent ageing retention of tensile strength. For the compound CB06, hardness is excellent and after ageing it does not change so much, but it shows weak tensile strength and elongation at break.

In the next series, the carbon black percentages of the compound CB04, NR, and SBR blend compound (HAF: FEF: SRF = 25:45:20) show excellent physical properties. Some amount of carbon black can be replaced by silica to improve the mechanical properties and also NR by polybutadiene (BR) rubber to reduce HBU and anti-mould sticking [17–20].

**Table 5.6:** Physical properties of the RR series rubber compounds.

| Sample name | Condition | Tensile strength in MPa | Retention of TS (%) | E B % | Modulus in MPa | | | Hardness (IRHD) |
|---|---|---|---|---|---|---|---|---|
| | | | | | 50% | 100% | 200% | |
| RR01 | Initial | 14.7 | 101 | 269 | 2.7 | 5.2 | 11.5 | 82 |
| | Ageing | 14.8 | | 216 | 3.5 | 6.6 | 14.0 | 81 |
| RR 02 | Initial | 12.8 | 119 | 209 | 3.1 | 5.8 | 11.6 | 83 |
| | Ageing | 15.2 | | 225 | 3.6 | 6.8 | 14.0 | 80 |
| RR03 | Initial | 14.7 | 97 | 213 | 3.3 | 6.3 | 13.2 | 82 |
| | Ageing | 14.3 | | 188 | 3.7 | 7.7 | 15.0 | 83 |
| RR04 | Initial | 14.7 | 98 | 217 | 3.5 | 6.8 | 13.9 | 83 |
| | Ageing | 14.4 | | 160 | 4.1 | 8.2 | – | 83 |
| RR05 | Initial | 14.5 | 97 | 239 | 3.0 | 5.8 | 12.4 | 82 |
| | Ageing | 14.1 | | 178 | 4.1 | 8.1 | – | 82 |

The HBU ($\Delta T$ in °C) results are as follows.

**Table 5.7:** Heat buildup data of all the rubber compounds for Case 1.

| Sample | 0 min | 5 min | 10 min | 15 min | 20 min | 25 min |
|---|---|---|---|---|---|---|
| CB 01 | 52 | 19 | 23 | 24 | 25 | 25 |
| CB 02 | 52 | 21 | 24 | 25 | 26 | 26 |
| CB 03 | 52 | 19 | 23 | 24 | 25 | 26 |
| CB 04 | 52 | 18 | 22 | 23 | 23 | 24 |

**Table 5.7** (continued)

| Sample | 0 min | 5 min | 10 min | 15 min | 20 min | 25 min |
|---|---|---|---|---|---|---|
| CB 05 | 52 | 21 | 26 | 26 | 27 | 27 |
| CB 06 | 52 | 21 | 26 | 27 | 27 | 27 |
| RR 01 | 52 | 15 | 21 | 21 | 22 | 24 |
| RR 02 | 52 | 16 | 20 | 21 | 22 | 24 |
| RR 03 | 52 | 19 | 23 | 25 | 25 | 25 |
| RR 04 | 52 | 17 | 21 | 22 | 22 | 22 |
| RR 05 | 52 | 21 | 24 | 24 | 24 | 24 |

The Heat build-up (HBU) data tabulated in the Table 5.7, shows that as the lower particle size carbon black is decreasing, HBU also decreases; the reason may be the better dispersion for higher particle size carbon black. Again for the next series of compounds for the addition of BR, the HBU decreases [34].

**Case 2: To increase the adhesion strength**
Here, two different resins can be used to increase the adhesion strength between the compound and fabric and compound and cord. Earlier there was one resin. Another variation is that the amount of silica is increased to augment the strength and to decrease the HBU and the total carbon black amount can be reduced from 80 to 75 PHR [21–24]. Formulation of the case study 2, CF series tabulated in the Table 5.8 [34].

**Table 5.8:** Formulations of the rubber compounds for Case 2.

| Ingredients | CF | | |
|---|---|---|---|
| | 01 | 02 | 03 |
| NR | | 35.0 | |
| SBR | | 55.0 | |
| BR | | 10.0 | |
| Silica | 15.0 | 20.0 | 25.0 |
| Silane | | 1.0 | |
| HAF | | 23.0 | |
| FEF | | 33.0 | |
| SRF | | 19.0 | |
| Wood resin | | 2.0 | |
| CI resin | | 2.0 | |
| Sulphur | | 3.0 | |
| Accelerator (MBTS) | | 2.0 | |
| Accelerator (TMTD) | | 0.25 | |

**Table 5.9:** Physical properties of the CF series compounds.

| Sample name | Condition | TS in MPa | Retention of TS (%) | EB (%) | Modulus in MPa | | | Hardness (IRHD) |
|---|---|---|---|---|---|---|---|---|
| | | | | | 50% | 100% | 200% | |
| CF01 | Initial | 14.4 | 87 | 285 | 2.7 | 4.8 | 10.6 | 82 |
| | Ageing | 12.5 | | 188 | 3.6 | 7.1 | – | 82 |
| CF02 | Initial | 14.7 | 88 | 277 | 2.9 | 5.3 | 11.3 | 83 |
| | Ageing | 12.9 | | 168 | 3.9 | 7.6 | – | 83 |
| CF03 | Initial | 13.2 | 94 | 263 | 2.9 | 5.1 | 10.6 | 85 |
| | Ageing | 12.4 | | 176 | 3.9 | 7.3 | – | 86 |

**Table 5.10:** Heat buildup data of all the compounds for Case 2.

| | 0 min | 5 min | 10 min | 15 min | 20 min | 25 min |
|---|---|---|---|---|---|---|
| CF 01 | 52 | 19 | 20 | 20 | 22 | 22 |
| CF 02 | 52 | 19 | 20 | 19 | 21 | 21 |
| CF 03 | 52 | 19 | 20 | 19 | 19 | 19 |

Table 5.9 shows that CF01 and CF02 compounds show good tensile strength but CF03 shows less. The reason maybe that higher filler agglomeration presents in CF03 compound because of a higher amount of silica. And it is easily visible that, as the amount of silica increases, elongation at break also decreased. On the other hand, modulus (50%, 100%, and 200%) and hardness (IRHD) do not vary significantly with respect to the other properties [34].

In the compound series CF, it shows that here due to addition of silica and increasing the amount of the silica by replacing carbon black, the HBU also decreases, but changes are very less as shown in Table 5.10. The effect of the addition of silica in the compound is that the adhesion strength also increased as illustrated in Table 5.11. And in the last series, resin increased and the adhesion strength also improved [26–28].

**Table 5.11:** Adhesion strength results for Case 2.

| Properties | CF01 | CF02 | CF03 |
|---|---|---|---|
| Fabric adhesion strength (180° Pell test) [ASTM D3330] in kgf | 17.12 | 14.80 | 19.27 |
| Cord adhesion strength [ASTM D855] in kgf | 5.76 | 3.42 | 4.39 |

## 5.17 DMA results

In Table 5.12, it is observed that the tan $\delta$ value is decreasing and if the tan $\delta$ value is less HBU also will be less, which is the confirmation of the HBU data discussed earlier [29, 30].

From the Figure 5.25, it can be observed that tan $\delta$ value is shifting downwards from **CB04** > **RR04** > **CF01**. This means that the tan $\delta$ value is decreasing and as the tan $\delta$ value is less HBU also will be less [31].

**Table 5.12:** DMA data of the selected compounds.

| Samples | tan $\delta_{max}$ | Temperature |
|---|---|---|
| CB 04 | 0.438 | −31.68 |
| RR 04 | 0.375 | −34.84 |
| CF 01 | 0.351 | −29.04 |

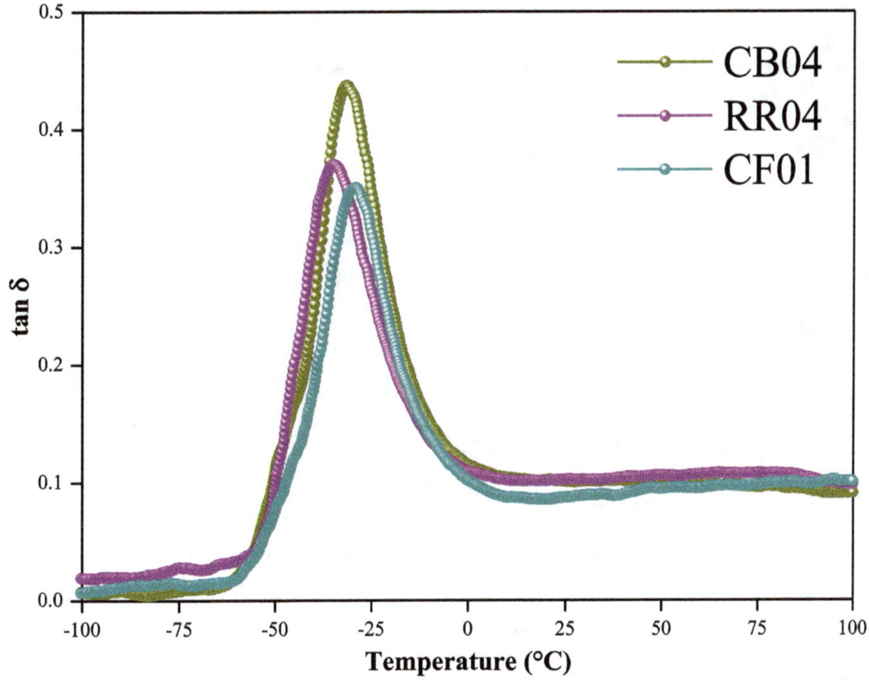


**Figure 5.25:** DMA data of the selected compounds.

## 5.18 Cost of the compounds

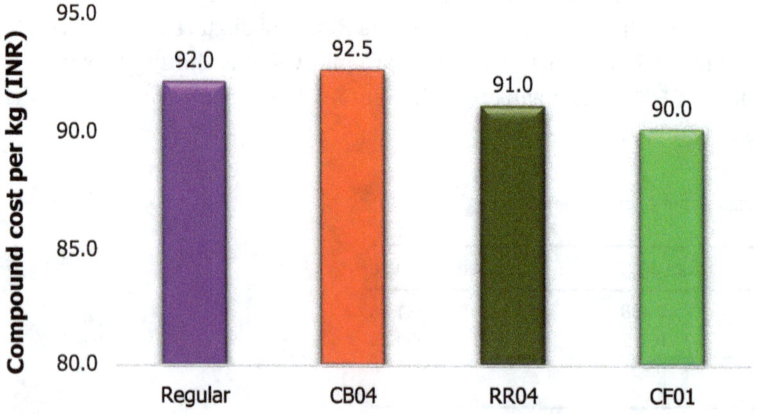


**Figure 5.26:** Compound costs comparisons.

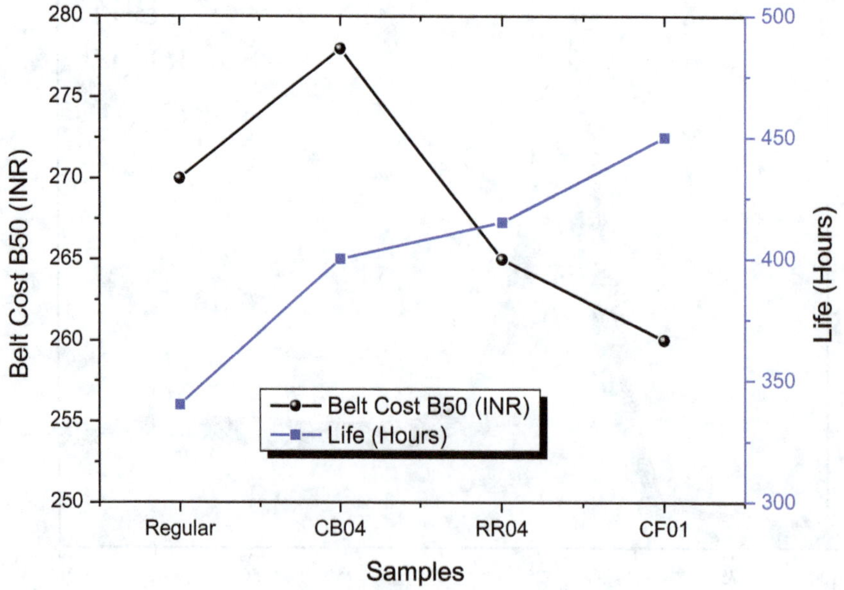


**Figure 5.27:** Comparison between belt cost per kg and belt life.

The compound cost has been optimised and calculated by variation of ingredients featured in Figure 5.26. Figure 5.27 depicts the belt cost and belt life comparison. It can be seen that the belt manufactured with CF01 base compound shows the best life as well as the hidden cost per piece. It is not only these two parameters; we have seen that the other properties also improved for this compound [32–34].

## References


[1] Erickson, W. D. Belt Selection and Applications for Engineers, Denver: GATES Rubber Company, Marcel Dekker Inc. (1987).
[2] Bhowmick, A. K., Hall, M. M., Benerey, H. A. Rubber Products and Manufacturing Technology, New York: Marcel Dekker, Inc. (1994).
[3] Gent, A. N. Engineering with Rubber, 3rd edition, HANSER Publishers (2012).
[4] Morton, M. Rubber Technology. Van Nostrand Reinhold Company, Inc. (1973).
[5] De, S. K., White, J. R. Rubber Technologist's Handbook, vol. 1, Rapra Technology Limited (2001).
[6] White, J., De, S. K., Naskar, K. Rubber Technologist's Handbook, Vol. 2, Rapra Technology Limited (2009).
[7] Donnet, J. B. Carbon Black: Science and Technology, CRC Press (1993).
[8] Rodgers, B. Rubber Compounding: Chemistry and Applications, CRC Press.
[9] Wootton, D. B. The Application of Textiles in Rubber. Rapra Technology Ltd. (2001).
[10] Roberts, A. D. Natural Rubber Chemistry and Technology, New York: Oxford University Press (1988).
[11] Fern, M. J., et al. Thermochimica Acta Elsevier 444:65–70 (2006).
[12] Beatty, J. R. Fatigue of Rubber. B. F. Goodrich Research Centre, OHIO, Rubber Chemistry and Technology.
[13] Mars, V., Fatemi, A. Rubber Chemistry & Technology 77:391–412 (2004).
[14] Patent: Power Transmission Belts, US 3122934, (1960).
[15] Patent: V-belt mechanism, US 4604082 A, (1986).
[16] Singh, G., et al. IOSR-JMCE 12(5):Ver. I 60–65 (2015).
[17] Sundararaman, S. Temperature-dependent fatigue-failure analysis of V – Ribbed serpentine belts. International Journal Fatigue 31(8):1262–1270 (2009).
[18] Chandrashekhara, K., et al. Asme 127:Nov-2005.
[19] Wongwitthayakool, P., Saeoui, P. Prediction of heat build-up behaviour under high load by use of conventional viscoelastic results in carbon black filled hydrogenated nitrile rubber. Plastics Rubber & Composites 40: (2011).
[20] Fanzhu, L., Liu, J. Numerical simulation and experimental verification of heat build-up for rubber compounds. Polymer 101(199–207): (2016).
[21] Medalia, A. I. Heat generation in Elastomer compounds: Causes and effects. Rubber Chemistry and Technology 64:481–492 (1991).
[22] Le Saux, V., Marco, Y. An energetic criterion for the fatigue of rubbers: An approach based on a heat build-up protocol and µ-tomography measurements. Procedia Engineering 2(1):949–958 (2010).
[23] Okado, Y., et al., U.S. Patent No. 4,360,627. 23 Nov. 1982.
[24] Sato, S., et al., U.S. Patent No. 4,477,621. 16 Oct. 1984.0.
[25] Saikrishna,, et al. Finite elements in analysis and design. 870–878 (2007).
[26] Srivastava, et al. Mechanism and Machine Theory 43(4):459–479 (2008).
[27] Milan,, Bani, S. Prediction of heat generation in rubber or rubber-metal springs. Thermal Science 16(2):S527–S539 (2012).
[28] Yohan Et, A. ECCMR 2007, London: Taylor and Francis Group (2007).
[29] Sarkar, P. P., Ghosh, S. K., Gupta, B. R., Bhowmick, A. K. International Journal of Adhesion and Adhesives (1):26–32 (1989).

[30] Sundararaman, S., Saikrishna. Mode-I fatigue crack growth analysis of V-ribbed belts. Finite Elements in Analysis and Design 43(11):870–878 (2007).
[31] Kar, K. K., Bhowmick, A. K. Hysteresis loss in filled rubber vulcanizates and its relationship with heat generation. Journal of Applied Polymer Science 64(8):1541–1555 (1997).
[32] Rajesh, C., et al. Cure characteristics and mechanical properties of short nylon fibre-reinforced nitrile rubber composites. Journal of Applied Polymer Science 92(2):2004 (1023–1030).
[33] Andriyana, A., Verron, E. International Journal of Solids and Structures 44:2079–2092 (2007).
[34] Barlow, F. Rubber Compounding, 2nd edition, New York: Marcel Dekker (1994).

Timir Baran Bhattacharyya

# Chapter 6
# The Science of Rubber Conveyor Belt: A Comprehensive Guide'

## 6.1 Introduction

Conveyor belts are the backbone of modern heavy industry. It transports material from one place to another. From the perspective of economy, conveyors perform the redundant tasks historically done by animals, people, handcarts, wagons, and trucks. From conveying integrated circuits over only a few inches to millions of tons of bulk material over many miles, conveyor systems are the backbone of global product handling [1].

Nowadays, conveyor belt is one of the most important and essential items that can be seen in most industries. In any modern plants, we can see that many conveyors are carrying different materials from one place to another, crossing one or more conveyors. This can be compared to the human body. Human body is full of veins that circulate blood from the heart to every part of the body and also carry all essential body nutrients and supply them to that particular part of body that needs the same. Likewise, a conveyor belt carries material from one position to another position inside plant where it is required. Hence, a conveyor belt can be called as the lifeline of a modern plant. This can be very easily understood from the photo of some modern plants given in Figure 6.1.

In this chapter, we have discussed construction, selection, and design of conveyor belts. Testing, splicing, and maintenance also have been discussed in brief.

**Figure 6.1:** Conveyors in modern industry.

https://doi.org/10.1515/9783110668537-006

## 6.2 History of the conveyor belt

As far as knowledge goes, the first conveyor was a manually operated small-distance conveyor used to shift husk of a husking mill. This was a single ply fabric stitched with its ends together. The pulley was rotated by hand. The name of the inventor of the first conveyor belt has sadly been lost to time.

After the Industrial Revolution in Great Britain during the eighteenth century, and then a little bit later in America, manual labour began to be phased out in many production environments and steam-powered machineries were favoured. Steam power was the 'hot' new tech at that time! The first steam-operated conveyor was put into use by the British Navy in the year 1804.

One may think that this new, machine-driven conveyor would have been used for loading ships but it was actually located in a bakery operation that produced biscuits for sailors to eat! In any case, this improvement meant that conveyors no longer had to be hand-cranked, which made them useful for more applications.

Seeing the advantage of this belt, people started developing various kinds of conveyors. In the early twentieth century, Henry Ford used conveyors in his automobile assembly lines. He did not invent them (as many people incorrectly believe) – he just improved upon the old technology. In 1844, Charles Goodyear patented vulcanisation of rubber. This invention brought a big change in the history of conveyor belt. People started making conveyor belts with rubber on top to protect the carcass.

In 1892, Thomas Robins introduced a conveyor belt for bulk material handling and since then slowly the conveyor belt took its position in each and every industry [2, 3]. Machine-driven conveyors caught on quickly and began appearing in all sorts of industries, though it would still be almost 100 years before they would be put to work in mining operations. Railcars were still the preferred method of moving aggregate and coal from within mines to surface operations for much of the 1800s.

This began to change though as new belt materials like rubber and steel appeared. In 1905, a conveyor belt was first used in coal mines by Richard Sutcliffe, which ultimately revolutionised the mining industry [4]. With the development of new man-made synthetic fibre and also because of development of rubber technology, the strength and simultaneously the length of a conveyor belt increased day by day.

Till now, the longest single haulage international conveyor belt is a steel cord conveyor belt between Meghalaya of India and Chhatak, Sylhet of Bangladesh; the distance is 17 km [5–7]. The belt carries limestone from one quarry of Meghalaya to a cement plant in Bangladesh. Seven kilometers of the belt is in India and 10 km is in Bangladesh. With multiple transfer points, the longest conveyor in the world is 98 km [8], situated in Western Sahara. It was built in 1972 by Friedrich Krupp GmbH (now ThyssenKrupp) from the phosphate mines of BouCraa to the coast, south of El-Aaiun.

Today, conveyors continue to improve and adapt as technology advances. As the industry applies new technology, conveyors have become more capable of handling complex and customised applications with the help of computerisation and modern technology. Nowadays, a conveyor belt is the most important and essential equipment in any industry.

## 6.3 Construction of a conveyor belt

A conveyor belt is basically a composite that consists of two parts: (i) the reinforcing material [9, 10] and (ii) the cover material [11, 12]. The reinforcing material or the carcass takes the tension or load and the cover material is to protect the carcass. The reinforcing material can be single ply or multiply – mostly three or four ply. The reinforcing material is basically a woven fabric having both warp and weft made of cotton, nylon, polyester, aramid or steel cord. Steel cord belts are mostly single ply. The carcass materials are bound through a skim compound placed between plies to construct the carcass. The cover material is then placed and fused together through the process of vulcanisation [13]. This can be understood from the cross section of conveyor belt as illustrated in Figure 6.2.

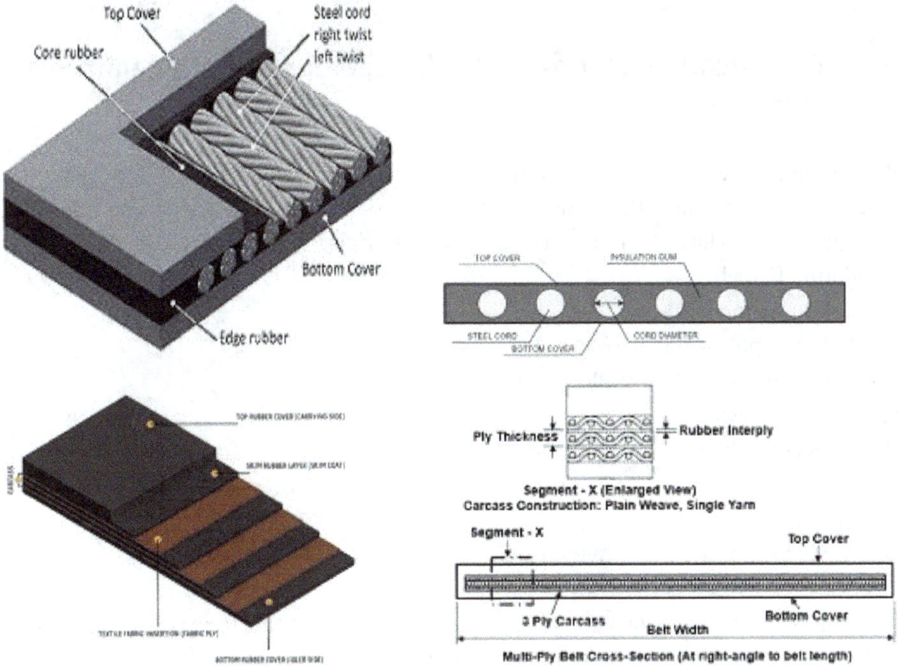


**Figure 6.2:** Cross section of both textile and steel cord conveyor belts with identification of different parts of the belt.

The width of a conveyor belt can be made as minimum as possible. However, generally, industrial conveyor belts have width between 400 and 3,600 mm. The length depends upon the requirement of the industry. The length of a conveyor can be few kilometres as mentioned above. However, while manufacturing, there are limitations in length, depending on the plant capacity. In general, conveyor belts are made in lengths from 100 to 500 m, depending on the customer's requirements. Longer lengths can reduce the number of splice joints in a long-length conveyor.

## 6.4 Classification of conveyor belts

Conveyor belts can be classified into different categories in different ways, viz. (i) Depending on the reinforcing material or the carcass material, (ii) depending on the protecting material, (iii) depending on the grade of the cover, and (iv) depending on the contour or configuration of the conveyor. We shall discuss below those classifications in detail. Here, in this chapter, we shall discuss mainly the classification depending on the carcass material and depending on the grades of the cover rubber.

### 6.4.1 Classification of conveyor belt depending on the reinforcing material

Reinforcing material is the most important part of a conveyor belt [14]. It bears the tension load and it also takes the impact [15]. Depending on the usage, the following varieties of materials are used as reinforcing materials:

(a) Cotton (C or B)
(b) Cotton–nylon (CN)
(c) Nylon or polyamide (NN or PP)
(d) Polyester–polyamide (EP)
(e) Aramid (Kevlar)
(f) Steel cord (SC)

#### 6.4.1.1 Cotton fabric for conveyor belt

The first modern conveyor belt was produced using cotton fabric as reinforcing material. Two or three layers of cotton fabric were used. A thin layer of rubber was used as a skim compound between the plies and a rubber layer was used on the top and bottom of the carcass [16]. The life of those conveyor belts was not very long because cotton was susceptible to attack by moisture and atmosphere [17]. Moreover, the adhesion between the plies and between the cover and the plies was not very high because there was no chemical bonding between the rubber and the cotton fabric [18]. In the case of cotton belting, the bonding between the rubber and the fabric was mechanical bonding. Cotton is a staple fibre [19].

It has so many anchoring points. This is demonstrated in Figure 6.3. Those anchors helped for mechanical bonding. Cotton carcass-made conveyor belt or so-called cotton conveyor belts were very much predominant in the Indian industry till the seventh decade of the twentieth century. Different kinds of fabrics were used, depending on the weight and strength of the fabric, viz., 28 oz, 32 oz, 34 oz, 36 oz, 42 oz, etc. Depending upon the number of plies, belts were designated as 4 ply 32 oz or 4 × 32 oz, 4 × 42 oz, etc.

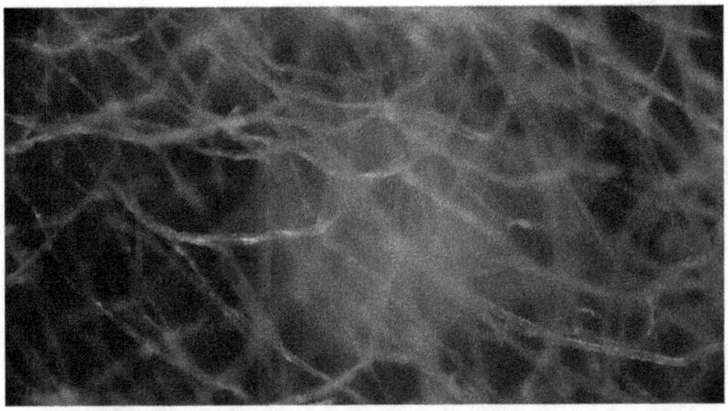

**Figure 6.3:** Cotton fibres under a compound microscope.

### 6.4.1.2 Nylon fabric for conveyor belt

Since the life of a cotton carcass-made conveyor belt was not very high [20], scientists were in search of a better carcass that will give more life than that of a cotton carcass-made belting. At this point, newly developed synthetic fibre came up – this was a synthetic polyamide fibre. The name of this polyamide was NYLON [21]. Two kinds of nylon were developed: one was called Nylon 6 and the other was called Nylon 66. Nylon 6 is made from one monomer that has six carbon atoms and Nylon 66 is made of two monomers, with each having six carbon atoms [22], which are illustrated in Figure 6.4.

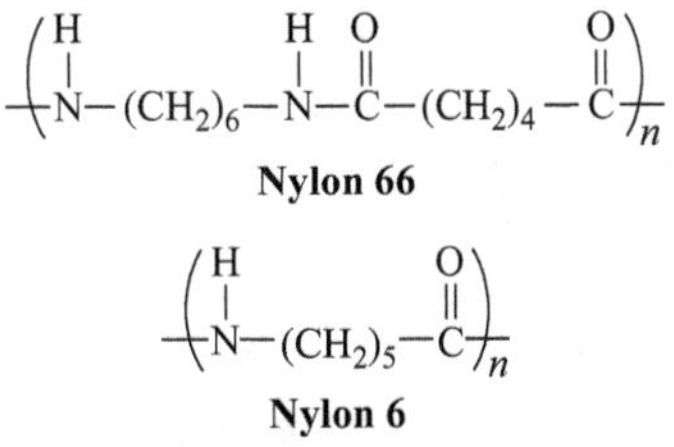


**Figure 6.4:** Structure of Nylon 6 and Nylon 66.

**Figure 6.5:** Nylon fibre under a compound microscope.

Nylon fibre has a higher strength and it does not lose its strength due to moisture attack [23]. The synthetic fibre was continuous and hence the surface was smooth, and there was no possibility of mechanical anchorage with rubber. Also, the fibre surface did not have any property to make chemical bonding with rubber. This is illustrated in Figure 6.5.

#### 6.4.1.3 Cotton–nylon fabric for conveyor belt

Although, there were so many advantages as far as nylon fibre is concerned, this newly developed fibre could not find its use in conveyor belts due to its low adhesion property with rubber [24]. Utilising the advantages of the newly developed nylon yarn, scientists started developing a new fabric, which was called CN fabric. In this CN fabric, nylon threads were twisted together with cotton thread in the warp direction. Generally, in the weft direction, no nylon threads were used. An image of a CN fabric is shown in Figure 6.6.

The advantage of use of nylon thread along with cotton thread is its higher strength. Hence, with the use of cotton nylon fabrics, the carcass was of lower thickness, lesser weight, and of higher strength. But with the newly developed CN fabric, adhesion was not improved. Hence, the life of the conveyor belt was also not improved.

#### 6.4.1.4 RFL-dipped nylon fabric

The newly developed nylon yarn had many improved properties like high strength, low weight, high temperature stability, low elongation, and high resistance to moisture attack. Scientists started searching for a method to improve the adhesion between rubber and the nylon fabric. Finally, a new method called RFL treatment of nylon fabric was developed. RFL stands for "resorcinol formaldehyde latex"" [25]. After a nylon fabric is dipped into an RFL solution and then dried, the nylon fabric

**Figure 6.6:** Cotton nylon fabric.

achieves bonding property with rubber. This bonding between the RFL-treated fabric and rubber being a chemical bonding [26], its strength is much higher than that of the bonding strength between the cotton fabric and rubber. Photos of the undipped and RFL-dipped nylon fabrics are depicted in Figures 6.7 and 6.8.

Therefore, a long pending problem was solved. A conveyor belt having nylon fabric made from carcass has less weight, less thickness, higher strength, higher adhesion, is not affected by moisture, has higher heat resistance property, etc.

**Figure 6.7:** Nylon fabric for a conveyor belt.

Due to so many advantages, nylon or polyamide replaced cotton fabric-made conveyor belts very fast and within one or two decades, most of the cotton conveyor belts were replaced by polyamide or nylon conveyor belts.

#### 6.4.1.5 RFL-dipped polyester fabric

After the successful use of nylon yarn in conveyor belts, people started using conveyor belts in every sphere of bulk material handling. However, few more problems were yet to be solved. Since nylon yarn has a higher intermediate elongation, conveyor belts made with nylon fabric as reinforcing material used to undergo higher elongation dur-

**Figure 6.8:** Dipped nylon fabric.

ing starting and during running. So, large take-up travels were required for long-length conveyor belts. Hence, when long-length (say 2–3 km) conveyors were required, one or two transfer points were required to be made. In order to solve this problem, a new yarn called polyester yarn came into the market.

Like nylon yarn, polyester yarn also did not have any inherent property of bonding with rubber. Even only RFL treatment did not give the desired adhesion. Therefore, polyester fabrics are dipped in a dual solution called isocyanate added with RFL. After the polyester-nylon or EP or PN fabrics are dipped in this solution, its colour changes to brown, which is identical to nylon RFL-dipped fabric colour and it achieves adhesion properties with rubber. The chemical structure of polyester is shown in Figure 6.9.

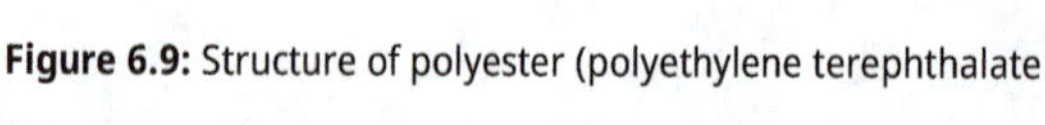


**Figure 6.9:** Structure of polyester (polyethylene terephthalate – PET).

The main advantage of this yarn was low elongation [27]. But it had some disadvantages also like low flexibility and low impact strength compared to nylon yarn. Hence, the fabric for a conveyor belt was made using polyester yarn in the longitudinal direction and with nylon 66 yarn in the transverse direction.

The main reason for using nylon yarn in weft was due to the higher flexibility requirement in the weft direction. This fabric is called the EP fabric. E stands for ester and P stands for polyamide. Although it is a combination of polyester and nylon yarn, this fabric is generally called polyester fabric.

After the introduction of polyester yarn-made fabric, the length of the conveyor belt increased 2/3-fold compared to nylon fabric-made belts. The take-up distances also came down by 2–3 times. Because of the high tenacity of polyester yarn, the strength of belt could also go as high as 2,500 kN/m.

#### 6.4.1.6 RFL-dipped aramid fabric

Here comes a new fabric called aramid fabric, which has so many advantages over nylon or polyester fabric [28]. Aramid (Aromatic Polyamide) is a high-performance fibre material with molecules that are characterized by rigid polymer chains. These molecules are linked by strong hydrogen bonds that transfer mechanical stress very efficiently, making it possible to use chains of low molecular weight. These materials are well-known and are extensively used in bullet-proof vests and car glasses [29]. Aramid's general molecular structure is shown in Figure 6.10.

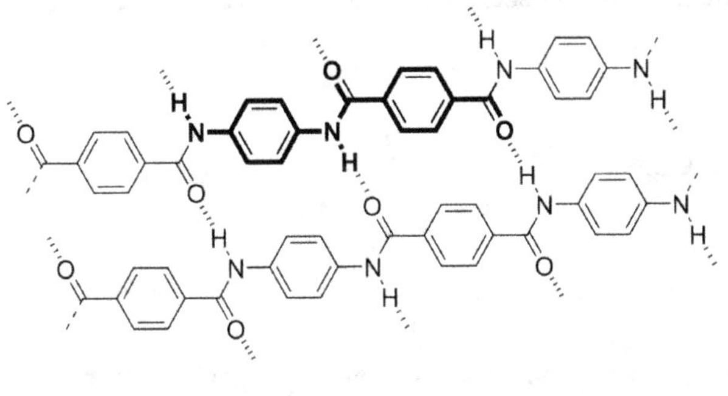

**Figure 6.10:** Structure of aromatic polyamide (Kevlar or aramid).

Nowadays, these materials are also used in the carcass material of the conveyor belt as a tension member due to its excellent strength that is comparable to steel and due to its less weight like a textile fabric. The main advantages of Kevlar-reinforced conveyor belt are listed as:

1. It is five times stronger than steel on a weight-for-weight basis.
2. It has very high strength than polyester and is comparable to steel.
3. Very low permanent elongation.
4. Excellent rip tear and impact resistance.
5. High heat and flame resistance

6. Anticorrosive.
7. Low energy consumption due to its less weight.

Comparative elongation of different fibres is shown in Figure 6.11.

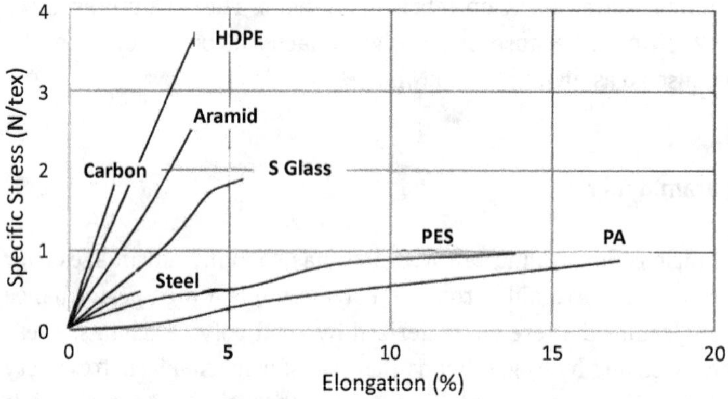


**Figure 6.11:** Comparative elongation of different fibres.

Although it has a lot of advantages, it rules itself out due to some disadvantages [30]:
1. Very high cost.
2. Very sensitive to moisture.
3. Poor resistance to UV.
4. Difficult to cut the belt at the time of splicing.

#### 6.4.1.7 Steel cord-reinforced conveyor belt

At this point of time, 'steel cord conveyor belt' was developed. In a steel cord conveyor belt, the carcass is made of steel wires of special construction.

A cross sectional image of a steel cord conveyor belt is shown in Figure 6.12.

The steel cords used in steel cord conveyor belts are made with high carbon steel. Different constructions of steel cord are used, viz., 7 × 7, 7 × 19, 19 + 7 × 7, etc.

A cross section sketch of a 7 × 7 cord is portrayed in Figure 6.13.

Each filament or wire is zinc coated to achieve bonding property with rubber.

Cords are placed alternatively (right-hand lay and left-hand lay alternatively) so that the belt is balanced properly and it does not sway towards any side during running. Construction of steel cords used for conveyor belts are different from normal steel ropes or steel wires [31]. The outer strand's diameter is always less than the inner strand's diameter. Likewise, in each strand, the outer filament diameter is always less than inner filament diameter (Figure 6.13).

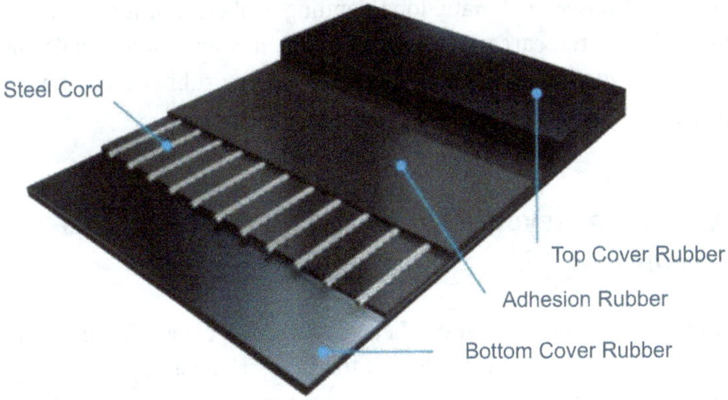


**Figure 6.12:** Cross section of steel cord conveyor belt.

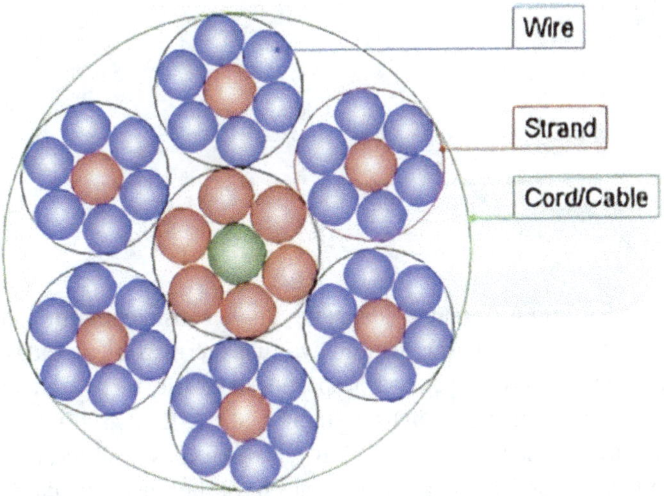


**Figure 6.13:** Cross section of a 7 × 7 steel cord.

This is to facilitate rubber penetration inside the cord. Higher the penetration of rubber into the cord, higher will be the bonding of rubber with the cord. Moreover, better rubber penetration results in less vacant space inside the cord and hence less chances of rusting of cord during the running of the belt.

After the introduction of the steel cord-reinforced conveyor belt, many long pending issues were solved, viz., length of conveyor could be increased to even 25–30 km, splicing efficiency increased to 100%, down time reduced a lot, take-up length also reduced, and over all, the life of conveyor belt increased many folds. There are many steel cord conveyor belts that are running for more than 15 years [32].

Steel cord conveyor belt has solved many long-pending problems of the industry, especially the increased life of the carcass (steel cord) demands enhanced quality of the protecting material and hence it was the responsibility of the rubber technologist to increase the quality of covers of the conveyor belts.

### 6.4.2 Classification of conveyor belt, depending on the grades of the conveyor belt

Apart from all the above varieties, a conveyor belt with rubber cover (Rubber conveyor belt) is most important and more than 95% of the industrial conveyor belts are rubber conveyor belts. Hence, in this chapter, we shall mainly provide emphasis on the various kinds of rubber conveyor belts.

Depending upon the usage and application, rubber conveyor belts are classified into the following categories:

i) General purpose
ii) Heat resistant
iii) Fire resistant
iv) Abrasion resistant
v) Oil resistant
vi) Rolling resistant
vii) Cut and gouge resistant

Each category is also divided into different grades. Like, if we consider the Indian standards for conveyor belt [33] for general purpose, we find there are two grades, viz., M24 and N17, each having different specified properties. Likewise, there are two different grades of heat-resistant conveyor belts – HR T1 and HR T2; two different grades of fire-resistant conveyor belts, FRAS and FRUG, one for surface application and the other for underground application. Most countries have their own standards for conveyor belts and of course, they have their own abbreviations. However, slowly ISO standards are becoming more popular.

A chart that shows the different national grades with their application and their physical properties are detailed in Table 6.1.

**Table 6.1:** Different grades of cover rubber and their properties.

| **Selection of cover rubber quality** | | | | | | | |
|---|---|---|---|---|---|---|---|
| **Cover type** | **Physical properties** | | | | | **Material** | **Characteristics** |
| | **Standard and grade** | **Min. tensile strength ($N/mm^2$)** | **Min. elongation @ break (%)** | **Max. abrasion loss ($mm^3$)** | **Max. temp. (°C)** | | |
| General purpose | DIN X | 25 | 450 | 120 | 70 | Iron ore, copper ore, stone, rock, etc. | Extra abrasion, cut, and gouge resistance. Suitable for transporting sharp, large lumps & rugged material under adverse loading conditions. |
| | AS M | 24 | 450 | 125 | 70 | | |
| | SANS M | 25 | 450 | 120 | 70 | | |
| | ISO H/JIS H | 24 | 450 | 120 | 70 | | |
| | IS M-24 | 24 | 450 | 150 | 70 | | |
| | DIN Y | 20 | 400 | 150 | 70 | Fine coal, ash, cement, earth, coal, salt, etc. | Abrasion resistance for normal service; suitable for transporting moderately abrasive materials. |
| | AS N | 17 | 400 | 200 | 70 | | |
| | SANS N | 17 | 400 | 150 | 70 | | |
| | IS N-17 / RMA – I | 17 | 400 | 200 | 70 | | |
| | ISO L / JIS L | 15 | 350 | 200 | 70 | | |
| | RMA-II / DIN-Z | 15 | 400 | 250 | 70 | | |
| Superior abrasion resistant | DIN W | 18 | 400 | 90 | 70 | Quarries, sandpits, limestone, coal, ash, ore, phosphate, raw material for glass works, etc. | Superior abrasion resistance for the heaviest service conditions; suitable for abrasive materials with a large proportion of fines. |
| | ISO D/JIS D | 18 | 400 | 100 | 70 | | |
| | AS A | 17 | 400 | 70 | 70 | | |

**Table 6.1** (continued)

| Selection of cover rubber quality | | | | | | | |
|---|---|---|---|---|---|---|---|
| **Cover type** | **Physical properties** | | | | | **Material** | **Characteristics** |
| | **Standard and grade** | **Min. tensile strength ($N/mm^2$)** | **Min. elongation @ break (%)** | **Max. abrasion loss ($mm^3$)** | **Max. temp. (°C)** | | |
| Heat resistant | DIN T | 12.5 | 350 | 250 | 125 | Hot sinter, hot cement, hot powder, chemicals and fertilisers, etc. | Moderate heat resistance; suitable for abrasive materials with a large proportion of fines. |
| | IS T1 | 12.5 | 350 | 250 | 125 | | |
| | IS T2 | 12.5 | 350 | 250 | 150 | | |
| Fire resistant | AS F | 14 | 300 | 200 | 70 | Materials having fire hazards, e.g., sulphur, coal, etc. | Resistance to flame propagation and burning; suitable for surface applications. |
| | CAN CSA-C | 17 | 350 | 200 | 70 | | |
| | ISO 340 | 17 | 350 | 175 | 70 | | |
| | IS-FRAS | 17 | 350 | 200 | 70 | | |
| | DIN K | 20 | 400 | 200 | 70 | | |
| | SANS F | 14 | 400 | 180 | 70 | | |
| | MSHA | 12 | 400 | 250 | 70 | Explosive and fire hazardous materials, e.g., coal powder | Resistance to flame propagation, extremely low burning rate, suitable for underground operation. |
| | DIN V | 15 | 350 | 200 | 100 | | |
| Oil and grease resistant | DIN G | 12 | 250 | 175 | 100 | Chemicals and fertiliser, wood, paper & pulp, recycling plants | Resists penetration of oil and therefore resists the damaging effect of oil and fat. |
| | AS Z | 12 | 250 | 175 | 100 | | |
| | IS OR | 12 | 250 | 175 | 100 | | |

### 6.4.3 Classification of conveyor belts depending on the contour or configuration of the conveyor

Depending on the contour or configuration, conveyor belts can be classified into the following categories:

i) Trough conveyor belt
ii) Chevron conveyor belt
iii) Side-wall conveyor belt
iv) Pipe conveyor belt
v) Sandwich conveyor belt
vi) Cable conveyor belt
vii) Bucket elevator belt

#### 6.4.3.1 Trough conveyor belt

Trough conveyor belts or generally called conveyor belts is the most popular and widely used type of conveyor belt. Out of the many different varieties of conveyor belts (depending on the contour of the conveyor), trough conveyor belts are most widely used. More than 80% conveyor belts that are used in industries are trough conveyor belts. Therefore, our discussion will concentrate on trough conveyor belt only. An image of a trough conveyor belt is shown in Figure 6.14.

**Figure 6.14:** Trough conveyor belt.

A cross-sectional diagram of a textile and steel cord trough conveyor belt is portrayed in Figure 6.15.

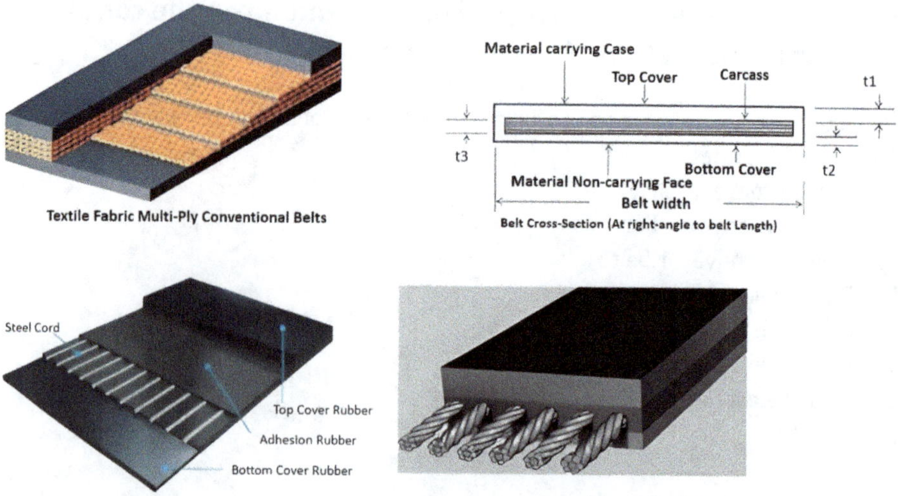


**Figure 6.15:** Cross section of a textile and steel cord trough conveyor belt.

#### 6.4.3.1.1 Designation of a trough conveyor belt with example

It is very important to designate a conveyor belt properly so that the requirement of users is properly communicated to the conveyor belt manufacturer and a correct material is received by the user. A guideline to designate conveyor belts is detailed in Table 6.2.

### 6.4.3.2 Chevron conveyor belt

Chevron conveyor belts are the right belts to meet requirements when higher angles of inclinations are required [34]. The Chevron profiles are vulcanised together with the base belt. This is very much useful for ready-mix and vegetable processing industry as well as for packaging industries. Few images with profiles are shown in Figure 6.16.

### 6.4.3.3 Side-wall conveyor belt

Side-wall conveyors are suitable for an infinite variety of applications in many industries, from food processing to mining. They can handle virtually any bulk material, and offer the advantage of being able to convey horizontally and elevate at a steep angle of incline (up to 60° and higher) on a single, continuous belt. They may be built in horizontal, inclined, L or Z configurations for lateral movement of materials. Few images and sketches of side-wall belts are portrayed in Figure 6.17.

**Table 6.2:** Designation of conveyor belts.

| Requirement of customer | | | | | | | | | | Final designation |
|---|---|---|---|---|---|---|---|---|---|---|
| Width (mm) | Stren-gth (kN/m) | No. of ply | Type of yarn | Grade of cover | Thickness of top cover (mm) | Thickness of bottom cover (mm) | Edge construction | Open-end or close-end | Length (m) | |
| 1,000 | 315 | 3 | NN | M24 | 5.0 | 2.0 | Moulded | Open end | 200 m | 1,000 × NN315/3 ME × (5 + 2) M24 × L-200 m × Open-end |
| 1,200 | 630 | 4 | EP | FR | 5.0 | 3.0 | Cut edge | Open End | 300 m | 1,200 × EP 630/4 CE × (5 + 3) FR × L-300 m × Open-end |
| 1,600 | 800 | 4 | EE | HR T2 | 6 | 3 | Cut edge | Open End | 300 m | 1,600 × EE 800/4 CE × (6 + 3) HRT2 × Cut-edge × L-300 m × Open-end |
| 1,600 | 1,250 | 4 | EE | SAR | 8 | 3 | Cut edge | Open End | 250 m | 1,600 × EE 1250/4 CE × (8 + 3) SAR × Cut-edge × L-250 m × Open-end |
| 2,000 | 1,600 | 4 | EP | M24-LRR | 8 | 3 | Cut edge | Open End | 250 m | 2,000 × EP1600/4 CE × (8 + 3) M24-LRR × Cut-edge × L-250 m × Open-end |

**Figure 6.16:** Few photographs of Chevron conveyor belt.

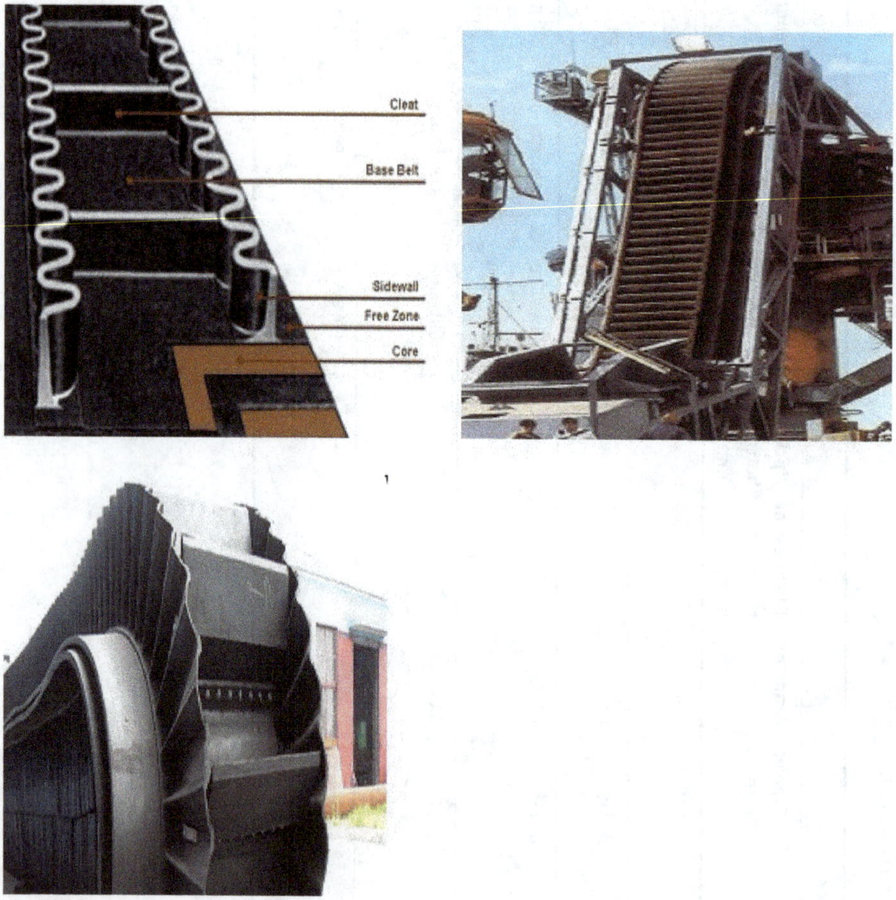


**Figure 6.17:** Few images of side-wall conveyor belt.

#### 6.4.3.4 Pipe conveyor belt

Pipe conveyors were invented in Germany in 1956, but have been utilised only in the last three decades. In fact, pipe conveyor has revolutionised the bulk material handling procedure by adopting an enclosed material transfer principle [35]. An illustrated image of a pipe conveyor belt is shown in Figure 6.18.

**Figure 6.18:** Illustrated image of pipe conveyor belt.

**Figure 6.19:** A closer look at the pipe conveyor belt.

With an overriding stress on maintaining a clean environment and the need for higher flexibility in routing in arduous and difficult terrains, the solution lies in carrying the material by a pipe conveyor.

The principles: The pipe conveyor is loaded like a conventional conveyor.

The material moves within the tubular profile towards the discharge or intermediate loading point. When the materials reach their final destination, the pipe opens up for discharging the carried materials. These operations are demonstrated in Figures 6.19–6.23.

**Figure 6.20:** Discharging materials by a pipe conveyor.

The pipe conveyor belts are utilised successfully in a wide variety of industries for carrying sized coal, limestone, lignite, fertilisers, etc.

#### 6.4.3.4.1 Advantages of pipe conveyor

(i) The environment is adequately protected against pollution. (ii) The material carried is protected against rain, dust, and wind. (iii) Wastage or spillage is greatly reduced. (iv) Transfer points are reduced compared to a conventional conveyor. (v) The belt can undergo vertical, horizontal, and three-dimensional curves. (vi) Best for areas where space is a constraint (Figures 6.22 and 6.23).

**Figure 6.21:** Unique idler arrangement maintains the desired pipe diameter throughout the conveyor length.

**Figure 6.22:** A space-saver modular system ensures a seamless movement through terrain curves.

**Figure 6.23:** Specially arranged idlers force the conveyor belt to give a pipe shape.

### 6.4.3.5 Cable belts

A cable belt's design differs from historically accepted conveyor designs. The cable conveyor belt is supported by two endless wire rope cables, one on each side of the belt. Here, the driving and carrying mechanisms are totally separated. The driving forces are engaged by two steel wire cables (ropes) and the material-carrying medium is a rubber conveyor belt. These cables are supported by grooved pulleys at well-ordered intervals along the length as depicted in Figure 6.24.

The conveyor belt is designed with a special V-shaped groove at the top and bottom of the surface. This engages and grips the underlying rope. The belt has almost no tension in the warp or longitudinal direction as it simply moves on the wire ropes. At the same time, the transverse direction has enough stiffness to withstand its own weight and material load. The principal advantage of the cable belt is its ability to work on tight curves even on long distance with minimal operating cost [36].

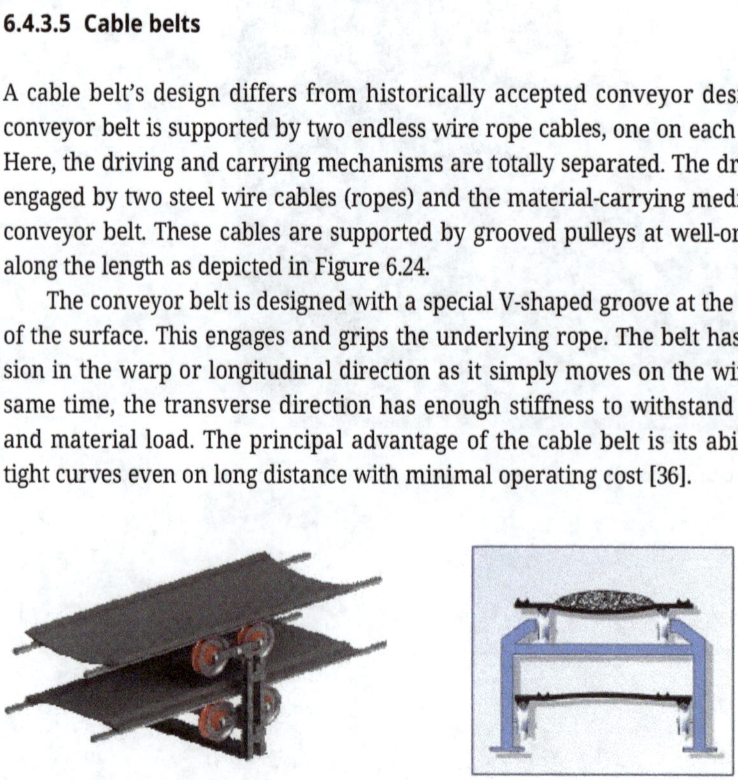

**Figure 6.24:** Cable belt.

### 6.4.3.6 Sandwich belt

As the name indicates, sandwich belt conveyors convey the material between two rubber belts, before it is inclined at angles up to 90 degrees. The sandwich belt conveyor uses two conveyor belts that share a common load-carrying path, face-to-face, to gently but firmly contain the product being carried; hence making the steep incline and even vertical-lift runs easily achievable. The top and bottom belts are independently driven and tensioned. Along the carrying path, the top and bottom belts are alternately supported against closely spaced troughing idlers. Here, the material is 'hugged' by the belts throughout the inclined section to ensure that it does not slide back down the incline even if the conveyor trips. This hugging pressure is achieved by either the belt tension or by means of pressing assemblies, which force the belts together; it is illustrated in Figure 6.25.

A fundamental requirement of any proposed application of sandwich belt conveyors is that the material to be transported must have a reasonable internal friction angle [37].

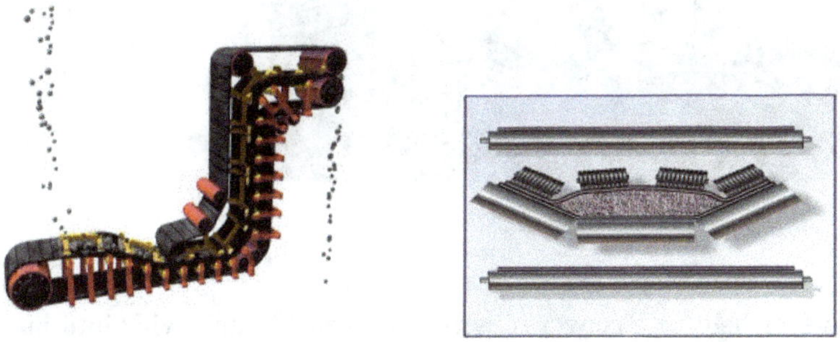

**Figure 6.25:** Sandwich belt with idler arrangements.

### 6.4.3.7 Bucket elevator belt

Bucket elevators are designed to move flowing powders or bulk solids vertically. The typical elevator consists of a series of buckets mounted on a belt operating over a sprocket or pulley. A steel casing encloses the bucket line. Material is fed into an inlet hopper. Buckets are filled with the material and conveyed up to and over the head sprocket/ pulley. This is shown in Figures 6.26 and 6.27. After the discharge, the emptied buckets then continue back down to the boot to continue the circle [38].

## 6.5 Selection of conveyor belt

The conveyor belt is a valuable component of a conveyor. Hence, it is very important to select the correct conveyor belt for any conveyor. In order to select or design

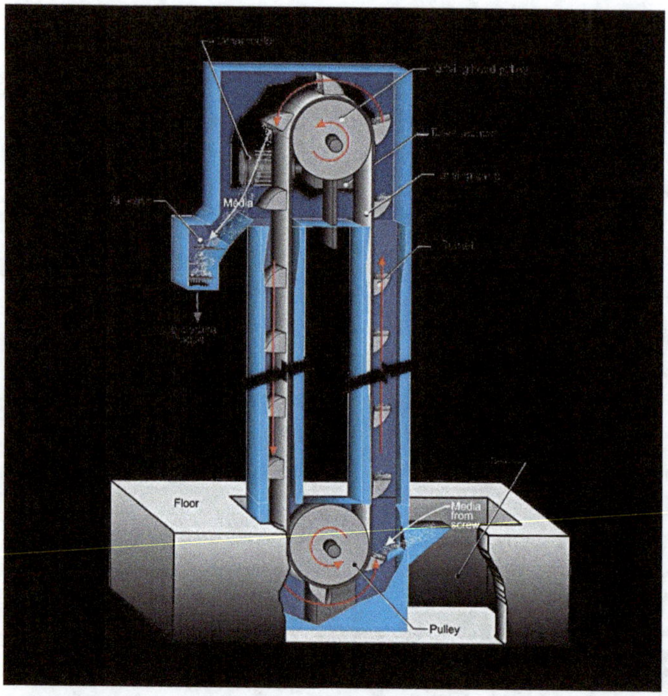


**Figure 6.26:** Bucket elevator.

the correct conveyor belt for a conveyor system, it needs the following information [39]:

1. Load surface of the belt that is supported by flat or troughed idlers; if troughed, then the troughing angle
2. Drive data: Single or multiple drive pulleys, motor horse power, angle of wrap, and if drive pulleys are wrapped or bare
3. Environmental condition
4. Vertical lifts, if any
5. Distance between the head and the tail pulley
6. Discharge per hour in tons/h
7. Material type, temperature, density, size, and percentage of maximum lumps
8. Pulley diameter, speed of belt, and width of belt
9. Take-up type (gravity or screw), location, and total amount of movement

First of all, effective belt tension (TE) is to be calculated. TE is the sum of the tension required to move the empty belt (TC), the tension required to move the belt horizontally (TL), and the tension required to lift the load (TH):

$$TE = TC + TL + TH$$

**Figure 6.27:** Bucket filled with material.

Calculations:

$$TC = F1 \times L \times CW$$

where $F1 = 0.035$ (normal friction factor to move empty belt), $L$ is the belt length between terminal pulleys, and CW is the weight of the conveyor belt and components, expressed in kg/m of conveyor length:

$$TL = F2 \times L \times MW$$

$F2 = 0.04$ (normal friction factor to move the load horizontally), $L$ is the belt length between terminal pulleys, and MW is the material weight per meter of belt = 0.278TPH/belt speed (m/s)

$$TH = H(\text{Difference in elevation of terminal pulleys in m}) \times MW$$

Slack-side tension (TS) is also to be calculated from the following formulation:

$$TS(T2) = D \times TE; D \text{ is the drive factor} = 0.42 \text{ for } 200° \text{ lagged snub pulley}$$

Total tension, which is called the operating tension (TO), is known as the allowable working tension; it is the value used to select the belt construction:

$$TO(T1) = TS(T2) + TE$$

TE is the effective belt tension at drive and TS is the slack-side tension.

Rough estimation to calculate the gravity take-up = 2TS.

Motor horsepower may be calculated using the following formula:
HP = TE × Belt speed/75
kw = TE × Belt speed/100
Example:
Belt width = 1,200 mm
Material = Limestone
TPH = 800
Lump size = −200 mm
Lift = 25 m
Speed = 2 m/s
C–C distance = 400 m
Take-up = gravity
Belt tension = to be calculated

Calculation:
Before we calculate the effective tension, we should calculate the effective length ($L_e$) using the following formulation:

**Effective length table**

| Length between terminal pulleys | $L_e$ (effective length) |
|---|---|
| Less than 1,000 m | 0.665L + 30 |
| 1,000–1,500 m | 0.60L + 32 |
| Above 1,500 m | 0.55L + 35 |

In our case, $L_e$ = (400 × 0.665) + 30 = 296
TC = $F1 \times L_e$ × CW = 0.035 × 296 × 92 (for 150 mm idler) = 953 kg.
TL = $F2 \times L_e$ × MW = 0.04 × 296 × 111 = 1,314 kg
TH = $H$ × MW = 25 × 111 = 2,775 kg
Now, TE = TC + TL + TH = 5,042 kg
TS ($T2$) = D × TE = 0.42 × 5,042 = 2,118 kg
TO ($T1$) = TS ($T2$) + TE = 7,160 kg
Unit tension = TO/width (kg/cm) × 0.9806 = 58.4 kN/m
Factor of safety = 8 times the splice strength for a textile belt
= 6.7 times the splice strength for a steel cord belt.

Splice strength = 80% of belt strength for a 5-ply belt, 75% of belt strength for a 4-ply belt, 67% of belt strength for a 3-ply belt.
For steel cord belt, splice strength = 100% of belt strength.
Now $T1$ = Splice strength/factor of safety. So, splice strength = 58.4 × 8 = 467 kN/m
Therefore, belt rating (considering 4-ply belt) = 467/0.75 ~630 kN/m
Hence, select a 4-ply belt having strength 630 kN/m.

## 6.5.1 Selection of cover grade and thickness

Selection of grade of cover depends on the application of the conveyor belt [40].

During selection of the grade of cover, it is necessary to see whether the belt is used for:

1. General-purpose application
2. Super abrasion resistant
3. Heat resistant
4. Fire resistant
5. Oil resistant.

Considering the above, the cover grade can be selected from Table 6.1.

Thickness of the cover can be selected from Table 6.3.

**Table 6.3:** Cover thickness selection table.

| **Cover rubber thickness selection table** | | | | | | |
|---|---|---|---|---|---|---|
| **Material handled** | **Top cover for belt length** | | | **Bottom cover for belt length** | | |
| | **Below 50 m** | **50–150 m** | **Over 150 m** | **Below 50 m** | **50–150 m** | **Over 150 m** |
| Nonabrasive: fine coal, wood chips, grain chips, ash cement, etc. | 4.0 mm | 3.0 mm | 3.0 mm | 1.5 mm | 1.5 mm | 1.5 mm |
| Slightly abrasive: sand, earth, clay, salt, etc. | 5.0 mm | 4.0 mm | 3.0 mm | 1.5 mm | 1.5 mm | 1.5 mm |
| Very abrasive: undressed coal, crushed stone, gravel coke, etc. | 6.5 mm | 6.0 mm | 5.0 mm | 3.0 mm | 2.0 mm | 2.0 mm |
| Extremely abrasive: limestone, metal ores, slag iron, glass, etc. | 9.0 mm | 8.0 mm | 6.5 mm | 3.0 mm | 3.0 mm | 3.0 mm |

Hence, the final construction of our belt shall be

$$1,200\,\text{mm} \times \text{EP}\,630/4 \times (6.5\;+\;3.0)\,\text{SAR} \times \text{L-800}\,\text{m}$$

## 6.6 Testing

It is necessary to test any product after manufacturing to check whether the product has met the required parameters or not. Testing the products gives a clear indication to predict the performance of the material in the actual application. It also helps the customers/end users to define the quality in a quantitative way. Like other products, both textile and steel cord-reinforced conveyor belts undergo internationally defined destructive and nondestructive type tests to assess the conveyor belt quality. We will discuss the tests conducted for the conveyor belt, in brief, here.

Most of the test parameters are common for both textile and steel cord-reinforced conveyor belts, and those are:
- Dimension (width, cover rubber thickness, and total thickness)
- Cover rubber tensile strength and elongation
- Abrasion loss
- Troughability
- Volume swelling (for oil-resistant grade belting)
- Electrical resistivity (for flame-resistant grade belting)
- Flame resistance property (for flame-resistant grade belting)
- Drum friction test (for flame-resistant grade belting).

Other test parameters followed for textile-reinforced conveyor belts are:
- Full thickness breaking strength
- Adhesion test

Similarly, the other tests for steel-reinforced conveyor belts are:
- Static and dynamic cord pull-out strength
- Peeling strength
- Air penetration test

A full width × 500 mm long sample from one end of a conveyor belt is cut and send to a lab for testing. All tests are conducted from this sample. We do not have the scope for discussing all the test procedure in detail in this chapter. Here we shall discuss few important tests of the conveyor belt in brief.

### 6.6.1 Troughability

The transverse flexibility (troughability) of a conveyor belt is a very important test parameter. It depicts the ability of the belt to sit on different conveying troughing idlers. A full width (*W*) belt sample of 150 mm on the longitudinal side is taken from the mother belt and it is suspended freely by holding both edges. The test sample starts to deflect due to its own eight and its stiffness [41].

Troughability is determined by measuring the maximum deflection (*F*) and it is expressed as the ratio of deflection and the belt width (*F*/*W*). An image of troughability testing is shown in Figure 6.28.

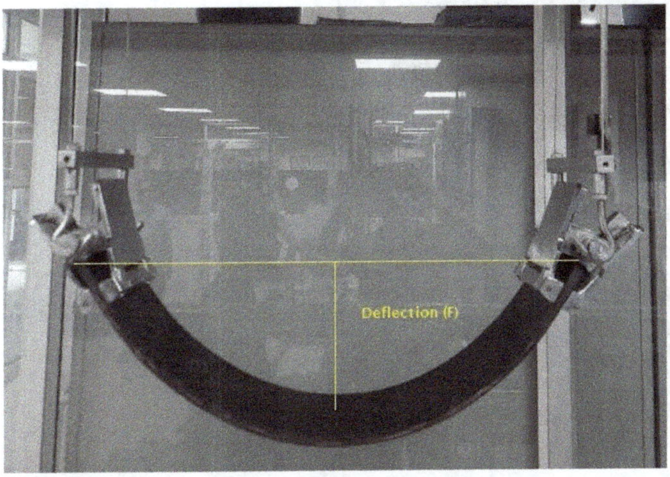


**Figure 6.28:** Measurement of troughability.

## 6.6.2 Flame resistivity, electrical resistivity, and drum friction test

These tests are conducted only for fire resistance grade belting, considering its critical nature of application.

### 6.6.2.1 Flame resistivity

A total of six test pieces (three from warp and three from weft) in 25 mm width × 200 mm length are cut from the belt and suspended vertically above the gas flame of 1,000 ± 20 °C.

During the test period, the gas flame burner is positioned 45° to the central vertical axis of the test piece. The distance between the burner tube and the bottom tip of the test piece in the 45° position is 50 mm. The flame is applied for 45 s and the duration of time taken to self-extinguish the test piece after the removal of flame is recorded and reported [42].

The sum of the self-extinguishing times of the six pieces shall be less than 45 s and no individual value shall be greater than 15 s. An image of a flame test is depicted in Figure 6.29.

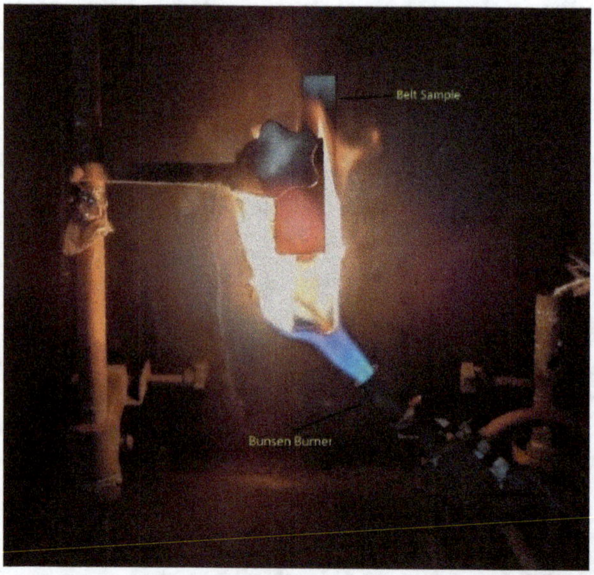


**Figure 6.29:** Flame test.

#### 6.6.2.2 Electrical resistivity

This test is conducted to check the ability of the rubber compound to conduct static charge accumulated during the continuous operation of the conveyor without major resistance. Sometimes, the static charges ignite and combust the conveyor [43].

A full thickness belt sample of 300 mm × 300 mm is cut from the belt sample and an electric potential of 1,000 V is applied such that not more than 1 W of energy is passed through the electrodes in the test piece, and the resistance between the electrodes is measured. The electrical resistance of the belt surface shall not exceed 300 M-Ohm. An image of an electrical resistance test is detailed in Figure 6.30.

#### 6.6.2.3 Drum friction test

Drum friction test is conducted to determine the ignitability and the maximum surface temperature of belt due to friction.

A 150 mm width × 1,500 mm long sample is passed around a drum pulley and the belt is held stationary under 35 kg load. While testing, the drum pulley is rotated at 200 rpm and the test continued until the sample breaks or for maximum 2 h, whichever is earlier. Due to the continuous friction between the rotating drum and the high-tension belt surface, the belt surface temperature starts to increase [44].

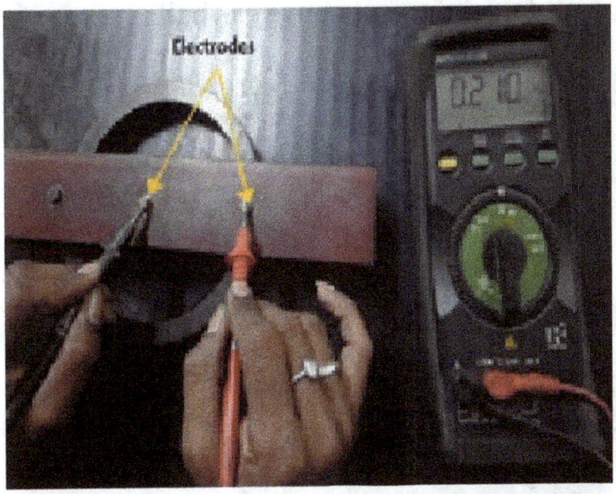


**Figure 6.30:** Electrical resistance test.

The temperature is recorded at regular intervals and the maximum surface temperature is reported. The temperature shall not exceed 325 °C temperature and there shall not be any sign of visible flame or glow. A drum friction test procedure is detailed in Figure 6.31.

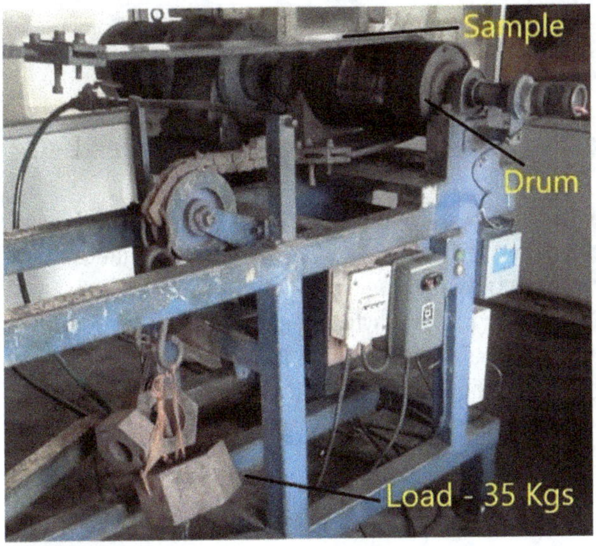


**Figure 6.31:** Drum friction test.

## 6.6.3 Adhesion test for textile-reinforced conveyor belt

The force required to separate the cover from the carcass and also between the plies in a standard known width of 25 mm is measured with a tensile testing machine at a constant rate of strain. Adhesion strength is the ratio of the average force to the width of the sample and it is expressed in N/mm [45]. An adhesion test procedure is portrayed in Figure 6.32.

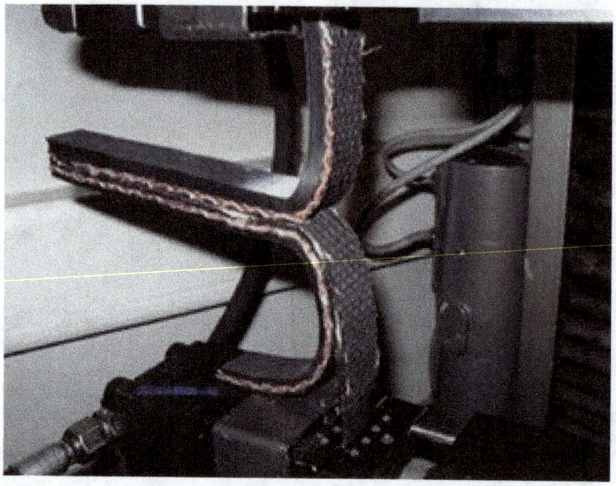

**Figure 6.32:** Adhesion test.

## 6.6.4 Full thickness breaking strength for textile-reinforced belt

Full thickness breaking strength helps to determine the total belt rating of the conveyor and the intermediate elongation at 10% load helps to give a clear indication of the actual belt elongation during running condition.

A dumbbell specimen of a standard known width is cut from the belt and it is stretched in a tensile testing machine at a constant rate of 100 mm/min until it breaks, which can be observed in Figure 6.33. The tensile strength per unit width, Elongation at predetermined load (10% of belt rating multiplied by sample width) and elongation-at-break are recorded [46].

**Figure 6.33:** Full thickness breaking strength test.

## 6.6.5 Static and dynamic cord pull-out test for steel-reinforced belt

### 6.6.5.1 Static

A full thickness sample containing five cords is cut from the belt and it is prepared as illustrated in Figure 6.34.

The prepared sample is mounted on a tensile testing machine and the force per unit length to pull out the centre cord is measured at a constant rate of strain.

A sample similar to the above figure is prepared. A cyclic load, ranging from 3.6% to 36% of the specified static pull-out strength, is applied in a suitable dynamic tester until the cord comes out from the sample and it is noted, as described in Figures 6.35 and 6.36. In this test, the bonding between the cords to rubber is measured to prevent any premature failure [47, 48].

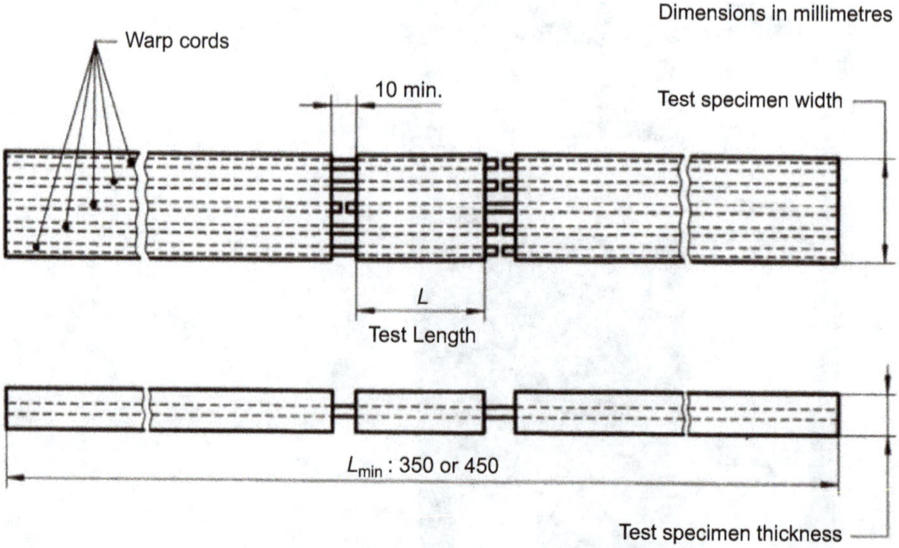


**Figure 6.34:** Sketch of a pull-out test sample.

**Figure 6.35:** Static pull-out test.

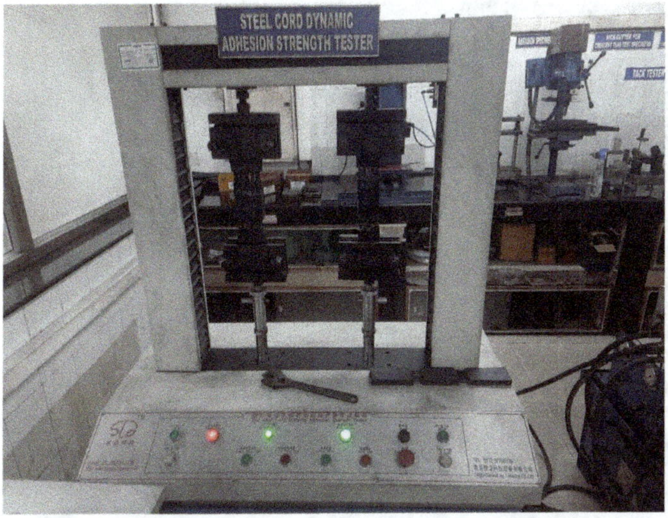


**Figure 6.36:** Dynamic pull-out test.

### 6.6.6 Peeling test for steel-reinforced belt

Full thickness sample containing at least two cords (25 mm width minimum) is cut from the belt [49]. The force per unit width to strip the cover and the core layer is measured with a tensile testing machine at a constant rate of strain – shown in Figure 6.37.

### 6.6.7 Air penetration test for steel-reinforced belt

This test is conducted to check the penetration of the rubber compound inside the steel cord. A pressure differential of 100 kPa is passed at one end of the 400 mm-long belt sample embedded with steel cord for 60 s and its output pressure difference is measured at the other end of the cord [48].

The loss in pressure should not be more than 5 kPa as depicted in Figure 6.38.

## 6.7 Splicing

A conveyor belt is not a conveyor belt until it is joined. As it is manufactured and supplied in open-end for ease of handling and transportation, the two ends are joined by the user to make it a closed loop. The splice joint must retain the maximum possible belt strength to avoid snapping in the joint area. Generally, there are two types of splicing methods in the current scenarios [50].

**Figure 6.37:** Peeling resistance test.

**Figure 6.38:** Air permeability test.

1. Mechanical splicing – joins belt ends by metal hinges or plates; and
2. Vulcanised splicing – joins belt ends through heat and/or chemicals.

Both methods have certain advantages and disadvantages. It is important to understand these while making a comprehensive decision on adopting the splicing method. The environmental factors, downtime, and criticality of the application are the main factors to consider and they are discussed in the respective sections.

## 6.7.1 Mechanical splicing

Mechanical splice installation is a fast and simple method. It is very effective and consumes very less time compared to other methods. It is not necessary to wait for a professional splicer, if any unexpected failures happen. In addition, mechanical splices can be made without considering any environmental factors like temperature and moisture or contaminants.

The mechanical splice is visible; wear and deterioration are apparent and can be taken care of before a complete belt failure. There are several types of mechanical fastener, each created for use with different belt widths, lengths, thicknesses, speeds, tensions, and belt cleaners.

Identifying the correct fastener for the application is essential to ensure maximum splice life and performance. Mechanical fasteners are available in two types: hinged and solid plate [51].

### 6.7.1.1 Solid-plate fasteners

Solid-plate fasteners consist of one plate that is installed on each end of the belt and fastened together. Solid-plate fasteners are used with larger pulleys and for applications that require a sift-free splice. These fasteners are installed to the belt using bolts, rivets, or staples. The solid-plate fastener typically is considered the longer-lasting and stronger splice because it has no hinge pin to wear out, shown in Figure 6.39 [52].

### 6.7.1.2 Hinge-type fasteners

Hinged plate fasteners are often used in small pulleys. Here, each end of the belt is hammered/fixed with hinge-plate fasteners, and then joined together by inserting a pin. It can significantly shorten the conveyor downtime compared to any other splice methods. For example, the spare belts can be prepared in advance with both ends pre-spliced (fixed with hinge-plate fasteners), requiring only the hinge pin to be inserted at the job site [53]. It also permits to join the two different belt thicknesses by

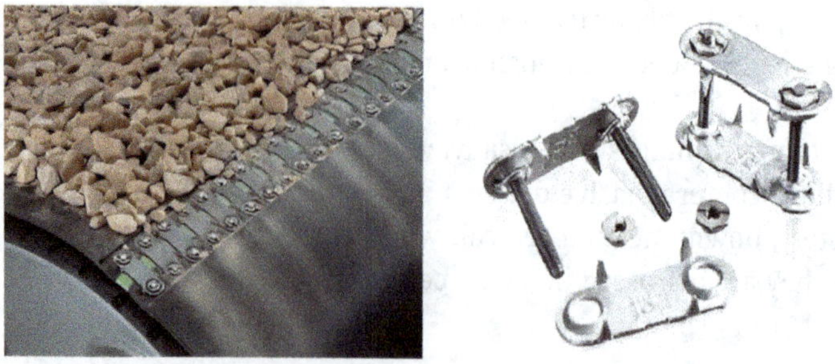

**Figure 6.39:** Images of solid-plate fastener.

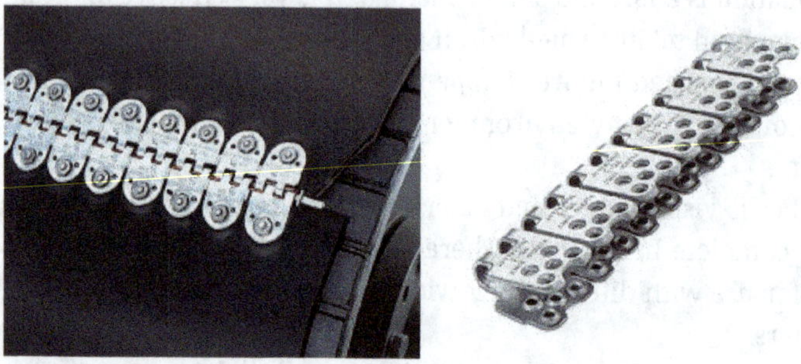

**Figure 6.40:** Hinge-type fastener.

using two different fastener halves fixed to the belt and joined by a hinge pin, depicted in Figure 6.40.

Every splicing method has its limitations and they are:

1) Mechanical fasteners cannot be used with higher-tension belts (i.e., not more than 1,400 N/mm).
2) Mechanical fasteners are noisy, incompatible with belt cleaners and scrapers, and generally damage the belt. If mechanical splices are properly installed, maintained and countersunk by skiving the belt, there should be no problem with noise or damage to the belt or belt cleaners.

## 6.7.2 Vulcanised splices

In favour of long conveyors and permanent installations, this kind of splicing method is adopted by all conveyor belt end users. It offers smooth splice with minimal risk of sagging, tearing, and other harmful wear to the belt. There are several types of vulcanised splice, including stepped splices, finger splices, and overlap splices. It is done by two types of vulcanising processes: hot splice and cold splice. Each process requires specific tools and an intimate knowledge of the rubber bonding process.

The preparation of a splice is similar for both hot splice and cold splice. But the vulcanisation process varies between two methods and plays a vital role. With hot vulcanisation, the splice portions are cured under specified time, temperature, and pressure with a vulcanising press. Whereas, cold vulcanisation uses a rubber bonding solution that provides a chemical reaction and holds the two belt ends together. In the cold process, vulcanisation presses are not required.

Based on the number of plies in the textile belt, the splice type is selected. For multiply belting, step joint is preferable and for monoply belting and steel cord belting, finger splices are preferable. The step joints are recommended at a bias angle of 16–22° to avoid the splice failure as the belt exhibits greater tension while running around the pulley [54].

#### 6.7.2.1 Hot vulcanisation

##### 6.7.2.1.1 Stepped bias splice for multiply belt

The rubber covers and plies are stripped out at a bias angle of approximately 16–22°, depending on the bias angle of the press. A schematic diagram is detailed in Figure 6.41.

**X-step length**

The step length of the fabric is selected based on the belt strength and the number of plies. Usually about 25 mm of the cover is also removed at the splice. Both ends of the belt are prepared in this way so that the two ends overlap and match up. The stripped areas are buffed lightly and cleaned with a suitable solvent. First, one coat of rubber solution made out of skim compound is applied on both ends of the spliced area and allowed to dry [55].

A layer of unvulcanised skim compound is applied over the exposed fabric and suitable rubber compounds are filled in the 25 mm vacant cover area. When the splice has been built up, it is then vulcanised in a portable splicing press, as depicted in Figure 6.42. Depending on the length and width of the splice area, the vulcanising press size is selected [56].

##### 6.7.2.1.2 Single-ply finger splice (solid-woven and steel cord)

Here, the carcass is cut across the width in a zigzag pattern. The length of the fingers and their width depends on the actual belt strength. The fingers are cut out, taking care that the trailing end of the belt has complete fingers at the edges and the covers are removed at the splice area. Then, the carcass zigzag pattern is matched with one another and the cover layers are applied.

A sketch of a finger splice of a textile belt is illustrated in Figure 6.43. The joint is then pressed and vulcanised with application of heat and allowed to cool under pressure.

The most popular and effective means of splicing steel cord belts is with the finger splice method, but in this case, the steel cords are stripped of rubber and alternate

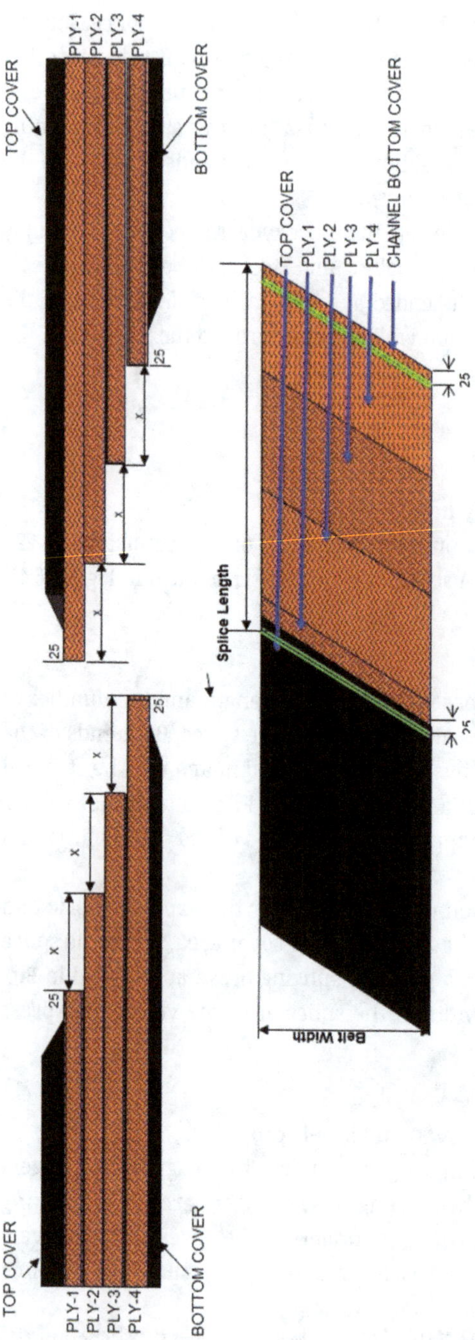


**Figure 6.41:** Sketch of a splice of a 4-ply textile belt.

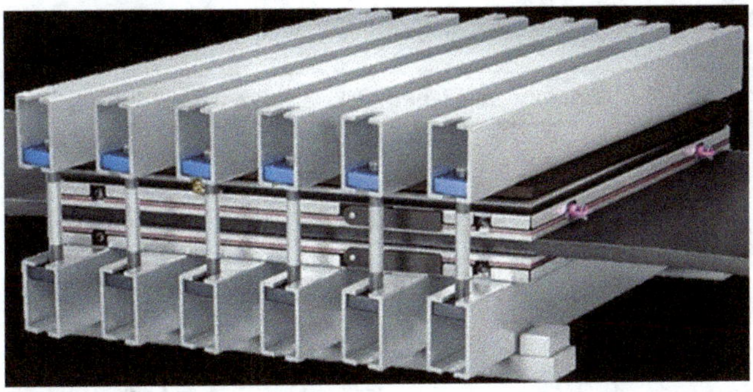

**Figure 6.42:** A schematic sketch of a portable vulcanising press.

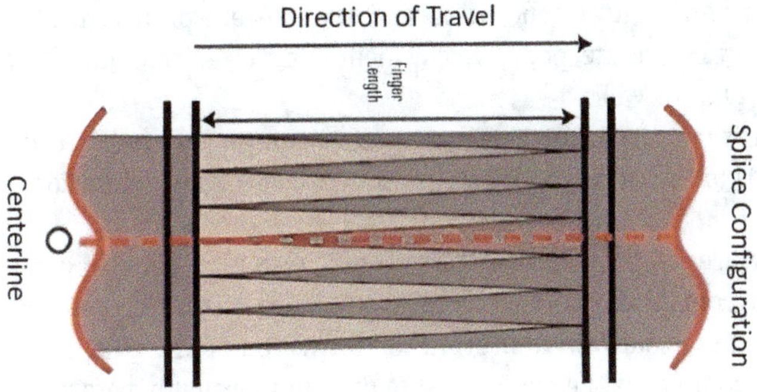


**Figure 6.43:** Schematic sketch of a finger splice.

cords cut back, so that the two ends of the belt intermesh. Rubber compound is then applied to build up the correct thickness. The joint is then vulcanised under optimum temperature, pressure, and time.

Different splice-making protocols are employed, depending on the strength of the belt, and these different types are referred to as Type I, Type II, and so on up to Type 'V' splice joints for very high-strength belts [57]. Standard splice length, number of steps, and minimum step lengths against belt rating are detailed in Table 6.4 and in Figure 6.44.

#### 6.7.2.1.3 Cold vulcanisation method of splicing

In cold vulcanisation, specific grade of synthetic rubber polychloroprene-based adhesive solution with hardener is used instead of unvulcanised rubber to join the two belt ends. This process is mostly adopted in shorter distance and in low-tension belts. The preparation process for splice is the same like a stepped bias joint. Cold vulcanising solution does not require any vulcanising press or temperature for joining.

**Table 6.4:** Belt strength, no. of steps, and step length.

| Belt type | Number of steps | Minimum step length (ls) | Splice length (lv) |
|---|---|---|---|
| ST-1250 and below | 1 | 350 | 650 |
| ST 1600 | 1 | 450 | 750 |
| ST 2000 | 2 | 400 | 1,150 |
| ST 2500 | 2 | 500 | 1,350 |
| ST 3150 | 2 | 650 | 1,650 |
| ST 3500 | 3 | 650 | 2,350 |
| ST 4000 | 3 | 750 | 2,650 |
| ST 4500 | 3 | 800 | 2,800 |

Lv represents splice length, Ls represents step length, *s* represents min. 3 × cord diameter.

It is completely dependent on chemical bonding. After both ends are prepared fully, cold solution is prepared. Just before use, pour 1 L of cement into a clean container. Mix 1 bottle of hardener with it and stir well.

When 1 L of container is procured, one bottle of hardener may be poured directly into it. Lay down both ends of the spliced belt on each other to check for dimensional accuracy.

Coat the entire splice area, fabric steps, cut edges of the solid rubber, and the 50 mm cover strip with the cold joining cement by means of a soft bristled brush.

Take care to spread out the coat well to prevent the formation of any globules of cement that fail to dry out completely. Leave the coat to dry out thoroughly before the next coat is applied.

Apply two coats for rubber surface and for finely structured fabric, but at least three coats for coarse fabric. Apply the last coat of cement simultaneously on both halves to ensure that the two halves of the splice are ready for bonding at the same time, as detailed in Figure 6.45.

Ensure that the last coat is not completely dried out before jointing. Join the two halves of the splice with care so that the fabric steps butt precisely and at the same time, stitch them down with a hand roller. Hammer the entire splice area with a rubber hammer.

Leave the splice untensioned for about 4–6 h before putting it back into operation [58].

## 6.8 Calculation of the life of conveyor belts

It has been found that the number of tons of material conveyed to wear out 1 mm of cover is directly proportional to the cycle time and the full cross sectional load area. Thus, values could be derived from the measured results for the number of tons that would wear 1 mm of top cover for a standard cycle time of 60 s.

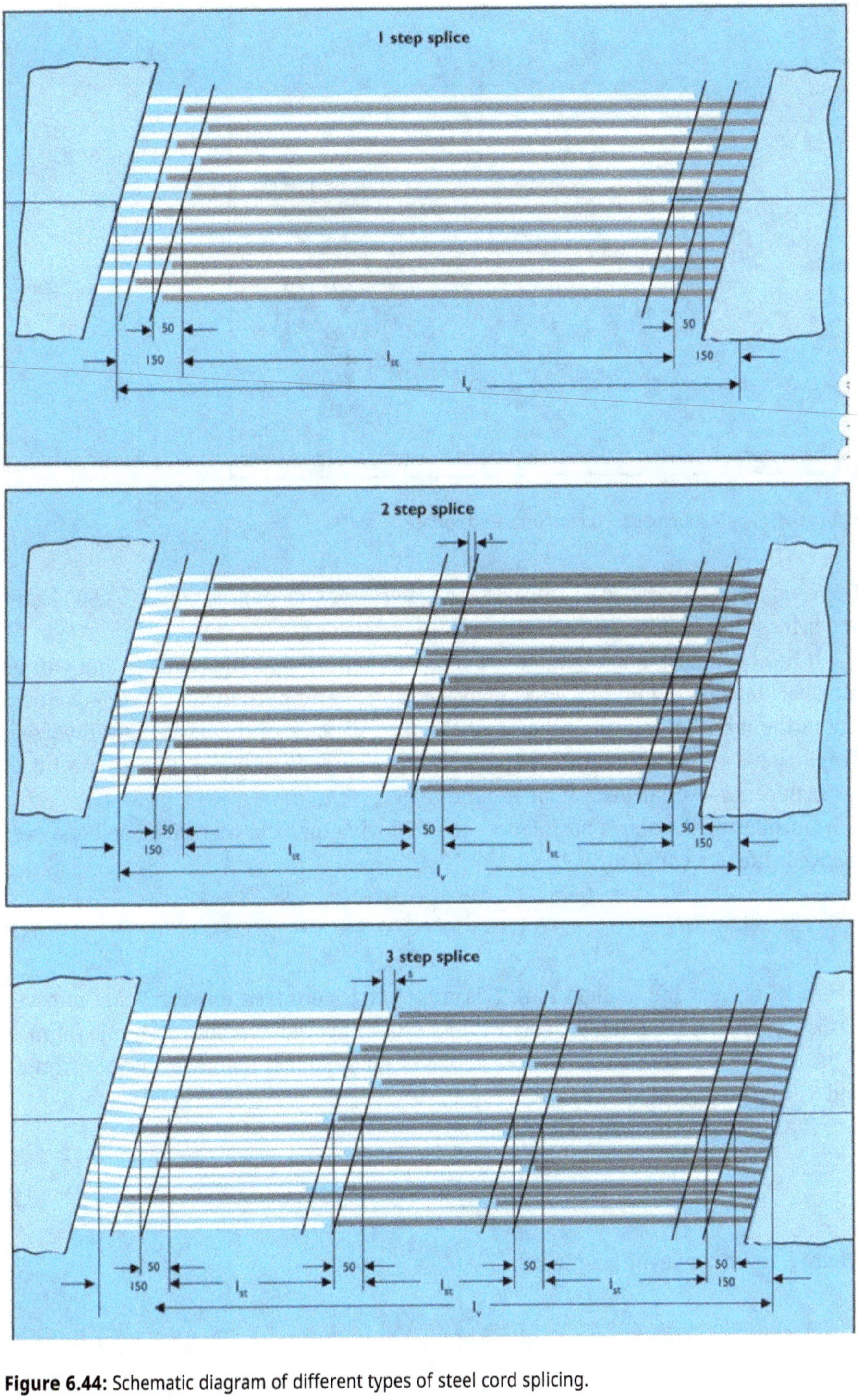


**Figure 6.44:** Schematic diagram of different types of steel cord splicing.

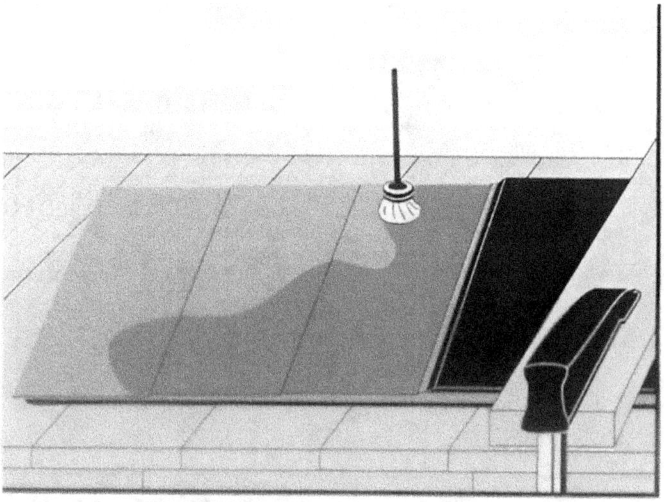

**Figure 6.45:** Application of glue on the prepared surface.

These values are described in Table 6.5 for material of bulk density equal to 1,500 kg/m$^3$ and belt speed of 5 m/s.

It has been postulated that, all other things being equal, the tonnage that will be conveyed to wear a unit thickness of cover is directly proportional to the relative density of the material and the relative abrasion resistance of the cover, and inversely proportional to the relative belt speed. It has also been assumed that each unit of cover thickness would wear at an equal rate [59].

Considering the above postulation, the following formula was developed to assess the wear life of a conveyor belt:

$$W_l = \frac{T_f \cdot D \cdot C_t \cdot A \cdot t}{S} \tag{6.1}$$

where $W_l$ is wear life (million ton), $T_f$ is million tons conveyed to wear 1 mm of cover for 60 s cycle, $D$ is the relative density of material conveyed, $C_t$ is the cycle time (min), $A$ is the relative abrasion resistance of belt cover material, $t$ is the cover thickness (mm), and $S$ is the belt speed.

The approximate cycle time is given by

$$C_t = \frac{2L}{60S} \tag{6.2}$$

where $L$ is the conveyor length (m).

Substituting for $C_t$, formula (6.1) can be rewritten as follows:

$$W_1 = \frac{T_f \cdot D \cdot A \cdot L \cdot t}{30S} \tag{6.3}$$

**Table 6.5:** $T_f$ (tonnage factor) Million tonnage to wear 1 mm of cover for a 60 s belt cycle time, material bulk density of 1,500 kg/m$^3$ and a belt speed is of 5 m/s.

| Belt width (mm) | Quantity of material to wear 1 mm of cover (million ton) |
|---|---|
| 750 | 0.50 |
| 900 | 0.75 |
| 1,050 | 1.10 |
| 1,200 | 1.35 |
| 1,350 | 1.85 |
| 1,500 | 2.30 |
| 1,800 | 3.35 |
| 2,100 | 4.55 |

Example: Let us consider a conveyor belt having the following conditions:
Width = 1,200 mm. Hence, $T_f = 1.35$
$D = 1.1$ g/cc; $A = 150$ mm$^3$; $L = 500$ m; $t = 5$ mm; $s = 3.5$ m/s
$W_1 = 1.35 \times 1.1 \times 150 \times 500 \times 5 \div 30 \div 3.5$
= 5,303 million ton.
When a 1,200 mm wide belt is carrying coal and running at 3.5 m/s, its tph is 1,700. Hence, the expected life will be
5,303 × 10,000,00 ÷ 1,700 h
= 3,119,411 h
= (3,119,411 ÷ 1,440 ÷ 365) years
= 5.9 years.

## 6.9 Packing and storing

Transportation, storing, and maintenance exert fundamental influences on the performance capability of rubber conveyor belts.

### 6.9.1 Packing

Irrespective of the length, thickness, width, and weight, rubber conveyor belts are packaged and wound on to wooden squares, wooden drums, and steel drums. The maximum shipping length will be governed by the kind of transportation, the maximum drum diameter, which depends on the technical conditions of the plant, and the maximum weight of the belt.

On delivery, the rubber conveyor belts will be corded five times. The cut edge will be provided with a foil cover in order to ensure its long-term preservation.

### 6.9.2 Storage

In order to avoid negative influences on the serviceability of the conveyor belts in case of long-term storing (longer than 3 months), emphasis should be given to the following details.

Belt should be kept in an upright position until it is commissioned. The belts are to be stored in cool, dry, covered, and well-ventilated places, away from direct sun light. Other factors like exposure to ozone in sea area, contact with all types of solvents, oils, greases, heat and moisture, as well as of other substances having detrimental effects on rubber should be avoided.

In case of belts weighing in excess of 1,500 kg, and not installed within 2 or 3 months and beyond, they should be supported off the ground on 'A' frames.

Belts kept as spares should definitely be stored off the ground and rotated ¼ turn every 3 months. Whenever lengths are cut off from the belting roll, the end should be sealed with a conveyor belting repair solution to prevent ingress of moisture into the carcass.

The drums in storage should be examined once in 2 or 3 months to ensure that wooden packing is not failing due to white ants and similar insects. Spraying of DDT over packing will prevent this type of damage [60].

## 6.10 Maintenance of conveyor belt

The belt tension has to be checked at regular intervals. Slippage of the belt around the drive pulley and too much belt sag between the idlers indicates a lack of belt tension. Especially in the case of take-up stations with fixed take-up pulleys, the take-up pulley has to be adjusted. During the initial operation period, if a new conveyor belt is employed, the belt tension has to be examined even after a longer time of operation.

The conveyed material should be loaded centrally on to the belt. Depositing the material from one only side will cause off-track movement of the belt and lead to damaging of the conveyor belt. When discharging bulk materials, their free flow on leaving the belt should be provided for.

Above are the main damages that are encountered by conveyor belts during running [61].

**Table 6.6:** Most frequently occuring damages to the belt, their causes, and prevention.

| Troubles | Causes | Correction |
|---|---|---|
| A. Excessive edge wear/broken/frayed due to rubbing | 1. Off-centre loading | 1. Adjust the chute to place the load on the centre of belt; discharge material in the direction of the belt travel. Check the inclination of the idler stations. Adjust for adequate belt tension. |
| | 2. Material spillage | 2. Improve loading and transfer conditions. Install cleaning devices and improve maintenance. |
| | 3. Belt-hitting structure | 3. Install training idlers on carrying and return run. |
| | 4. Off-track running of the belt in the drive station. | 4. Eliminate tilt of take-up equipment. Check the pulley alignment. |
| | 5. Inadequate edge clearance | 5. Minimum recommended clearance between the belt edge and structure is 75 mm. |
| B. Excessive top cover wear | 1. Defective or standing idler in return run. | 1. Replace idlers. |
| | 2. Missing idlers on the return side | 2. Fit the missing idlers. |
| | 3. Wear marks due to skirt boards that exert too much pressure on the belt. | 3. Replace and adjust skirt board. |
| | 4. Defective scrappers. | 4. Repair/replace scrappers. |
| | 5. Cover quality too low. | 5. Replace with better quality cover belt or higher cover thickness belt. |
| C. Grooving, gouging, or stripping of cover | 1. Skirt board wrongly adjusted or made of improper material. | 1. Adjust skirt board support to a minimum of 26 mm between the metal and the belt. |
| | 2. Material hanging up in or under the chute. | 2. Change the design of the chute and also increase the gap between the chute and the belt. |
| | 3. Excessive impact on belt | 3. Reduce impact by improving the chute design. |

Apart from these, there are many other causes of damage that can be observed by users during the running of a belt in a conveyor system such as severe bottom cover wear, cover blisters, lengthwise or transverse carcass breaks with cover intact, splice failure, belt breaks just after the mechanical splice area, ply separation, etc.; these are listed in Table 6.6.

For all these causes of damage, the remedy of the conveyor and its design need to be looked into.

## 6.11 Conclusion

Economists estimate an industrial growth of 6% for the years 2022–2030.The growth of the market can be attributed to the increasing use of conveyor belts in various end-user industries, such as mining and construction.

The growth in these end-user industries is anticipated to positively impact the growth of the conveyor belt market. Higher production demands across all bulk handling segments require increased efficiency at the lowest cost of operation in the safest and most effective manner possible.

As conveyor systems become wider, faster, and longer, more energy output and more controlled throughput will be needed.

With the push towards wider and higher-speed belts, bulk handlers will need substantial development in more reliable components, such as idlers, impact beds, and chutes.

The life of a conveyor belt depends on many factors but most important is designing of the correct conveyor with a perfect belt and of course, proper maintenance of the conveyor.

Selection of a proper belt is very important to get the optimum life of a conveyor belt. Nowadays, many new rubber compound grades have come up and many new designs of textiles are coming up. By selecting a proper grade and a proper reinforcing material, one can enhance the life of a conveyor belt to many folds.

## References

[1] McGuire, P. M. Conveyors: Application, Selection, and Integration. 1st edition. Boca Raton, FL:CRC Press, 35 (2009-08-05). ISBN 9781439803905.

[2] Thomas Robins, Inventor, 89, Dies. Developer of Heavy-Duty Conveyor Belt Had Headed Hewitt-Robins Company. The New York Times. November 5, 1957. Archived from the original on December 25, 2013. Retrieved 2013-12-18.

[3] Rines, G. E. (ed). Robins, Thomas. Encyclopaedia Americana (1920).

[4] Sutcliffe, R. J., Sutcliffe, E. Richard Sutcliffe – The Pioneer of Underground Belt Conveying. Privately printed, 3rd edition. T and A. Constable Ltd., University of Edinburgh (1955).

[5] Phoenix Conveyor Belt Systems GMBH Hannoversche Strasse 88 D-21079 Hamburg, Germany Phone +49-40-7667-03 Fax +49-40-7667-2411 E-mail info@phoenix-cbs.com www.phoenix-conveyor-belts.com

[6] Asian Development Bank description of the Lafarge Surma project. Pid.adb.org. Archived from the original on 2014-01-11. Retrieved 2013-03-27.

[7] Stoppage of limestone supply to Lafarge Surma Indian SC issues show cause notice on central, Meghala. Mines and Communities (2007-06-20). Archived from the original on 2014-01-11 Retrieved 2013-03-27.
[8] Morocco's fish fight: High stakes over Western Sahara. BBC News. 15 December 2011. Archived from the original on 16 December 2011.
[9] Forech Textile Belt catalogue (2016)
[10] Afolabi, D. I., Mayungbe, O. E., Funmilayo, A. D., Bolaji, O. O., Development of a belt conveyor for small scale industry. Journal of Advancement of Engineering and Technology (2017).
[11] Forech Steel Cord Belt Catalogue (2016).
[12] Dunlop, F. Conveyor Handbook. Fenner Dunlop Conveyor Belting Australia, 1–70 (2009).
[13] NIIR Board of Consultants and Engineer. The Complete Book on Rubber Processing and Compounding Technology.
[14] https://patents.google.com/patent/US4769202A/enProcess of making a conveyor belt
[15] Dunlop, F. Conveyor Hand Book. Australia: Fenner Dunlop Conveyor Belting (June 2009).
[16] http://slurrymaster.co.za/_Documents/Conveyor.pdf
[17] https://digibuo.uniovi.es/dspace/bitstream/10651/38442/3/TFMPelayoLopezGRUO.pdf
[18] Rao, D. V. S. (ed). The Belt Conveyor: A Concise Basic Course. Boca Raton, FL, USA: CRC Press (2020).
[19] Fayed, M. E., Skocir, T. Mechanical Conveyors: Selection and Operation.
[20] https://allstateconveyors.com/conveyor-spare-parts/conveyor-belt/
[21] Palmer, R. J. Polyamides, plastics. In: Polyamides, Plastics. Encyclopaedia of Polymer Science and Technology (4th edition). John Wiley & Sons, Inc. (2001). doi: 10.1002/0471440264.pst251.
[22] Viers, B. D. Polymer Data Handbook, Oxford University Press, Inc, 189 (1999). ISBN 978-0195107890.
[23] https://bcjplastics.com.au/nylon-plastic-benefits-and-uses/
[24] Rezaeian, I., Zahedi, P., Rezaeian, A. Rubber adhesion to different substrates and its importance in industrial applications: A review. Journal of Adhesion Science and Technology 26(6):721–744 (March 2012).
[25] Pocius, A. V. Adhesion and Adhesives Technology: An Introduction, Ohio: Hanser, Cincinnati, (1997).
[26] McBain, J. W., Hopkins, D. G., On adhesive and adhesive actions. Journal of Physical Chemistry 29:188 (1925).
[27] http://www.meroller.com/index.php
[28] https://news.bulk-online.com/bulk-solids-handling-archive/aramid-in-conveyor-belts-for-extended-lifetime-energy-savings-and-environmental-effects.html
[29] Bandaru, A. K., Chavan, V. V., Ahmad, S., Alagirusamy, R., Bhatnagar, N. Ballistic impact response of Kevlar reinforced thermoplastic composite armours.
[30] https://www.guilford.edu/original/academic/chemistry/current_courses/chem110/sloan.html
[31] https://conveyorbeltguide.com/steel-cords.html
[32] Kumar, U., Installation and maintenance of belt conveyor system in a thermal plant. IJESS 2(6) (June 2012). ISSN: 2249–9482.
[33] Conveyor and Elevator Textile Belting, Part 1: General Purpose Belting, IS 1891-11994.
[34] https://www.conveyorbeltworld.com/chevron-conveyor-belts.html
[35] Jiotode, A. S., Raut, A. A. Advancement in conveyor system: Pipe conveyor. IJSRD – International Journal for Scientific Research and Development 5(09) (2017). ISSN (online): 2321-0613 All rights reserved by www.ijsrd.com, 352.
[36] JLV Industries PTY LTD. Catalogue of Cable Belt. Harvey, Western Australia 6220.
[37] Dos Santos, J. A. Sandwich belt high angle conveyors – Applications in open pit mining. Bulk Solids Handl (1984).
[38] Belt Bucket Elevator Design. ESBN:C60-592B-1b38-20A1.
[39] Ananth, et al. Design and selecting the proper conveyor-belt. International Journal of Advanced Engineering Technology IV(II):43–49 (April-June, 2013). E-ISSN 0976-3945 IJAET.

[40] IS 11592:2000 Selection and design of Belt Conveyors-Code of Practice.
[41] ISO 703: 2017 – Conveyor Belts – Transverse Flexibility( Troughability)-Test Method.
[42] ISO 340:2022 – Conveyor Belts – Laboratory scale flammability characteristics-Requirements and characteristics.
[43] ISO 284:2012 – Electrical Conductivity – Specification and Test Method.
[44] ISO 20238:2018 – Conveyor Belts – Drum Friction Testing.
[45] ISO 252:2007 – Conveyor Belts Adhesion Between Constitutive elements-Test Methods.
[46] ISO 283:2007 – Textile conveyor belts-Full thickness tensile strength, elongation at break and elongation at reference load- Test method.
[47] ISO 7623:2022 – Steel cord conveyor belts – Cord-to-coating bond test – Initial test and after thermal treatment.
[48] AS 1333: 1994 – Conveyor Belting of elastomeric and steel cord construction.
[49] ISO 8094: 2013 – Steel cord conveyor belts — Adhesion strength test of the cover to the core layer.
[50] https://www.coveya.co.uk/belt-joints-mechanical-vs-vulcanised/#:~:text=A%20critical%20area%20of%20importance,through%20heat%20and%2For%20chemicals
[51] https://www.flexco.com/NA/EN/Flexco/Products/Mechanical-Belt-Fastening-Systems.htm
[52] https://www.fastenersweb.com/proddetail/26751/solid-plate-conveyor-belt-fastener-2-inch
[53] https://www.abec-apollo.co.uk/products/bolt-solid-plate-system
[54] https://foundations.martin-eng.com/learning-center/learning-center/vulcanized-conveyor-belt-splices-part-3
[55] Hardygora, M., Bajda, M., Blazej, R. Laboratory testing of conveyor textile belt joints used in underground mines. Mining Science 22:161–169 (2015).
[56] Chuen-Shii, C., Ching-Liang, L., Wei-Chung, C. Optimum conditions for vulcanizing a fabric conveyor belt with better adhesive strength and less abrasion. Materials and Design 44:172–178 (2013).
[57] Miller, D. Bulk Material conveyor belt installation, vulcanizing and maintenance. Bulk Solids Handling 18(4): (1998).
[58] https://www.rematiptop.com/assets/tech/ind/manuals/REMA-TIP-TOP-Cold-Splice-Instructions.pdf
[59] Simplified life cycle assessment of a return belt conveyor idler, Proceedings of 11th International conference on accomplishments in Electrical and Mechanical Engineering and Information Technology – “DEMI 2013”, Banja Luka, BiH, May 30th–June 1st, University of Banja Luka, Faculty of Mechanical Engineering, ISBN 978-99938-39-46-0, 201–206 (2013).
[60] https://www.dunlopcb.com/conveyor-belting-storage/
[61] https://goodyearrubberproducts.com/2018pdfs/Contitech_Conveyor_Belt_Installation_Manual_Goodyear_Belting/pdf/Contitech_Conveyor_Belt_Installation_Manual_Goodyear_Belting.pdf

Saikat Das Gupta, Hirak Satpathi, Tirthankar Bhandary, and Rabindra Mukhopadhyay

# Chapter 7 Reverse engineering: a tool for the chemical composition analysis of finished rubber products

## 7.1 Introduction

Rubber is a versatile material. Special characteristics like incompressibility, stretchability, viscoelasticity, and low-temperature flexibility make rubber a significant material for wide application. From a very simple product like a rubber band to very complex products like tyres, rubber finds its application successfully. Different types of rubbers are available to suit different applications. These rubbers are categorised into two major groups: natural rubber and synthetic rubber. Rubber products are basically a complex homogeneous mixture of a number of different ingredients. These ingredients include rubber, filler, processing aids, oligomeric resins, antidegradants, accelerators, activators, pigments for colour products, chopped fibres, and special chemicals [1–3]. Each ingredient has typical characteristics and provides a synergistic effect to the rubber compound. The combination of these ingredients varies depending on the application of rubber products [4–8]. Characterisation of finished rubber products with respect to material composition and performance parameters are critical to any manufacturer. This helps the product manufacturer to gain confidence about their products. Understanding of material composition from finished rubber products is generally termed as material reverse engineering. Reverse engineering is a well-known technique in every industry; it helps industry:

- understand the edge of its product or service over its competitor;
- create better marketing strategy;
- support its technical service.

Chemical reverse engineering of rubber products is one of the most critical as well as complex processes. Rubber industry generally deals with two major groups of materials: (1) reactive material and (2) nonreactive materials. Reactive materials in a rubber compound changes its form and chemical structure during different processing operations, whereas nonreactive materials are stable materials and do not change its chemical nature. Due to this chemical change, characterisation of some of the ingredients in rubber product becomes difficult. This chapter will discuss some of the tools and techniques used for characterisation of chemical ingredients in rubber products.

https://doi.org/10.1515/9783110668537-007

Before knowing about reverse engineering, it is very important to have an idea about the different components of finished rubber products. From the name, it is pretty obvious that the major component of finished rubber products is rubber or elastomer.

## 7.2 Ingredients of rubber finished product

### 7.2.1 Rubber

Rubbers are mainly classified into two categories, natural and synthetic. Based on the application, synthetic rubbers are classified into two categories, general purpose rubber and special purpose rubber. Classification of rubber is shown in Figure 7.1.

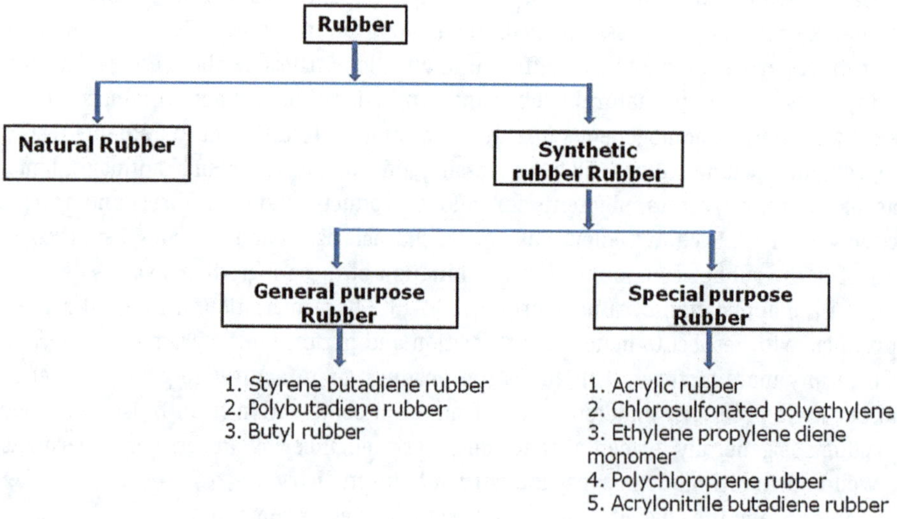


**Figure 7.1:** General classification of rubber.

### 7.2.2 Natural rubber

Natural rubber is produced from the isoprene compound present in latex obtained from the tapping process of rubber tree, *Hevea brasiliensis*. Latex is coagulated using formic acid and purified to prepare different types of technically specified natural rubber such as TSR-10, TSR-20, etc. Based on the dirt content in rubber the grade of natural rubber is defined. The polyisoprene structure is depicted in Figure 7.2.

Natural rubber is being used in several types of products because of its unique properties such as high tensile and tear strength, excellent fatigue resistance, out-

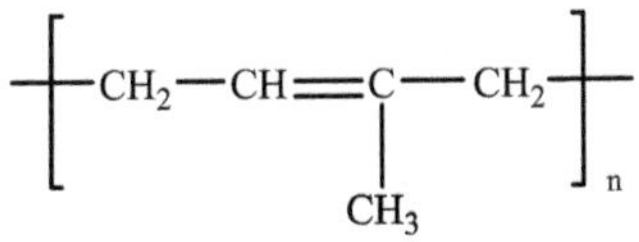

**Figure 7.2:** 1,4-Polysoprene.

standing tack, and green strength. It shows low hysteresis loss, which leads to low heat buildup [9–17]. Due to its strain-induced crystallisation, it is resistant to cutting, chipping, and tearing [18]. One major drawback of natural rubber is its poor resistance to oxygen, ozone, and heat.

## 7.2.3 Styrene-butadiene rubber

Styrene–butadiene rubber (SBR) is a copolymer of styrene and 1,3-butadiene. A typical SBR contains 23% styrene and 67% butadiene. Based on the manufacturing technique, SBR is classified into emulsion SBR (ESBR) and solution SBR (SSBR). Figure 7.3 shows the SBR structure.

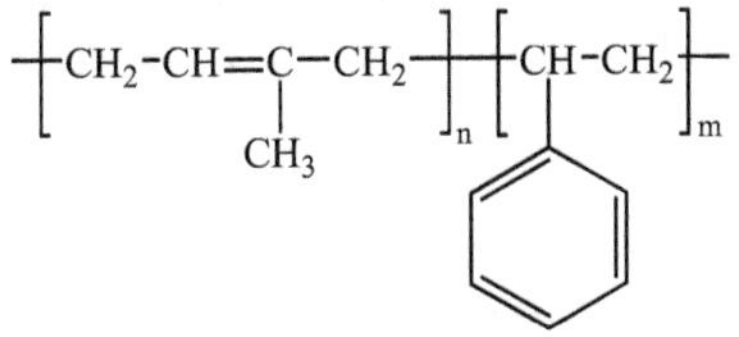

**Figure 7.3:** Styrene–butadiene rubber.

Due to its inherent properties, SBR is becoming a substitute of NR [19–24]. Unlike NR, raw SBR is very weak unless reinforced with carbon black filler. With increase in styrene content in the copolymer backbone of SBR [25–28], tensile strength, and grip, abrasion properties improve in SBR [29–33]. Low-temperature property of SBR is inferior with respect to NR.

## 7.2.4 Polybutadiene rubber

Polybutadiene rubber (PBR) is a synthetic rubber [34–37] manufactured by polymerisation of 1,3-polybutadine monomer. High cis (~96%) content PBR is manufactured using Nd, Co, Ni, or Ti catalyst whereas low cis (~20%) content PBR is manufactured by Li catalysis process.

Due to its low glass transition temperature, PBR exhibits low rolling resistance, high rebound resilience, and excellent abrasion resistance. Due to its inherent properties, apart from tyres industries, PBR is also being used in conveyor belt, hose, shoe soles, etc. [38–45] Figure 7.4 portrays the structure of polybutadiene rubber.

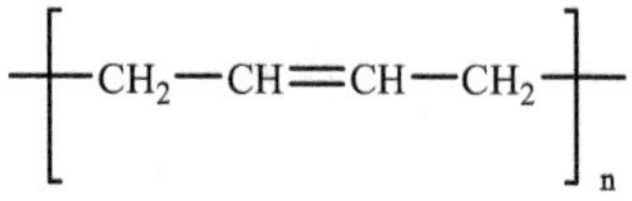

**Figure 7.4:** Polybutadiene rubber.

## 7.2.5 Butyl rubber

Butyl rubber, isoprene isobutylene rubber (IIR), is a copolymer of isobutylene and small quantities of isoprenemonomer [46–54]. Butyl rubber is widely known for its air retention and good flex properties [55–61].

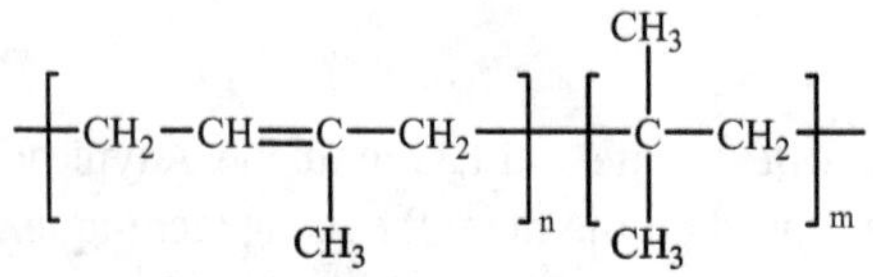

**Figure 7.5:** Butyl rubber.

Due to its good air retention property, butyl rubber is extensively used in tyres inner liners, tyres tubes, and in the medical sector [62–64]. The absence of double bond in the backbone makes butyl rubber weather-resistant. Figure 7.5 shows the structure of butyl rubber.

## 7.2.6 Acrylic rubber

Acrylic rubber, alkyl acrylate copolymer (ACM), is a copolyme, manufactured by emulsion as well as solution polymerisation of butyl acrylate and ethyl acrylate. Due to the polar group presence, ACM is resistant to oil. ACM manifests good resistance to weathering [65–74].

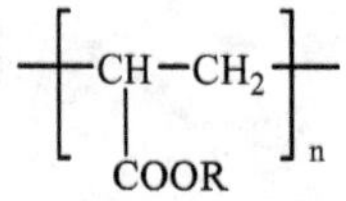

**Figure 7.6:** ACM rubber.

Due to its inherent properties, ACM is being used where combined resistance of oil as well as heat is required [75–82]. Acid and water resistivity of ACM rubber is very poor. Figure 7.6 depicts the structure of ACM rubber.

## 7.2.7 Chlorosulfonated polyethylene rubber

Chlorosulfonated polyethylene rubber (CSM) [83–91] is a synthetic polymer manufactured by treatment of polyethylene with chlorine and sulphuric acid under UV radiation.

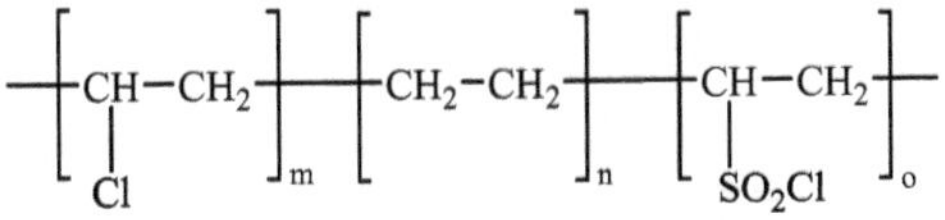

Figure 7.7: Chlorosulfonated polyethylene rubber.

Due to its extraordinary weather resistance, CSM is being used in various automobile, hose, rubber gasket, and seal industries [92–98]. But poor compression set property resists usage of CSM in dynamic seal applications. Fuel resistivity of CSM is very poor unlike ACM and NBR. Figure 7.7 illustrates the structure of CSM.

## 7.2.8 Ethylene propylene diene monomer

Ethylene–propylene–diene monomer (EPDM), a copolymer of ethylene, propylene, and dienes such as ethylidene norbornene, dicyclopentadiene, and vinyl norbornene, is considered as the most water-resistant rubber [99–106]. The absence of unsaturation in backbone makes EPDM rubber resistant to heat, ozone, and UV radiation. EPDM rubber exhibits very good electrical resistivity and resistance to abrasion and tearing [107–114]. Due to its inherent characteristics, EPDM is being used in several industries like tyres, seal, conveyor belt, O-ring, electrical insulator, vibrator, etc. Figure 7.8 structure of EPDM rubber.

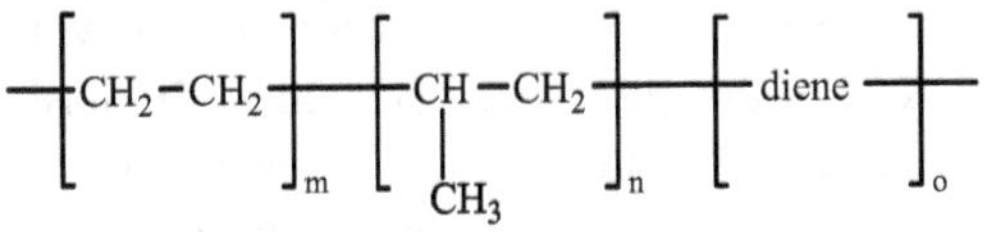

Figure 7.8: EPDM rubber.

## 7.2.9 Polychloroprene rubber

Polychloroprene rubber (CR) or chloro-butadiene rubber is commonly known as neoprene rubber. It is manufactured by free radical emulsion polymerisation of 2-chloro-1,3-butadiene.

The presence of chlorine in the backbone of the polymer reduces the reactivity towards oxygen and ozone. Chloroprene rubber is well known for its self-extinguishing property and good metal-to-rubber bonding. The major applications of chloroprene rubber are in cable, transmission belt, gloves, and gasket industries [115–122]. Figure 7.9 illustrates the structure of polychloroprene rubber.

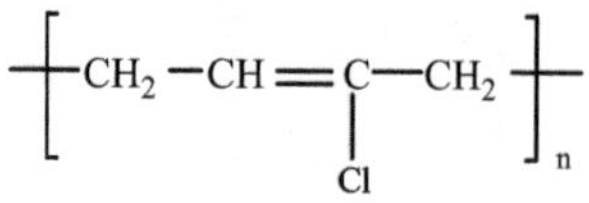

**Figure 7.9:** polychloroprene rubber.

## 7.2.10 Acrylonitrile butadiene rubber

Acrylonitrile rubber, famous as nitrile rubber (NBR) is produced by free radical polymerisation of butadiene monomer and 10–50% of acrylonitrile monomer. The presenceof polar acrylonitrile group makes NBR oil-resistant. With increase in acrylonitrile group, oil resistivity and abrasion resistivity increase and the low temperature property decreases. NBR with suitable acrylonitrile content is used in a wide variety of applications such as roller, gloves, fuel hose, and some automotive parts [122–132]. Figure 7.10 depicts the structure of NBR.

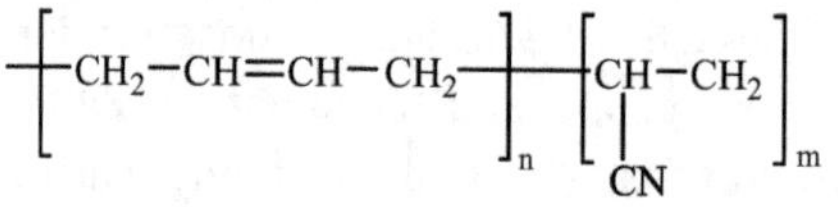

**Figure 7.10:** Acrylonitrile butadiene rubber.

At low temperature rubber exhibits brittleness and in high temperature it tends to flow. To avoid this difficulty, cross-linking of polymeric chain is done by vulcanisation process. To get the desired properties of application, several types of chemicals, viz., process aids, vulcanising agents, fillers, softeners, antioxidants, antiozonants, blowing agent, etc. are used. Even though the nature of the elastomer determines the basic properties of the products, it can be enhanced significantly by using different kinds of ingredients and by varying its amount in the rubber compound. In general, the materials used in the rubber compound, apart from elastomer, are classified as follows.

## 7.2.11 Fillers (carbon black)

Various ASTM and non-ASTM-grade furnace blacks are extensively used as reinforcement fillers for rubber compounding. Using ASTM D1765, the nomenclature of carbon black is done based on its particle size and structure. The name of carbon black consists of one letter (N or S) and three numbers, where 'N' and 'S' denote normal cure and slow cure, respectively. The first number expresses the particle size of the carbon black [133] whereas the remaining numbers demonstrate the structure of the black. Physical properties of rubber vulcanisates depend on the particle size and structure of the carbon black [134–141].

With increase in particle size or surface area of a carbon black, tensile strength, abrasion resistance, tread wear, and tear resistance for a rubber compound improve while deterioration in viscosity, dispersion, rebound, and dynamic properties takes place. Physical properties of a rubber compound such as viscosity, modulus, stiffness, hardness, conductivity, and dispersibility meliorate, whereas swelling during extrusion and tear strength are decreased [142–151].

The conductivity of a carbon black depends on the particle size as well as structure of a carbon black. Conductivity of carbon black increases with decrease in particle size and increase in structure. The typical properties of carbon black are shown in Table 7.1

Physical properties of a rubber vulcanisate depend on the loading of carbon black as well. With increase in carbon black loading in NR, SBR, or NR-SBR blended rubber compounds, hardness increases with decrease in resilience and wear rate. Modulus at 100% elongation, tensile strength, and elongation at break decrease after increasing to an optimum level [152–160]. A similar study with increase in carbon black loading in NBR rubber vulcanisate displays that tensile strength, modulus at 100% elongation, hardness, and compression set increases with decrease in elongation at break.

**Table 7.1:** Key properties of various ASTM-grade carbon blacks.

| S. no. | Carbon black | Primary particle size (nm) | Surface area ($m^2/g$) | Structure (OAN, Oil absorption no., mL/100 g) |
|---|---|---|---|---|
| 1 | N110 | 15–18 | 139–151 | 108–118 |
| 2 | N134 | 15–18 | 137–147 | 122–132 |
| 3 | N220 | 20–25 | 116–126 | 109–119 |
| 4 | N234 | 28–36 | 115–125 | 120–130 |
| 5 | N326 | 28–36 | 77–87 | 68–76 |
| 6 | N330 | 28–36 | 77–87 | 97–107 |
| 7 | N339 | 28–36 | 86–94 | 115–125 |
| 8 | N550 | 39–55 | 38–48 | 116–126 |
| 9 | N660 | 56–70 | 31–41 | 86–94 |
| 10 | N770 | 71–96 | 24–34 | 67–77 |
| 11 | N990 | 250–350 | 7–9 | 33–43 |

## 7.2.12 Plasticiser

For better processability, based on the solubility with rubber, two categories of plasticiser oils are being used in rubber industries. Those are mostly petroleum-based and ester-type. Based on the basic constituent such as aromatic, naphthenic, and paraffinic content, petroleum-based plasticisers are broadly classified into three categories: aromatic type, naphthenic type, and aromatic type. Polarity of the paraffinic oil is the least whereas aromatic oil shows maximum polarity. Polar rubbers containing hetero atoms, such as NBR and ACM are not compatible with petroleum oil. To enhance the plasticising effect for such rubbers, ester-type plasticisers, viz., dioctylsebacate, dibutylsebacate, dioctyl phthalate, dibutyl phthalate, trioctyl phosphate, and triethylene glycol dimethacrylate are added [161–168].

### 7.2.13 Antidegradants

Ageing of rubber occurs through reaction between oxygen and ozone with rubber backbone in the presence of light or heat [169–173]. Some of the antidegradants such as microcrystalline wax, paraffinic wax, etc. make a layer on the surface of the products to protect oxygen and ozone attack. Another type of antidegradant, such as *para*-phenylenediamine-type, quinolone-type, and phenolic type is used. These types of antioxidants bloom to the surface of the products and react with oxygen and ozone before it attacks the rubber [174–181]. Figure 7.11 shows details of different antidegradants used in the rubber industry.

### 7.2.14 Processing aids

Processing aids are materials used to modify rubber during the mixing or processing steps such as extrusion, calendering, or moulding operation. This includes peptisers and plasticiser. The purpose of processing aids is to enhance the processability of rubber compounds by achieving the desired viscosity.

### 7.2.15 Vulcanising agent

Vulcanising agents are essential as cross-link between the polymeric chains to enhance the physical properties. In case of unsaturated rubber such as NR, SBR, or BR, sulphur is used as a vulcanising agent. For saturated rubber such as EPDM and IIR, peroxides are used to cross-link the polymer backbone. In some cases, metal oxides are used to cure the halogenated rubber such as CIIR, BIIR, and CR [182–189].

### 7.2.16 Accelerator

To increase the rate of sulphur vulcanisation reaction, sulfonamide, guanidine, or thiazole types of accelerator are commonly used [190–194]. A detailed list of accelerators is shown in Table 7.2.

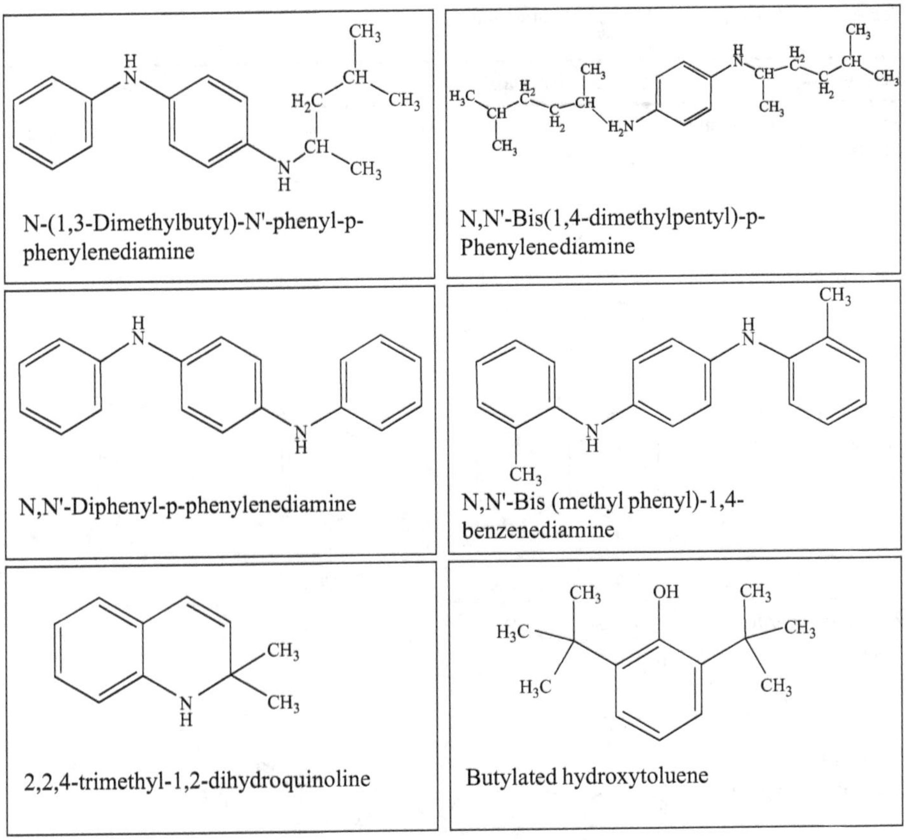


**Figure 7.11:** Different antidegradants used in rubber industries.

### 7.2.17 Accelerator activator

This ingredient produces chemical complexes with accelerator and thus obtains the maximum benefits from an acceleration system by increasing vulcanisation rates and improving the properties. A combination of zinc oxide and stearic acid is most commonly used in sulphur vulcanisation [195–199]. The stepwise vulcanisation process is detailed in Table 7.3.

## 7.3 Reverse engineering concept

It has already been noticed from the previous discussion that to achieve the desired processability and physical properties, several chemicals are added to the rubber matrix. Some of the chemicals such as accelerator, vulcanising agent, etc. change their

**Table 7.2:** Commonly used rubber accelerator with chemical formula.

| Accelerator | Chemical structure | Remarks |
|---|---|---|
| Hexamethylenetetramine (HMT) | 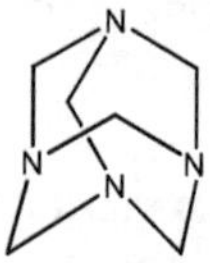 | Group: amines |
| Diphenyl guanidine (DPG) | 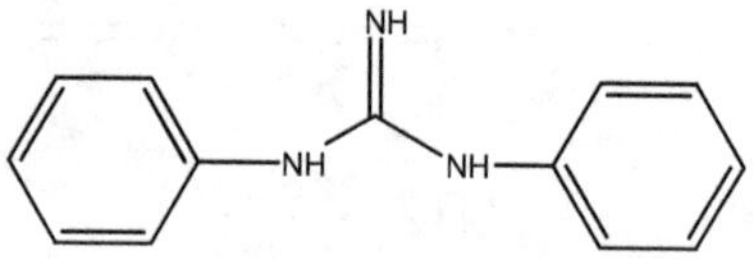 | Group: guanidines |
| *N, N'*-Diorthotolyl guanidine (DOTG) | 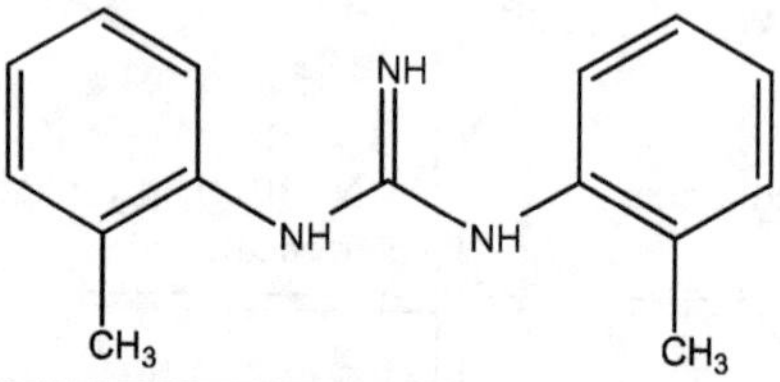 | Group: guanidines |
| 2-Mercaptobenzothiazole (MBT) | 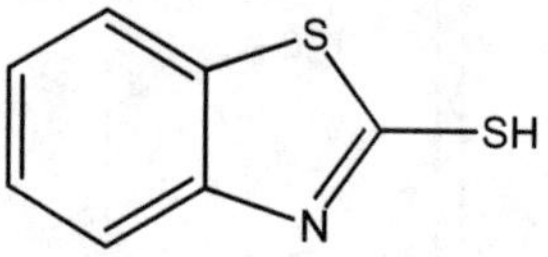 | Group: thiazoles |
| 2-2′-Dithiobis (benzothiazole) (MBTS) | 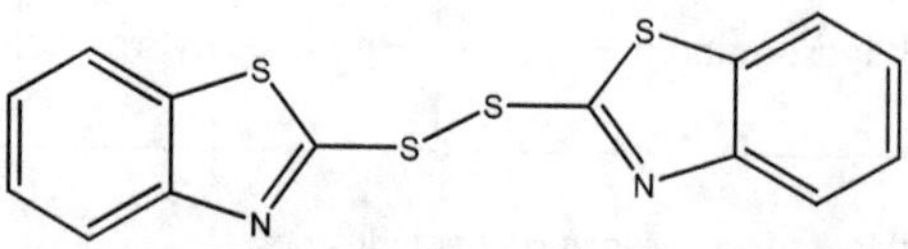 | Group: thiazoles |
| Zinc-2-mercaptobenzothiazole (ZMBT) | 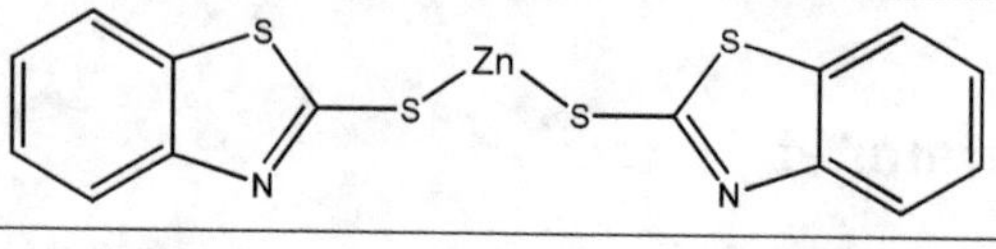 | Group: thiazoles |
| *N*-Cyclohexyl-2-benzothiazole sulfenamide (CBS) | 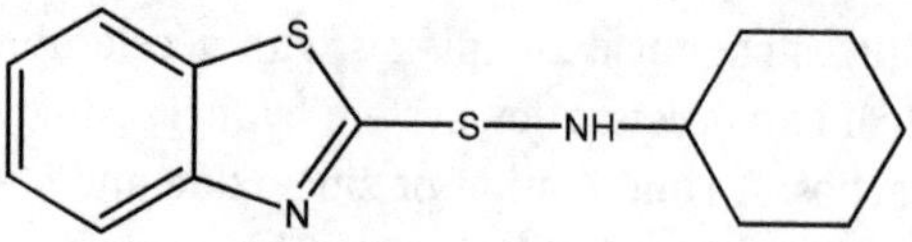 | Group: sulfenamides |
| N-tert-butyl-2-benzothiazole sulfenamide (TBBS) | 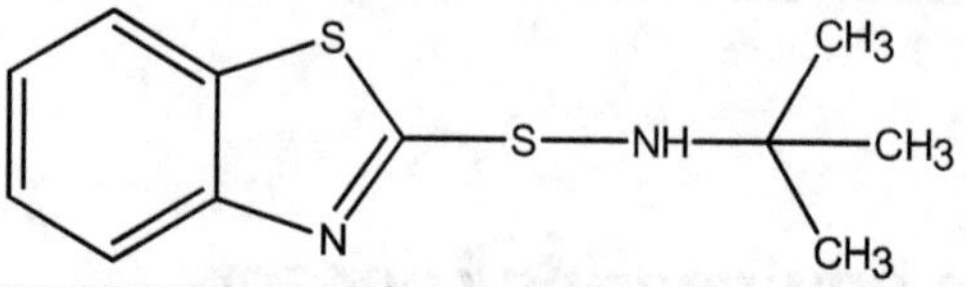 | Group: sulfenamides |

**Table 7.2** (continued)

| Accelerator | Chemical structure | Remarks |
|---|---|---|
| 2-(4-Morpholinothio)-benzothiazole (MBS) | 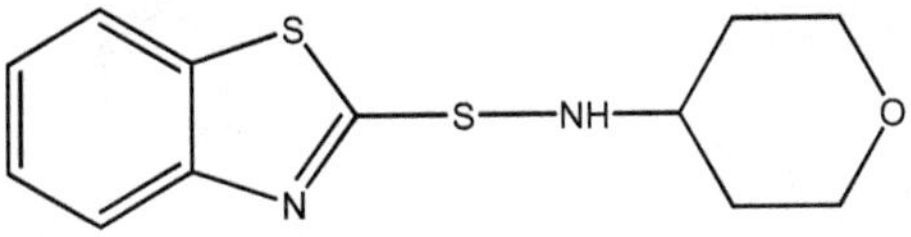  | Group: sulfenamides |
| *N,N′*-Dicyclohexyl-2-benzothiazole sulfenamide (DCBS) | 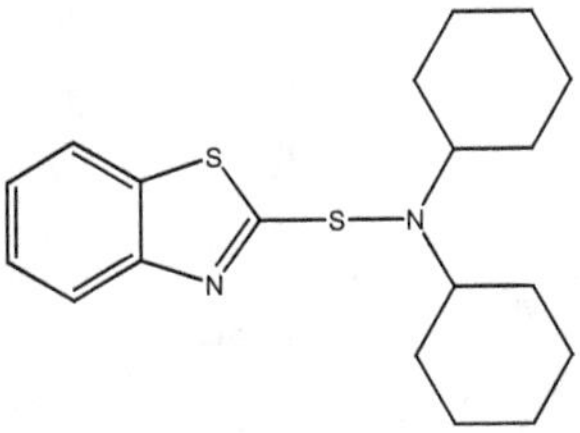  | Group: sulfenamides |
| Ethylene thiourea (ETU) | 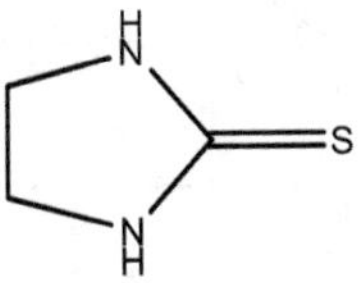  | Group: thioureas |
| Dibutyl thiourea (DBTU) | 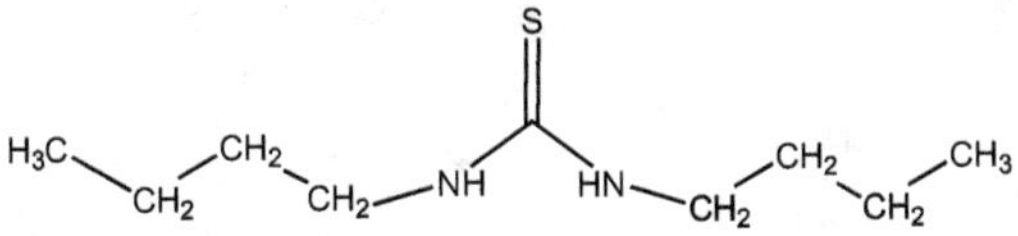  | Group: thioureas |
| Tetramethylthiuram monosulfide (TMTM) | 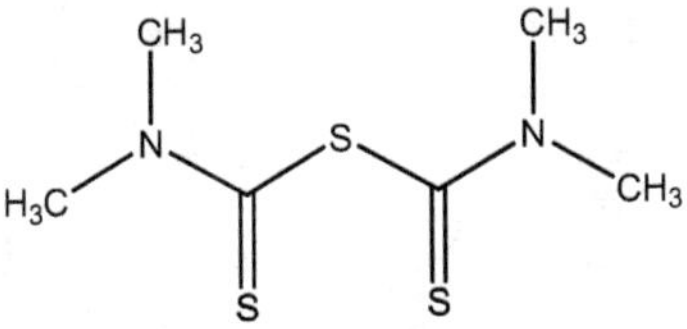  | Group: thiurams |
| Tetramethylthiuram disulfide (TMTD) | 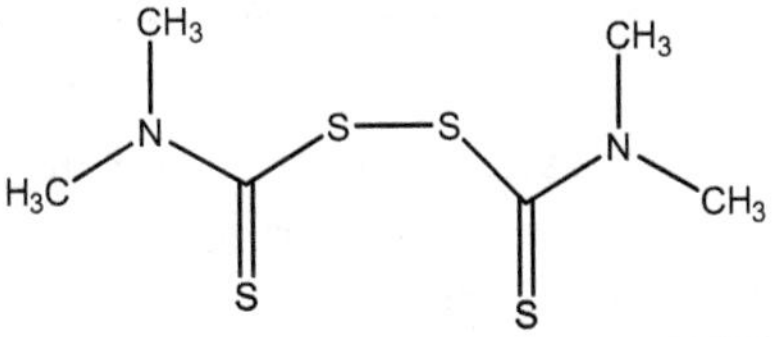  | Group: thiurams |
| Dipentamethylenethiuram tetrasulfide (DPTT) | 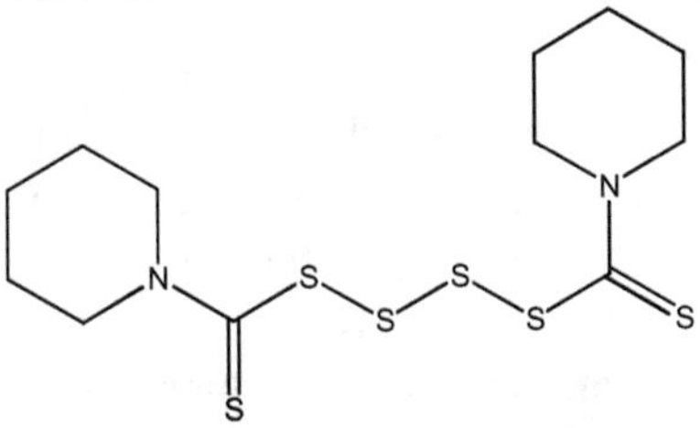  | Group: thiurams |

**Table 7.2** (continued)

| Accelerator | Chemical structure | Remarks |
|---|---|---|
| Tetrabenzylthiuram sisulfide (TBzTD) | 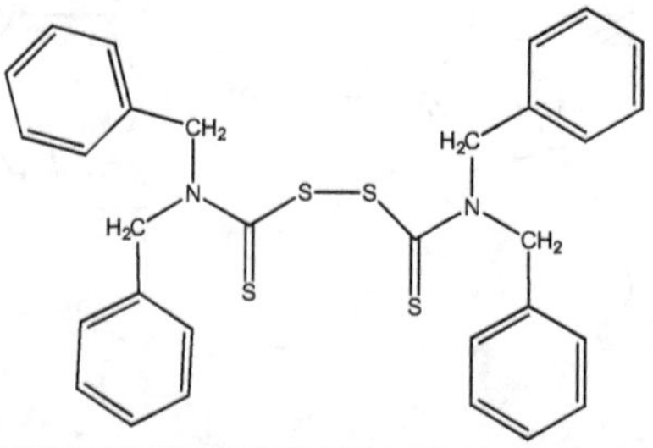 | Group: thiurams |
| Zinc dimethyldithiocarbamate (ZDMC) | 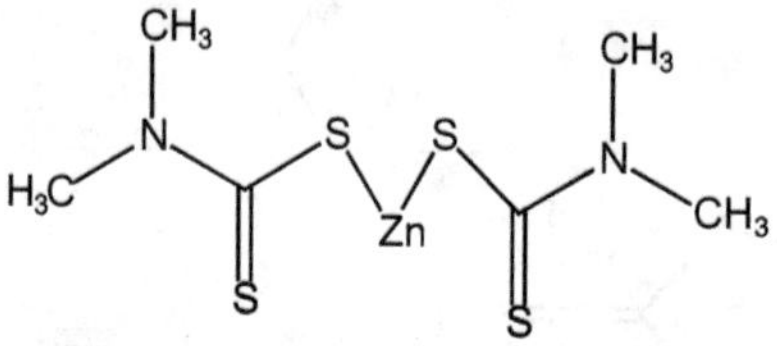 | Group: dithiocarbamates |
| Zinc diethyldithiocarbamate (ZDEC) | 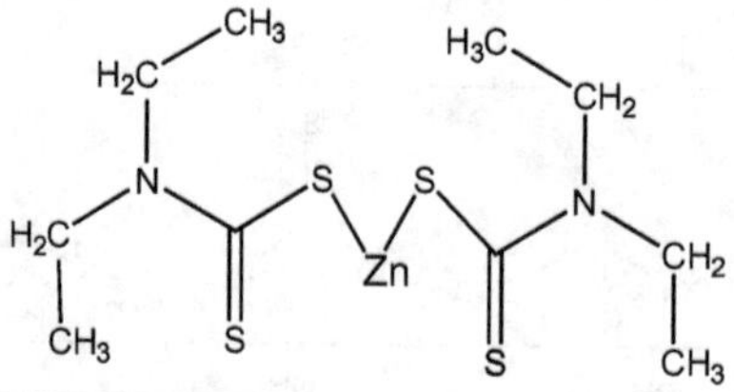 | Group: dithiocarbamates |
| Zinc dibutyldithiocarbamate (ZDBC) | 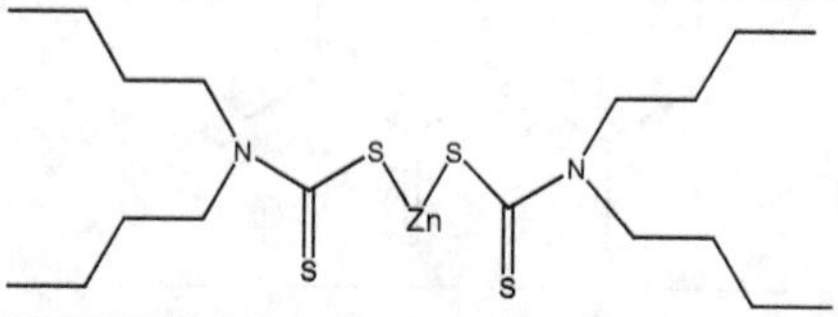 | Group: dithiocarbamates |

chemical properties during vulcanisation reaction. It is a challenge for an analyst to identify those chemicals from a composite. The process of identifying the raw materials used in finished product is known as reverse engineering.

Reverse engineering is a very common process in several industries to survive in the competition and to identify the efficient way of achieving excellence by introducing new products in competitive markets. The three pillars of reverse engineering in rubber industry are design reverse engineering, process reverse engineering, and material reverse engineering. Due to convolution in identifying the materials used in rubber products, samples undergo several chemical tests along with advanced instrumental techniques.

Reverse engineering of polymeric products includes chemical analysis to estimate extractable content, metal oxide content, separation of extractable to estimate oil, antidegradants, and wax content. A simple graphical abstract of all the techniques in-

**Table 7.3:** Stepwise vulcanisation process of rubber.

| | |
|---|---|
| Formation of active accelerator complex | 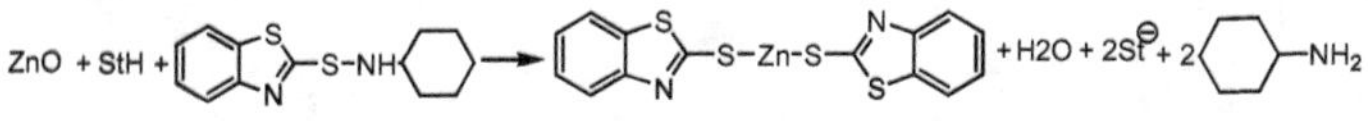 |
| Formation of active sulfurating agent | 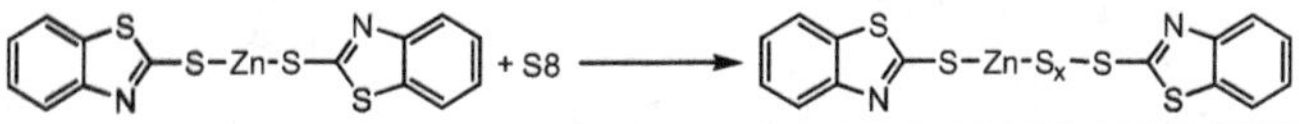 |
| Formation of cross-linking intermediate |  |
| Formation of polysulfide bond |  |
| Final vulcanisation | 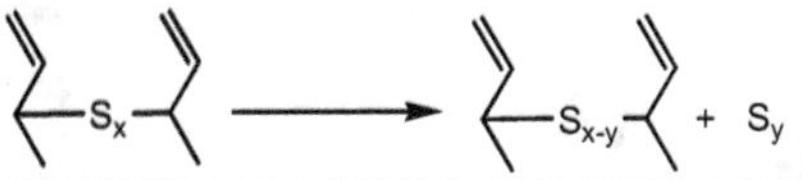 |

volved in reverse engineering of finished rubber products is illustrated in Figure 7.12 with the following discussions.

## 7.3.1 Extractable material content

A portion of vulcanised sample was crushed in a two roll mill and extracted in Soxhlet apparatus using suitable solvent. Plasticiser, antidegradant, unreacted accelerator, and unreacted sulphur leached out in the solvent [200]. Extractable material content can be calculated by the following equation:

$$\text{Extractable material content}(\%) = \frac{\text{Initial weight of the sample} - \text{final weight of the sample}}{\text{Initial weight of the sample}} \times 100$$

## 7.3.2 Separation on extractable material

The lower portion of the solvent extract is evaporated using rotary vacuum evaporator and the solid mass in the round-bottom flask has been separated through column chromatography. A 60–120 mesh silica filled column has been used in hexane medium. Solid mass obtained from evaporation, dissolved in hexane and poured into the column. After collection of hexane part, toluene has been poured into the column. Once the toluene part is collected, in the next stage, methanol is poured into the top of the column. The hexane part obtained from column chromatography consists of wax and is further ana-

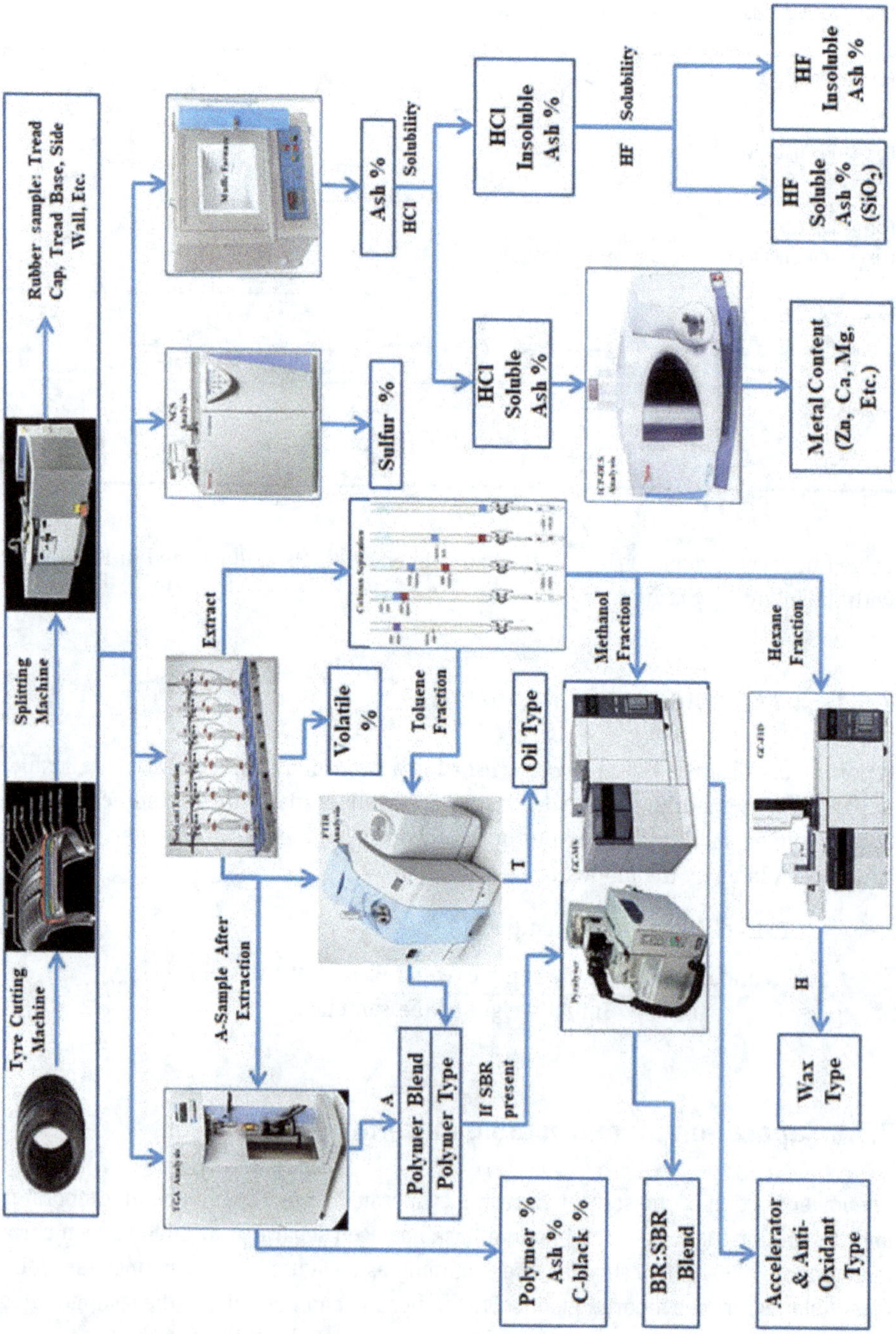


**Figure 7.12:** Chemical composition analysis process with the entire techniques involved for finished rubber product step-by-step.

lysed using gas chromatography to identify the type of wax used in the product. The toluene part is evaporated to obtain the dissolved plasticiser and carbon type analysis of the masses done to identify the type of oil used. Methanol part is analysed using gas chromatography–mass spectroscopy (GC-MS) to identify the presence of antidegradants, for example, DPPD, 6PPD, and TMQ, and accelerator used in the vulcanisation process.

### 7.3.3 Analysis of metal oxide

A portion of the sample is taken in a silica crucible having known weight and burnt at 600 °C inside a muffle furnace [200]. Total metal oxide, commonly known as ash content is calculated using the following equation:

$$\text{Total metal oxide content } (\%) = \frac{\text{Weight of cruible with ash} - \text{initial weight of crucible}}{\text{Initial weight of sample}} \times 100$$

The residue obtained from ashing process is dissolved in 1:1 dilute hydrochloric acid and filtered through a sintered glass crucible (G4 grade). Different metal contents in the filtrate can be estimated through titration with ethylenediamine tetra-acetoacetate (EDTA) or instrumental techniques, such as inductively coupled plasma (ICP-OES) or atomic absorption spectroscopy (AAS). A portion of the residue obtained on the sintered glass crucible is digested in hydrofluoric acid and evaporated at 550 °C. Silica present in the residue reacts with hydrofluoric acid to form silicon tetrafluoride ($SiF_4$) and is evaporated at 550 °C. Silica content can be obtained from the following equation:

$$\text{Silica content in ash } (\%) = \frac{\text{Sample weight} - \text{weight of residue}}{\text{Sample weight}} \times 100$$

Throughout reverse engineering of polymeric composites, instruments such as thermogravimetric analyser (TGA), differential scanning calorimeter, elemental analyser, inductive coupled plasma, gas chromatography coupled with mass spectroscopy, and Fourier-transform infrared spectrometer (FTIR) are essential.

### 7.3.4 Thermogravimetric analyser (TGA)

TGA is the basic equipment used to perform composition analysis of polymeric products. Mass change of a specimen is continuously monitored with change in temperature and time. A portion of the specimen is heated at a constant heating rate under inert atmosphere till the organic parts decompose out. Switching the gas into oxygen or air is done to burn the carbon black. During the process, metal oxide portion present in the sample remains as residue [201–207]. Figure 7.13 shows a typical TGA thermogram for rubber compounds.

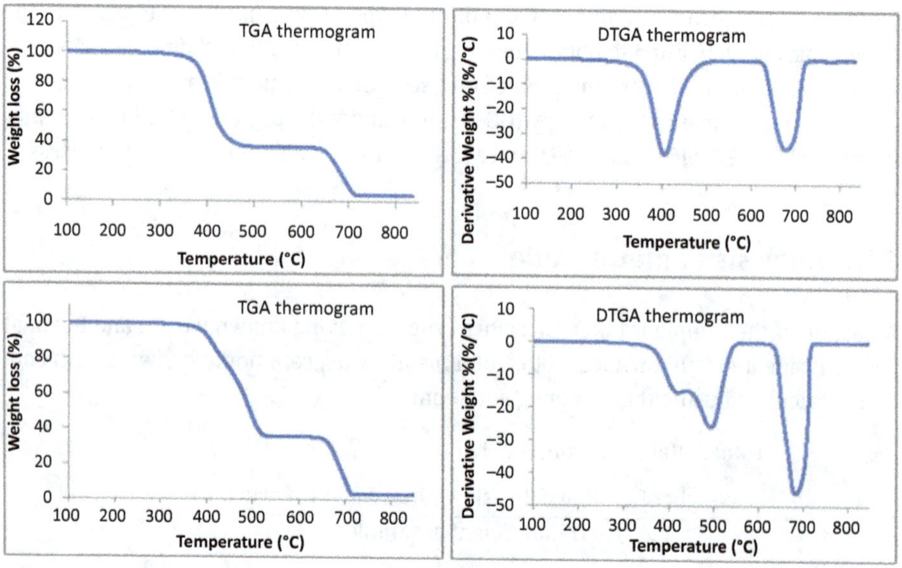


**Figure 7.13:** Typical TGA thermogram of rubber products.

## 7.3.5 Differential scanning calorimeter (DSC)

During the DSC study, a rubber specimen is analysed with reference to a specimen blank. During the temperature scan, specimen and reference holder are heated. Enthalpy change due to transformation in physical and chemical properties as a function of temperature is measured. Glass transition temperature, melting point, crystallisation temperature, oxygen induction time, reaction kinetics, etc. can be measured using DSC. During the reverse engineering process, glass transition temperature of the sample is measured to get an idea of the polymer used in the product [208–213]. Figure 7.14 shows a typical DSC thermogram for rubber compounds.

## 7.3.6 Elemental analyser

An elemental analyser is used to measure nitrogen, carbon, hydrogen, and sulphur content in the sample. After burning under controlled conditions, a portion of the rubber specimen passes through a layer filled with copper to convert nitrogen oxide ($NO_X$) into nitrogen ($N_2$). Individual gases, separated while passing through gas chromatography column, provide signals based on their thermal conductivity [214]. Figure 7.15 shows a typical graph obtained from an elemental analyser.

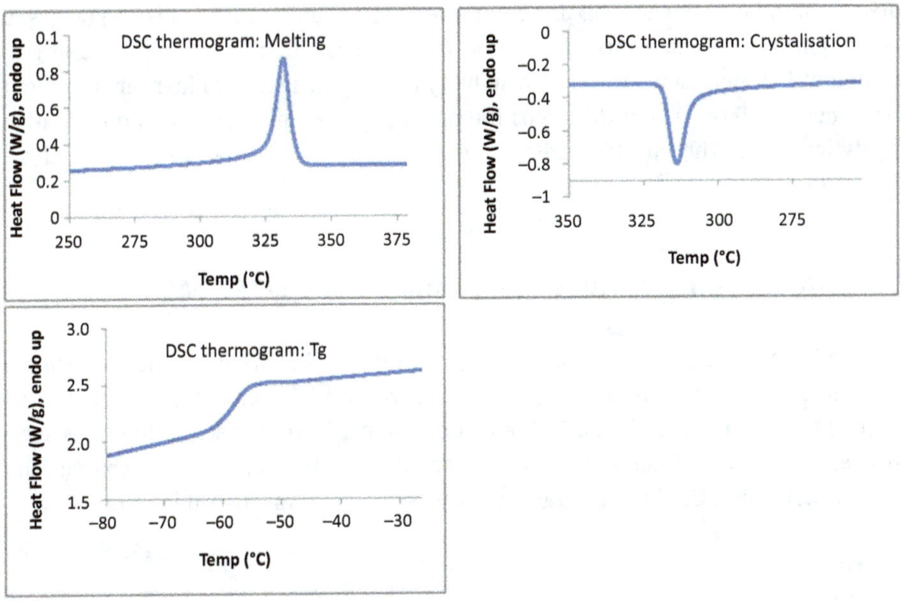


**Figure 7.14:** Typical DSC thermogram of rubber products.

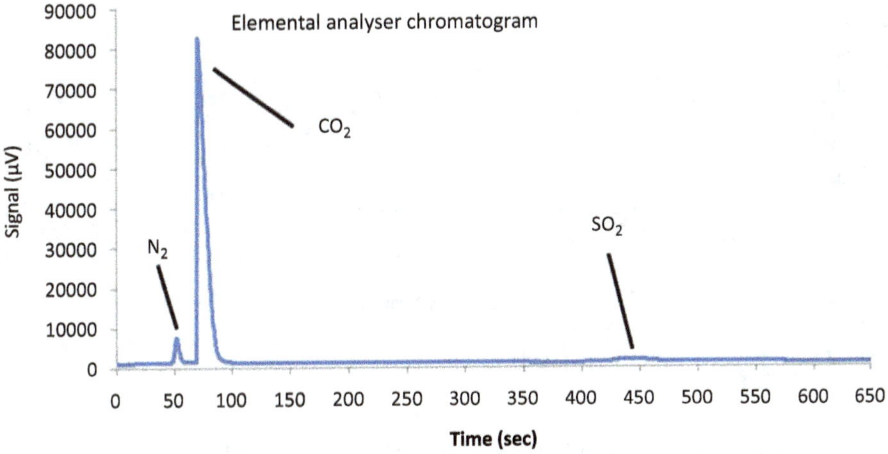


**Figure 7.15:** Typical elemental analyzer diagram of rubber products.

## 7.3.7 Inductive coupled plasma (ICP-OES)

During the reverse engineering process, ICP is used to determine the metal ion content in finished products. Metal oxide obtained from the ashing process of sample is solubilised in dilute hydrochloride acid and analysed through ICP. The energy generated from radio frequency coil present inside the instrument helps ionise argon gas to create plasma. Plasma excites metal atoms, which leads to photon emission and ionisation of the metal ions. It has high stability and is considered as inert environment with low interference and this helps ICP produce better qualitative and quantitative analytical data. The emitted radiation from the plasma is then used for analysis [215–220].

### 7.3.8 Gas chromatography–mass spectroscopy (GC-MS)

The stationary phase of GC columns mostly consists of silica. While passing through the column, the mixture of materials injected to the preheated GC column is separated based on their boiling point and polarity. Separated chemicals are ionised in the mass spectrometer and the molecules are identified based on their unique mass-to-charge ratio [221–228]. A typical GC chromatogram and mass spectrum is mentioned in Figure 7.16.

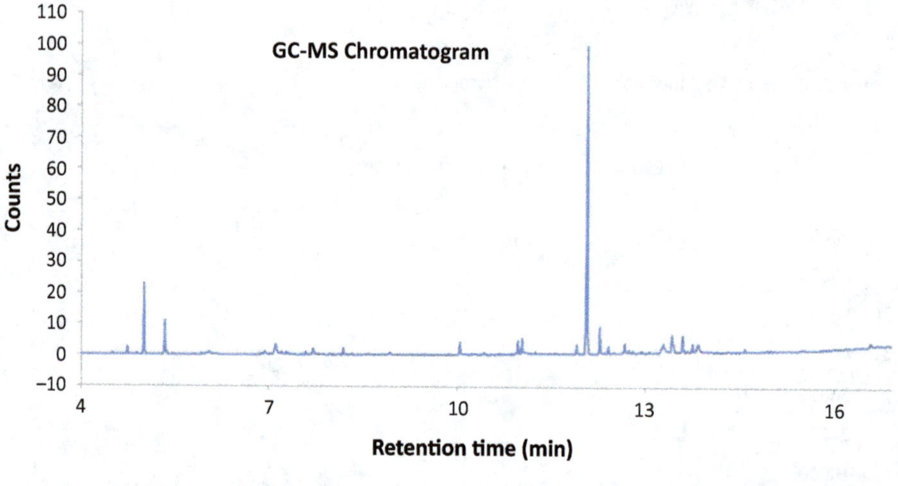


**Figure 7.16:** A typical GC chromatogram and mass spectrum.

### 7.3.9 Fourier-transform infrared spectrometer (FTIR)

Atoms in every molecule have a periodic motion called molecular vibration. The pattern of vibration can be stretching, scissoring, rocking, wagging, or twisting. When infrared light passes through the sample, molecules absorb energy similar to the wavelength of molecular vibration. From the absorbed light spectra, characterisation of a chemical is done by FTIR [109, 200, 229–234]. FTIR peaks of different rubber functional groups are tabulated in Table 7.4.

**Table 7.4:** FTIR peaks of different rubbers.

| S. No. | Peak at ($cm^{-1}$) | Functional group |
|---|---|---|
| 1 | 699 | 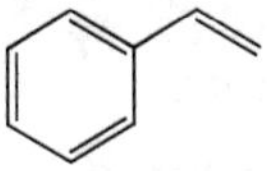 |
| 2 | 722 | —CH2— |
| 3 | 770 | —C—Cl (with bonds above and below the C) |
| 4 | 887 | >C=$CH_2$ |
| 5 | 909 | —CH=$CH_2$ (vinyl) |
| 6 | 965 | —CH=CH— (trans) |
| 7 | 990 | —CH=$CH_2$ |
| 8 | 1,365 | >CH—$(CH_3)_2$ |
| 9 | 1,375 | —$CH_3$ |
| 10 | 1,385 | >CH—$(CH_3)_2$ |
| 11 | 1,494 | 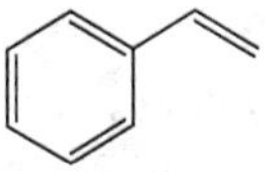 |
| 12 | 1,740 | >C=O |
| 13 | 2,239 | —CN |

## 7.4 Case studies of formulation reconstruction

Some case studies for analysing data obtained from instruments, viz., TGA, FTIR, DSC, and GC-MS during the reverse engineering process are demonstrated below.

### 7.4.1 Case study 1

A rubber part (bias tyres tread) was sliced and crushed in two roll mill to increase the surface area. 2.598 g of crushed sample was wrapped in filter paper and extracted for 16 h. The extracted sample was dried. Then the sample was cooled to room temperature and the noted weight was 2.338 g. The calculated extractable content present in the sample was:

$$\text{Extractable content in the tread sample} = \frac{2.598 - 2.338}{2.598} \times 100 = 10.0\%$$

A portion of the crushed sample was analysed in TGA to get the carbon black content and residue content. During TGA study, the sample was heated from 100 to 850 °C at 40 °C/min heating rate. From 100 to 600 °C, to get decomposition characteristics of the polymer, the sample was heated under nitrogen atmosphere and switched to oxygen at 600–850 °C to get the details of carbon black content. Once the atmosphere changed to oxygen, the sample started losing weight due to burning of carbon black. Weight loss from 600 to 850 °C was noted as carbon black present in the sample and unburnt residue at 850 °C was noted as ash content in the sample.

Another portion of the sample was analysed in DSC to get glass transition temperature of the specimen. While measuring glass transition temperature through DSC the sample was heated from (−)90 °C to (+)20 °C at 10 °C/min heating rate.

A portion of the extracted sample was pyrolysed and analysed through FTIR in transmittance mode and the obtained spectrum was analysed to identify the polymer type.

From TGA study of the sample, it is found that the degradation temperature of the polymer is ~410 °C (Figure 7.17). Difference in weight percentage from 600 to 850 °C is 25.0 and residue at 850 °C is found as 5.5, which denote that the sample contains 25.0% carbon black and 5.5% ash. From DSC study of the sample, glass transition temperature of the sample is obtained at (−)60 °C (Figure 7.17). From the FTIR analysis it is found that two characteristic peaks at 887 $cm^{-1}$ and 1,375 $cm^{-1}$ are present, which confirms that the sample contains $>C=CH_2$ and $-CH_3$ functional group. So, the polymer is confirmed to be isoprene rubber, that is, NR.

From the above study, the extractable content in the sample is found to be 10.0%, carbon black content 25.0%, ash content 5.5%, and the base polymer is NR. So, the NR content is (100–10.0–25.0–5.5)% = 59.5%.

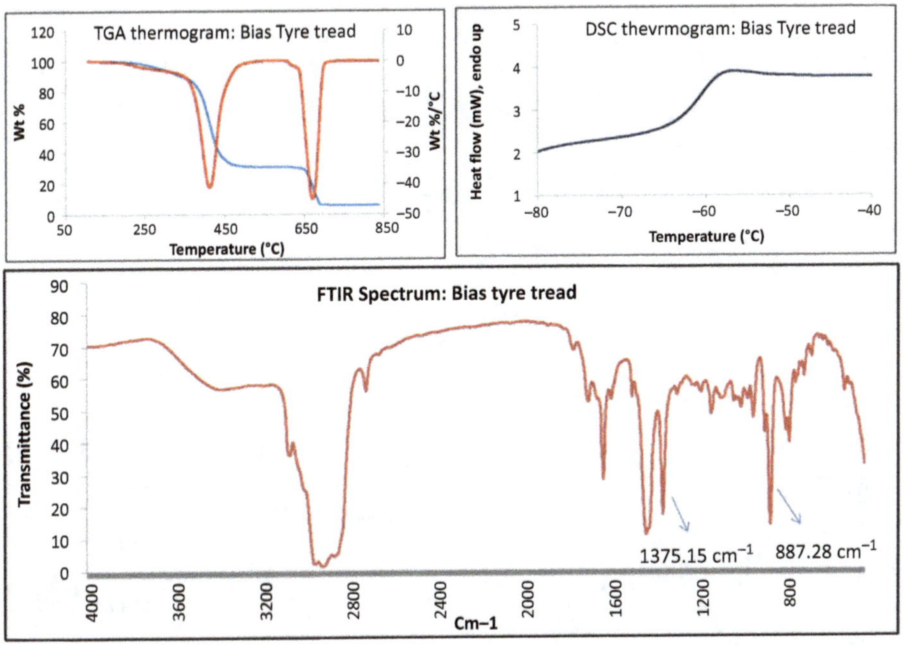


**Figure 7.17:** TGA, DSC, and FTIR analysis for the bias tyres tread (Case study 1).

The composition of bias tyres tread is as follows:

| Parameter | Result |
|---|---|
| Polymer content (%) | 59.5 |
| Carbon black content (%) | 25.0 |
| Ash content (%) | 5.5 |
| Extractable content (%) | 10.0 |
| Polymer type | Natural rubber |

## 7.4.2 Case study 2

A PCR tyres tread sample was collected and analyzed in the same way as briefed in Case study 1 and the obtained results are mentioned in Figure 7.18. Extractable content in the sample obtained was 17.9%.

TGA thermogram (Figure 7.18) obtained from analysis of the PCR tyres tread sample displays degradation peak of polymer at ~485 °C. The difference in weight from 600 to 850 °C is 38.0% and residue at 850 °C is 1.8%. Glass transition temperature from Figure 7.18 is obtained at (−)52.9 °C. FTIR spectrum of the pyrolysed extract indicates major characteristic peaks at 699 cm$^{-1}$, 909 cm$^{-1}$, 965 cm$^{-1}$, 1,375 cm$^{-1}$, and 1,494 cm$^{-1}$

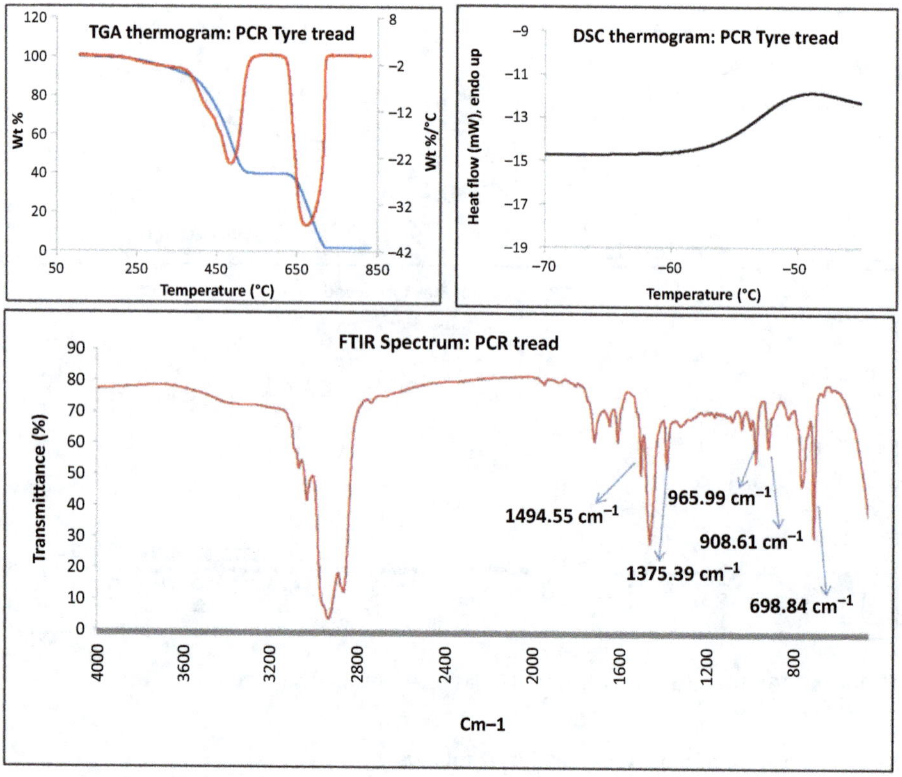


**Figure 7.18:** TGA, DSC, and FTIR analysis for the passenger car radial (PCR) tyres tread (Case study 2).

which validates the presence of styrene, trans —CH=CH—, vinyl, and $-CH_3$ functional group.

From the above study, one may conclude that the basic composition of a PCR tyres tread is as follows:

| Parameter | Result |
|---|---|
| Polymer content (%) | 42.3 |
| Carbon black content (%) | 38.0 |
| Ash content (%) | 1.8 |
| Extractable content (%) | 17.9 |
| Polymer type | Styrene-butadiene rubber |

### 7.4.3 Case study 3

A PCR inner liner sample was collected and analysed to get the composition details. All the testing parameters were kept the same as in previous case studies and percent extractable found in the inner liner sample was 8.9%.

TGA study (Figure 7.19) of PCR inner liner confirmed that the degradation temperature of the sample is ~450 °C and weight loss in oxygen atmosphere is 33.6% and the obtained residue at 850 °C is 1.9%.

DSC study of the inner liner sample (Figure 7.19) reveals that the glass transition temperature of the sample is (−)63.5 °C. From FTIR spectra, characteristic peaks at 888 $cm^{-1}$, 1,365 $cm^{-1}$, and 1,385 $cm^{-1}$ confirm that the sample contains isoprene, isobutylene, and $-CH_3$ functional group.

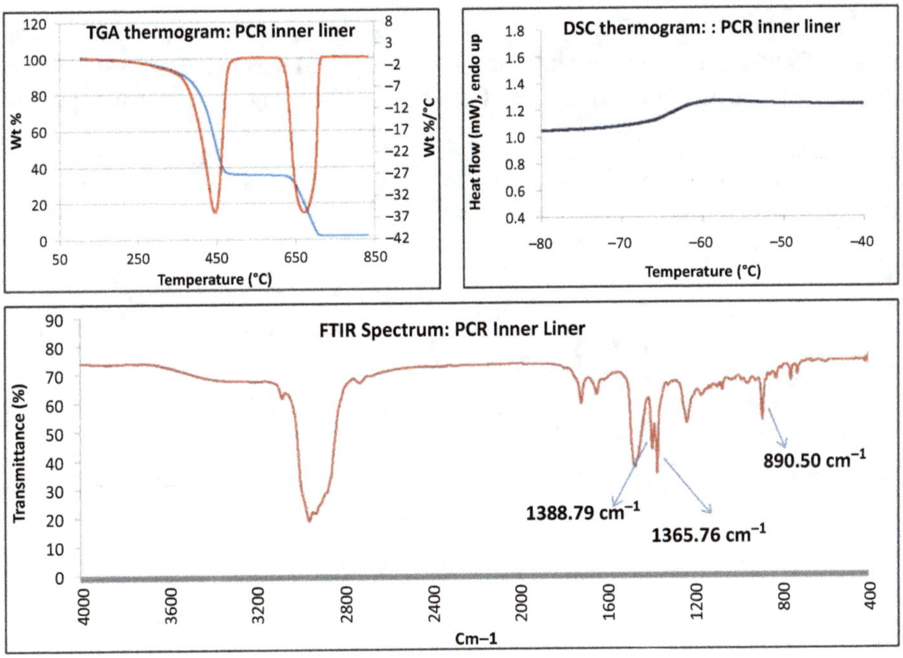


**Figure 7.19:** TGA, DSC, and FTIR analysis for the passenger car radial (PCR) inner liner (Case study 3).

From the above study, one may conclude that the PCR inner liner sample composition is as follows:

| Parameter | Result |
|---|---|
| Polymer content (%) | 55.6 |
| Carbon black content (%) | 33.6 |
| Ash content (%) | 1.9 |
| Extractable content (%) | 8.9 |
| Polymer type | Halobutyl type |

### 7.4.4 Case study 4

Composition analysis of rubber roller used in textile industries was performed following the same conditions used in previous case studies. Extractable content of the sample was found to be 16.7%.

From Figure 7.20, TGA study of the sample exhibits that the sample is degrading at ~450 °C and difference in weight from 600 to 850 °C is 4.9%, and residue at 850 °C is 10.2%. Glass transition temperature of the sample from Figure 7.20 is obtained at

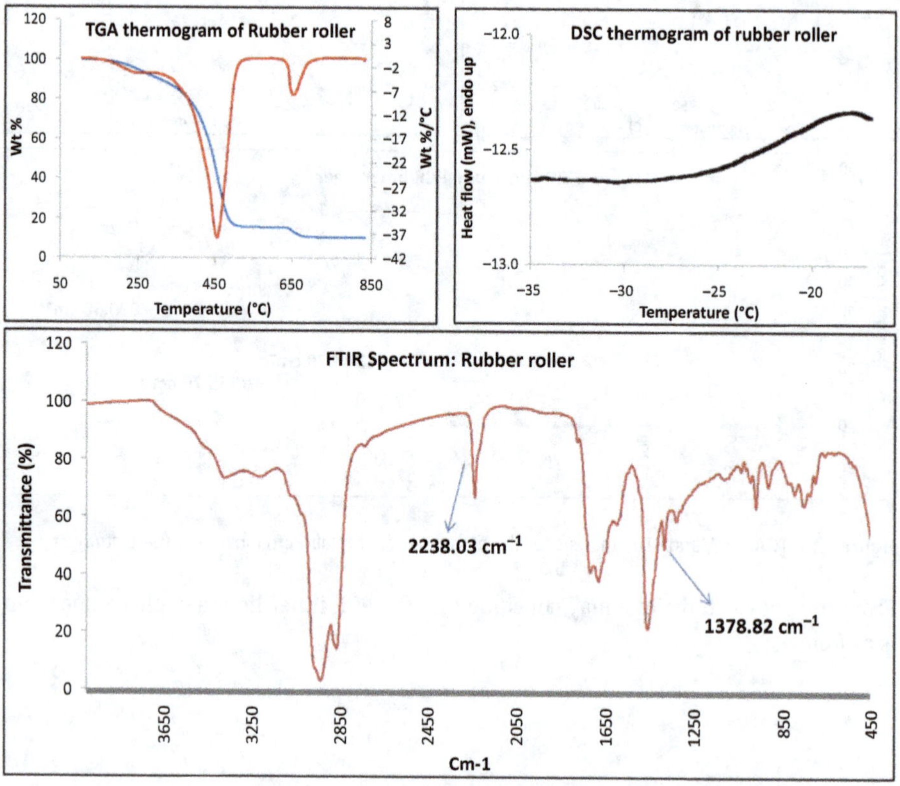


**Figure 7.20:** TGA, DSC, and FTIR analysis for rubber roller compound (Case study 4).

(–)21.67 °C. From the FTIR spectra (Figure 7.20), characteristic peaks at 2,239 $cm^{-1}$ and 1,375 $cm^{-1}$ confirm the presence of –CN and $-CH_3$ functional group.

The above study confirms composition of the sample is as follows:

| Parameter | Result |
|---|---|
| Polymer content (%) | 68.2 |
| Carbon black content (%) | 4.9 |
| Ash content (%) | 10.2 |
| Extractable content (%) | 16.7 |
| Polymer type | Acrylonitrile butadiene rubber |

### 7.4.5 Case study 5

One O-ring sample was analysed using the same conditions as above. Extractable content of the sample was 10.9%.

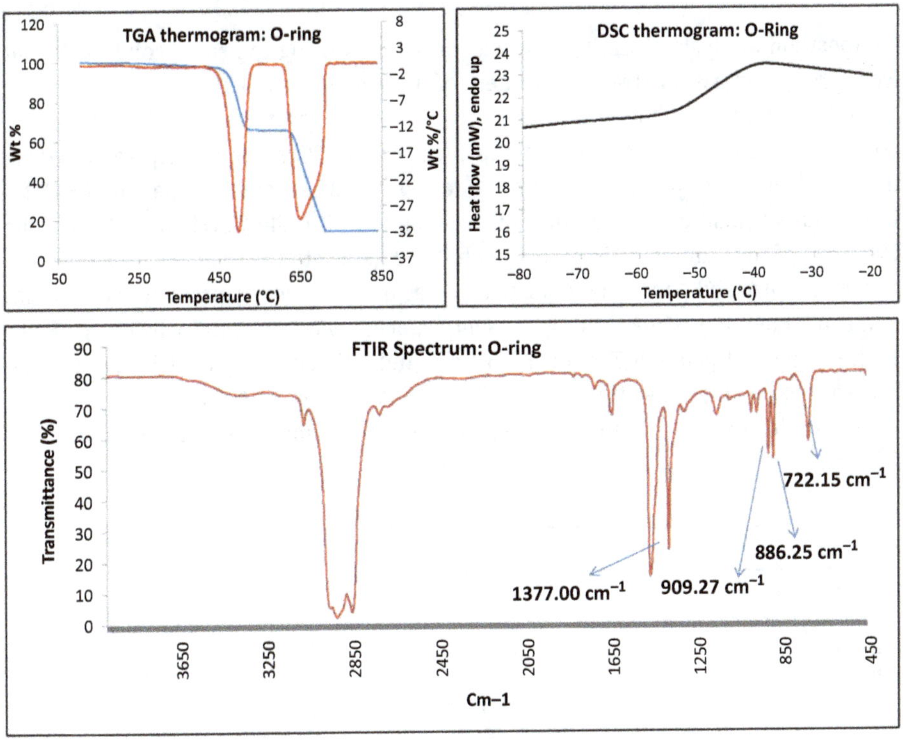


**Figure 7.21:** TGA, DSC, and FTIR analysis of O-ring compound (Case study 5).

From TGA study of the O-ring sample (Figure 7.21) it is noticed that the degradation temperature of the sample is ~500 °C. Delta in weight percentage from 600 to 850 °C is 51.3 and residue at 850 °C is 13.7%. DSC study (Figure 7.21) of the O-ring sample confirms the glass transition temperature is (−)48.1 °C. From the FTIR spectra, peaks at 722 $cm^{-1}$, 887 $cm^{-1}$, 909 $cm^{-1}$, and 1,375 $cm^{-1}$ confirm the presence of $-CH_2-$, $>C=CH_2$, $-CH=CH_2$, and $-CH_3$ functional group.

The above study indicates that the composition of the O-ring is as follows:

| Parameter | Result |
|---|---|
| Polymer content (%) | 24.1 |
| Carbon black content (%) | 51.3 |
| Ash content (%) | 13.7 |
| Extractable content (%) | 10.9 |
| Polymer type | EPDM rubber |

### 7.4.6 Case study 6

Composition analysis of timing belt was performed using the same conditions that were previously used. Extractable content of the sample was 19.9%.

While performing the TGA analysis, the sample degraded at ~265 °C and ~450 °C (Figure 7.22). The first degradation temperature at 265 °C is due to dihydro halogenation. Delta in weight percentage from 600 to 850 °C is 24.5% and residue after 850 °C is 12.4%. Glass transition temperature of the sample is at (−)40 °C (Figure 7.22). During pyrolysis of the sample, it is observed that the generated vapour was turning red-coloured Congo red paper into blue, which was not observed in previous case studies and confirmed that while heating the sample, acidic vapour of HCl was being generated. Presence of peaks at 770 $cm^{-1}$, 965 $cm^{-1}$, and 1,375 $cm^{-1}$ (Figure 7.22) confirm the presence of $C-Cl$, $-CH=CH-$, and $-CH3$ functional group in the sample.

From the above data, one may conclude that the composition of the sample is as follows:

| Parameter | Result |
|---|---|
| Polymer content (%) | 43.2 |
| Carbon black content (%) | 24.5 |
| Ash content (%) | 12.4 |
| Extractable content (%) | 19.9 |
| Polymer type | Neoprene rubber |

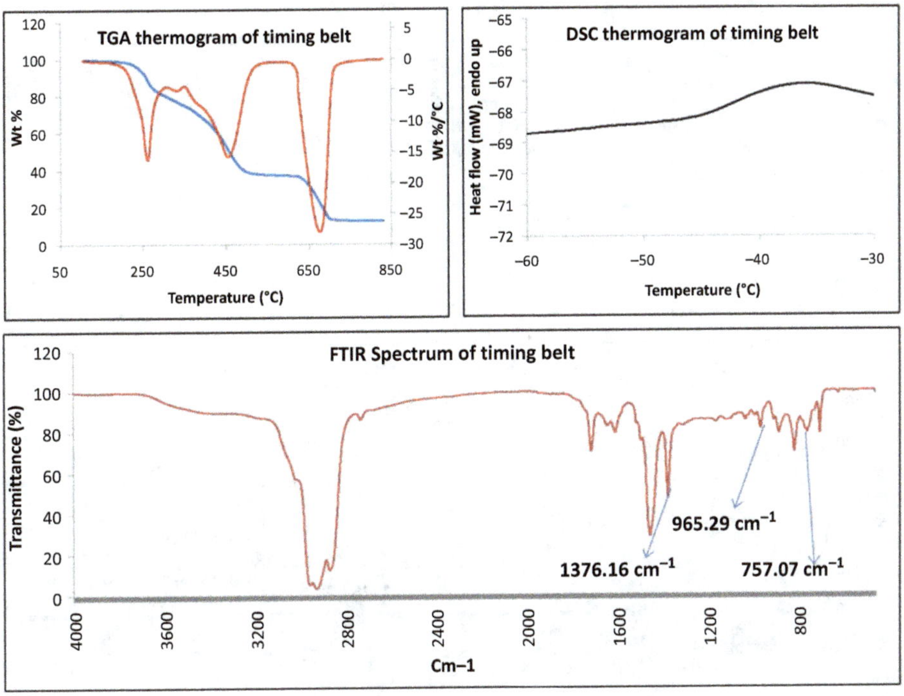


**Figure 7.22:** TGA, DSC, and FTIR Analysis of timing belt compound (Case study 6).

### 7.4.7 Case study 7

In the last six case studies, the basic composition of the rubber part was discussed. In this case study, a truck bus radial (TBR) tyres tread has been analysed. TGA and FTIR analysis of the rubber part has been done using the same conditions followed in previous case studies.

Once extraction completed, solvent was dried in rotary evaporator and column separation of the extracted mass was done. The hexane, methanol, and toluene parts were collected and dried in a rotary evaporator. The hexane and methanol parts were analysed using GC-MS to know about the wax type and chemical type, respectively.

From the TGA study (Figure 7.23), it is observed that the sample exhibits two degradation peaks at ~410 °C and ~480 °C. The ratio of peak area is 60:40. Delta in weight loss from 600 to 850 °C is 20.5% and residue at 850 °C is 12.7%. The percent extractable of the TBR tread sample is 8.6%. From the glass transition temperature study (Figure 7.23), it is observed that the sample shows two glass transition temperatures at (−)60 °C and (−)50 °C, which confirms that the sample contains two different types of polymer. During pyrolysis FTIR (Figure 7.23), peaks at 1,495 $cm^{-1}$, 1,375 $cm^{-1}$, 965 $cm^{-1}$, 909 $cm^{-1}$, and 888 $cm^{-1}$ are observed, which confirms the sample contains styrene

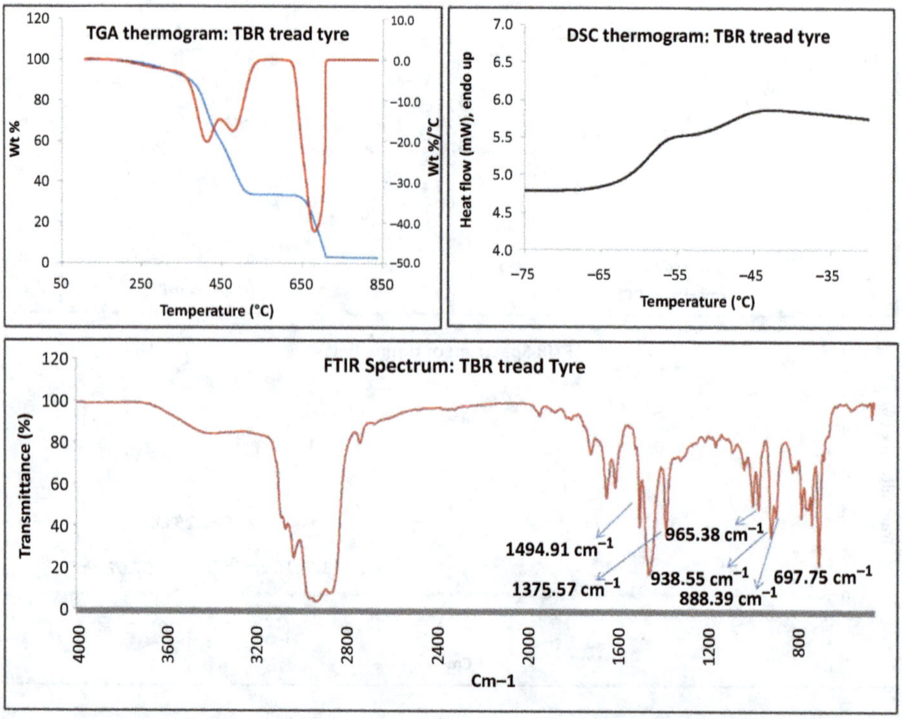


**Figure 7.23:** TGA, DSC, and FTIR analysis of truck bus radial (TBR) tread tyres compound (Case study 7).

functional group, ⊣CH3, ⊣CH = CH⊣, ⊣CH = $CH_2$, and 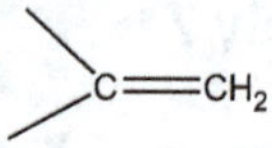 functional group, respectively. The basic composition of the sample is as follows:

| Parameter | Result |
|---|---|
| Polymer content (%) | 58.2 |
| Carbon black content (%) | 20.5 |
| Ash content (%) | 12.7 |
| Extractable content (%) | 8.6 |
| Polymer type | NR:SBR (60:40) |

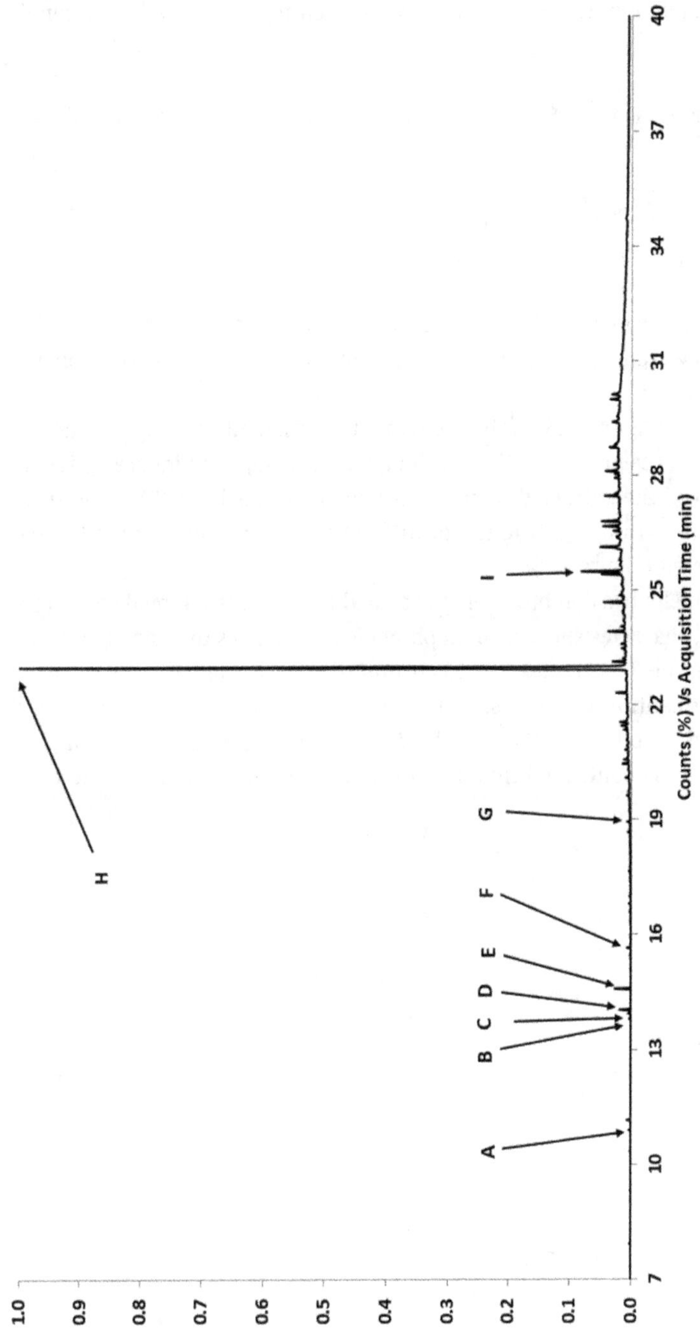


**Figure 7.24:** Key components separated (from A to I) during the GC-MS analysis of the methanol fraction.

GC-MS study of the methanol extract has been performed following the below conditions:

- Inlet temp: 350 °C
- Oven parameter: Heat from 50 to 325 °C at 10 °C/min heating rate and hold for 10 min
- Injection volume: 2 μL
- MS condition: scan mode (50–550 amu)
- Transfer line temp.: 275 °C

During GC-MS study of the methanol fraction, components were separated (Figure 7.24) and key components were marked from A to I. Mass spectra of individual components are mentioned in Figure 7.25.

The GC-MS spectra (Figure 7.25) of the methanol fraction confirms that the sample contains thiazole type accelerator, TMQ, 6PPD, PVI, stearic acid, and hindered phenol.

The hexane part of the extracted material has been analysed in GC-MS and from the observed chromatogram, the fraction of different carbon number in extracted wax has been estimated (Figure 7.26).

FTIR study of toluene fraction has been performed in transmission mode and from the obtained spectra; characteristic peak of naphthenic oil confirms that the naphthenic type of oil is used in the TBR tread tyres. A portion of the sample has been burnt at 550 °C for 16 h and quantitatively HCl soluble content has been measured and found that 18% of total ash is getting dissolved. ICP-OES study of the acid soluble part has been performed and it is confirmed that 2.1% of ZnO is present in it. Approximately

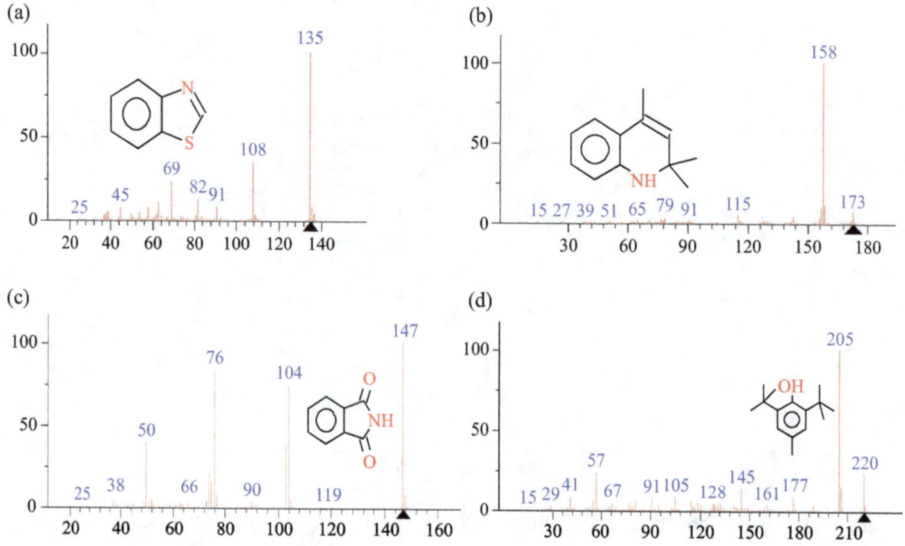


**Figure 7.25:** Mass spectra of the individual component.

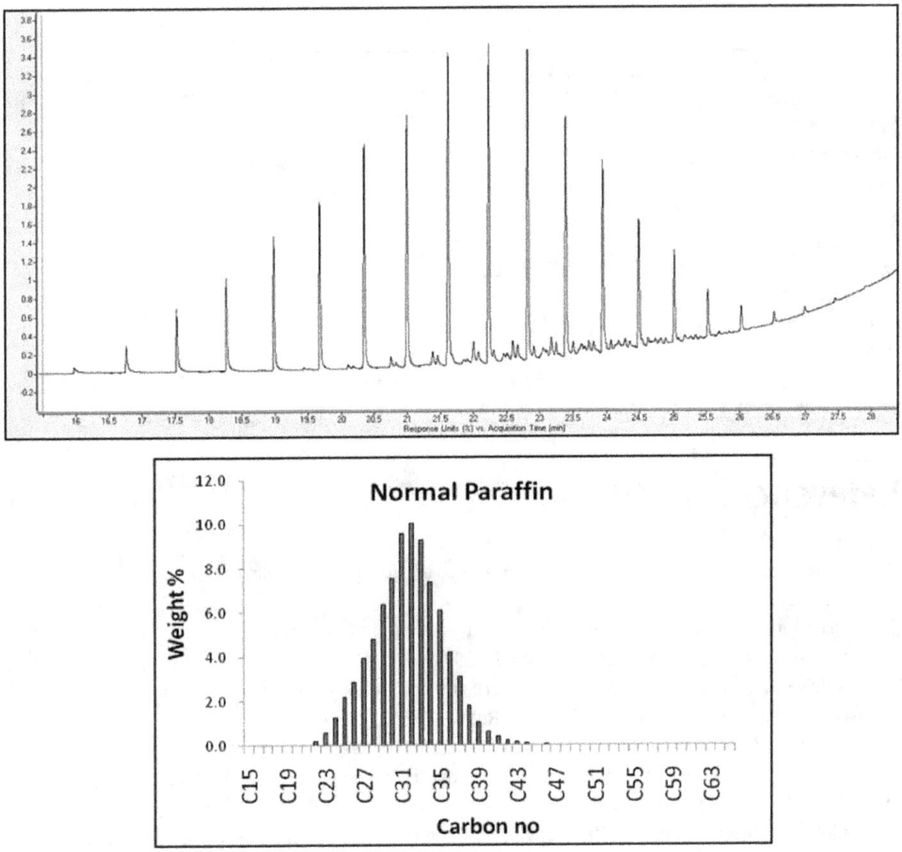


**Figure 7.26:** GC-MS analysis of the hexane fraction.

0.2 g HCl insoluble part has been taken in platinum crucible and treated with hydrofluoric acid followed by heating at 500 °C. It is found that all the HCl insoluble ash has been reacted with hydrofluoric acid and evaporated. This study confirms that the HCl insoluble part contains silica. A portion of the crushed sample has been tested in NCS analyser and found that total organic sulphur in the sample is 1.6%.

From the above study, it is concluded that the detail composition of TBR tyres tread is:

| Parameter | Result |
|---|---|
| Polymer content (%) | 58.2 |
| Carbon black content (%) | 20.5 |
| Ash content (%) | 12.7 |
| Extractable content (%) | 8.6 |
| Polymer type | NR:SBR (60:40) |
| Accelerator | Thiazole type |
| Antioxidant/antidegradant | TMQ, 6PPD, wax |
| ZnO content (%) | 2.1 |
| Plasticiser type | Naphthenic type |
| Sulphur content | 1.6 |
| Other chemicals | Stearic acid, PVI, hindered phenol |

# References


[1] Moore, J. Some chemical and physical properties of rubber. British Journal of Applied Physics 1:6 (1950).

[2] Lichtman, J. Z., Chatten, C. K. Physical properties of natural and synthetic rubber materials at low temperatures. Analytical Chemistry 24:812 (1952).

[3] Kohjiya, S., Ikeda, Y. Chemistry, manufacture and applications of natural rubber. In: Chemistry, Manufacture and Applications of Natural Rubber, 1–502 (2014).

[4] Kim, D. Y., Park, J. W., Lee, D. Y., Seo, K. H. Correlation between the crosslink characteristics and mechanical properties of natural rubber compound via accelerators and reinforcement. Polymers (Basel) 12:1 (2020).

[5] Mukhopadhyay, R., De, S. K. Effect of vulcanization temperature and different fillers on the properties of efficiently vulcanized natural rubber. Rubber Chemistry and Technology 52:263 (1979).

[6] Zhao, F., Bi, W., Zhao, S. Influence of crosslink density on mechanical properties of natural rubber vulcanizates. Journal Of Macromolecular Science Part B-Physics 50:1460 (2011).

[7] Studebaker, M. L. Effect of curing systems on selected physical properties of natural rubber vulcanizates. Rubber Chemistry and Technology 39:1359 (1966).

[8] Chukwu, M. N., Madufor, I. C., Ayo, M. D., Ekebafe, L. O. Effect of stearic acid level on the physical properties of natural rubber vulcanisate. Pacific Journal of Science and Technology 12:344 (2011).

[9] Ulfah, I. M., Fidyaningsih, R., Rahayu, S., Fitriani, D. A., Saputra, D. A., Winarto, D. A., Wisojodharmo, L. A. Influence of carbon black and silica filler on the rheological and mechanical properties of natural rubber compound. Procedia Chemistry 16:258 (2015).

[10] Thaptong, P., Sirisinha, C., Thepsuwan, U., Sae-Oui, P. Properties of natural rubber reinforced by carbon black-based hybrid fillers. Polymer-Plastics Technology and Engineering 53:818 (2014).

[11] South, J. T., Case, S. W., Reifsnider, K. L. Effects of thermal aging on the mechanical properties of natural rubber. Rubber Chemistry and Technology 76:785 (2003).

[12] Hinchiranan, N., Lertweerasirikun, W., Poonsawad, W., Rempel, G. L., Prasassarakich, P. Cure characteristics and mechanical properties of hydrogenated natural rubber/natural rubber blends. Journal of Applied Polymer Science 111:2813 (2009).

[13] Nasir, M., Teh, G. K. The effects of various types of crosslinks on the physical properties of natural rubber. European Polymer Journal 24:733 (1988).

[14] Nimpaiboon, A., Amnuaypornsri, S., Sakdapipanich, J. Influence of gel content on the physical properties of unfilled and carbon black filled natural rubber vulcanizates. Polymer Testing 32:1135 (2013).
[15] Sombatsompop, N. Analysis of cure characteristics on cross-link density and type, and viscoelastic properties of natural rubber. Polymer-Plastics Technology and Engineering 37:333 (1998).
[16] González, L., Rodríguez, A., Valentin, J. L., Marcos-Fernández, A., Posadas, P. Conventional and efficient crosslinking of natural rubber effect of heterogeneities on the physical properties. KGK Kautschuk Gummi Kunststoffe 58:638 (2005).
[17] Helaly, F. M., El Sabbagh, S. H., El Kinawy, O. S., El Sawy, S. M. Effect of synthesized zinc stearate on the properties of natural rubber vulcanizates in the absence and presence of some fillers. Materials and Design 32:2835 (2011).
[18] Toki, S. The effect of strain-induced crystallization (SIC) on the physical properties of natural rubber (NR). Chemistry, Manufacture and Applications of Natural Rubber 135 (2014).
[19] Naebpetch, W., Junhasavasdikul, B., Saetung, A., Tulyapitak, T., Nithi-Uthai, N. Influence of filler type and loading on cure characteristics and vulcanisate properties of SBR compounds with a novel mixed vulcanisation system. Plastics, Rubber and Composites 46:137–145 (2017).
[20] Choi, S. S., Park, B. H., Song, H. Influence of filler type and content on properties of styrene-butadiene rubber (SBR) compound reinforced with carbon black or silica. Polymers for Advanced Technologies 15:122–127 (2004).
[21] Ward, A. A., Khalf, A. I. Electrical and mechanical properties of SBR filled with carbon black-silica blends. Journal of Elastomers and Plastics 60:623 (2007).
[22] Malas, A., Pal, P., Das, C. K. Effect of expanded graphite and modified graphite flakes on the physical and thermo-mechanical properties of styrene butadiene rubber/polybutadiene rubber (SBR/BR) blends. Materials and Design 55:664 (2014).
[23] Mostafa, A., Abouel-Kasem, A., Bayoumi, M. R., El-Sebaie, M. G. Insight into the effect of CB loading on tension, compression, hardness and abrasion properties of SBR and NBR filled compounds. Materials and Design 30:1785 (2009).
[24] Zhao, J., Ghebremeskel, G. N. A review of some of the factors affecting fracture and fatigue in SBR and BR vulcanizates. Rubber Chemistry and Technology 74:409 (2001).
[25] Yang, S., Liang, P., Peng, X., Zhou, Y., Hua, K., Wu, W., Cai, Z. Improvement in mechanical properties of SBR/Fly ash composites by in-situ grafting-neutralization reaction. Chemical Engineering Journal 354:849 (2018).
[26] Liu, Z., Zhang, Y. Enhanced mechanical and thermal properties of SBR composites by introducing graphene oxide nanosheets decorated with silica particles. Composites Part A: Applied Science and Manufacturing 102:236 (2017).
[27] Esmaeeli, R., Farhad, S. Parameters estimation of generalized Maxwell model for SBR and carbon-filled SBR using a direct high-frequency DMA measurement system. Mechanics of Materials 146:2020 (2020).
[28] Alkadasi, N. A. N., Sarmade, B. D., Kapadi, U. R., Hundiwale, D. G. Effect of bis (3-triethoxy silylpropyl) tetrasulphide on the mechanical properties of flyash filled styrene butadiene rubber. Journal of Scientific and Industrial Research (India) 63:287 (2004).
[29] Ajam, A. M., Al-Nesrawy, H., Al-Maamori, M. Effect of Reclaim Rubber Loading on the Mechanical Properties of Sbr Composites. International Journal of Chemical Science 14:2439 (2016).
[30] Hwang, E. H., Ko, Y. S. Comparison of mechanical and physical properties of SBR-polymer modified mortars using recycled waste materials. Journal of Industrial and Engineering Chemistry 14:644 (2008).
[31] Zhang, J., Wang, J., Wu, Y., Wang, Y., Wang, Y. Evaluation of the improved properties of SBR/weathered coal modified bitumen containing carbon black. Construction and Building Materials 23:2678 (2009).

[32] Demirhan, E., Kandemirli, F., Kandemirli, M., Kovalishyn, V. Investigation of the physical and rheological properties of SBR-1712 rubber compounds by neural network approaches. Materials and Design 28:1737 (2007).
[33] Han, S. C., Han, M. H. Fracture behavior of NR and SBR vulcanizates filled with ground rubber having uniform particle size. Journal of Applied Polymer Science 85:2491 (2002).
[34] Yoshioka, A. Structure and physical properties of high-vinyl polybutadiene rubbers and their blends. International Union of Pure and Applied Chemistry 58 (1986).
[35] Railsback, H. E. Cis-Polybutadiene-Natural rubber blends. Rubber Chemistry and Technology 32:308 (1959).
[36] Gopi Sathi, S., Stoček, R., Kratina, O. Reversion free high-temperature vulcanization of cis-polybutadiene rubber with the accelerated-sulfur system. Express Polymer Letters 14:823 (2020).
[37] Rodríguez Garraza, A. L., Mansilla, M. A., Depaoli, E. L., Macchi, C., Cerveny, S., Marzocca, A. J., Somoza, A. Comparative study of thermal, mechanical and structural properties of polybutadiene rubber isomers vulcanized using peroxide. Polymer Testing 52:117 (2016).
[38] Posadas, P., Fernández, A., Brasero, J., Valentín, J. L., Marcos, A., Rodríguez, A., González, L. Vulcanization of polybutadiene rubber with dipentamethylene thiuram tetrasulfide. Journal of Applied Polymer Science 106:3481 (2007).
[39] Alkadasi, N. A. N., First, K. Effect of coupling agent on the mechanical properties of fly ash–filled polybutadiene rubber. Journal of Applied Polymer Science 91:1322 (2003).
[40] Wang, S., Zhang, Y., Peng, Z., Zhang, Y. New method for preparing polybutadiene rubber/clay composites. Journal of Applied Polymer Science 98:227 (2005).
[41] Marzocca, A. J., Rodriguez Garraza, A. L., Sorichetti, P., Mosca, H. O. Cure kinetics and swelling behaviour in polybutadiene rubber. Polymer Testing 29:477 (2010).
[42] Lin, J. P., Chang, C. Y., Wu, C. H., Shih, S. M. Thermal degradation kinetics of polybutadiene rubber. Polymer Degradation and Stability 53:295 (1996).
[43] Hamed, G. R., Kim, H. J., Gent, A. N. Cut growth in vulcanizates of natural rubber, cis-polybutadiene, and a 50/50 blend during single and repeated extension. Rubber Chemistry and Technology 69:807 (1996).
[44] Bellander, M., Stenberg, B., Persson, S. Crosslinking of polybutadiene rubber without any vulcanization agent. Polymer Engineering and Science 38:1254 (1998).
[45] Mishra, S., Shimpi, N. G. Studies on mechanical, thermal, and flame retarding properties of polybutadiene rubber (PBR) nanocomposites. Polymer-Plastics Technology and Engineering 47:72 (2008).
[46] Fusco, J. V., Hous, P. Butyl and Halobutyl rubbers. Rubber Technology 284 (1999).
[47] Gunter, W. D. Butyl and Halogenated Butyl rubbers. Development of Rubber Technology 155 (1981).
[48] Vukov, R. Halogenation of Butyl Rubber – a model compound approach. Rubber Chemistry and Technology 57:275 (1984).
[49] Kulbaba, K., Adkinson, D., Beilby, J. Examining functionalized butyl rubber. Rubber and Plastics News 14 (2015).
[50] Dubey, V., Pandey, S. K., Rao, N. B. S. N. Research trends in the degradation of butyl rubber. Journal of Analytical and Applied Pyrolysis 34:111 (1995).
[51] Tikhomirov, S. G., Polevoy, P. S., Semenov, M. E., Karmanov, A. V. Modeling of the destruction process of butyl rubber. Radiation Physics and Chemistry 158:205 (2019).
[52] Jiang, K., Shi, J., Ge, Y., Zou, R., Yao, P., Li, X., Zhang, L. Complete devulcanization of sulfur-cured butyl rubber by using supercritical carbon dioxide. Journal of Applied Polymer Science 127:2397 (2012).
[53] Chen, F., Qian, J. Studies on the thermal degradation of polybutadiene. Fuel Processing Technology 67:53 (2000).

[54] Cong, L., Shi-Ai, X., Fang-Yi Xiao, C.-F. W. Dynamic mechanical properties of EPDM rubber blends. European Polymer Journal 42:2507 (2006).
[55] Chang, V. S. C., Kennedy, J. P. Gas permeability, water absorption, hydrolytic stability and air-oven aging of polyisobutylene-based polyurethane networks. Polymer Bulletin 8:69 (1982).
[56] Yang, X., Zhang, Y., Xu, Y., Gao, S., Guo, S. Effect of octadecylamine modified graphene on thermal stability, mechanical properties and gas barrier properties of brominated butyl rubber. Macromolecular Research 25:270 (2017).
[57] Lian, H., Li, S., Liu, K., Xu, L., Wang, K., Guo, W. Study on modi ed graphene/butyl rubber nanocomposites. I. Preparation and characterization. Polymer Engineering and Science 51:2254 (2011).
[58] Formela, K., Haponiuk, J. T. Curing characteristics, mechanical properties and morphology of butyl rubber filled with ground tire rubber (GTR). Iranian Polymer Journal (English Edition) 23:185 (2014).
[59] Kar, K. K., Ravikumar, N. L., Tailor, P. B., Ramkumar, J., Sathiyamoorthy, D. Performance evaluation and rheological characterization of newly developed butyl rubber based media for abrasive flow machining process. Journal of Materials Processing Technology 209:2212 (2009).
[60] Xia, L., Li, C., Zhang, X., Wang, J., Wu, H., Guo, S. Effect of chain length of polyisobutylene oligomers on the molecular motion modes of butyl rubber: Damping property. Polymer (Guildf) 141:70 (2018).
[61] Flory, P. J. Effects of molecular structure on physical properties of Butyl Rubber. Rubber Chemistry and Technology 19:552 (1946).
[62] Hungate, R. E., Smith, W., Clarke, R. T. J. Suitability of Butyl Rubber stoppers for closing anaerobic roll culture tubes reciprocal recombination of chromosome and F- Merogenote in escherichia coli methyl dipicolinate monoester from spores of Bacillus cereus var. glob igii. Microbiology 91:908 (1966).
[63] Kumar, A., Modi, S., Panwar, A., Schmidt, D., Barry, C. M. F., Mead, J. L. Highly impermeable nanocomposites of brominated butyl rubber with modified montmorillonite clay. Rubber Chemistry and Technology 87:579 (2014).
[64] Takahashi, S., Goldberg, H. A., Feeney, C. A., Karim, D. P., Farrell, M., O'Leary, K., Paul, D. R. Gas barrier properties of butyl rubber/vermiculite nanocomposite coatings. Polymer (Guildf) 47:3083 (2006).
[65] Li, Y., Iwakura, Y., Zhao, L., Shimizu, H. Nanostructured poly(vinylidene fluoride) materials by melt blending with several percent of acrylic rubber. Macromolecules 41:3120 (2008).
[66] Kader, M. A., Bhowmick, A. K. Acrylic rubber-fluorocarbon rubber miscible blends: Effect of curatives and fillers on cure, mechanical, aging, and swelling properties. Journal of Applied Polymer Science 89:1442–1452 (2003).
[67] Balakrishnan, S., Start, P. R., Raghavan, D., Hudson, S. D. The influence of clay and elastomer concentration on the morphology and fracture energy of preformed acrylic rubber dispersed clay filled epoxy nanocomposites. Polymer (Guildf) 46:11255 (2005).
[68] Shakun, A., Poikelispää, M., Das, A., Vuorinen, J. Improved electromechanical response in acrylic rubber by different carbon-based fillers. Polymer Engineering and Science 58:395 (2018).
[69] Levita, G., Marchetti, A., Lazzeri, A. Toughness of epoxies modified by preformed acrylic rubber particles. Makromolekulare Chemie. Macromolecular Symposia 41:179 (1991).
[70] Hesami, M., Jalali-Arani, A. Cold crystallization behavior of poly(lactic acid) in its blend with acrylic rubber; the effect of acrylic rubber content. Polymer International 66:1564 (2017).
[71] Abolhasani, M. M., Jalali-Arani, A., Nazockdast, H., Guo, Q. Poly(vinylidene fluoride)-acrylic rubber partially miscible blends: Crystallization within conjugated phases induce dual lamellar crystalline structure. Polymer (Guildf). 54:4686 (2013).
[72] Abolhasani, M. M., Naebe, M., Jalali-Arani, A., Guo, Q. Influence of miscibility phenomenon on crystalline polymorph transition in poly(vinylidene fluoride)/acrylic rubber/clay nanocomposite hybrid. PLoS One 9 (2014).

[73] Martinez-Martinez, D., Schenkel, M., Pei, Y. T., De Hosson, J. T. M. Microstructural and frictional control of diamond-like carbon films deposited on acrylic rubber by plasma assisted chemical vapor deposition. Thin Solid Films 519:2213 (2011).

[74] Dao, T. D., Lee, H. il, Jeong, H. M. Alumina-coated graphene nanosheet and its composite of acrylic rubber. Journal of Colloid and Interface Science 416:38 (2014).

[75] Kocevski, S., Yagneswaran, S., Xiao, F., Punith, V. S., Smith, D. W., Amirkhanian, S. Surface modified ground rubber tire by grafting acrylic acid for paving applications. Construction and Building Materials 34:83 (2012).

[76] Wong-On, J., Wootthikanokkhan, J. Dynamic vulcanization of acrylic rubber-blended PVC. Journal of Applied Polymer Science 88:2657 (2003).

[77] Kader, M. A., Bhowmick, A. K. Effect of filler on the mechanical, dynamic mechanical, and aging properties of binary and ternary blends of acrylic rubber, fluorocarbon rubber, and polyacrylate. Journal of Applied Polymer Science 90:278–286 (2003).

[78] Leong, Y. C., Lee, L. M. S., Gan, S. N. The viscoelastic properties of natural rubber pressure-sensitive adhesive using acrylic resin as a tackifier. Journal of Applied Polymer Science 88:2118–2123 (2003).

[79] Bandyopadhyay, A., De Sarkar, M., Bhowmick, A. K. Effect of reaction parameters on the structure and properties of acrylic rubber/silica hybrid nanocomposites prepared by sol-gel technique. Journal of Applied Polymer Science 95:1418–1429 (2005).

[80] Sun, Y., Sun, G. Effects of a hindered phenol compound on the dynamic mechanical properties of chlorinated polyethylene, acrylic rubber, and their blend. Journal of Applied Polymer Science 80:2468–2473 (2001).

[81] Wootthikanokkhan, J., Rattanathamwat, N. Effects of a hindered phenol compound on the dynamic mechanical properties of chlorinated polyethylene, acrylic rubber, and their blend. Journal of Applied Polymer Science 102:248 (2006).

[82] Wu, C., Wei, C., Guo, W., Wu, C. Dynamic mechanical properties of acrylic rubber blended with phenolic resin. Journal of Applied Polymer Science 109:2065 (2008).

[83] Tanrattanakul, V., Bunchuay, A. Microwave absorbing rubber composites containing carbon black and aluminum powder. Journal of Applied Polymer Science 105:2036 (2007).

[84] Maiti, S. N., Das, R. Mechanical properties of talc filled i-PP/CSM rubber composites. International Journal of Polymeric Materials and Polymeric Biomaterials 54:835 (2005).

[85] Boonsong, K., Seadan, M., Lopattananon, N. Compatibilization of natural rubber (NR) and chlorosulfonated polyethylene (CSM) blends with zinc salts of sulfonated natural rubber. Songklanakarin Journal of Science and Technology 30:491 (2008).

[86] Flauzino Neto, W. P., Mariano, M., da Silva, I. S. V., Silvério, H. A., Putaux, J. L., Otaguro, H., Pasquini, D., Dufresne, A. Mechanical properties of natural rubber nanocomposites reinforced with high aspect ratio cellulose nanocrystals isolated from soy hulls. Carbohydrate Polymers 153:143 (2016).

[87] Sae-oui, P., Sirisinha, C., Thepsuwan, U., Hatthapanit, K. Dependence of mechanical and aging properties of chloroprene rubber on silica and ethylene thiourea loadings. European Polymer Journal 43:185 (2007).

[88] Anyszka, R., Bieliński, D. M., Pędzich, Z., Szumera, M. Influence of surface-modified montmorillonites on properties of silicone rubber-based ceramizable composites. Journal of Thermal Analysis and Calorimetry 119:111 (2015).

[89] Maiti, S. N., Das, R. Mechanical properties of impact i-PP/CSM rubber blends. International Journal of Polymeric Materials and Polymeric Biomaterials 54:467 (2005).

[90] Nanda, M., Tripathy, D. K. Physico-mechanical and electrical properties of conductive carbon black reinforced chlorosulfonated polyethylene vulcanizates. Express Polymer Letters 2:855 (2008).

[91] Marković, G., Radovanović, B., Marinović-Cincović, M., Budinski-Simendić, J. The effect of accelerators on curing characteristics and properties of natural rubber/chlorosulphonated polyethylene rubber blend. Materials and Manufacturing Processes 24:1224 (2009).
[92] Sisanth, K. S., Thomas, M. G., Abraham, J., Thomas, S. General introduction to rubber compounding. In: Progress in Rubber Nanocomposites, 1–39 (2017), doi: 10.1016/B978-0-08-100409-8.00001-2.
[93] Janowska, G., Kucharska-Jastrząbek, A. The effect of chlorosulphonated polyethylene on thermal properties and combustibility of butadiene-styrene rubber. Journal of Thermal Analysis and Calorimetry 101:1093 (2010).
[94] Zhang, M. Q., Rong, M. Z. Self-Healing polymers and polymer composites. Self-Healing of Polymers and Polymer Composites (2011).
[95] Tabaei, T. A., Bagheri, R., Hesami, M. Comparison of cure characteristics and mechanical properties of nano and micro silica-filled CSM elastomers. Journal of Applied Polymer Science 132 (2015).
[96] Phiriyawirut, M., Luamlam, S. Influence of Poly(vinyl chloride) on natural Rubber/Chlorosulfonated polyethylene blends. Open Journal of Organic Polymer Materials 03:81 (2013).
[97] Tanrattanakul, V., Petchkaew, A. Mechanical properties and blend compatibility of natural rubber -Chlorosulfonated polyethylene blends. Journal of Applied Polymer Science 99:127 (2006).
[98] Roychoudhury, A., De, P. P., Bhowmick, A. K., De, S. K. Self-crosslinkable ternary blend of chlorosulphonated polyethylene, epoxidized natural rubber and carboxylated nitrile rubber. Polymer (Guildf) 33:4737 (1992).
[99] Basfar, A. A., Abdel-Aziz, M. M., Mofti, S. Accelerated aging and stabilization of radiation-vulcanized EPDM rubber. Radiation Physics and Chemistry 57:405 (2000).
[100] Akhlaghi, S., Kalaee, M., Mazinani, S., Jowdar, E., Nouri, A., Sharif, A., Sedaghat, N. Effect of zinc oxide nanoparticles on isothermal cure kinetics, morphology and mechanical properties of EPDM rubber. Effect of zinc oxide nanoparticles on isothermal cure kinetics, morphology and mechanical properties of EPDM rubber. Thermochimica Acta 527:91 (2012).
[101] Eom, Y., Choi, B., Park, S. I. A study on mechanical and thermal properties of PLA/PEO blends. Journal of Polymers and the Environment 27:256 (2019).
[102] Choudhary, V., Varma, H. S., Varma, I. K. Polyolefin blends: effect of EPDM rubber on crystallization, morphology and mechanical properties of polypropylene/EPDM blends. 1. Polymer (Guildf). 32:2534 (1991).
[103] Choudhary, V., Varma, H. S., Varma, I. K. Effect of EPDM rubber on melt rheology, morphology and mechanical properties of polypropylene/HDPE (90 10) blend. 2. Polymer (Guildf). 32:2541 (1991).
[104] Hamza, S. S. Effect of aging and carbon black on the mechanical properties of EPDM rubber. Polymer Testing 17:131 (1998).
[105] Poltabtim, W., Wimolmala, E., Saenboonruang, K. Properties of lead-free gamma-ray shielding materials from metal oxide/EPDM rubber composites. Radiation Physics and Chemistry 153:1 (2018).
[106] Shaw, S., Singh, R. P. Studies on impact modification of polystyrene (PS) by ethylene–propylene–diene (EPDM) rubber and its graft copolymers. I. PS/EPDM and PS/EPDM-g-styrene blends. Journal of Applied Polymer Science 40:685–692 (1990).
[107] Amin, A. R. LDPE/EPDM multilayer films containing recycled LDPE for greenhouse applications. Journal of Polymers and the Environment 9:25 (2001).
[108] Khattak, A., Amin, M. Accelerated aging investigation of high voltage EPDM/silica composite insulators. Journal of Polymer Engineering 36:199 (2016).
[109] Gunasekaran, S., Natarajan, R. K., Kala, A. FTIR spectra and mechanical strength analysis of some selected rubber derivatives. Spectrochimica Acta, Part A: Molecular and Biomolecular Spectroscopy 68:323 (2007).
[110] Vazquez, A., Dominguez, V. A., Kenny, J. M. Journal of thermoplastic composite. Journal of Thermoplastic Composite Materials 12:477 (1999).

[111] Zhao, Q., Li, X., Hu, J., Ye, Z. Degradation characterization of ethylene-propylene-diene monomer (EPDM) rubber in artificial weathering environment. Journal of Failure Analysis and Prevention 10:240 (2010).
[112] Sutanto, P., Picchioni, F., Janssen, L. P. B. M., Dijkhuis, K. A. J., Dierkes, W. K., Noordermeer, J. W. M. EPDM rubber reclaim from devulcanized EPDM. Journal of Applied Polymer Science 102:5948 (2006).
[113] Zhao, Q., Li, X., Gao, J., Jia, Z. Evaluation of ethylene-propylene-diene monomer (EPDM) aging in UV/condensation environment by principal component analysis (PCA). Materials Letters 63:1647 (2009).
[114] Paeglis, A. U. A simple model for predicting heat aging of EPDM rubber. Rubber Chemistry and Technology 77:242 (2004).
[115] Kang, G.-H., Kim, C.-S. Nonlinear analysis of rubber bellows for the high speed railway vehicle. Journal of the Korea Academia-Industrial Cooperation Society 14:3631 (2013).
[116] Ito, M., Okada, S., Kuriyama, I. The deterioration of mechanical properties of chloroprene rubber in various conditions. Journal of Materials Science 16:10 (1981).
[117] Arruda, E. M., Boyce, M. C. A three-dimensional constitutive model for the large stretch behavior of rubber elastic materials. Journal of the Mechanics and Physics of Solids 41:389 (1993).
[118] Balasubramanian, P., Ferrari, G., Del Prado, Z. J. G. N., Amabili, M. Theoretical and experimental study on large amplitude vibrations of clamped viscoelastic plates. ASME's International Mechanical Engineering Congress and Exposition Proceedings 4B (2016).
[119] Goette, D. K. Raccoon-like periorbital leukoderma from contact with swim goggles. Contact Dermatitis 10:129–131 (1984).
[120] Bardy, E. R., Mollendorf, J. C., Pendergast, D. R. Thermal conductivity and compressive strain of aerogel insulation blankets under applied hydrostatic pressure. Journal of Heat Transfer 129:232 (2007).
[121] Sabura Begum, P. M., Mohammed Yusuff, K. K., Joseph, R. Preparation and use of nano zinc oxide in neoprene rubber. International Journal of Polymeric Materials and Polymeric Biomaterials 57:1083 (2008).
[122] Garu, P. K., Chaki, T. K. Acoustic & Mechanical properties of neoprene rubber forencapsulation of underwater transducers. International Journal of Engineering, Science and Technology 1:231 (2012).
[123] Botros, S. H., Tawfic, M. L. Preparation and characteristics of EPDM/NBR rubber blends with BIIR as compatibilizer. Polymer-Plastics Technology and Materials 44:209 (2005).
[124] Zhu, L., Cheung, C. S., Zhang, W. G., Huang, Z. Compatibility of different biodiesel composition with acrylonitrile butadiene rubber (NBR). Fuel 158:288 (2015).
[125] Zhao, X., Niu, K., Xu, Y., Peng, Z., Jia, L., Hui, D., Zhang, L. Morphology and performance of NR/NBR/ENR ternary rubber composites. Composites Part B: Engineering 107:106 (2016).
[126] El-sabbagh, S. H., Yehia, A. A. Detection of crosslink density by different methods for natural rubber detection of crosslink density by different methods for natural rubber blended with SBR and NBR. Egyptian Journal of Solids 30:157 (2007).
[127] George, S., Varughese, K. T., Thomas, S. Thermal and crystallisation behavior of isotactic polypropylene/nitrile rubber blends. Polymer (Guildf) 41:5485 (2000).
[128] Bhattacharjee, S., Bhowmick, A. K., Avasthi, B. N. Degradation of hydrogenated nitrile rubber. Polymer Degradation and Stability 31:71 (1991).
[129] Degrange, J. M., Thomine, M., Kapsa, P., Pelletier, J. M., Chazeau, L., Vigier, G., Dudragne, G., Guerbé, L. Influence of viscoelasticity on the tribological behaviour of carbon black filled nitrile rubber (NBR) for lip seal application. Wear 259:684 (2005).
[130] Mahmoud, W. E., Mansour, S. A., Hafez, M., Salam, M. A. On the degradation and stability of high abrasion furnace black (HAF)/acrylonitrile butadiene rubber (NBR) and high abrasion furnace black (HAF)/graphite/acrylonitrile butadiene rubber (NBR) under cyclic stress-strain. Polymer Degradation and Stability 92:2011–2015 (2007).

[131] Balachandran, M., Devanathan, S., Muraleekrishnan, R., Bhagawan, S. S. Optimizing properties of nanoclay-nitrile rubber (NBR) composites using face centred central composite design. Materials and Design 35:854 (2012).
[132] Kim, J. tae, Oh, T. su, Lee, D. ho. Preparation and characteristics of nitrile rubber (NBR) nanocomposites based on organophilic layered clay. Polymer International 52:1058 (2003).
[133] Stacy, C. J., Johnson, P. H., Kraus, G. Effect of carbon black structure aggregate size distribution on properties of reinforced rubber. Rubber Chemistry and Technology 48:538 (1975).
[134] Li, Q., Ma, Y., Wu, C., Qian, S. Effect of carbon black nature on vulcanization and mechanical properties of rubbe. Journal of Macromolecular Science Part B-Physics 47:837 (2008).
[135] Zhang, B. S., Lv, X. F., Zhang, Z. X., Liu, Y., Kim, J. K., Xin, Z. X. Effect of carbon black content on microcellular structure and physical properties of chlorinated polyethylene rubber foams. Materials and Design 31:3106 (2010).
[136] Gent, A. N., Hartwell, J. A., Lee, G. Effect of carbon black on crosslinking. Rubber Chemistry and Technology 76:517 (2003).
[137] Xu, Z., Song, Y., Zheng, Q. Payne effect of carbon black filled natural rubber compounds and their carbon black gels. Polymer (Guildf) 185 (2019).
[138] Dutta, N. K., Khastgir, D., Tripathy, D. K. The effect of carbon black concentration on the dynamic mechanical properties of bromobutyl rubber. Journal of Materials Science 26:177 (1991).
[139] Luheng, W., Tianhuai, D., Peng, W. Effects of conductive phase content on critical pressure of carbon black filled silicone rubber composite. Sensors Actuators, A Physics 135:587 (2007).
[140] Gupta, B. R., Maridass, B. Effect of carbon black on devulcanized ground rubber tire – natural rubber vulcanizates: Cure characteristics and mechanical properties. Journal of Elastomers and Plastics 291–302 (2006).
[141] Lamba, R. Effect of carbon black on dynamic properties. Rubber World 222:43 (2000).
[142] Song, J. Ping, Tian, K. Yan, Ma, L. Xiang, Li, W., Yao, S. The effect of carbon black morphology to the thermal conductivity of natural rubber composites. Chune International Journal of Heat and Mass Transfer 137:184 (2019).
[143] Luheng, W., Tianhuai, D., Peng, W. Influence of carbon black concentration on piezoresistivity for carbon-black-filled silicone rubber composite. Carbon NY 47:3151 (2009).
[144] Maiti, M., Sadhu, S., Bhowmick, A. K. Effect of carbon black on properties of rubber nanocomposites. Journal of Applied Polymer Science 96:443 (2005).
[145] Dannenberg, E. M. Bound rubber and carbon black reinforcement. Rubber Chemistry and Technology 59:512 (1986).
[146] Clarke, J., Clarke, B., Freakley, P. K., Sutherland, I. Compatibilising effect of carbon black on morphology of NR-NBR blends. Plastics, Rubber and Composites Processing and Applications 30:39 (2001).
[147] Caruthers, J. M., Cohen, R. E., Medalia, A. I. Effect of carbon black on hysteresis of rubber Vulcanizates: Equivalence of surface area and loading. Rubber Chemistry and Technology 49:1076 (1976).
[148] Medalia, A. I. Electrical conduction in carbon black composites. Rubber Chemistry and Technology 59:432 (1986).
[149] Chen, L., Gong, X. L., Li, W. H. Effect of carbon black on the mechanical performances of magnetorheological elastomers. Polymer Testing 27:340 (2008).
[150] Mohamed, R., Mohd Nurazzi, N., Huzaifah, M. Effect of carbon black composition with sludge palm oil on the curing characteristic and mechanical properties of natural rubber/styrene butadiene rubber compound. IOP Conference Series: Materials Science and Engineering 223 (2017).
[151] Smit, P. P. A. Glass transition in carbon black reinforced rubber. Rheologica Acta 5:277 (1966).
[152] Yeoh, O. H. Characterization of elastic properties of carbon-black-filled rubber vulcanizates. Rubber Chemistry and Technology 63:792 (1990).

[153] Lion, A. A constitutive model for carbon black filled rubber: Experimental investigations and mathematical representation. Continuum Mechanics and Thermodynamics 8:153 (1996).
[154] Savetlana, S., Zulhendri, Sukmana, I., Saputra, F. A. The effect of carbon black loading and structure on tensile property of natural rubber composite. IOP Conference Series: Materials Science and Engineering, Vol. 223 (2017).
[155] Ismail, R., Mahadi, Z. A., Ishak, I. S. The effect of carbon black filler to the mechanical properties of natural rubber as base isolation system. IOP Conference Series: Earth and Environmental Science, Vol. 140 (2018).
[156] Boonstra, B. B., Medalia, A. I. Effect of carbon black dispersion on the mechanical properties of rubber Vulcanizates. Rubber Chemistry and Technology 36: 115 (1963).
[157] Zhang, Y., Ge, S., Tang, B., Koga, T., Rafailovich, M. H., Sokolov, J. C., Peiffer, D. G., Li, Z., Dias, A. J., McElrath, K. O., Lin, M. Y., Satija, S. K., Urquhart, S. G., Ade, H., Nguyen, D. Effect of carbon black and silica fillers in elastomer blends. Macromolecules 34:7056 (2001).
[158] Hamed, G. R., Al-Sheneper, A. A. Effect of carbon black concentration on cut growth in NR vulcanizates. Rubber Chemistry and Technology 76:436 (2003).
[159] Medalia, A. I. Effect of carbon black on ultimate properties of rubber Vulcanizates. Rubber Chemistry and Technology 60:45 (1987).
[160] Ismail, H., Ramly, A. F., Othman, N. Effects of silica/multiwall carbon nanotube hybrid fillers on the properties of natural rubber nanocomposites. Journal of Applied Polymer Science 128:2433 (2013).
[161] Cataldo, F., Ursini, O., Angelini, G. Biodiesel as a plasticizer of a SBR-Based tire tread formulation. ISRN Polymer Science 2013:1 (2013).
[162] Cadogan, D. F., Howick, C. J. Ullmann's encyclopaedia of industrial chemistry. Plasticizers 27: (2000).
[163] Geiss, O., Tirendi, S., Barrero-Moreno, J., Kotzias, D. Investigation of volatile organic compounds and phthalates present in the cabin air of used private cars. Environment International 35:1188 (2009).
[164] Egorov, E. N., Ushmarin, N. F., Efimov, K. V., Sandalov, S. I., Spiridonov, I. S., Koltsov, N. I. The influence of functional ingredients on the technological properties of water-swelling rubber sealing elements. Thematic Section: Research into New Technologies 22: 146–147 (2019), doi: 10.37952/ROI-jbc-01/17-50-4-45.
[165] Flanigan, C., Beyer, L., Klekamp, D., Rohweder, D., Haakenson, D. Using bio-based plasticizers, alternative rubber. Rubber and Plastics News 11:15 (2013).
[166] Levin, M., Redelius, P. Determination of three-dimensional solubility parameters and solubility spheres for naphthenic mineral oils. Energy and Fuels 22:3395 (2008).
[167] Capponi, S., Alvarez, F., Racǩo, D. Free volume in a PVME polymer–water solution. Macromolecules 53:4770 (2020).
[168] Rodgers, B., Waddell, W. H., W. K. Rubber Compounding (2004). doi: 10.1201/9780203740385.
[169] Keller, R. W. Oxidation and ozonation of rubber. Rubber Chemistry and Technology 58:637 (1985).
[170] Tsurugi, J., Murakami, S., Goda, K. Charge transfer complexing mechanism of antioxidants. Fate of aromatic amines during thermal oxidation of natural rubber vulcanizates. Rubber Chemistry and Technology 44:857 (1971).
[171] Bernard, D., Cain, M. E., Cunneen, J. I., Houseman, T. H. Oxidation of vulcanized natural rubber. Rubber Chemistry and Technology 45:381–401 (1972).
[172] Layer, R. W., Lattimer, R. P. Protection of rubber against ozone. Rubber Chemistry and Technology 63:426 (1990).
[173] Datta, R. N., Huntink, N. M., Datta, S., Talma, A. G. Rubber vulcanizates degradation and stabilization. Rubber Chemistry and Technology 80:436 (2007).
[174] Cox, W. L. Chemical antiozonants and factors affecting their utility. Symposium on Effect of Ozone on Rubber, pp. 57 (2009).
[175] Huntink, N. M., Datta, R. N., Noordermeer, J. W. M. Addressing durability of rubber compounds. Rubber Chemistry and Technology 77:476 (2004).

[176] Li, G. Y., Koenig, J. L. A review of rubber oxidation. Rubber Chemistry and Technology 78:355 (2005).
[177] Waddell, W. H. Tire black sidewall surface discoloration and non-staining technology: a review. Rubber Chemistry and Technology 71:590 (1998).
[178] Ambelang, J. C., Kline, R. H., Lorenz, O. M., Parks, C. R., Wadelin, C., Shelton, J. R. Antioxidants and antiozonants for general purpose elastomers. Rubber Chemistry and Technology 36:1497 (1963).
[179] Kuczkowski, J. A. Stabilization of raw dienic synthetic rubber polymers. Rubber Chemistry and Technology 84:273 (2011).
[180] Baldwin, J. M., Bauer, D. R., Ellwood, K. R. Accelerated aging of tires, Part III. Rubber Chemistry and Technology 78:767 (2005).
[181] Baldwin, O. M., Bauer, D. R. Rubber oxidation and tire aging - A review. Rubber Chemistry and Technology 81:338 (2008).
[182] Yasin, U. Q., Kamarun, D., Said, C. M. S., Samsuri, A. Effect of compounding ingredients and crosslink concentration on blooming rate of natural rubber compounds. Advanced Materials Research 1134:50 (2015).
[183] Tamási, K., Kollár, M. S. Effect of different sulfur content in Natural Rubber mixtures on their thermo-mechanical and surface properties. International Journal of Engineering Research and Science 4:28 (2018).
[184] Jurkowski, B., Jurkowska, B. On the mechanism of sulfur behavior in rubber compounds. Journal of Macromolecular Science-Physics 37:135 (1998).
[185] Joseph, A., George, B., Madhusoodanan, K., Alex, R. Current status of sulphur vulcanization and devulcanization chemistry: process of vulcanization. Rubber Science 28:82 (2015).
[186] Ghosh, P., Katare, S., P, P. Sulfur vulcanization of natural rubber for benzothiazole accelerated formulations. Rubber Chemistry and Technology 76:592 (2003).
[187] Van Duin, M., Orza, R., Peters, R., Chechik, V. Mechanism of peroxide cross-linking of EPDM rubber. Macromolecular Symposia, Vol. 291–292, pp. 66, (2010).
[188] Kruzelák, J., Dosoudil, R., Sýkora, R., Hudec, I. Rubber composites cured with sulphur and peroxide and incorporated with strontium ferrit. Bulletin of Materials Science 40:223 (2017).
[189] Boonkerd, K., Deeprasertkul, C., Boonsomwong, K. Effect of sulfur to accelerator ratio on crosslink structure, reversion, and strength in natural rubber. Rubber Chemistry and Technology 89:450 (2016).
[190] Khang, T. H., Ariff, Z. M. Vulcanization kinetics study of natural rubber compounds having different formulation variables. Journal of Thermal Analysis and Calorimetry 109:1545 (2012).
[191] Marykutty, C. V., Mathew, E. J., Thomas, S. Studies on a new binary accelerator system in sulfur vulcanization of natural rubber. KGK Kautschuk Gummi Kunststoffe 61:383 (2008).
[192] Coran, A. Y. Vulcanization. Part VII. Kinetics of sulfur vulcanization of natural rubber in presence of delayed-action accelerators. Rubber Chemistry and Technology 38:1 (1965).
[193] Coran, A. Y. Chemistry of the vulcanization and protection of elastomers: A review of the achievements. Journal of Applied Polymer Science 87:24 (2003).
[194] Zhong, B., Jia, Z., Luo, Y., Jia, D. A method to improve the mechanical performance of styrene-butadiene rubber via vulcanization accelerator modified silica. Composites Science and Technology 117:46 (2015).
[195] Heideman, G., Noordermeer, J. W. M., Datta, R. N., Van Baarle, B. Effect of zinc complexes as activator for sulfur vulcanization in various rubbers. Rubber Chemistry and Technology 78:245 (2005).
[196] Heideman, G., Datta, R. N., Noordermeer, J. W. M., Van Baarle, B. Activators in accelerated sulfur vulcanization. Rubber Chemistry and Technology 77:512 (2004).
[197] Nieuwenhuizen, P. J. Zinc accelerator complexes. Versatile homogeneous catalysts in sulfur vulcanization. Applied Catalysis A: General 207:55 (2001).

[198] Zhong, B., Jia, Z., Hu, D., Luo, Y., Jia, D., Liu, F. Enhancing interfacial interaction and mechanical properties of styrene-butadiene rubber composites via silica-supported vulcanization accelerator. Composites Part A: Applied Science and Manufacturing 96:129 (2017).

[199] Akiba, M., Hashim, A. S. Vulcanization and crosslinking in elastomers. Progress in Polymer Science 22:475 (1997).

[200] American Society for Testing & Material Standard test methods for rubber products – chemical analysis. *ASTM D297* 9(01):1 (2011).

[201] American Society for Testing & Material Standard test method for rubber –compositional analysis by thermogravimetry. *ASTM D6370* 9(01):6 (2019).

[202] Fernández-Berridi, M. J., González, N., Mugica, A., Bernicot, C. Pyrolysis-FTIR and TGA techniques as tools in the characterization of blends of natural rubber and SBR. Thermochim Acta 444:65 (2006).

[203] Shi, S., Lei, B., Li, M., Cui, X., Wang, X., Fan, X., Tang, S., Shen, J. Thermal decomposition behavior of a thermal protection coating composite with silicone rubber: Experiment and modeling. Progress in Organic Coatings 143:1–15 (2020). doi: 10.1016/j.porgcoat.2020.105609.

[204] Kodal, M., Karakaya, N., Wis, A. A., Ozkoc, G. Thermal properties (DSC, TMA, TGA, DTA) of rubber nanocomposites containing carbon nanofillers. In: Carbon-Based Nanofillers and Their Rubber Nanocomposites: Fundamentals and Applications, Elsevier, 325–366 (2019). doi: 10.1016/B978-0-12-817342-8.00011-1

[205] Bystritskaya, E. V., Monakhova, T. V., Ivanov, V. B. TGA application for optimising the accelerated aging conditions and predictions of thermal aging of rubber. Polymer Testing 32:197 (2013).

[206] Harun, N. Y., Afzal, M. T., Azizan, M. T. TGA snalysis of rubber seed kernel. International Journal of Engineering 3:639 (2010).

[207] American Society for Testing & Material Standard test method for compositional analysis by Thermogravimetry. *ASTM E1131* 14(05):6 (2015).

[208] Mano, J. F., Gómez Ribelles, J. L., Alves, N. M., Salmerón Sanchez, M. Glass transition dynamics and structural relaxation of PLLA studied by DSC: Influence of crystallinity. Polymer (Guildf) 46:8258 (2005).

[209] Schawe, J. E. K. Principles for the interpretation of modulated temperature DSC measurements. Part 1. Glass transition. Thermochim Acta 261:183 (1995).

[210] American Society for Testing & Material Standard test method for assignment of the DSC procedure for determining Tg of a polymer or an elastomeric compound. *ASTM D7426* 9(01):1 (2013).

[211] Rahman, M. S., Al-Marhubi, I. M., Al-Mahrouqi, A. Measurement of glass transition temperature by mechanical (DMTA), thermal (DSC and MDSC), water diffusion and density methods: A comparison study. Chemical Physics Letters 440:372 (2007).

[212] Tseretely, G. I., Smirnova, O. I. DSC study of melting and glass transition in gelatins. Journal of Thermal Analysis 38:1189 (1992).

[213] Liu, P., Yu, L., Liu, H., Chen, L., Li, L. Glass transition temperature of starch studied by a high-speed DSC. Carbohydrate Polymers 77:250 (2009).

[214] Thermo Fisher Scientific Gmbh. User manual-organic elemental analysis eager Xperience For FLASH elemental analyzers. User Manual Part no 31711055 (2014).

[215] International Organization for Standardization Rubber, raw, vulcanised – determination of metal content by ICP-OES. *ISO 19050*. TC 45:1 (2009).

[216] Rezić, I., Steffan, I. ICP-OES determination of metals present in textile materials. Microchemical Journal 85:46 (2007).

[217] Kumaravel, S., Alagusundaram, K. Determination of mineral content in Indian spices by ICP-OES. Oriental Journal of Chemistry 30:631 (2014).

[218] Bocca, B., Forte, G., Petrucci, F., Costantini, S., Izzo, P. Metals contained and leached from rubber granulates used in synthetic turf areas. Science of the Total Environment 407:2183 (2009).

[219] Massadeh, A., Gharibeh, A., Omari, K., Al-Momani, I., Alomary, A., Tumah, H., Hayajneh, W. Erratum: Simultaneous determination of Cd, Pb, Cu, Zn, and Se in human blood of Jordanian smokers by ICP-OES (Biological Trace Element Research (DOI 10.1007/s12011-009-8405-y)). Biological Trace Element Research 133:120 (2010).
[220] American Society for Testing & Mater Standard test methods for steel tyre cords. *ASTM D2969* 7 (01):1 (2014).
[221] Lachowicz, T., Zięba-Palus, J., Kościelniak, P. Z Analysis of rubber samples by PY-GC/MS for forensic purposes. Zagadnien Nauk Sadowych 91:195 (2012).
[222] Jones, M. Gas Chromatography-Mass Spectrometry: A National Historic Chemical Landmark. https://www.acs.org/content/acs/en/education/whatischemistry/landmarks/gas-chromatographymass-spectrometry.html.
[223] Sparkman, O. D., Penton, Z., Kitson, F. G. Gas chromatography and mass spectrometry: A practical guide In: Gas Chromatography and Mass Spectrometry: A Practical Guide, Academic Press, (2011). doi: 10.1016/C2009-0-17039-3.
[224] Liu, X., Zhao, J., Liu, Y., Yang, R. Volatile components changes during thermal aging of nitrile rubber by flash evaporation of Py-GC/MS. Journal of Analytical and Applied Pyrolysis 113:193 (2015).
[225] Kamarulzaman, N. H., Le-Minh, N., Stuetz, R. M. Identification of VOCs from natural rubber by different headspace techniques coupled using GC-MS. Talanta 191:535 (2019).
[226] Li, X., Berger, W., Musante, C., Mattina, M. J. I. Erratum to characterization of substances released from crumb rubber material used on artificial turf fields. Chemosphere 80:1406 (2010).
[227] Gohlke, R. S., Mclafferty, F. W. Early gas chromatography/mass spectrometry. American Society for Mass Spectrometry 4:367 (1993).
[228] Hübschmann, H. J. Handbook of GC/MS: Fundamentals and applications. In: Handbook of GC/MS: Fundamentals and Applications, Wiley-VCH Verlag GmbH & Co. KGaA, 1–719, (2008). doi: 10.1002/9783527625215.
[229] Griffiths, P. R. Fourier transform infrared spectrometry. Science (80-.) 222:297 (1983).
[230] American Society for Testing & Material Standard practice for rubber chemicals – determination of infrared absorption. *ASTM D2702* 9(01):1 (2016).
[231] International Organization for Standardization Rubber – identification – infrared spectrometric method, ISO 4650. *ISO 4650*. TC 45:1 (2011).
[232] American Society for Testing & Material Standard test methods for rubber – identification by infrared spectrophotometry. *ASTM D3677* 9(01):1 (2019). doi: 10.1520/D3677-10R19.
[233] Chakraborty, S., Bandyopadhyay, S., Ameta, R., Mukhopadhyay, R., Deuri, A. S. FTIR spectra and mechanical strength analysis of some selected rubber derivatives. Polymer Testing 26(1):38–41 (2007), doi: 10.1016/j.polymertesting.2006.08.004.
[234] Azevedo, J. B., Murakami, L. M. S., Ferreira, A. C., Diniz, M. F., Silva, L. M., de Cássia Lazzarini Dutra, R. Application of FTIR in characterization of acrylonitrile-butadiene rubber (nitrile rubber). Polimeros 28(5):440–449 (2018), doi: 10.1590/0104-1428.00918.

Shambhu Lal Agrawal and Abhijit Adhikary

# Chapter 8 Thermal and mechanical analysis study of different rubber applications

## 8.1 Introduction

Rubbers are used for various applications to achieve the desired performance properties during their end use. General-purpose rubbers like natural rubber (NR), styrene–butadiene rubber (SBR), and polybutadiene rubber (PBR) may be used interchangeably in a majority of the rubber components, based on the cost factor. These rubbers have major usage in tyres, followed by retread, conveyor, footwear, hose, V-belt, rice polisher brake, rice roller, other moulded rubber products, etc. Special-purpose rubbers have characteristics to achieve specific properties during product performance, for example, butyl rubber (IIR) and halobutyl rubbers (CIIR and BIIR), are used due to their air retention characteristics, ethylene–propylene–diene rubber (EPDM) for weather resistance, chloroprene rubber (CR) for flame resistance, acrylonitrile rubber (NBR) for oil resistance, fluoro-elastomers (FKM) for high temperature resistance, silicone rubbers (Q) for biocompatibility and low temperature flexibility, etc. Special-purpose rubbers are used for many applications like tube and inner liners in tyres, curing bladder, pharma stopper, acid tank liner, vibration isolators, squash ball, ball bladder, seal, high temperature hose, etc.

While rubber products are used in a majority of applications, the largest of which is in automobile components, slightest quality variation might lead to malfunctioning of the product or complete halt of operation, depending on the criticality of its function. The tragic accident of the Space Shuttle Challenger, which exploded 73 s after lift-off on January 28, 1986, killing seven crew members, traumatised the world. The cause of the disaster was traced to an O-ring malfunctioning.

In general, there are two types of testing: (1) quality control and (2) performance evaluation. For quality control purposes, similarity of test equipment, standard test method, and uniformity of reference standard material are the requirements. Inter-Laboratory Correlation (ILC) is one such method to assess the variation of test results between the labs using *Z*-scores. On the other hand, performance evaluation tests remain a subjective area. Understanding the minute details of product performance and simulating those in close experimental conditions are the challenges.

Rubber sample preparation for testing involves the following steps: (a) recipe finalisation, (b) mixing and the subsequent processing, (c) vulcanisation, (d) maturation, and (e) sample preparation. These are tested for various mechanical properties in static as well as in dynamic modes of testing, such as specific gravity, Mooney viscosity, rheometric properties, compound processability tests, uniaxial stress–strain

https://doi.org/10.1515/9783110668537-008

(in tensile and compression modes), tear, hardness (IRHD and Shore A), and biaxial and triaxial tests. Dynamic mechanical analysis in various deformation modes (tension, compression, and shear), including effect of temperature, frequency and strain, and time-temperature superposition, rebound resilience, hysteresis, testing of flex properties, and fatigue properties are done to estimate the effect of accelerated test conditions. Swell index, volume fraction, cross-link density (including ammoniacal conditions), etc. are also measured to check the state of cure of rubber vulcanisates.

The source of variation in rubber products quality may come from following (a) incoming raw material quality and its suitability of use, (b) changes in material processing and handling conditions, and (c) pre- and post-vulcanisation conditions, and material contamination at various stages of material processing operations. There are currently 150 ASTM test methods for raw material testing, 15 ASTM test methods for testing processing behaviour of rubber compounds, and more than 100 ASTM test methods for testing vulcanisate properties.

To check the effect of elastomer and its blends, anti-degradants, cure system/cross-link density, other rubber compound ingredients on product performance during application, accelerated ageing properties (anaerobic-thermal oxidation and aerobic exposure) are performed. Thermal degradation mechanism and factors affecting mechanical properties like strain-induced crystallisation, reinforcement by filler, processing (mainly mixing/extrusion), process aid for filler dispersion, etc. are studied.

Thermal behaviour of rubbers and its vulcanisates are important to assess its performance in different application conditions. Each rubber has a specific glass transition temperature, which helps for its selection for application. Glass transition temperature is measured using differential scanning calorimeter (DSC) and dynamic mechanical analysis. In general, polymer decomposition temperature and degradation kinetics are specific to elastomers for a given environment and rate of heating, which is evaluated using thermogravimetric analyser (TGA). Other thermal properties such as specific heat, thermal conductivity, thermal diffusivity, coefficient of thermal expansion (CTE), and surface heat transfer coefficient are important for product performance requirements and product life assessment.

Test frequency determination to ascertain the quality of output in a manufacturing setup is a complex subject and is related to various factors like raw material procurement policy, material storage facility, plant process reliability, age of processing equipment, its maintenance strategy, reliability of test equipment, etc. It is part of the management strategy, mostly determined by the complexity of the product, the underlying manufacturing processes, process capability requirements, and technological superiority of the product in the marketplace.

# 8.2 Sample preparation

## 8.2.1 Sampling

The selected sample should truly represent the lot to be tested; hence, sample selection is very important before further analysis. Two terms that are often mixed up in rubber testing are 'sample' and 'specimen'. As per ASTM D1566 [1], sample is the portion selected to represent the lot and specimen is a piece of the material appropriately shaped and prepared so that it is ready to use for test. In other words, the sample is the material taken from a lot in a manner such that the properties or characteristics of the sample are essentially the same as those of the entire lot. On the other hand, a specimen is a small part of the sample that is tested after being suitably prepared for a test. Generally, several specimens are prepared and tested from a single sample. A sample may be directly taken from a product using various tools or it may be prepared after compound mixing.

## 8.2.2 Mixing and moulding

Mixing and moulding of rubber compounds are performed following ASTM D3182 [2].

### 8.2.2.1 Mixing

Rubber compound recipe is mixed through a two-roll mill, a miniature internal mixer, a Banbury mixer, or their combination. These recipes may be either as per the respective standard test methods that are available for most of the rubber and rubber ingredients or application formulation. Pictures of internal mixer and two-roll mill are portrayed in Figures 8.1 and 8.2. In a two-roll mill, rolls are rotated in opposite directions with a friction ratio between them. Rubber is masticated and a band is formed on the front roll. Once a band is formed, the ingredients are added as per sequence and time. Cuts and folds are applied on the rubber compound to achieve the desired homogeneity of ingredients in the rubber. In a Banbury internal mixer, the rotors (either tangential or intermeshing type) are rotated inside a closed chamber. Here too, the rubber is loaded first for mastication, followed by the ingredients as per the specific sequence and time. On achieving required mixing time, batch temperature, and mixing energy, the mixed compound is dumped and sheeted out using two roll mills. For masterbatch compounds, the dump temperature is generally maintained near about 150 °C, whereas it is maintained around 100 °C for the final batch mixing. For EPDM-based compound with a high filler and oil lading, upside down mixing is preferred in which at first the filler and oil are loaded in the mixer chamber, followed by the rubber. The respective standard method for evaluation covers the detailed mixing procedure including the sequence and timing of chemical addition, mixing condition like rotor speed, TCU setting, mixing time, etc. These test methods also cover the milling procedure for the mixed compounds and the maturation period before further steps or testing. Standard ingredients are used for the ASTM recipe; however, commercial grade ingredients can also be used for evaluation of application formulations. For application study, mixing equipment/conditions may be used in line with the actual production. Mixing may be performed in single or multiple stages, based on filler loading, compound properties requirement, etc.

Normally, the compounds are mixed in the first stage without curing the ingredients and it is called masterbatch mixing. Once the curing ingredients are added to the masterbatch compound, it is called the final batch compound. To reduce the compound's Mooney viscosity and to improve the filler dispersion/distribution, mainly in case of higher filler loading, the compounds are also mixed in a re-mill/re-pass stage in which no chemicals are added. All the compounding ingredients are taken as phr (parts per hundred parts of rubber), which means part per hundred parts of rubber. The main objective of the mixing is to achieve a compound with optimised dispersion and distribution of the filler and other ingredients in the rubber matrix.

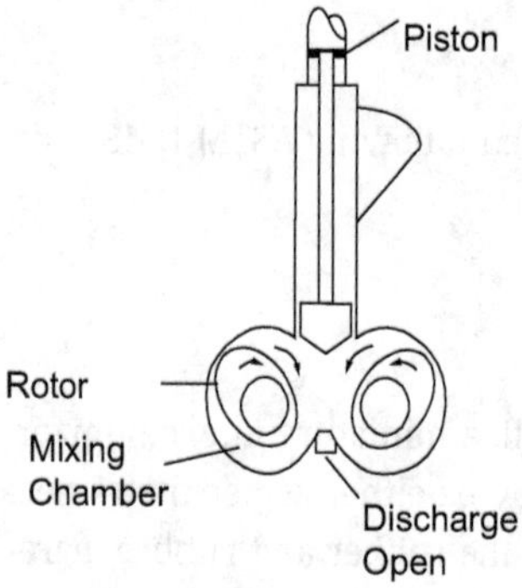


**Figure 8.1:** Internal mixer.

### 8.2.2.2 Moulding

Mixing is followed by moulding after a specified maturation period, which is required for a better dispersion of the ingredients and for homogenisation. Moulding is performed using various equipment like compression, injection, and transfer moulding at specific sets of conditions like time, temperature, and pressure. However, for lab test specimen preparation, compression moulding press is mostly used. A representative picture of a typical compression moulding press is shown in Figure 8.3. For ASTM standard recipe, fixed moulding temperature, time, and pressure conditions are followed as mentioned in the respective standard. However, in the case of application study, the

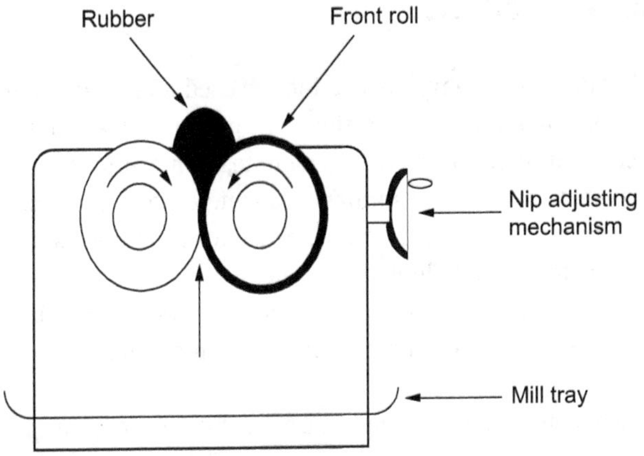


**Figure 8.2:** Two-roll mill.

analyst may use a moulding time based on rheometer test data. Normally, for less-thick sample preparation for stress–strain, fatigue-to-failure test, ozone, gas permeability, dynamic mechanical analysis, etc. moulding is performed at an optimum cure time ($t_{C90}$) plus 2 min at the specific temperature. Whereas, for high thickness samples like Goodrich Flexometer, Demattia, abrasion test, rebound, cut and chip, etc. moulding time is used as an optimum cure time plus 5 min. This is in line with the basic curing principle that thicker rubber products should be moulded at lower curing temperatures and for higher time, whereas thin rubber products should be moulded at higher curing temperatures and for lesser time. Moulded samples are stored in laboratory conditions for 16–96 h for maturation before final testing. Before moulding, sample performs are prepared by proper milling of the rubber compound to the desired thickness and the milling direction is marked. In the moulding process, the uncured final batch rubber compound is transformed to three-dimensional networks by the application of heat and pressure for a specified time. Moulded samples are used for further testing of failure properties. The sample weight for performs is normally taken as approx. 10% more than the calculated volume of the mould cavity to avoid any moulding defects like starvation, flow marks, etc. This leads to moulding flashes that are removed using a knife or scissors. Here, the moulds play a very important role in getting a smooth surface of the moulded sample. Hence, mould surfaces should be well polished and are fabricated using steel (tool/mild/stainless). These moulds should be further chrome plated; however, for moulding highly corrosive or adhesive materials, they may be coated with Teflon materials. Each mould has specific dimensions as mentioned in the respective standard test method. Sample dimensions are particularly important for accurate test results. Hence, any moulded sample not meeting the dimensions as per the requirement should be discarded.

### 8.2.3 Samples from finished products

This is part of destructive testing in which the product cannot be used after taking out the sample for testing. For all kinds of application development purposes, we require guideline formulation to reduce the number of trials. To facilitate, many rubber product manufacturers perform reverse engineering studies of a competitor's products. Sometime the finished product's sample testing is also performed to check the test parameters versus specification. This may also help in cost reduction if their product is overdesigned with respect to the specification. Few equipment like slicing machine, drilling machine, punching dies, etc. are used for preparing sample from finished products, following ASTM D3183 [3].

Products like tyres, conveyor belt, etc. are manufactured using multiple components to achieve the final performance properties. Slicing procedure is required to separate these multilayer components using a movable band cutting knife. This machine helps to slice the required thickness of sample from the finished product; however, good experience is required to identify and differentiate the various layers of the finished product. Once the samples are sliced with the desired thickness, they are punched to the required test specimen dimensions using hydraulic/pneumatic punching machine and specific dies. ISO-type dumbbells are punched due to the smaller specimen size.

These dumbbells may be used for measurement of stress–strain and hardness properties. Abrasion specimens may also be prepared using a drilling machine and a die. If needed, the surface of these samples may be buffed using a buffing machine. This operation removes any unevenness of the surface. These samples are allowed to stay for a minimum of 16 h for conditioning before testing.

### 8.2.4 Sample conditioning

As discussed above, samples prepared at various stages like mixing, moulding, slicing/punching, etc. are kept for maturation for a specific length of time at laboratory conditions. Normally, laboratories are maintained at a standard temperature of 25 ± 2 °C and 50 ± 5% relative humidity [4]. Any change in temperature and humidity of the laboratory may affect the properties of the specimen. Also, it may be one of the reasons for variation of the test results between two laboratories. Laboratory temperatures are maintained using an air-conditioner unit, and humidity is controlled using dehumidifier machines. Laboratory areas should be dust-proof, which may help to protect the testing apparatus and the samples from contamination for analytical tests. Uncured rubber compounds are conditioned for 1–24 h before further processing or testing and cured samples are conditioned for 16–24 h at standard lab conditions. However, this duration may be changed as per the test plan. This maturation will improve filler dispersion in the rubber matrix for vulcanisates.

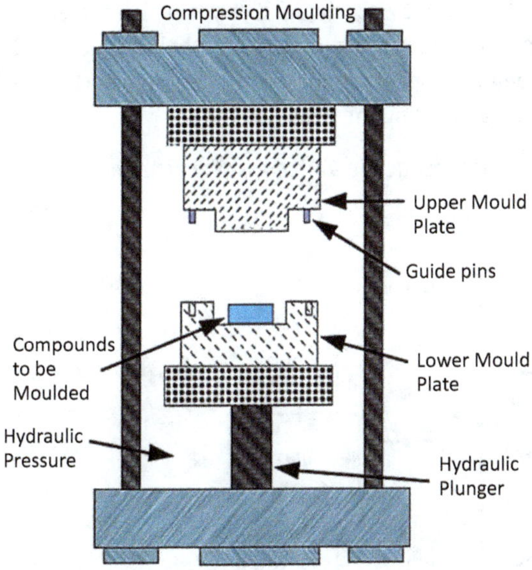


**Figure 8.3:** Compression moulding press.

# 8.3 Sample testing – uncured compound

## 8.3.1 Specific gravity

A compound's specific gravity is measured using a compressed volume densimeter for uncured compounds and the hydrostatic method is used for cured specimens. A representative picture of a compressed volume densimeter is depicted in Figure 8.4. It follows the ASTM D297 [5] method.

This method is used to evaluate the effect of ingredients, including the metal oxide level, on a rubber compound, mainly carbon black other fillers. This test also gives an idea about the missing ingredients, mainly filler/ metal oxide, during mixing. For an uncured compound sample, the mass of the sample is determined nearest to 0.01 g as also the final volume attained by sample is measured when it is subjected to a compressive force to take the shape of chamber after its flow. Approx. 87 psi air pressure is used to compress the sample in the machine chamber. A compressed volume densimeter has an air-operated piston cylinder test chamber of known dimensions. Sample volume is taken as approx. 100 cc to achieve repeatable test results. This method may also be used for raw rubber, whose specific gravity is normally less than 1. The hydrostatic method follows the Archimedes principle of buoyant force in which the sample weight is determined nearest to 0.1 mg and the specimen is taken such that its weight is between 1 and 10 gm. The specimen is immersed in water and its weight is again determined. Air bubbles on the specimen, if any, may lead to errors in the test results. Hence, the test specimen is first dipped in methyl/ethyl alcohol-water mixture, followed by blotting before suspending in water. The test is performed at 25 °C and the specific gravity is calculated using the below formula:

$$0.9971 \times \text{mass of specimen in air}/(\text{mass of specimen in air} - \text{mass of specimen in water}) \quad (8.1)$$

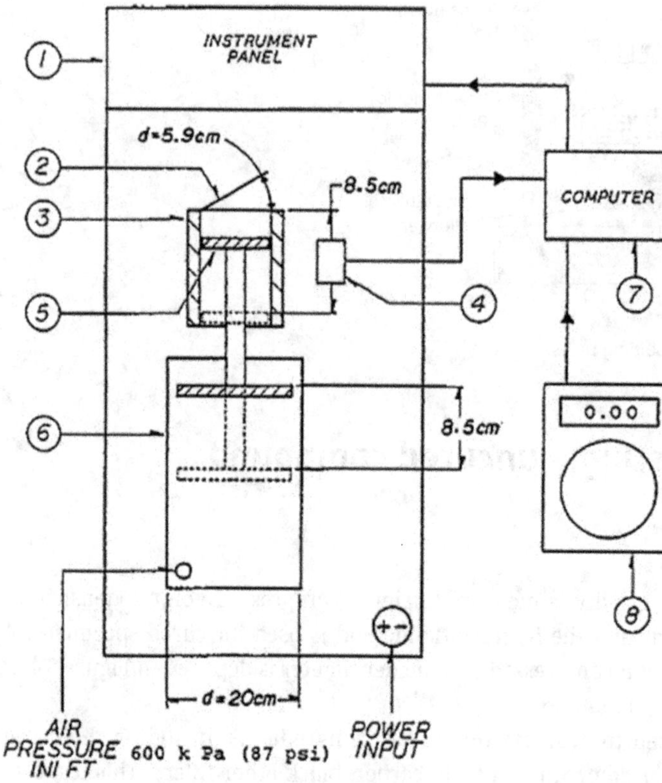


**Figure 8.4:** Compressed volume densimeter.

## 8.3.2 Bound rubber

It is measured for rubber compound masterbatch (without curatives) to determine the polymer-filler interaction [6]. When uncured rubber compound is immersed in a good solvent for a sufficiently long time (approx. 96 h), some part of rubber is not extracted by the solvent and the unextracted part is known as bound rubber. A higher bound rubber content indicates more polymer-filler interaction. This is desirable to achieve better mechanical and failure properties as it indicates higher reinforcement.

Around 0.2–0.3 g of sample is accurately weighed and cut to small pieces. It is then put to 350-mesh steel wire cage in excess toluene/suitable solvents for 120 h. After every 24 h, the solvent is changed until it becomes colourless. It is then dried at room temperature for 24 h and then dried at 100 °C in an air oven until constant weight. It is then cooled in desiccators and the final weight is taken after 24 h. The bound rubber content (%) is calculated using the below mentioned formula:

$$\text{Bound rubber content}(\%) = [(W_{\text{d}} - W_{\text{f}})/W_{\text{o}}] \times 100 \tag{8.2}$$

where $W_d$ is the final weight of the dried sample in g, $W_f$ is the weight of the filler in the sample in g, and $W_o$ is the weight of the rubber in the sample in g.

Total bound rubber has the following components: primary layer (or tightly bound rubber), secondary layer (or loosely bound rubber), and connecting filaments (bridges). It is possible to measure the individual components of bound rubber content. To measure the primary layer bound rubber, the sample needs to be extracted at 90 °C for 120 h, followed by sonication at 45 °C for 3 h. Secondary layer (or loosely bound rubber) is the difference of the total bound rubber content and the primary layer bound rubber.

For a silica-filled compound, the primary layer (or tightly bound rubber) is measured by solvent extraction of the sample at 90 °C for 120 h followed by bubbling with ammonia gas for 20 h at room temperature, and sonication at 45 °C for 3 h. Ammoniacal condition will break loose the chemical bond between the silica filler and rubber and hence provide information about the chemical polymer-filler interaction.

## 8.3.3 Mooney viscosity

Mooney viscosity of a raw rubber or compounded rubber is measured using Mooney viscometer following ASTM D1646 [7] or ISO289. Its value is expressed as Mooney units (MU). A representative picture of a typical shearing disc Mooney viscometer is portrayed in Figure 8.5. It measures the torque required to rotate a rotor at two revolutions per minute when it is embedded in the specimen and packed in a closed die cavity at a specific temperature. Two rotors with different diameters are used for Mooney viscosity measurement. The larger rotor has a diameter of 38.10 + 0.03 mm whereas the smaller rotor diameter is 30.48 ± 0.03 mm. Generally, ML (1 + 4)@100 °C is measured, which indicates that the large rotor (L) is used, one minute preheating time given, and the specimen is tested for 4 min. The small rotor is used for high-viscosity specimens like rubber latex coagulates.

The Mooney viscosity value depends on the molecular weight, molecular structure, and the nonrubber constituents present in the rubber. A compound's Mooney viscosity value determines its processing; both high and low Mooney viscosity compounds are difficult to process. Therefore, for every compound in a manufacturing setup, a narrow band of Mooney viscosity is preferred for ease of processing. A compound's Mooney viscosity increases with increasing filler loading and decreases with increasing process oil loading. For few raw rubbers other than NR, IIR, HIIR, EPDM, etc. massed specimen may also be used for Mooney viscosity test. A 250 ± 5 g sample is passed between hot rolls at 50 ± 5 °C in a two-roll mill following the procedure mentioned in ASTM D1646. For butyl and halo-butyl raw rubber samples, the temperature of the rolls is increased to 145 ± 5 °C and the nip gap is also slightly increased. Mooney viscometer may also be used for determining the stress relaxation of raw rubber and rubber compounds, and the scorch safety time of the final rubber compound, following the same ASTM method mentioned above.

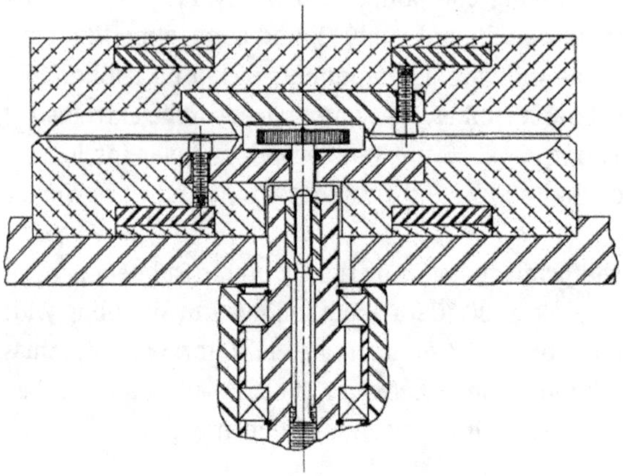

**Figure 8.5:** Shearing disk Mooney viscometer.

### 8.3.3.1 Mooney stress relaxation

For the stress relaxation test, the Mooney viscosity test is extended in which the rotor is stopped after completion of the 4 min test time. The torque experienced by the sample is recorded versus time for 2 min and the log of Mooney viscosity versus log time is plotted to measure the slope of the plot. A slow rate of relaxation indicates the presence of higher elastic components in the sample and vice versa. A representative picture of a Mooney viscosity and stress relaxation test are illustrated in Figures 8.6 and 8.7, respectively.

### 8.3.3.2 Mooney scorch

In the Mooney scorch test, Mooney viscosity is plotted against the test time. When the rubber compound attains a specific Mooney value, the rotor is automatically stopped. This method measures the onset of vulcanisation; so it can be used to understand ac-

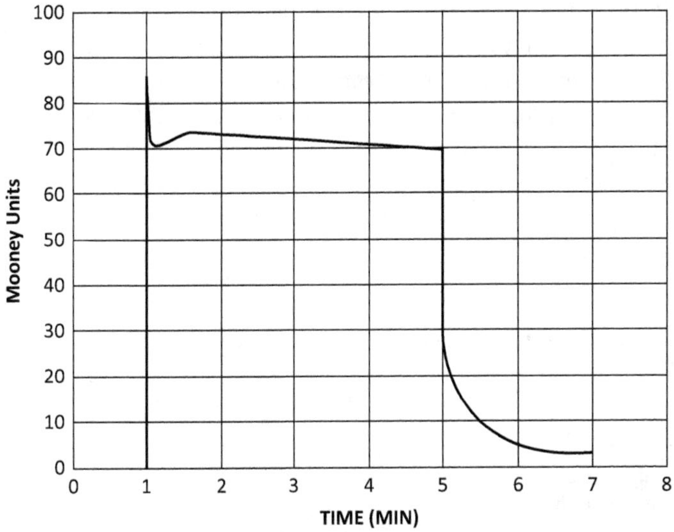


**Figure 8.6:** An example of Mooney viscosity and stress relaxation test.

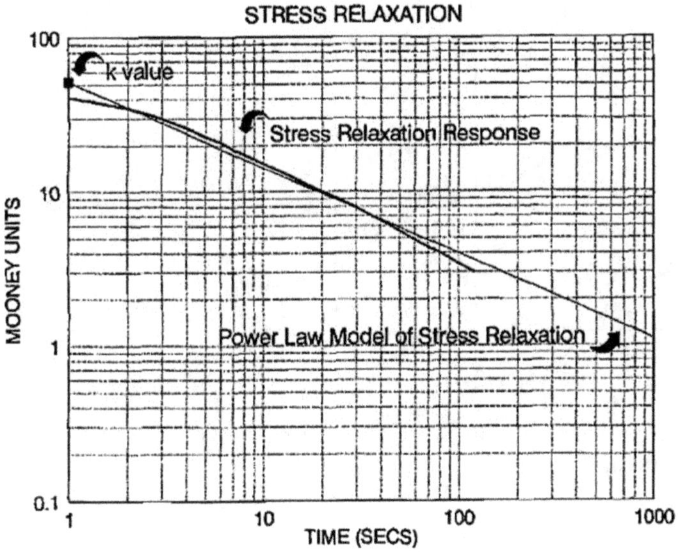


**Figure 8.7:** Plot of log Mooney viscosity versus log time for stress relaxation test.

cidental (scorch) cure time and cure rate during the initial curing stage. A higher Mooney scorch value is desirable for a better processing of the rubber compound. Mooney scorch depends on the cure system like dosage and the type of vulcanising chemicals. Normally, sulfenamide accelerators are used for better scorch safety.

A representative image of Mooney scorch test is shown in Figure 8.8. When a large rotor is used, the machine is stopped when the Mooney viscosity value increases by 35 MU to the minimum Mooney viscosity; however, in the case of the small rotor, it stops after an increase of 18 MU. The time corresponding to an increase of 5 MU for the large rotor and 3 MU for the small rotor indicates the scorch safety time, and cure index can be calculated using the below formula:

$$\text{For large rotor,}\quad t_{35} - t_5 \tag{8.3}$$

$$\text{For small rotor,}\quad t_{18} - t_3 \tag{8.4}$$

Here, $t_{35}$, $t_5$, $t_{18}$, and $t_3$ are the times corresponding to an increase of 35, 5, 18, and 3 MU, respectively, to the minimum Mooney viscosity. All values are expressed as minutes.

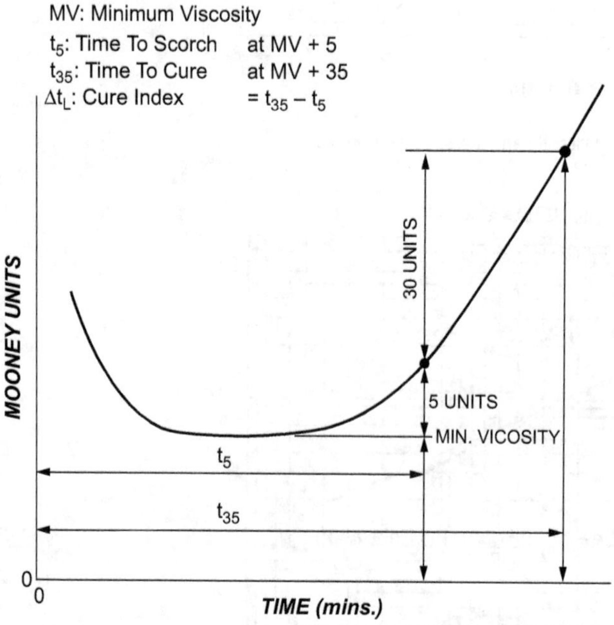


**Figure 8.8:** An example of a Mooney scorch test.

## 8.3.4 Rheometric properties

Rheometric properties can be measured using a moving die rheometer (MDR) and an oscillation disc rheometer (ODR), following ASTM D5289 and ASTM D2084, respectively [8, 9]. Both equipment are used with a test frequency of 100 cycles per minute in the isothermal condition. However, the strain level is maintained as 0.5° for MDR while there are three options – 1°, 3°, and 5° for ODR. Rheometric properties provide an idea about the cure characteristics of the final batch compound. The test is performed at an isothermal condition at a fixed strain and frequency. A specimen of approx. 5 g (4.75 ± 0.75 cc) weight is taken for MDR and 10 g (9.5 ± 1.5 cc) is taken for ODR. This equipment measures the torque change in the rubber compound with respect to test time. A rheograph is a plot of torque versus time. The parameters have five stages. Representative pictures of the ODR and rheometer cure test are depicted in Figures 8.9 and 8.10, respectively. In the warming up stage, the compound attains the set temperature and hence torque is reduced to its lowest value, known as the minimum torque (ML). Scorch safety time ($t_{S2}$) is calculated as time to achieve a torque rise of 2 units to minimum torque (ML). It is defined as the time interval from the beginning of the heat cycle to the beginning of cure, and it is expressed in minutes. However, scorch safety time measured using Mooney viscometer is more related to the actual plant conditions. In the curing region, the torque increases at a fast rate to attain the optimum torque level due to the increasing cross-linking. The optimum cure time ($t_{C90}$) is calculated as the time corresponding to the torque calculated using the below formula:

$$t_{C90} = 0.9\ (\mathrm{MH} - \mathrm{ML}) + \mathrm{ML}[\text{corresponding time of the calc torque}] \tag{8.5}$$

Here, $t_{C90}$ is the optimum cure time expressed in minutes.
ML is the minimum torque expressed in dN-m or lb-in.
MH is the maximum torque expressed in dN-m or lb-in.

The above optimum cure times ($t_{C90}$) decide the time required to cure different specimens for various physical properties. In the plateau region, the torque remains constant at the maximum level, and it indicates that the rubber compound's properties are also maintained at the maximum level. In the overcure region, either an increasing (marching) trend is observed for rubbers like SBR, NBR, etc. due to the presence of the vinyl structure in the polymer chain, which cross-links further or a decreasing trend is observed for rubbers like NR, IR, and IIR due to the presence of methyl ($-CH_3$) group, which activates the double bond and leads to chain breakage. The scorch safety time and the cure rate depend on the curative system, including the type and dosage. We can calculate rate of cure also using the below formula:

$$\text{Rate of cure} = 100/(t_{C90} - t_{S2}) \tag{8.6}$$

Here, $t_{C90}$ is optimum cure time expressed in minutes and $t_{S2}$ is the scorch safety time expressed in minutes.

## 8.3.5 Processability by rubber process analyser

A rubber process analyser (RPA) is used to check the processability of a rubber compound. RPA is used for comparative study with respect to a control sample. It is heated by a digitally controlled system in forced air cooling. It hosts a rotor-less bi-

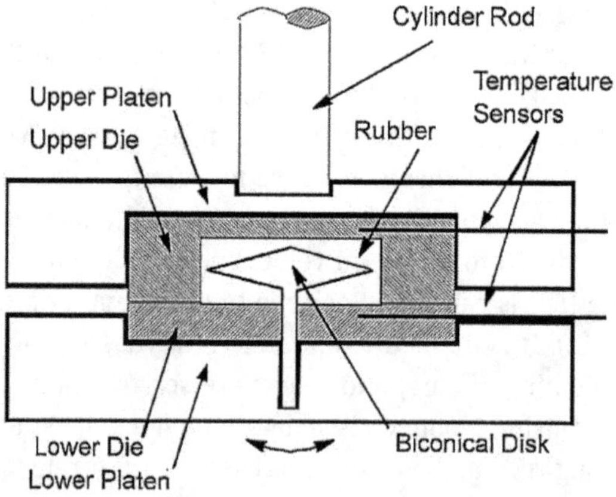


**Figure 8.9:** ODR rheometer assembly.

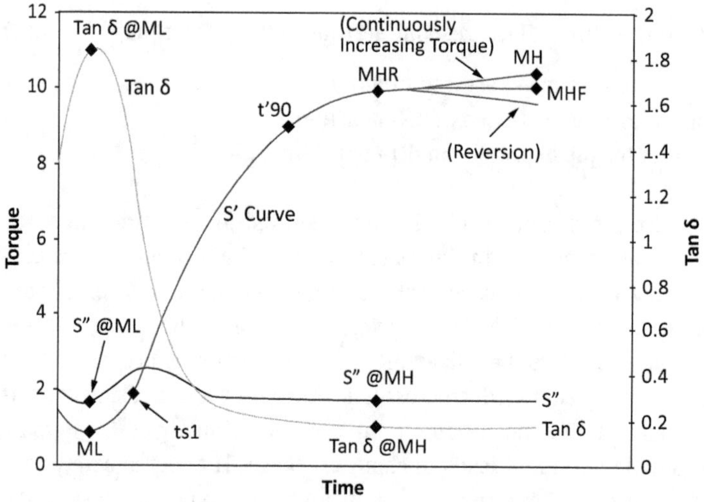


**Figure 8.10:** An example of a rheometer cure test.

conical die, which is formed when both the upper and lower parts of die contact each other. Excess sample extrudes through the spew channel and thus the cavity contains a constant volume of the sample for every test. The digitally controlled motor, which is attached to the lower die, can vary the frequency and strain (by oscillating the lower die) and a reaction torque transducer on the upper die measures the torque transmitted through the sample. It can test the viscoelastic materials over a wide range of temperatures, frequencies, and strains (single strain amplitude, SSA and double strain amplitude, DSA). It provides the properties of the viscoelastic material before, during, and after the curing of the rubber compound, providing information on the processing and curing properties of the green compound and performance properties of the rubber product. The RPA test requires approx. 5 cc sample for testing. The sample is exposed to cyclic deformation (torsional shear) and the response (torque) is measured. This torque is split into two components: elastic torque ($S'$) and viscous torque ($S''$) by applying the Fourier transform to complex torque ($S^*$) signals. Based on the die factor and other parameters, elastic modulus ($G'$), viscous modulus ($G''$), and complex viscosity ($\eta^*$) are also calculated; tan $\delta$ is calculated as the ratio of viscous torque ($S''$) to elastic torque ($S'$).

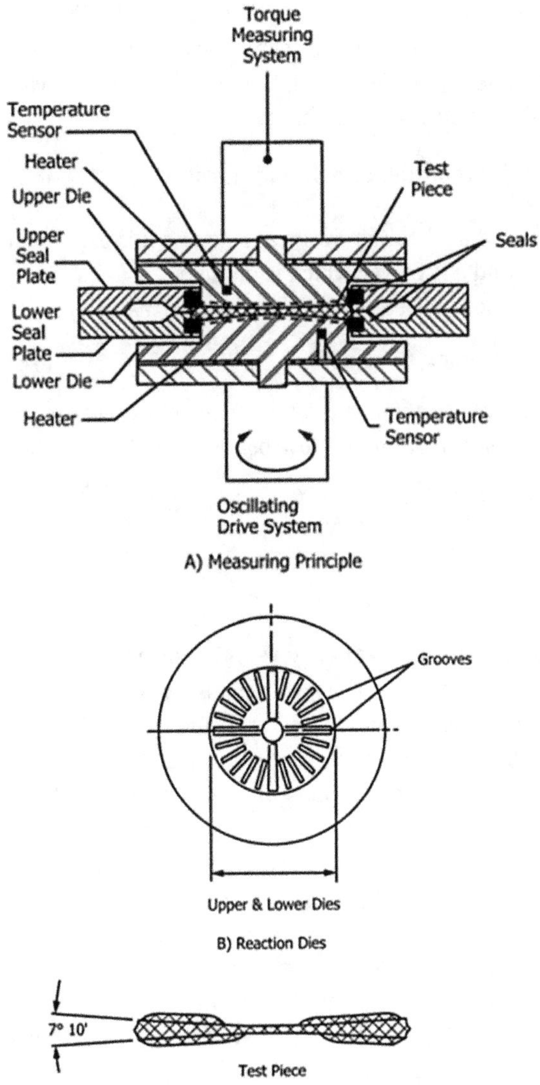


**Figure 8.11:** Torsional shear rotor-less rheometer, RPA (rubber process analyser).

ASTM D6204 [10, 11] is used for checking the rheological properties of the raw rubber and rubber compound using RPA. A representative image of RPA is shown in Figure 8.11.

#### 8.3.5.1 Frequency sweep test

In a frequency sweep test, the temperature and strain remain constant and the frequency changes. Raw elastomer on this mode of testing provides an idea about the relative molecular weight distribution; tan δ difference is checked at low and high frequencies; higher difference in values indicates a narrow molecular weight distribution and vice versa. Flow behaviour of rubber compounds can be determined by measuring the power law index in a frequency sweep test [12]. As per Newton's power law, shear stress is proportional to the (shear rate)$^n$ and hence,

$$\Rightarrow \text{Complex viscosity}(\eta^*)\ \text{is proportional to the (shear rate)}^{n-1} \tag{8.7}$$

Here, complex viscosity ($\eta^*$) is calculated as shear stress/shear rate and it is expressed in Pa-s, shear rate is expressed in $s^{-1}$, and $n$ is power which is unitless.

From eq. (8.7), the slope of log $\eta^*$ and log shear rate plot is equal to $n-1$:

$$\Rightarrow \text{Power law index, } n = \text{slope} + 1 \tag{8.8}$$

Power law index ($n$) close to zero or a relatively lower value means plastic nature of the rubber compound; hence, a better shear thinning (flow behaviour) of the rubber compound.

#### 8.3.5.2 Strain sweep test

In the strain sweep test, the temperature and frequency remain constant and the strain changes. Raw elastomer testing gives an idea about the relative branching tan $\delta$ difference is checked between low and high strains. A higher difference indicates a linear structure and a lower difference indicates the presence of a high long-chain branched structure in the elastomer. The filler networking or agglomeration of filler particles, which is developed mainly by filler–filler interactions in a rubber compound, is quantified from the strain dependence of the elastic modulus, $G'$. The filler network is gradually destroyed on increasing the strain (at strains well below 100%).

This results in a decrease in the elastic modulus $G'$ with increase in strain amplitude. Payne effect is measured as the difference between elastic modulus ($G'$) at low and high strain, and is a measure of the filler–filler interaction. Higher the Payne effect means more filler–filler interaction. Generally, high filler loading compounds and silica filler-based compounds show high Payne effect values [12, 13].

#### 8.3.5.3 Temperature sweep test

Activation energy of the rubber compound flow can be measured using the temperature sweep test in RPA, in which the frequency and strain (shear rate) remain constant and the temperature rises. Complex viscosity ($\eta^*$) is measured at intervals of 10 °C. Normally, the test starts at 70 °C and ends at 130 °C. Activation energy can be calculated using the Arrhenius–Eyring formula [12, 14]:

$$\eta^* = Be^{Ea/RT} \tag{8.9}$$

$$\Rightarrow \text{Log}\,\eta^* = \text{Log}\,B + Ea/RT \tag{8.10}$$

Here, complex viscosity ($\eta^*$) is measured at a shear rate and it is expressed in Pa-s, $B$ is a constant, $Ea$ is activation energy and is expressed in kcal/mol/g, $R$ is the gas constant, and $T$ is the absolute temperature.

The slope of $1/T$ against log $\eta^*$ plot is equal to $Ea/R$. Activation energy is calculated by multiplying the slope by the gas constant $R$.

#### 8.3.5.4 After-cure properties by RPA

ASTM D6601 [10, 15] is used to determine the cure and after-cure characteristics of the final batch rubber compounds using RPA. Both, isothermal and non-isothermal (variable temperature cure, and VTC) cure characterisations are possible. Isothermal cure is performed at fixed frequency, strain, and temperature. Minimum torque, maximum torque, final torque, scorch safety time, and optimum cure time are reported like the MDR and ODR tests. Non-isothermal cure is performed using the thermocouple data of different components of tyres or other composite product for cure simulation. Based on the results, the mould opening time is set for tyres and these products.

Here, for every component of these products, non-isothermal (VTC) cure test is run in RPA to get an idea about the actual cure state for every component. Based on these results, over-curing and under-curing of product locations can be avoided. Dynamic mechanical properties like elastic modulus ($G'$), viscous modulus ($G''$), and loss factor (tan $\delta$) of the final batch compound is checked after curing the specimen in the RPA itself. After a fixed cure time at a temperature, the temperature is reduced and the tan $\delta$ values measured at 30°, 60–700, and 100 °C, keeping a fixed frequency and strain level as per the test plan. These tan $\delta$ values can be related to dry traction, rolling resistance, and heat buildup properties, respectively.

A higher tan $\delta$ value at 30 °C indicates better dry traction of the tyres-tread compound to the road and lower tan $\delta$ values at 60–70 °C and 100 °C indicates lower rolling resistance and lower heat buildup, respectively, during tyres performance. Frequency, strain, and temperature sweep tests are also possible after curing the specimen in the die cavity to check its dynamic mechanical properties. Various RPA test configurations are detailed in Table 8.1.

**Table 8.1:** Test configuration in RPA.

| Parameters | Temperature (°C) | Strain (%) | Frequency (Hz) |
|---|---|---|---|
| Relative molecular weight distribution | | | |
| Frequency sweep (raw rubber) | 100 | 14 | 0.05, 0.1, 0.21, 0.43, 0.88, 1.81, 3.71, 7.61, 15.61, 32 |
| Flow behaviour study (Power law index) | | | |
| Frequency sweep (compounded rubber) | 120 | 15 | 0.1, 0.2, 0.5, 1, 2, 5, 10, 20, 30 |
| Filler-filler and polymer-filler interaction study (Payne effect) | | | |
| Strain sweep (compounded rubber) | 70 | 0.5, 1, 5, 10, 15, 20, 25, 30, 35, 40, 45, 50 | 0.2 |
| Activation energy study for compounded rubber by Arrhenius equation | | | |
| Temperature sweep (low shear rate, 1.26 $sec^{-1}$) | 70, 80, 90, 100, 110, 120, 130 | 100 | 0.2 |
| Temperature sweep (medium shear rate, 5.03 $sec^{-1}$) | | | 0.8 |
| Temperature sweep (high shear rate, 10.05 $sec^{-1}$) | | | 1.6 |

#### 8.3.5.5 Stress relaxation test by RPA

It is measured for raw and compounded rubbers following ASTM D6048 [16] and ISO13145, respectively [17]. Various techniques are available to check the stress relaxation of rubber and its compound. This test provides a measure of the viscoelastic response of a specimen over a period without affecting its structure. Also, the specimen can be tested at temperatures between room temperature and 225 °C and at various shear rates. In this test, the torque decrease is determined at constant strain and temperature. Stress relaxation is the time-dependent decrease in stress under constant strain at a temperature; it provides information about the viscoelasticity of the specimen.

Both processability and mechanical properties are related to viscoelasticity; so measurement of stress relaxation will help to predict both processing and service performance of a specimen. The test specimen is preheated for one minute at 100 °C for conditioning purpose. 150% strain at 0.1 Hz frequency is applied for six oscillations.

Elastic torque (G′), viscous torque (G″), and loss factor (tan $\delta$) are recorded for the last three cycles. High strain is applied for better repeatability. Then, static recovery is performed for one minute.

It is followed by stress relaxation at 150% for 35 s. The percentage torque reduction is measured after one second and after 20 s. The total test time is approx. 4 min. The percentage torque reduction indicates the presence of elastic and viscous components, relatively.

## 8.4 Mechanical analysis: static mode

### 8.4.1 Stress–strain properties in tensile mode

Stress–strain properties of rubber vulcanisates are determined, following ASTM D412 [18] or ISO37 for tensile mode (method A for straight specimen and method B for ring specimen) in the uniaxial direction. The tensile strength depends on the materials as well as the testing conditions like test speed, temperature, humidity, specimen dimension, etc. Natural rubber-based vulcanisates show higher tensile strength due to their high molecular weight and due to strain-induced crystallisation. In general, addition of reinforcing filler in a rubber compound improves its tensile strength with loading and the addition of process oil decreases its tensile strength value.

It is observed that conventional a vulcanisation system shows better stress–strain properties due to the higher proportion of polysulfide linkages. Pneumatic/other types of grips are used to clamp the test specimen to avoid slippage during testing. The load experienced by the specimen is measured by a load cell and elongation is measured through contact- or noncontact-type extensometers. In some equipment, the test chamber is equipped to perform tensile test at low and high temperatures.

This is to correlate the test results at actual performance conditions. A representative picture of a tensile test specimen for die type C is shown in Figure 8.12. In the tensile mode test, a dumbbell or a ring specimen is fixed in grip at one end and is stretched from the other end. Test speed is maintained as 500 mm/min for tensile test; however, other jaw separation speeds could also be used as per test plan.

These tests provide the modulus values, elongation at break, and tensile strength at rupture of the specimen, at various stain levels. Normally, five test specimens are tested, and the median value is reported. One can also calculate the area under the stress–strain curve to know the work needed to rupture the specimen. The parameters calculated for a dumbbell specimen are given in formula (8.11)–(8.13).

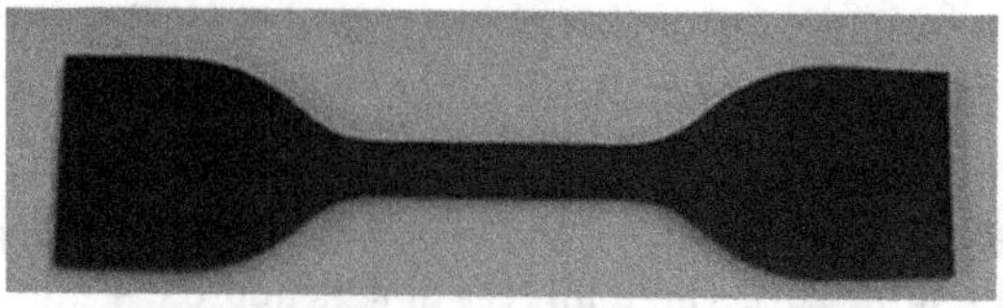

**Figure 8.12:** Tensile specimen for die type C.

$$\text{Tensile strength} = \text{Breaking load/cross-sectional area} \tag{8.11}$$

Where tensile strength is calculated in MPa, breaking load in N, and cross-sectional area (specimen thickness × width) in mm$^2$:

$$\text{Tensile modulus at specified elongation} = \text{Load at elongation/cross-sectional area} \tag{8.12}$$

$$\text{Elongation at break(\%)} = \text{(Length at break-original length)/Original length} \times 100 \tag{8.13}$$

Here, the original length is the length of the benchmark on the specimen, which is generally 25 mm for a specimen cut using die C and 17.78 mm for a specimen cut using the ISO-type die.

**Ring specimen**: Radial width and thickness of the ring-type (type 1 with inside circumference as 50.0 ± 0.01 mm and type 2 with inside circumference as 100.0 ± 0.2 mm) and ISO-type (normal with 4 mm thickness and small with 1 mm thickness) cut specimen is measured using a dial micrometer. The inside circumference can be measured using a stepped cone or by go-no go gauges. A ring-type specimen is fixed in a special test fixture of a universal testing machine. The test is performed at 500 mm/min for types 1 and 2 specimen and 100 mm/min for ISO-type specimen.

The average circumference is used to calculate the elongation properties. The tensile strength and tensile modulus at the specified elongations are calculated using eqs. (8.11) and (8.12), respectively. Elongations are calculated using formulas (8.14) and (8.15):

$$\text{Specified elongation(\%)} = 200 \times L1/C \tag{8.14}$$

Here, $L1$ is the increase in grip separation at the specified elongation, in mm, $C$ is the average circumference of the ring-test specimen in mm:

$$\text{Elongation at break(\%)} = 200 \times L2/C \tag{8.15}$$

Here, $L2$ is the increase in grip separation at break, in mm and $C$ is the average circumference of the ring-test specimen, in mm.

## 8.4.2 Stress–strain properties in compression mode

ASTM D575 [19] is used to measure the rubber properties in compression mode. Few rubber products are used under extensively compression deflection like the engine mount, shock absorbent, sealing, etc. Therefore, it is very important to evaluate the rubber vulcanisate performance in compression mode for these applications. This test method is useful for comparing the stiffness of the rubber vulcanisates in compression mode. The compression mode stress–strain test is complex due to the huge variation between the specimens, which arises due to the friction between the platens and the contact surfaces of the test specimens.

This leads to nonuniform load distribution over the test specimens, leading to variations in the test data. To avoid this variation, suitable lubricating gel (silicon oil) is used as a lubrication agent. In the compression mode test, the specimen height should be less than its diameter to avoid lateral bucking from the side. Since the test specimen is not clamped during compression, friction resistance due to lateral sliding at the end surface is very high.

This converts the test specimen into a cylindrical shape, leading to variation in stress distribution. The cylindrical test specimen may be prepared with 28.6 ± 0.1 mm diameter and 12.5 ± 0.5 mm thickness. Force is applied at the rate of 12 mm/min speed until the specified deflection is achieved. Deflection is the change in thickness of the specimen, on applying a compressive force. Post attainment of the deflection, the force is released at the same rate, and this loading cycle is repeated for the second and third time. Now, the force required to create a specified deflection is measured. The forces are measured for deflection at intervals of 5% and the stress–strain curve is plotted. Compressive deflection depends on the shape factor of the specimen.

A test specimen with a similar shape factor may give the same deflection and compressive force. The shape factor for various specimens is calculated using the following formula:

$$\text{Shape factor for block-type specimen} = a/4t \tag{8.16}$$

Here, $a$ is the width of specimen in mm and $t$ is height of the specimen in mm:

$$\text{Shape factor for cylinder-type specimen} = d/4t \tag{8.17}$$

Here, $d$ is the diameter of the specimen in mm and $t$ is the height of the specimen in mm.

## 8.4.3 Tear strength

Vulcanised rubber products fail during service due to the generation and propagation of a special type of rupture, known as tear. In this test method, resistance to tearing action is measured as tear strength. Tear strength depends on the strain-induced anisotropy, stress distribution, strain rate, and the size of the specimen. Natural rubber-based vulcanisates show higher tear strength values in comparison to that of synthetic rubber. Silica filler could be used in a compound recipe to improve its tear strength.

Tear strength is measured using a universal tensile testing machine, following ASTM D624 [20] or ISO34. Various types of specimens are used for testing purposes, like die types A, B, and C; trouser type; and constrained path type. A representative picture of the angle tear specimen for die type C is depicted in Figure 8.13, which is mostly used for testing purposes.

Tensile slab is cured by a compression moulding press at a specified temperature and time, based on its rheometric properties. A specimen of required dimensions as per the standard test method is punched using a sharp cutting die using a pneumatic cutter, and testing is done in a similar fashion for its tensile properties.

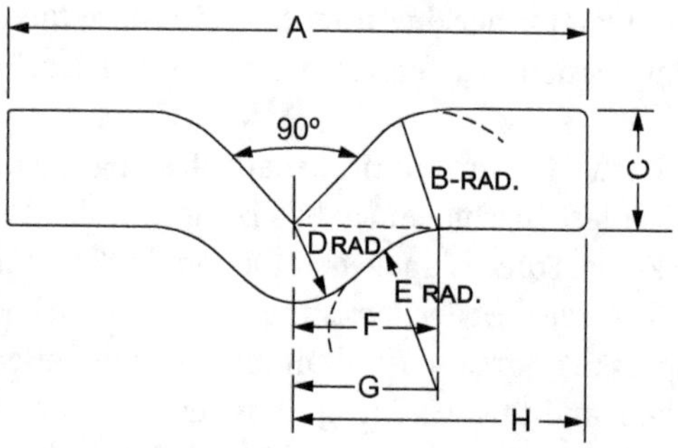


DIE C

| Dimension | Millimetres | | Inches | |
|---|---|---|---|---|
| | Value | Tolerance | Value | Tolerance |
| A | 102 | ±0.50 | 4.0 | ±0.02 |
| B | 19 | ±0.05 | 0.75 | ±0.002 |
| C | 19 | ±0.05 | 0.75 | ±0.002 |
| D | 12.7 | ±0.05 | 0.5 | ±0.002 |
| E | 25 | ±0.05 | 1.0 | ±0.002 |
| F | 27 | ±0.05 | 1.061 | ±0.002 |
| G | 28 | ±0.05 | 1.118 | ±0.002 |
| H | 51 | ±0.25 | 2.0 | ±0.01 |

**Figure 8.13:** Tear (angle) specimen for die type C.

Normally, five test specimens are tested per sample and the median value is reported. Tear strength value is calculated using the following formula:

$$\text{Tear strength (N/mm)} = \text{Maximum force in N/specimen thickness in mm} \quad (8.18)$$

## 8.4.4 Hardness

Hardness of a cured rubber test specimen is measured using International Rubber Hardness Degree (IRHD) and shore A-type hardness testers, following ASTM D1415 [21] or ISO48 and ASTM D2240 [22] or ISO7619, respectively. These instruments measure the resistance to indentation under conditions that do not puncture the specimen. The indentation hardness is inversely proportional to the penetration and is also dependent on the elastic modulus and viscoelastic behaviour of the material. This test gives an idea about the degree of curing, and it depends on the filler and the process oil loading. A higher hardness value indicates a high degree of curing and/or higher reinforcement. Each elastomer has a base hardness value; however, as a rule of thumb, addition of carbon black by 2 phr typically increases the hardness by one unit and the addition of process oil by 2 phr level decreases the hardness by one unit. The indentation force is applied for 30 s for both types of tester and the hardness value is recorded using a dial gauge. A specimen of minimum 6 mm thickness with smooth, flat, and parallel upper and lower surfaces is required for the hardness test.

The test specimen is conditioned at laboratory conditions for minimum 3 h, prior to testing. Specimens are tested in five different locations and the median value is reported.

## 8.4.5 Biaxial and triaxial tests

When stress is applied from one direction (either tension or compression), it is called uniaxial deflection. For a biaxial test, stress is applied on the specimen from both the directions. The stress–strain measurement in tension and compression mode and fatigue/fracture properties measurement using the conventional method are examples of uniaxial testing. For analysis of elastomers under close to real loading, biaxial measurement is required. Such data are primarily used by finite element analysis calculations for the calibration of the constitutive models.

However, the measurement of such biaxial deformation is a complicated method. If a specimen is loaded in two perpendicular directions with servo hydraulic systems, biaxial strain state is achieved at the middle of the specimen and due to the strain concentration at the edges, it is not possible to test specimens at high strains In biaxial measurements. Four separately controlled electromechanical linear motors with sensors for load and displacement are used. A representative picture of a clamping system is shown in Figure 8.14. The specimen is stretched by around 10% equi-biaxially and the temperature, as a function of time, is measured using an infrared camera.

This provides information about the stored and dissipated energy within the specimen. Load displacement data are also recorded to produce the stress–strain curve. This test may be performed in both static and dynamic conditions up to 150% strain levels. The specimen is prepared in a square shape with 77 mm side and 1.5 mm thickness, with a proper grip design [23].

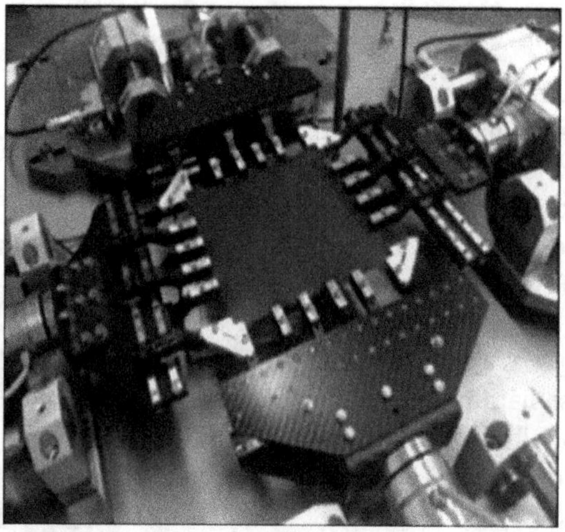

**Figure 8.14:** Clamping system for biaxial test.

This test may also be used to measure the fatigue and fracture properties in the biaxial direction. These test results are used for designing of rubber parts using the finite element methods. Under dynamic loading conditions, the temperature may be measured using an infrared camera to check the relationship between the mechanical energy, dissipative sample heating, and heat loss. It is observed that a major part of the mechanical energy is used internally, and only a small part is dissipated as heat. This testing setup can also be used to measure the crack propagation and tearing energy in the biaxial direction.

Triaxial deformation is also known as hydrostatic compression. In this test, compressive force is applied uniformly on all surfaces of the rubber test specimen. Under this condition, the volume of the rubber specimen decreases but its shape remains unchanged.

## 8.5 Mechanical analysis: dynamic mode

Dynamic properties of the rubber vulcanisates are instrument-specific and depend on the test method used for analysis. Information on the energy loss by rubber products in dynamic conditions, for example, vibration mounts and automotive tyres, is important for its performance evaluation parameters like, heat generation and fatigue life for vibration mounts and traction, and rolling resistance and heat generation for tyres. Dynamic properties of the rubber vulcanisates and their dependence on temperature may be affected by the elastomer type (macro and microstructure) as well as by other compounding ingredients, including filler (surface area, structure, and surface characteristics), filler loading, other ingredients like resins/oils, and cross-link density.

The definitions of few popular terms used in the dynamic mode test are given below:

- Resilience: In a rubber-like body subjected to and relieved of stress, the ratio of energy given up, on recovery from deformation, to the energy required to produce the deformation is resilience. Resilience for these materials is expressed as a percent.
- Hysteresis: It is the percent energy lost per cycle.
- Dynamic modulus: It is the ratio of stress-to-strain under vibratory conditions. It is expressed in Pascals or in pounds per square inch for unit strain.
- Damping: It refers to the progressive reduction of vibrational amplitude in a free vibration system. Damping is a result of hysteresis, and the two terms are frequently used interchangeably.

## 8.5.1 Dynamic mechanical analyser (DMA)

Dynamic mechanical analyser (DMA) is used to measure the elastic ($E'$) and viscous modulus ($E''$) in frequency sweep, strain sweep, and temperature sweep test, following ASTM D5992 [24]. Elastic modulus ($E'$) is a component of the complex modulus ($E^*$) in phase with dynamic deflection. Viscous modulus ($E''$) is a component of the complex modulus ($E^*$), 90° out of phase with dynamic deflection, which is responsible for the conversion of mechanical energy to heat. The ratio of the viscous modulus ($E''$) to elastic modulus ($E'$) is expressed as loss factor (tan $\delta$). The modulus is reported in kPa/MPa, whereas tan $\delta$ is a unitless parameter. Based on the equipment model, test configuration capabilities may vary; however, it is expected to test specimens from a low temperature (approx. −150 °C) to a high temperature (+450 °C). However, for rubber application, it is between −120 and +120 °C only.

Commonly, double shear-type and compression-extension-type specimens are used for testing. However, compression, torsional, and bending-mode testing is also possible in some models. In a double shear-type specimen, the height-to-thickness ratio is maintained between 8 and 10, which gives a constant strain throughout the specimen. The image of a double shear-type specimen is depicted in Figure 8.15.

The image of a compression-extension-type specimen is shown in Figure 8.16.

A rectangular or circular cross-sectional specimen is loaded in compression and tension mode. Normally, a 2 mm-thick specimen with 10 mm width and 10 mm height is used for this mode with pre-tension, to avoid bulging during testing. Different laboratories follow different test conditions and deformation modes for the measurement of dynamic test parameters; tan $\delta$ peak, measured for compounded and cured specimens, is termed as dynamic glass transition temperature ($T_g$).

A lower value of dynamic $T_g$ indicates better tyres durability during service; the tan $\delta$ peak height also indicates the filler–filler interaction – a lower value is desirable. Tan $\delta$, measured at various temperatures, correlates with the tyres performance properties. The tan $\delta$ value at 0 – (−10 °C) and 30 °C relates to the tyres tread in wet

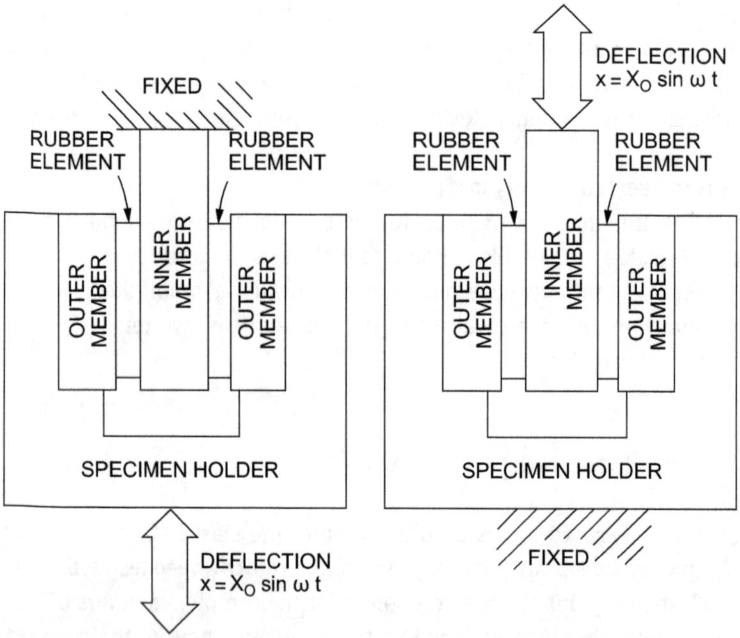


**Figure 8.15:** Double shear-type specimen.

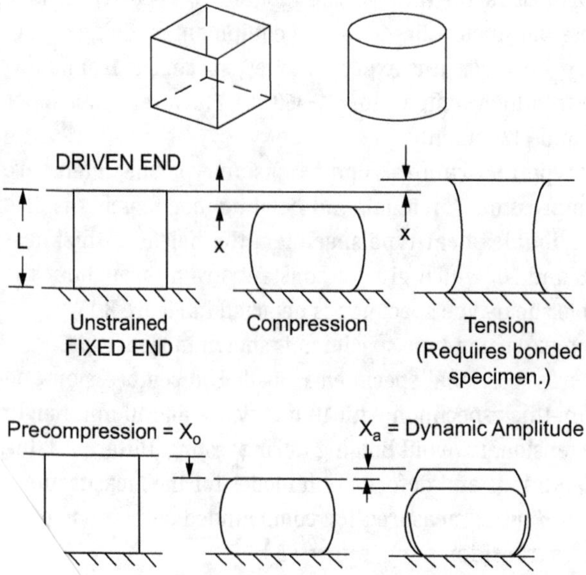


ompression-extension-type specimen.

and dry traction; at 70 and 100 °C, it relates to the tyres tread's rolling resistance and heat buildup, respectively.

Tyres tread, based on styrene butadiene rubber, shows a higher tan $\delta$ value at 0 and 30 °C, whereas the same tread prepared with polybutadiene and natural rubber shows a lower tan $\delta$ at 70 and 100 °C. The temperature versus tan $\delta$ curve to correlate tyres performance is illustrated in Figure 8.17.

Since it is not possible to experiment at high frequencies using the available DMAs, time and temperature superposition principles are used, as proposed by William, Landel, and Ferry (WLF). Thus, high frequency performances like tyres tread dry ($10^2$–$10^4$ Hz) and wet ($10^5$–$10^9$ Hz) traction could be related using existing DMA by testing specimens at lower temperatures, at 10 Hz.

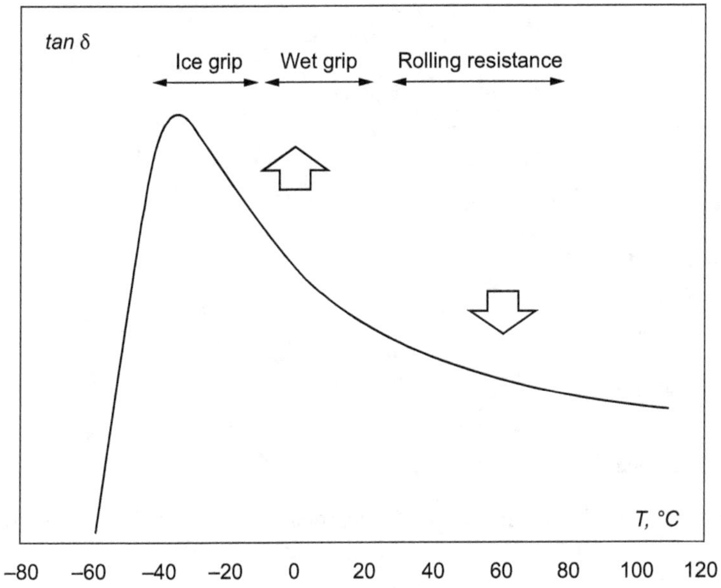


**Figure 8.17:** Temperature versus tan $\delta$ curve to correlate tyres performance.

## 8.5.2 Rebound resilience test

Rebound resilience is the ratio of the energy returned to the energy applied. It is measured at room as well as at high temperature using a rebound resilience tester, following ASTM D2632 [25] for vertical type and ISO 4662 [26] for pendulum-type machine. It is a function of both dynamic modulus and internal friction within a rubber matrix. The value of rebound increases with increase in the compound storage modulus and decreases with higher hysteresis loss. Rebound resilience at room temperature or 30 °C correlates with dry traction of the tyres. A low value of rebound resilience is desirable at this test temperature. Rebound resilience value at 70 °C correlates with the rolling resistance of a tyres. A high rebound resilience value is desirable at this test temperature.

As the test temperature is increased, the value of rebound resilience increases due to decrease in the viscous modulus of the rubber vulcanisates. It is reported as the average or median of three specimens. Rubber industries prefer the pendulum-type machine for rebound testing. The temperature is controlled using a heating system attached to a test piece holder. The test specimen of thickness 12.5 ± 0.5 mm is fixed on the test piece holder. A pendulum of a specified mass and dimensions strikes the specimen at a fixed velocity. A representative picture of the equipment is shown in Figure 8.18.

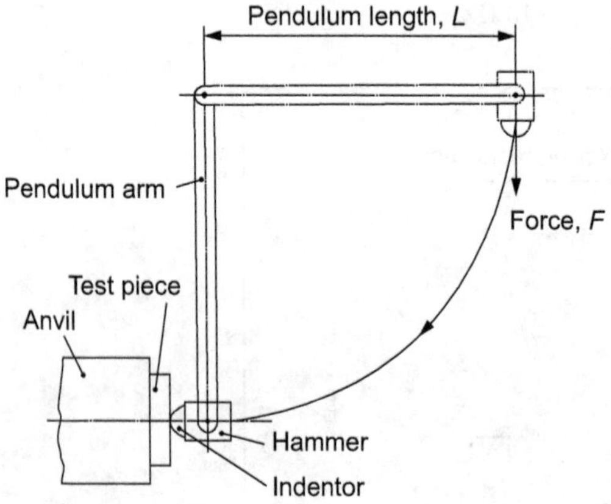


**Figure 8.18:** Pendulum-type rebound resilience tester.

The calculation is given below:

$$\text{Percentage rebound resilience [27]} = \{1 - \cos(\text{angle of rebound})\}/\{1 - \cos(\text{angle of fall})\} \times 100 \quad (8.19)$$

In the vertical-type machine, the resilience is determined as a ratio of the rebound height to the drop height of a metal plunger of a prescribed mass and shape, which is allowed to fall on the rubber specimen. The plunger mass shall be 28 ± 0.5 g and drop height is 400 ± 1 mm.

The test specimen shall have thickness of 12.5 ± 0.5 mm. The calculation of rebound resilience is given as follows:

$$\text{Percentage rebound resilience} = h_R/h_O \times 100 \quad (8.20)$$

$h_R$ is the rebound height and $h_O$ is the height of fall.

### 8.5.3 Hysteresis loss

Energy is absorbed when a viscoelastic material is deformed, and part of the energy is released when the force is removed again. Rubber vulcanisates absorb more energy during loading; it releases less during unloading. The process of energy transfer to its neighbouring molecules results in energy loss.

The work consumed during the reversible deformation is stored in the solid partially as potential energy and is partially used for the reduction of entropy due to deformation. When a large stress is applied or the specimen is subjected to cycles of high strain, the secondary network of filler aggregates the break. The breakdown and reformation of these secondary networks result in hysteresis loss.

The lost energy is reflected as temperature rise. Rubber vulcanisates, based on linear structure elastomers, show lower hysteresis loss as compared to the branched ones. Higher filler loading may lead to higher hysteresis loss due to the breakage of the filler–filler networks during cyclic deformation. A typical hysteresis loss curve for viscoelastic materials (rubber) is illustrated in Figure 8.19.

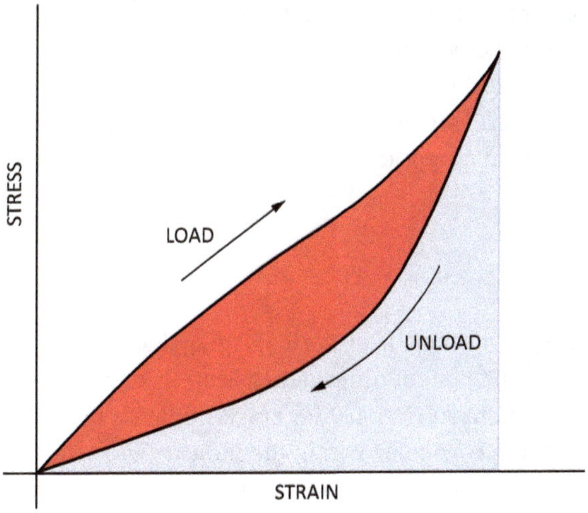


**Figure 8.19:** Hysteresis loss curve for viscoelastic materials (rubber).

### 8.5.4 Goodrich flexometer test

It is used to measure the heat buildup parameter for vulcanised rubber specimen, following ASTM D623 [28]. In this instrument, a fixed compressive load, equivalent to 143 psi, is applied on a cylindrical specimen of diameter 17.8 ± 0.1 mm and height 25 ± 0.15 mm. A stroke of 4.45 ± 0.03 mm with frequency, 30 ± 0.2 Hz under cyclic compression is used for testing. Goodrich flexometer has two anvils, lower and upper. The specimen is kept on the lower anvil and the cyclic deformation, as specified, is applied by the hammering action of the upper anvil. The lower anvil has a thermocouple to measure the base surface's temperature generation.

This test may be performed at 50 or 100 °C for a test time up to 30 min, based on the compound recipe. For measuring the temperature generation at the centre and at the end of the test, a needle-type pyrometer is inserted at the centre of the specimen. In both cases, heat buildup is calculated as the difference of final and initial temperatures. Before starting the test, the specimen is conditioned at the test temperature for a minimum of 30 min. Heat buildup data measured through this method can be used to get an idea about the relative performance of rubber vulcanisates under dynamic applications.

High heat generation is not desirable for products under dynamic application. Heat generation depends on the elastomer's macro and microstructure, filler particle size, and loading. A smaller particle size or a higher loading leads to high heat generation during dynamic applications. It is also affected by change in the cross-link density of the rubber vulcanisates. Normally, a tighter cross-link gives a lower heat buildup value. The change in specimen's height before and after the test is also measured to get an idea about the compression set of the rubber vulcanisates during dynamic loading.

It gives indication about the degree of softening of the rubber vulcanisates. For a blow out test, a higher stroke height of 6.35 mm is applied and the specimen is cut at the end of the test to check whether or not air pockets are generated.

## 8.5.5 Flex properties

Demattia flex testing using Wallace-type machine is performed, following ASTM D430 [29] for crack initiation and ASTM D813 [30] or cut propagation. Scott flexing machine and Dupont apparatus is also mentioned in ASTM D430 for crack initiation measurement; however, Wallace flexing machine is popular within the industry. A Scott flexing machine is used to measure rubber to fabric separation in flexing mode whereas a Dupont apparatus can be used to measure crack initiation and growth as well as rubber to fabric separation.

In the Wallace Demattia machine, a moulded specimen of length 150 mm, width 25 mm, and thickness 6.4 mm with semicircular groove of specific dimensions at centre is fixed for the crack initiation test or the crack propagation test. For the crack initiation test, no cut on specimen is applied, whereas for the crack propagation test, a 2 ± 0.1 mm cut is inserted with a specific piercing tool at the centre of the circular groove in a single attempt. The test can be performed at room or at elevated temperature conditions. At elevated temperatures, crack initiation may appear faster than at room temperatures, and the crack growth rate may be higher due to degradation of the rubber vulcanisates at elevated temperatures.

The specimen is conditioned at laboratory temperature for a minimum of 12 h before testing. Three specimens are tested and the average of the observed values is reported. In the Wallace-type machine, test speed is fixed as 300 ± 10 cycles/min, whereas the frequency can be changed in the new variable speed Demattia tester. For the bend flexing test, the grips approach each other to 19.0 ± 0.1 mm and separate to 75.9 + 0.3–0.0 mm.

To measure crack initiation, the specimen is evaluated for a crack of 2 mm length after every 1,000 cycles of testing and the values are reported as the number of cycles. For the crack growth rate test, the length of the crack is measured using Vernier calipers after every 1,000 cycles of testing for a crack length developed to a minimum of 12.5 mm – the values are reported in mm – against the number of cycles in multiple of 1,000.

During the measurement of crack growth and observation of crack initiation, the specimen should be folded by keeping both specimen-holding stations close to each other. The Demattia test gives an estimate of the ability of the rubber vulcanisates to resist crack initiation and crack growth when the specimen is subjected to bending or flexing. In general, natural rubber-based vulcanisates show poor crack initiation, which is due to the presence of natural flaws, and better crack propagation due to its strain-induced crystallisation characteristics. Synthetic rubber-based vulcanisates show just the opposite, meaning, crack initiation is difficult for synthetic rubber-based vulcanizates, but once the crack is initiated, its propagation is fast. Hence, to achieve improved crack initiation and crack propagation resistance in tyres tread and sidewall compounds, a blend of natural and synthetic rubber is used in the compound formulation. A representative picture of a Wallace Demmatia flex tester is shown in Figure 8.20.

For footwear applications, a Ross flex tester is used to measure the crack growth properties, following ASTM D1052 [31]. In this test, a specimen of dimensions similar to the one use in the Demattia test is pierced for a cut of 2 mm length and the specimen is then bent freely over a 10 mm diameter rod during the test, at a 90° angle. One end of the specimen is clamped to a holder arm and the pierced end is placed between two rollers, which should facilitate free bending movement of the specimen during testing.

The test is performed at 100 ± 5 cycles/min speed. The crack length is measured using Vernier calipers after every 1,000 cycles, and the crack growth rate is reported as the average of three values. The specimen may also be punched from the finished product and tested after proper buffing. This test may be performed for the original (unaged) specimen after conditioning at laboratory conditions for a minimum of 16 h or for an aged specimen, after exposure at 100 °C for 24 h, followed by laboratory conditioning. In general, a faster crack growth rate is observed for an aged specimen as against an original specimen due to the degradation of rubber due to exposure at high temperature.

Fatigue-to-failure tester (FTFT) is used to relate the fatigue life of the rubber vulcanizates, mainly tyres sidewall and tread, following ASTM D4482 [32]. Two stations of 12 specimens each can be tested at a time. The specimens are gripped at both the ends. All specimens in each station are subjected to the same extension.

**Figure 8.20:** Wallace Demmatia flex tester.

The 'top station' and the 'bottom station' are driven by separate drive cams, which help each station to run different extensions, simultaneously. The dumbbell-shaped specimens are cut using a punching die from moulded slabs with 'end ribs' to prevent slippage from the test clamps. This test is normally performed at 100 ± 10 cycles/min speed under 100% extension mode using a dumbbell-shaped specimen. The test may also be performed at other extension ratios on changing the cam for the required extension ratio. The moulded specimens are conditioned at laboratory temperature for a minimum of 24 h before the test. Set is developed in the specimens on initial cyclic deformation is taken up using 'adjustable screws' as necessary after 1, 10, or 100 kilocycles. The test may be performed at room temperature as well as at a high temperature. For a high temperature test, the machine should have an environmental chamber to attain the desired test temperature. At the high temperature, the fatigue life value is drastically reduced because of rubber degradation and because the strain effect vanishes fast. During testing, cracks are initiated on the specimen due to naturally occurring flaws.

The growth of cracks under repeated deformation, known as fatigue, leads to catastrophic failure. This fatigue failure is initiated as minute flaws where the stress is high. Mechanical rupture at such points will lead to the development of cracks and ultimately to the failure of the specimen. Attack by ozone may also initiate cracks at the surface, and these will grow as a result of mechanical fatigue. Fatigue life is influenced by polymer type, filler, type and degree of cross-linking, type of protective agents, and quality of dispersion. Natural rubber-based vulcanisates show a higher fatigue life as compared to synthetic rubber-based vulcanisates.

A conventional vulcanisation system helps in improving the fatigue life due to the flexibility of the S-S bond in polysulfidic linkage. The fatigue tester should be capable of providing a specimen extension ratio of 1.6 to 2.4. The extension ratio is calculated using the following formula:

$$\lambda = L/L_0 \tag{8.21}$$

where $\lambda$ is the extension ratio, $L$ is the extended length of the specimen, and $L_0$ is the unextended length of the specimen.

To take care of the variations in fatigue-to-failure test, six specimens are tested per sample to check the fatigue life of the rubber vulcanisates, out of which four close values are selected for calculation purpose, eliminating the extreme values. Fatigue life is calculated using the JIS average formula as given below or by determining the geometric mean of all values. It is expressed in kilocycles unit:

$$\text{Fatigue life in kilocycles} = 0.5A + 0.3B + 0.1(C + D) \tag{8.22}$$

where $A$ is the highest value out of close four values, $B$ is the second highest value, $C$ is the third highest value, and $D$ is the lowest value.

The fatigue behaviour of the specimens may be compared at constant strain or constant strain energy. The stain energy for a particular extension is obtained from the area under the stress–strain curve up to a specific extension using the universal tensile testing machine. The specimen's width and thickness are measured with the help of Vernier calipers and dead load thickness gauge meter, respectively. The specimen is stretched in a tensile testing machine 30 times to cause maximum extension so as to be used in fatigue testing. The specimen is elongated at 50 mm/min and the stress at every 10% elongation increment is recorded until the extension ratio is 2.5. It is expressed in kJ/m$^3$. Using the strain energy basis, rubber vulcanisates of different modulus can be compared.

## 8.6 Physicochemical properties

### 8.6.1 Swell index

It is measured for rubber vulcanisate specimens using the test method in line with ASTM D3616 [33]. The value of the swell index gives a relative idea about the cross-link density of the rubber vulcanisate. A higher swell index value indicates a lower cross-link density and vice versa. An accurately weighed specimen of 0.2–0.3 g (0.39–0.41 g as per ASTM D3616) from vulcanised rubber sheet is immersed in a suitable solvent for

48 h at room temperature. Normally, toluene is used for general-purpose rubber (NR, SBR, BR)-based vulcanisates and cyclohexane is used for butyl or halo-butyl rubber based vulcanisates. After 48 h, the specimen is taken out from the solvent and the excess solvent is then blotted from the specimen. Immediately, the specimen is put in a weighing bottle, its lid is closed, and the swollen weight of the specimen is taken, nearest to four decimal places. The swell index is calculated using the following formula:

$$\text{Swell index} = S/T \tag{8.23}$$

where $S$ is the weight of the swollen specimen in g and $T$ is the original dry weight of the specimen in g.

## 8.6.2 Volume fraction of rubber and cross-link density

For a vulcanised rubber specimen, the volume fraction of rubber ($V_r$) in the swollen state is the preferred way of expressing the amount of the rubber component present at equilibrium swelling [34]. When a vulcanised rubber is immersed in a suitable solvent, the elastomer absorbs the solvent and undergoes swelling to the extent determined mainly by the cross-link density, the nature of the polymer, and the solvent [35]. The volume fraction of rubber is calculated to get an indication of the apparent cross-link density.

This test is a continuation of the swell index test as mentioned above. The swollen specimen is dried in an oven at 100 °C to a constant weight. The dried weight of the specimen is measured after cooling the specimen in desiccators for a sufficient duration.

The volume fraction, $V_r$, of the rubber specimen is calculated using the following formula [34–36]:

$$V_r = [(D - FT)/\rho_r]/[(D - FT)/\rho_r + A_O/\rho_s] \tag{8.24}$$

where $D$ is the weight of the de-swollen specimen, $F$ is the weight fraction of the insoluble nonrubber ingredients, $T$ is the original dry weight of the specimen, $A_o$ is the weight of solvent absorbed; it can be calculated as $(S - T)$, $\rho_r$ is the density of the raw rubber, and $\rho_s$ is the density of the solvent.

The cross-link density, $\nu$, of the rubber specimen is calculated using the Flory–Rehner relationship as follows [37–39]:

$$\nu = (-)[\ln(1 - V_r) + V_r + \chi(V_r)^2]/[2 \times \rho_r \times V_s \times (V_r)^{1/3}] \tag{8.25}$$

where $X$ is the cross-link density (g mol/g RH), $V_r$ is the volume fraction of rubber in the specimen, $\chi$ is the rubber–solvent interaction parameter, $\rho_r$ is the density of the raw rubber, and $V_s$ is the molar volume of solvent.

## 8.7 Accelerated ageing properties (anaerobic and aerobic exposure) in stress–strain mode

Rubber products require resistance to oxidation and thermal ageing with time. Stress–strain and tear strength properties are checked after accelerated ageing of the test specimen to predict the performance of the rubber vulcanisate to anaerobic and aerobic thermal exposure. Normally, anaerobic ageing is performed on rubber specimens that are inner components of products, like tyres-casing compound, etc., where the components are not exposed to air during application. During anaerobic ageing, the rubber compound is moulded as a tensile slab using a compression moulding press and it is kept inside the mould or inside a heated closed cell filled with nitrogen for anaerobic ageing.

The temperature is set as per the test plan and the ageing continues for a fixed duration. In air ageing, the vulcanised rubber specimens are exposed to air at a fixed temperature for a specified time. Post the ageing, the specimens are conditioned at laboratory conditions for 24 h before the stress–strain properties are measured to calculate the retention of these properties with respect to its original properties. Aerobic ageing is preferred for rubber specimens of components of products that are exposed to an open environment, like tread, sidewall of a tyres, etc. General purpose synthetic rubber-based vulcanisates are exposed at high temperature, at around 100 °C, for 72 h and the NR-based vulcanised specimens are exposed at a relatively low temperature, at around 70 °C, for 168 h.

Two types of equipment are used to expose the specimen at elevated temperatures to perform accelerated air ageing. Test tube enclosure or multicell ageing oven are used, following ASTM D865, and an air oven, following ASTM D573, to check the deterioration of rubber specimens by exposing in air at elevated temperatures [40, 41]. Air ovens with various interior sizes, from 30 × 30 × 30 cubic cm to 90 × 90 × 120 cubic cm, may be used for ageing purposes. Rubber specimens should be hanged in such way that they do not touch the walls of the oven chamber. Air circulation is required at atmospheric pressure.

The air oven should maintain the temperature within ±1 °C range of the set value uniformly throughout chamber. In the test tube enclosure, the rubber specimens are heated in individual test tubes with circulating air to prevent cross contamination of the antioxidant, etc. from other specimens. Borosilicate glass test tube with 3.8 cm diameter and 30 mm length are used for holding the specimen. The temperature should be maintained within ±1 °C range of the set value by thermostatic control. If any liquid medium is used, then suitable circulation is required to ensure a uniform temperature.

When an oil bath is used for ageing, provision should be made to remove fumes. Before keeping the specimen for ageing, it should be kept at room temperature for at least 24 h after curing.

The percentage of change in each of the stress–strain properties (tensile strength, elongation at break, and modulus) is calculated using the following formula:

$$P = (A - O)/O \times 100 \quad (8.26)$$

where $P$ is the percentage change in property, $A$ is the value after ageing, and $O$ is the original value.

Retention of the stress–strain properties after air ageing depends on the elastomer/blend type/ratio, antidegradant type and their dosage, cure system, cross-link density, presence of other ingredients like oil, etc.

## 8.8 Thermal analysis

Thermal analysis provides a rapid method for measuring the transition due to the thermochemical changes in a polymer when it is heated or cooled through a specified temperature range. Changes in specific heat capacity, heat flow, and temperature are measured for these transitions. A DSC is used to measure the thermal transition in the polymeric materials/compounds.

These techniques are used in the rubber industry for applications like material characterisation, material development, and quality control. Thermogravimetric analysis techniques are getting prominence in the rubber industry, mainly for composition analysis. For composition analysis, the test specimen should be taken out very carefully to avoid its contamination with another component.

Physical changes like thermal stability, oxidative stability, extent of polymerisation, and curing can be studied using thermal analysis techniques. This information is helpful to understand the environment conditions for use of the material.

Thermomechanical analysis (TMA) techniques are used to measure the CTE and other transitions like glass transition temperature ($T_g$), etc. It can also provide information about the compatibility of the materials, suitability of materials in specific environment conditions, mechanical properties of materials, etc.

### 8.8.1 Thermal analysis by TGA, DSC, and TMA

Two types of temperature-dependent transition could happen for rubber, viz., first-order and other is second-order transition. In the first-order transition, properties like volume, specific heat, modulus, etc. change drastically. In the second-order transition, there is not much change observed in properties like volume, but the thermal expansion coefficient changes significantly. The first-order changes lead to crystallisation, whereas the second-order changes result in the material becoming glassy.

**TGA** measures the amount and the rate of weight change against temperature at a defined heating rate under controlled conditions. It is used to determine the composition and thermal stability of rubber and its compounds. It characterises the materials that exhibit weight loss or gain due to decomposition, oxidation, etc. Weight loss occurs for thermal/thermo-oxidative decomposition due to bond breakage, evaporation of volatiles, reduction, and desorption. Weight gain happens for oxidation and absorption. The TGA instrument should be capable of continuously weighing a test specimen at an accuracy of ±2 µg and it should record the change in mass of the specimen under controlled environment throughout defined temperature range.

When a polymer sample degrades, its mass decreases due to the production of gaseous products like carbon monoxide, carbon dioxide, water vapour, and other gases, depending on the chemical composition of the polymer/compound. The test is done following ASTM D6370 [42]. A typical TGA thermogram is shown in Figure 8.21. Time, temperature, heating rate, and environment could be programmed as per the test plan before the experiment. It requires only a specimen of 5–15 mg weight for this test. In one of these programs, the specimen is heated to 50 °C for 2 min for instrument equilibrium. Then, it is heated to 560 °C at the rate of 10 °C/min in a nitrogen atmosphere. The % weight loss during this heating shows the presence of various organic materials, including polymers in the specimen. The specimen is then cooled to 300 °C and kept for 2 min for instrument equilibrium. Then, it is heated to 800 °C at the rate of 10 °C/min in an air or oxygen atmosphere. The % weight loss during this heating will give the presence of carbon black and ash content in the tested material.

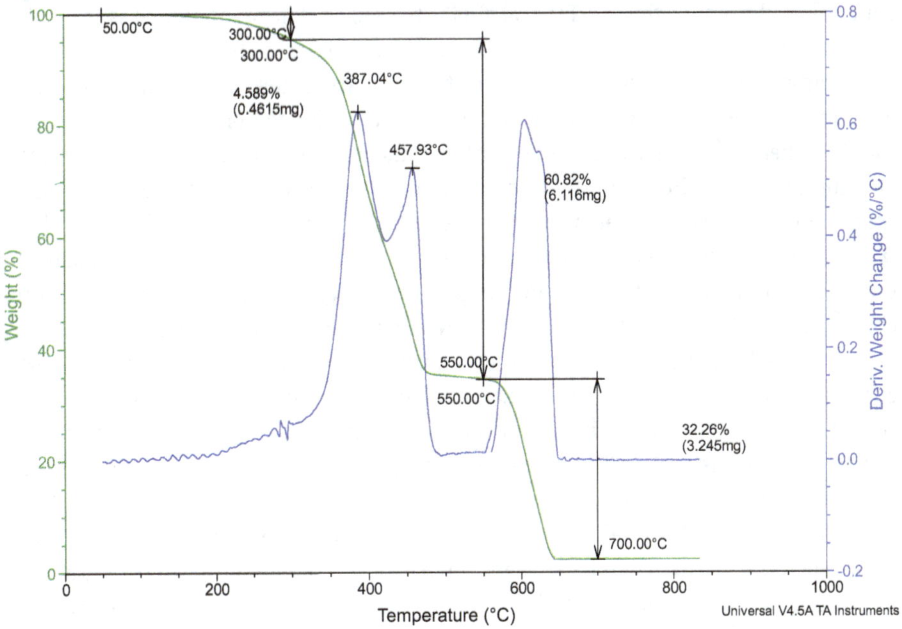


**Figure 8.21:** Typical thermogram by TGA.

**DSC** analyses the effect of heating on the polymer during the physical changes in terms of glass transition, melting, and so on. It also measures the heat flow associated with thermochemical changes like oxidation. Differential thermal analysis is a technique used to study the thermal behaviour of rubber when it undergoes physical and chemical changes due to heating and cooling. It is named as differential because one thermocouple is placed in the specimen while the other thermocouple is placed in an inert reference material. The reference material will not undergo any thermal transformation.

The temperature difference between the specimen and the reference material is calculated based on the net voltage difference at a point in time. Both first-order and second-order transition properties are studied using this technique. The specimen and the reference materials are cooled to a temperature below the expected transition temperature and the temperature is then increased at a given rate. At the point of transition, the specimen will change with the heat of the chamber without changing its temperature till the transition is complete.

DSC measures the exothermic and endothermic transitions with respect to temperature. In the endothermic, heat is absorbed by the specimen whereas in exothermic, heat is released by the specimen. It measures various transitions like glass transition temperature $T_g$, melting, crystallisation, curing and cure kinetics, onset of oxidation, and heat capacity. Endothermic events are glass transition temperature, melting, evaporation, and few decompositions; and exothermic events are crystallisation, cure reactions, oxidation, and decomposition. ASTM D7426 [43] is used to determine the glass transition temperature of the elastomer or other polymeric materials. To check the glass transition temperature, the difference in heat flow or temperature is measured between the specimen and the reference material when they are either heated or cooled at a specified rate. The thermal curve is analysed to calculate the glass transition temperature. Generally, the heating and cooling rates are maintained as 10 °C/min under an inert atmosphere and the specimen quantity is between 10 and 40 mg. The test chamber of the DSC instrument should be capable of providing controlled heating of the specimen and the reference material at a constant temperature or a constant rate for temperature change within the range of −120 to 500 °C. Normally, two cycles of heating and cooling are performed; the first cycle is to get the thermal history information and the second cycle is to provide information with the previous thermal history erased. The specimen should be homogeneous as it is used in very small quantities. The $T_g$ is not a fixed material constant but is partly affected by the heating and cooling rate. Slow temperature changes lower the $T_g$ of the tested material.

ASTM D3418 [44] is used to measure the transition temperatures and enthalpies of fusion and crystallisation of the polymers. A typical DSC graph is illustrated in Figure 8.22. In this test method, the test specimen is heated and cooled at a controlled rate under a specific purging gas at a controlled flow rate. For the programmed test conditions, a base line is generated and stored in computer memory. The difference in heat input between the 'sample' and the 'reference' pan due to energy change in the specimen is measured continuously. Any absorption or release of energy by the specimen, resulting in endothermic or exothermic peak, is considered as transition. For first-order transition like melting and crystallisation, around 5 mg weight of specimen is taken and nitrogen as purge gas is used during testing. There should be intimate contact between the pan and the specimen for better test results. The initial thermal cycle is performed by heating the specimen from a minimum of 50 °C below and 30 °C above the melting temperature, with heating rate like the test heating rate to erase the previous thermal history.

For crystallisation, cool the specimen to 50 °C below the crystallisation peak temperature at a heating rate similar to the test heating rate. The heating rate may be 10 °C or 20 °C/min. Record the cooling curve and hold the temperature for 5 min. This curve can be used to the calculate enthalpy of transition.

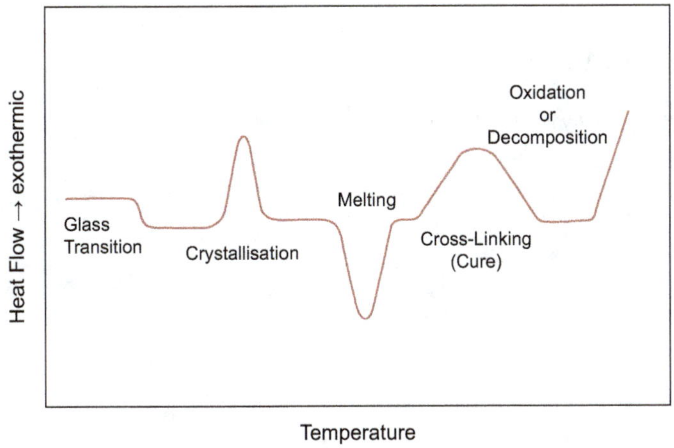


**Figure 8.22:** Thermal analysis by DSC.

The melting endotherm (Figure 8.23) or freezing exotherm (Figure 8.24) are shown for reference. Integrate the area under the graph as a function of time to get the enthalpy or heat (mJ) of the transition. Further, calculate the mass-normalised enthalpy or heat (J/g) of the transition by dividing the above enthalpy with the mass of the specimen.

**TMA** studies the dimension change of a material over a predefined temperature range. Stress is applied on the sample and the strain is measured while it is exposed to the programmed temperature condition. Various ASTM test methods are used to determine different parameters. ASTM E831 is used to check the coefficient of linear thermal expansion of solid materials. This coefficient is used for design purposes and to understand the failure of the solid body, which is composed of two different materials, under temperature variations. ASTM E1545 is used to measure $T_g$ of a material under compression conditions, which depends on the thermal history of the material to be tested. The $T_g$ of amorphous and semicrystalline materials may provide information about the thermal history, processing conditions, stability, progress of chemical reactions, and mechanical and electrical behaviour. ASTM E2092 is used to check the

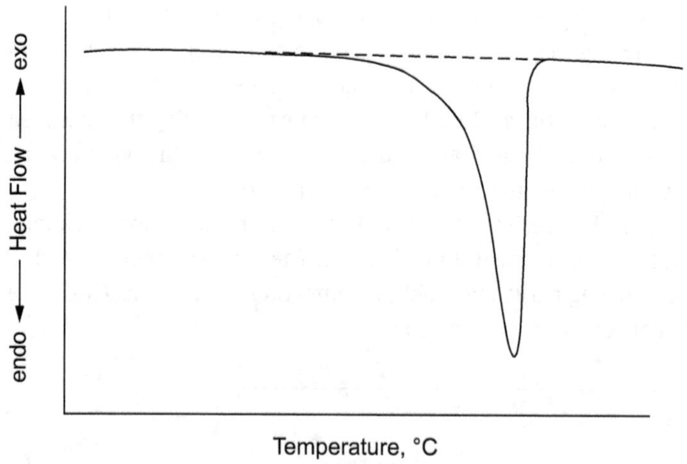


**Figure 8.23:** Thermal analysis by DSC (typical heating curve).

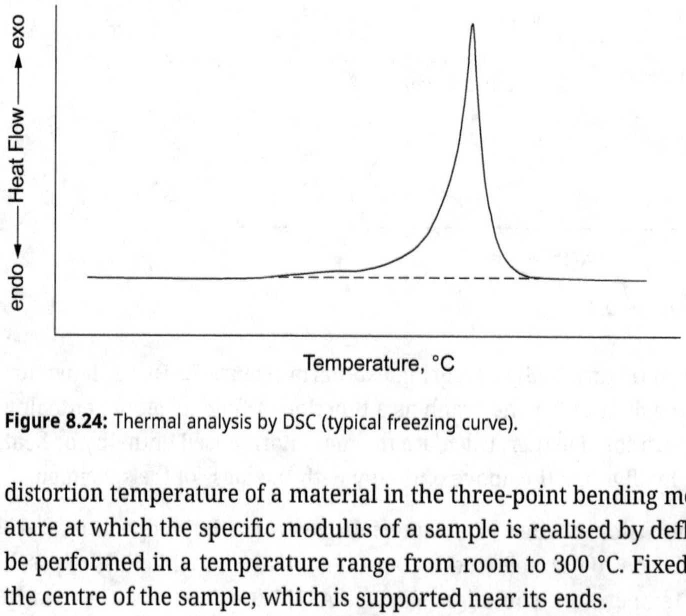


**Figure 8.24:** Thermal analysis by DSC (typical freezing curve).

distortion temperature of a material in the three-point bending mode. It is the temperature at which the specific modulus of a sample is realised by deflection. The test may be performed in a temperature range from room to 300 °C. Fixed stress is applied on the centre of the sample, which is supported near its ends.

The sample is heated at a constant heating rate. The deflection of the test sample is recorded as a function of temperature. The temperature at which the predetermined level of strain is observed is considered as the distortion temperature.

# References


[1] ASTM D1566, Standard Terminology Relating to Rubber.
[2] ASTM D3182, Standard Practice for Rubber – Materials, Equipment, and Procedures for Mixing Standard Compounds and Preparing Standard Vulcanized Sheets.
[3] ASTM D3183, Standard Practice for Rubber – Preparation of Pieces for Test Purposes from Products.
[4] Brown, R. Short-Term Mechanical Tests. In: Handbook of Polymer Testing, RAPRA Technology Limited, (2002). ISBN: 1-85957-324-X.
[5] ASTM D297, Standard Test Methods for Rubber Products – Chemical Analysis.
[6] El-Sabbagh, S. H., Ahmed, N. M., Turky, G. M., Selim, M. M. Rubber nanocomposites with new core-shell metal oxides as nanofillers. Progress in Rubber Nanocomposites 253 (2017). doi: http://dx.doi.org/10.1016/B978-0-08-100409-8.00008-5
[7] ASTM D1646, Standard Test Methods for Rubber – Viscosity, Stress Relaxation, and Pre-Vulcanization Characteristics (Mooney Viscometer).
[8] ASTM D5289, Standard Test Method for Rubber Property – Vulcanization Using Rotorless Cure Meters.
[9] Dick, J. S. Basic Rubber Testing: Selecting Methods for a Rubber Test Program, ASTM D2084, Standard Test Method for Rubber Property – Vulcanization Using Oscillating Disk Cure Meter, (2003). ASTM Stock Number: MNL39, ISBN: 0-8031-3358-8.
[10] Dick, J. S. Basic Rubber Testing: Selecting Methods for a Rubber Test Program (2003). ASTM Stock Number: MNL39, ISBN: 0-8031-3358-8.
[11] ASTM D6204, Standard Test Method for Rubber – Measurement of Unvulcanized Rheological Properties Using Rotorless Shear Rheometers.
[12] Dasgupta, S., et al. Characterization of eco-friendly processing aids for rubber compound. Polymer Testing 26(4):489–500 (2007).
[13] Sierra, C. A., Galan, C., Fatou, J. M. G., Quiteria, V. R. S. Rubber Chemistry and Technology 68:259 (1995).
[14] Kumar, N. R., Bhowmick, A. K., Gupta, B. R. Kautschuk Gummi Kunstsoffe 5:531 (1992).
[15] ASTM D6601, Standard Test Method for Rubber Properties – Measurement of Cure and After-Cure Dynamic Properties Using a Rotorless Shear Rheometer.
[16] ASTM D6048, Standard Practice for Stress Relaxation Testing of Raw Rubber, Unvulcanized Rubber Compounds, and Thermoplastic Elastomers.
[17] ISO13145, Rubber – Determination of viscosity and stress relaxation using a rotorless sealed shear rheometer.
[18] ASTM D412, Standard Test Methods for Vulcanized Rubber and Thermoplastic Elastomers – Tension.
[19] ASTM D575, Standard Test Methods for Rubber Properties in Compression.
[20] ASTM D624, Standard Test Method for Tear Strength of Conventional Vulcanized Rubber and Thermoplastic Elastomers.
[21] ASTM D1415, Standard Test Method for Rubber Property – International Hardness.
[22] ASTM D2240, Standard Test Method for Rubber Property – Durometer Hardness.
[23] Dedova, S., Schneider, K. Kautschuk Gummi Kunststoffe 06:85 (2018).
[24] ASTM D5992, Standard Guide for Dynamic Testing of Vulcanized Rubber and Rubber-Like Materials Using Vibratory Methods.
[25] ASTM D2632, Standard Test Method for Rubber Property – Resilience by Vertical Rebound.
[26] ISO4662, Rubber, vulcanized or thermoplastic – Determination of rebound resilience.
[27] Morton, M. Chapter 5: Physical testing of vulcanisates. In: Maurice Morton (eds) Rubber Technology (3rd edition). Akron Ohio: Kluwer Academic Publishers (1999).
[28] ASTM D623, Standard Test Methods for Rubber Property – Heat Generation and Flexing Fatigue In Compression.

[29] ASTM D430, Standard Test Methods for Rubber Deterioration – Dynamic Fatigue.
[30] ASTM D813, Standard Test Method for Rubber Deterioration – Crack Growth.
[31] ASTM D1052, Standard Test Method for Measuring Rubber Deterioration – Cut Growth Using Ross Flexing Apparatus.
[32] ASTM D4482, Standard Test Method for Rubber Property – Extension Cycling Fatigue.
[33] ASTM D3616, Standard Test Method for Rubber – Determination of Gel, Swelling Index, and Dilute Solution Viscosity.
[34] Hergenrother, W. L., Hilton, A. S. Use of $\chi$ as a function of volume fraction of rubber to determine crosslink density by swelling. Rubber Chemistry and Technology 76:832–845 (2003).
[35] Mark, H. F. Composites, fabrication to die design. In: Charles G. Overberger (eds) Encyclopedia of Polymer Science and Technology (IInd edition). UK: Wiley–Blackwell, Vol. 4, 356 (1986).
[36] Gent, A. N., Hartwell, J. A. Effect of carbon black on crosslinking. Rubber Chemistry and Technology 76:517 (2003).
[37] De, D., Gent, A. N. Tear strength of carbon black-filled compounds. Rubber Chemistry and Technology 69:834 (1996).
[38] Hamed, G. R., Rattanasom, N. Effect of crosslink density on cut growth in gum natural rubber vulcanizates. Rubber Chemistry and Technology 75:323 (2002).
[39] Dasgupta, S., Agrawal, S. L., Bandyopadhyay, S., Chakraborty, S., Mukhopadhyay, R., Malkani, R. K., Ameta, S. C. Characterisation of eco-friendly processing aids for rubber compound: Part II. Polymer Testing 27:277 (2008).
[40] ASTM D865, Standard Test Method for Rubber – Deterioration by Heating in Air (Test Tube Enclosure).
[41] ASTM D573, Standard Test Method for Rubber – Deterioration in an Air Oven.
[42] ASTM D6370, Standard Test Method for Rubber – Compositional Analysis by Thermogravimetry (TGA).
[43] ASTM D7426, Standard Test Method for Assignment of the DSC Procedure for Determining Tg of a Polymer or an Elastomeric Compound.
[44] ASTM D3418, Standard Test Method for Transition Temperatures and Enthalpies of Fusion and Crystallization of Polymers by Differential Scanning Calorimetry.

Dipankar Mondal, Soumyajit Ghorai, Dipankar Chattopadhyay, and Debapriya De

# Chapter 9 Devulcanisation of discarded rubber: a value-added disposal method of waste rubber products

## 9.1 Introduction

The challenge of waste disposal and management is one of a number of issues facing people in the twenty-first century. In 2016, global municipal solid waste production was approximately 2.01 billion tonnes. By 2050, it is projected to reach around 3.4 billion tons per year [1]. Although on a global scale, per capita waste production averages around 0.74 kg/day, it varies considerably between 0.11 kg/day and 0.45 kg/day [1], and the rate of waste production per inhabitant will increase significantly in the years to come [2]. Tyres from used vehicles play a significant role in waste generation. Approximately 1.4 billion tyres are manufactured annually throughout the globe [3]. But after long service life when these tyres are discarded, only a very small amount of rubber (<1%) gets abraded from the tyre. Most of the rubber from end-of-life tyres is discarded and disposed of in landfills; it is estimated that 4 billion tyres are landfilled worldwide [4]. It is further predicted that annually over 1.5 billion used tyres weighing approximately 17 million tons are thrown around the world [5–7]. Only 15–20% of the waste tyres from this enormous quantity of discarded tyres are reused [8].

It is anticipated that this number is expected to increase, as tyre demand rises each year. However, it is important to note that developed nations such as the United States, Japan, European Union (EU) countries, Australia, and South Korea also contribute significantly to the accumulation of scrap tyres.

These countries have a higher prevalence of automobile usage. India alone produces about a million tons of tyre scrap annually [9]. Both developed and developing countries pay great attention to the disposal of these large quantities of rejected tyres. Due to the cross-linked nature of rubbers and the inclusion of stabiliser and other additives, these materials cannot be biologically degraded, hydrolysed, or decomposed by organisms like plants or animals to return to the natural environment. Therefore, towards attaining the objective of preserving the environment and recycling resources, it is necessary to dispose of used tyres in a way responsibly [10, 11].

https://doi.org/10.1515/9783110668537-009

## 9.2 Present waste rubber generation scenario

The current world scenario for residual rubber production is presented in Figure 9.1. North America (31%) and Europe (29%), for example, are major producers of waste rubber due to their large automotive and manufacturing industries, as well as their high levels of consumption of consumer products. In contrast, some countries in Asia (excluding Japan – only 19%), such as China and India, are also significant producers of waste rubber due to their rapidly growing economies and industrial activities. Nevertheless, waste management practices and recycling efforts can also differ widely across regions and countries, which can impact the amount of waste rubber that is generated and how it is managed.

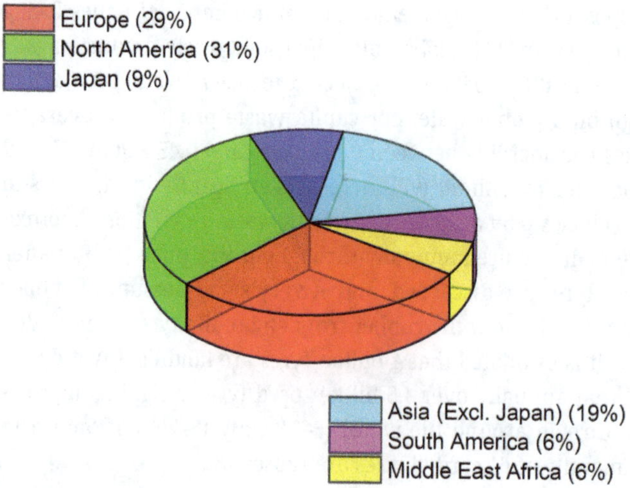


**Figure 9.1:** The current world scenario for waste rubber production.

## 9.3 Different disposal methods of waste rubber products

There are various disposal methods for waste rubber products, ranging from traditional methods such as landfilling to more sustainable options such as recycling and repurposing. Here are some common methods:

**Landfilling**: This involves burying waste rubber products in landfills, which is a common disposal method but not a sustainable one. Rubber products take a long time to break down in landfills, which can lead to environmental pollution and potential health hazards.

**Incineration**: This involves burning waste rubber products, which can generate energy but also releases toxic pollutants into the air. Incineration is not considered a sustainable or environmentally-friendly disposal method.

**Recycling**: This involves processing waste rubber products into new materials that can be used for various purposes. For example, waste tyres can be shredded and processed into rubber crumb, which can be used in the construction of playgrounds, sports fields, and other surfaces. Rubber can also be repurposed into products like footwear, automotive parts, and other consumer goods.

**Pyrolysis**: In order to create oil, gas, and carbon black, waste rubber products must be heated in the absence of oxygen. The oil and gas can be used as fuel, while the carbon black can be used as a raw material in the production of new rubber products.

**Devulcanisation**: This involves breaking down the molecular structure of waste rubber products to reclaim the rubber for reuse in new products. Devulcanisation is a promising method for reducing waste and increasing the sustainability of rubber products.

**Civil engineering application**: Overall, sustainable disposal methods such as recycling and repurposing are preferable to traditional methods such as land filling and incineration. These methods can reduce the amount of waste that ends up in landfills or pollutes the environment, while also creating economic opportunities and conserving natural resources.

### 9.3.1 Landfilling

Landfilling is one of the most common methods of waste disposal, but it has significant environmental impacts. The main problem with landfilling is that the waste rubber does not decompose quickly and can take hundreds of years to degrade. Landfills can also lead to pollution of groundwater and soil, and can create odor and nuisance problems. Around 4 billion tyres are estimated to be landfilled around the world [4]. However, due to the decreasing number of suitable sites and the accompanying cost increase caused by the implicit cost of transporting the trash to the landfill site and establishing and maintaining landfills to meet environmental requirements, it is no longer feasible to dispose of scrap rubber in this way [12]. The statistics of used tyres management over the last 15 years in the EU is represented in Figure 9.2. When compared to total used tyre treatment, it can be shown that end-of-life tyre material and energy recovery increased from 31% to 78%, while landfilling reduced to 4% (from nearly 50% in 1996) throughout that time. While reuse and export of part-worn tyres have mostly remained at the same level over time, retreading has dropped over time from 12% to 8%. Due to environmental concerns, landfilling discarded tyres is not an acceptable choice in the coming future. Since tyres are not disposed of properly, they can provide a breeding site for mosquitoes and rodents [13, 14].

Additionally, when rubber is being compounded, a number of additives were added, including stabilisers, plasticisers, colourants, flame retardants, etc. Low-molecular-weight additives are likely to leach from the bulk phase of the polymer to the surface phase of tyres and then potentially leach from the surface phase into the environment due to the concentration gradient after disposing of tyres for landfilling. These additives with low molecular weight are harmful to the environment and may be responsible for killing beneficial bacteria in soil.

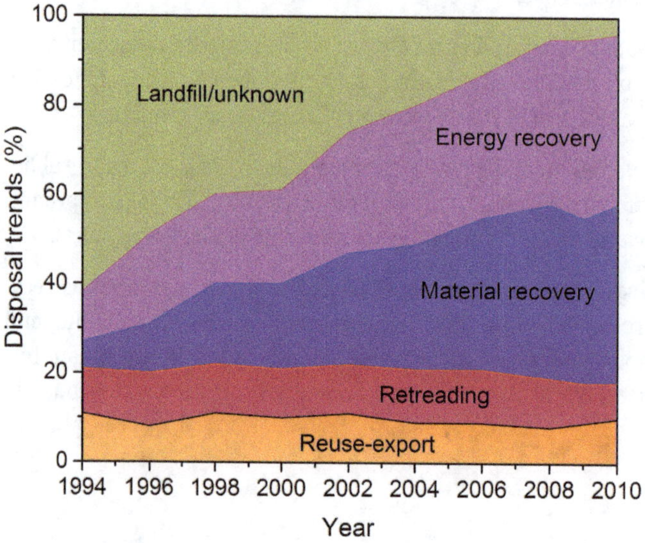


**Figure 9.2:** Statistics of used tyres management.

## 9.3.2 Incineration

Incineration is a waste treatment process that involves the combustion of organic substances (more than 90%) contained in waste tyres. It is a method of waste management that is used to reduce the volume of waste disposal as well as to generate fuel energy through the burning of these materials. The heat value of coal can vary depending on the type of coal, but it is generally between 24 and 28 MJ/kg (ca. 8,000–12,000 Btu/lb) [15, 16]. In contrast, the heat value of waste tyres typically ranges from about 32–36 MJ/kg (ca. 14,000 Btu/lb).

Incineration has several advantages over other waste disposal methods. For one, it can significantly reduce the volume of waste, which can help to conserve landfill space. It can also reduce the environmental impact of waste disposal by reducing the amount of greenhouse gases emitted by landfills. Additionally, incineration can generate electricity and heat, which can be used to power homes and businesses [17]. The ash and other byproducts of the combustion process can also be used as a source of raw materials for the production of construction materials, such as concrete and asphalt. Germany, one of the major tyre producers in the EU, ranks as the leading nation with the highest recycled energy from scrap tyres [18].

The calorific value of different combustible materials as an energy resource is depicted in Figure 9.3, which demonstrates that tyres have a calorific value comparable to that of coal and crude oil [19]. While incineration can be an effective method of waste management, there are also concerns associated with the process. These include the emission of pollutants such as dioxins, furans, and heavy metals, as well as emission of polycyclic aromatic hydrocarbon, which consists of alkylated naphthalene's, fluorenes, and phenanthrenes formed in this process. It may have an adverse effect on both the environment and human health. There are also concerns about the release of greenhouse gases, for instance, carbon dioxide, which may be a factor in climate change. Additionally, incineration can create a perception that waste is being 'taken care of' without addressing the underlying issues of waste generation and consumption.

Overall, incineration is a waste treatment process that has both benefits and potential drawbacks. Its effectiveness and suitability depend on various factors, including the type and quantity of waste being treated, the technology used, and the specific environmental and regulatory conditions of the area where it is being carried out.

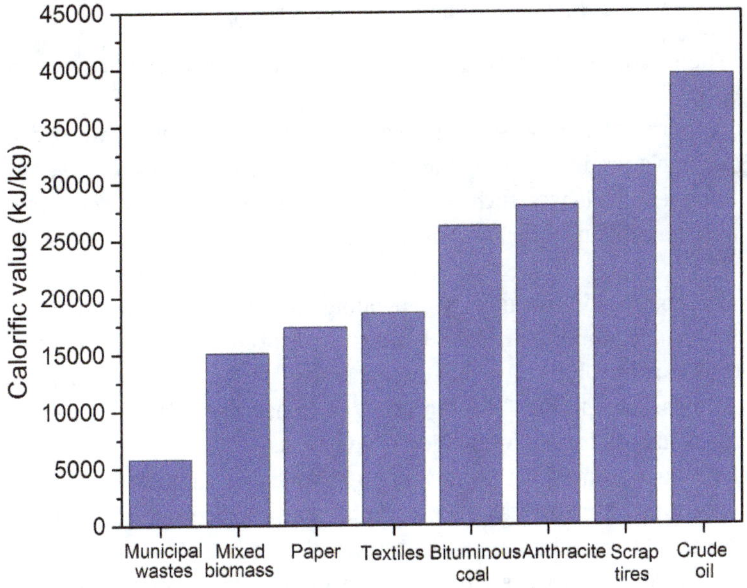


**Figure 9.3:** The calorific value of different combustible materials.

### 9.3.3 Pyrolysis

Pyrolysis of waste rubber is a process that involves the thermal degradation of rubber in the absence of oxygen to produce useful products. The process typically involves heating the waste rubber to high temperatures (typically between 400 and 600 °C) in a closed vessel, causing the rubber to break down into its constituent compounds.

During pyrolysis, the rubber undergoes a series of complex chemical reactions, resulting in the formation of a range of products such as gas, liquid, and solid residues. The gas phase typically contains methane, hydrogen, carbon monoxide, and other hydrocarbons, which can be used as fuels. The liquid phase consists of oils that can be further refined into various chemicals such as benzene, toluene, and xylene. The solid residue is a char that can be used as a fuel or as carbon black filler.

The pyrolysis process typically involves the following steps:

**Pretreatment**: Waste rubber is typically shredded into smaller pieces to increase the surface area and improve the efficiency of the pyrolysis process.

**Heating**: The shredded rubber is then heated in a pyrolysis reactor to temperatures ranging from 300 to 500 °C in the absence of oxygen. The heating process causes the rubber to break down into smaller molecules.

**Collection and separation**: The resulting products are then collected and separated into various fractions, such as fuel oil, carbon black, and syngas.

**Posttreatment**: The fractions are then further processed to remove impurities and improve their quality.

The pyrolysis of waste rubber has several advantages, including reducing waste disposal, producing valuable products, and reducing greenhouse gas emissions. However, the process also has a few challenges, such as the need for high capital investment, complex technology, and the potential for environmental pollution, if not properly controlled.

Products of waste rubber pyrolysis can vary depending on the specific process conditions, but typically include a mixture of liquid and gaseous products. The liquid product is often referred to as pyrolysis oil or bio-oil and can be used as a fuel in a variety of applications. The gaseous product is often referred to as syngas and can also be used as a fuel or as feedstock for the production of chemicals.

Carbon black produced during pyrolysis can also be used as a valuable commodity in a variety of applications, including as filler in rubber and plastic products, as a pigment in ink and paint, and as a reinforcing agent in tyres.

The pyrolysis of tyres uses a variety of reactor technologies, including fixed bed, screw kiln, rotary kiln, vacuum, and fluidised bed reactors [20]. Researchers looked into the effects of stepwise co-pyrolysis temperature and different ratios of palm shells and recycled tyres in the feedstock to find any enhancement in both the quantity and the quality of the liquid fuel [21]. Outcomes showed that the liquid and gas contents somewhat increased and the char level reduced with an increase in co-pyrolysis temperature to 800 °C from an optimal temperature of 500 °C. Both the quality and quantity of the liquid were successfully increased by adding scrap tyres to the pyrolysis of biomass. Considering the aforementioned factors, in a different study, discarded tyres were co-pyrolysed with biomass using various reactors, including fixed bed and continuous auger reactors, to improve the liquid biofuel produced during pyrolysis [22]. The results revealed that adding waste tyres to biomass feedstock enhances the properties of the bio-oil in both systems and highest improvement is observed for auger reactor.

Furthermore, radical interactions between by-products of biomass pyrolysis and waste tyres can encourage the production of stable oil with better properties. In situ and ex situ desulphurisation of discarded tyres was carried out by a two-stage pyroliser with an auger and fluidised bed and extremely affordable additives such as calcium oxide, iron, calcined olivine, and iron oxide to yield pyrolysis oil with extremely low sulphur content. Environmental pollution would be reduced by reducing the sulphur in the tyre pyrolysis oil due to its combustion [23]. In a typical pyrolysis process, before being pyrolysed, scrap tyres received a pretreatment of waste coal tar [24]. Here, waste coal tar was utilized as a modifier for scrap tyres with the aim of enhancing both the pyrolysis efficiency of the tyres and the quality of the resulting tar. To optimise the pyrolysis temperature, the waste tyre as well as the pretreated tyre was pyrolysed at a variety of temperatures such as 400, 450, 500, 550, and 600 °C. The results showed that for scrap tyres and pretreated tyres, the char yield steadily decreased and the tar and gas production gradually increased with an increase in temperature, as shown in Figure 9.4a and b, respectively. Conversely, the extent of increase of tar yield was larger and gas production was lower for pretreated tyres and the extent of decrease of char yield was lower for pretreated tyres compared to that of the scrap tyres.

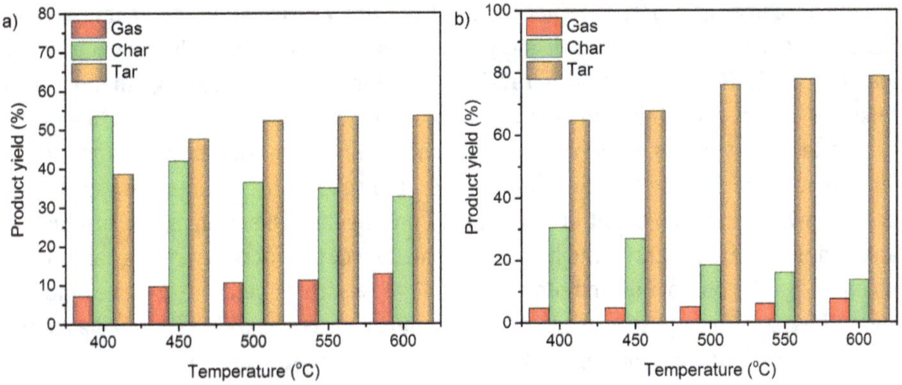


**Figure 9.4:** The effect of temperature on the gas, char and tar yield of (a) scrap tyres and (b) pretreated tyres.

Co-pyrolysis of scrap tyres and pistachio seed was performed at 500 °C with a catalyst to improve the quality and quantity of the pyrolysed products derived from waste tyres and biomass [25]. The findings showed that catalytic co-pyrolysis of waste tyres and biomass had a synergistic effect to enhance the standard of the pyrolytic oil. Additionally, the co-pyrolysis with 50% mixed biomass improved heat value and carbon content while reducing the pyrolytic oil's oxygen content. Thus, co-pyrolysis of scrap tyres in the presence of biomass may be considered an effective technique for producing high-grade pyrolysis oil simultaneously as well as this process of scrap tyre disposal may be chosen as a partially value-added disposal technique of discarded tyre products.

### 9.3.4 Asphalt-rubber pavement

Crumb rubber obtained from the grinding of end-of-life tyres can be used as a bitumen modifier in the production of bituminous mixtures used for road pavement construction and maintenance [26]. This is a sustainable approach to recycling end-of-life tyres, which would otherwise end up in landfills or be incinerated, creating environmental problems.

The addition of crumb rubber to bituminous mixtures has several benefits. Firstly, it improves the elasticity and durability of the asphalt, resulting in better resistance to cracking and deformation under heavy traffic loads. This prolongs the service life of the road, reducing maintenance costs and enhancing safety [27–29]. Crumb rubber-modified bitumen can be prepared in two different processes such as wet process and dry process. In the wet process, crumb rubber is mixed with bitumen to form a modified binder known as asphalt rubber. This modified binder is then combined with aggregates in a hot mix plant to produce gap-graded and open-graded types of bituminous mixtures, which are used to build surface courses and displayed satisfactory field performance [30]. Actually, in the dry process, crumb rubber is added to the production flow of bituminous mixtures as a partial replacement of the fine aggregate. When crumb rubber particles act as a filler material, it increases the elastic response of the mixture under loading [31, 32]. However, the early raveling and moisture-related damage that occur in mixtures could be attributed to several factors, including the quality and grading of the aggregates, the type and amount of asphalt binder, the compaction process, and the environmental conditions [33]. The addition of crumb rubber in bituminous mixtures has been shown to have potential environmental and health risks. During the production and laying of these mixtures, the heating and melting of the rubber particles can release volatile organic compounds and other harmful gases, such as benzene, toluene, and styrene. These gases can pose a risk to the health of construction workers, as well as to the general public living or working in close proximity to the construction site [34].

Different treating agents can be used to improve the properties of crumb rubber modifier (CRM) in bituminous mixtures [35]. These agents are typically added during the production process to enhance the compatibility and interaction between the rubber particles and the asphalt binder, thereby improving the overall performance of the mixture. The study found that hydrated lime is the best anti-stripping additive for normal asphalt concrete. According to the results of the study, the use of hydrated lime in the asphalt concrete mixture resulted in improved performance of the CRM mixture compared to the normal mixture in terms of fatigue life, deformation strength, and indirect tensile strength, but to varying degrees at different temperatures.

Specifically, the study found that the CRM mixture showed the most significant improvement in fatigue life at 20 °C, which refers to its ability to resist damage caused by cyclic loading over time. The improvement in deformation strength at 60 °C, which refers to its ability to resist deformation under load, was moderate. On the other hand, the improvement in indirect tensile strength at 25 °C, which measures the tensile strength of a material, was the least significant compared to that of the normal mixture. It seems that the study also found that the addition of 10% by weight of CRM to the asphalt concrete mixture resulted in a significant improvement in fatigue resistance compared to the normal mixtures without CRM. Fatigue resistance refers to the ability of a material to resist damage caused by cyclic loading over time, and the improved fatigue resistance with the addition of CRM suggests that it can help to extend the service life of the asphalt pavement.

Furthermore, the study found that there was no significant difference in fatigue resistance between the dry and wet processed CRM mixtures. This suggests that both dry and wet processing methods can be effective in producing asphalt concrete mixtures with improved fatigue resistance when using CRM as an additive. It appears that the wet-processed CRM mixtures showed improved resistance against deformation at high temperatures and tensile strength at ambient temperature compared to the dry mixtures [36].

Slurry oil, which is a by-product of the FCC process in refineries, can be a suitable emulsifier for CRMA. The slurry oil contains hydrocarbons with high boiling points and a significant amount of catalyst residue, making it an effective emulsifying agent. When blended with CRMA, slurry oil can help to improve the road performance of the asphalt at high temperatures [37]. The results revealed that rotational viscosity of CRMA decreases as the amount of FCC slurry increases, as evident from Figure 9.5(a) and as a result the viscosity of CRMA is comparable with asphalt. Furthermore, it is depicted from Figure 9.5(b) that the increment of ductility at 45 °C is 22 cm at 3 wt% of slurry content, whereas its decrement of softening point is 45 °C, which is only about 1 °C lower than that of the CRMA-0, but same as that of the softening point (45 °C) of the asphalt binder. The developed emulsion can meet the requirements of slurry or chip seal and sealcoat pavements for sealing and repairing roads, parking lots, and blacktop surfaces. Slurry seal and chip seal are both surface treatments that are applied to asphalt pavements to protect them from weathering and improve their skid resistance. Improved uniformity and adhesion ability are important benefits of using an emulsifier like slurry oil in asphalt binders [38].

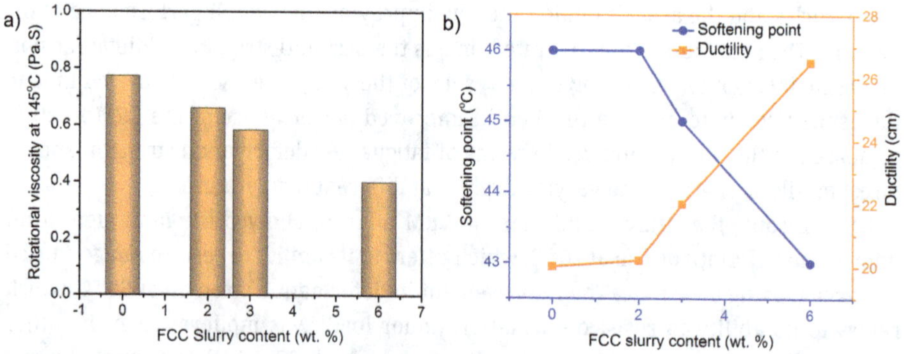


**Figure 9.5:** The influence of FCC slurry content on (a) rotational viscosity and (b) softening point and ductility of crumb rubber.

The rheological behaviour of a polymer refers to its response to an applied force or deformation. The addition of *trans*-polyoctenamer rubber (TOR) to chlorinated rubber-modified asphalt (CRMA) can have a significant impact on the rheological behaviour of the material. The modification mechanism between CRMA and TOR can be understood by examining the changes in the material's rheological properties [39]. Figure 9.6(a)–(c) shows how the rubber content and TOR in CRMA affect the physical qualities.

Based on the given statement, it appears that the study examined the effect of crumb rubber content on the penetration of asphalt. The results suggest that as the crumb rubber content in asphalt increases, the penetration of the asphalt decreases. Furthermore, the study found that there is a significant decrease in penetration when the crumb rubber content increases from 15% to 20%. However, when the crumb rubber content is further increased from 20% to 25%, the extent of decrease in penetration is comparatively less than the previous range (15–20%). Yet, the penetration of crumb rubber in asphalt decreases in the presence of TOR, but the decrease in penetration is less significant compared to the use of CRMA. On the other hand, specifically, the softening point of CRMA is increased to a greater extent when TOR is added to the mixture, which may be beneficial for improving the high-temperature performance of the material. The decrease in penetration and increase in softening point observed in CRMA with the addition of crumb rubber and TOR may be attributed to the formation of a new equilibrium state between crumb rubber and asphalt in the presence of TOR. The elastic recovery is the most important parameter of CRMA.

From Figure 9.6(c), it is evident that increasing the content of crumb rubber (CR) in the material leads to an increase in its elastic recovery, which means that the material is better able to return to its original shape after being deformed. This could be due to the properties of CR itself, which is known for its elasticity and resilience and the incorporation of TOR into the CRMA further improves its elastic recovery. This could be due to TOR interacting with the CR in a way that enhances its properties or

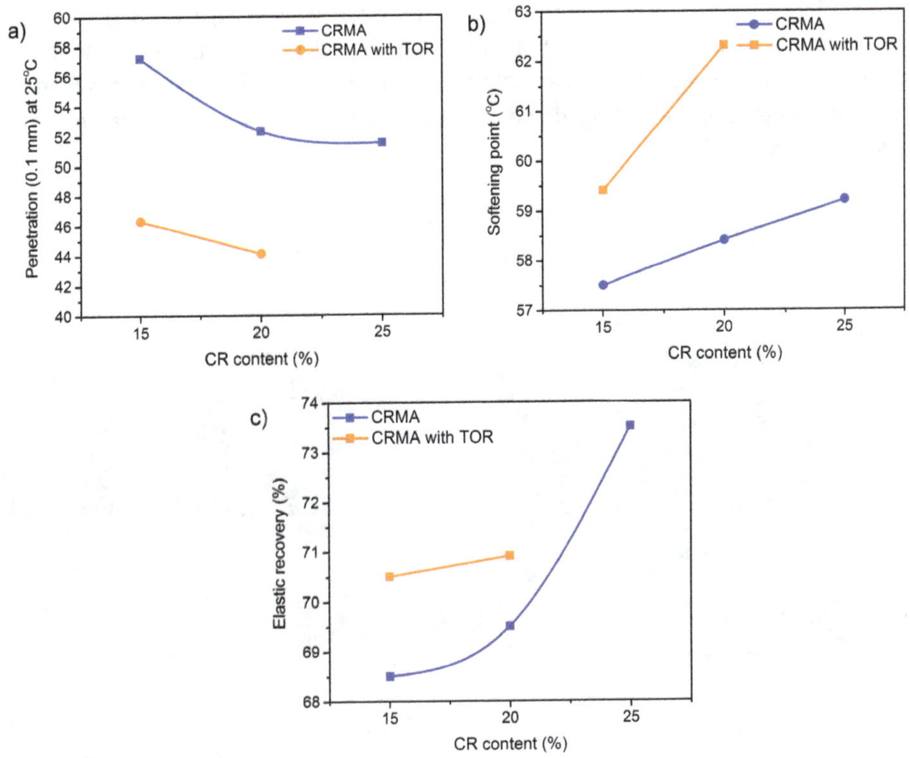


**Figure 9.6:** Effect of crumb rubber content on (a) penetration (0.1 mm), (b) softening point, and (c) elastic recovery.

provides additional elasticity to the material. The improved elastic recovery of CRMA could potentially help reduce high-temperature rutting and low-temperature cracking of asphalt [40].

The addition of cross-linking agents to asphalt rubber can indeed help improve its workability, among other properties [41]. Crumb rubber, and TOR were used to produce asphalt rubber binders. Specifically, two different base asphalt binders (PG67-22 and PG64-22) were used, along with three different proportions of crumb rubber (8%, 10%, and 12% of the weight of binder) and three different proportions of TOR (0%, 3%, and 6% of the weight of binder) as a cross-linking agent. To evaluate the high temperature properties and stability of a material, likely an asphalt rubber binder, evaluation was carried out using two tests: the dynamic shear rheometer (DSR) and the tube method in conjunction with DSR tests.

The results indicate from Figure 9.7(a)–(d) that an increase of 14% and 20% in the $G^*/\sin\delta$ of asphalt rubber binders in original state was found when the proportion of TOR was 3% and 6%, respectively. Rolling thin film oven (RTFO) aged residue showed a decreased rate of growth. TOR thereby improved the high-temperature properties of the binder when it was added to asphalt rubber. However, Figure 9.8(a)–(d) indicates that the phase angle of asphalt rubber binders decreased up to 1.8% with increasing proportion of TOR. TOR added to the asphalt rubber binder (TOR) resulted in improved elastic properties, as demonstrated by the testing of both unaged binders and RTFO test residue.

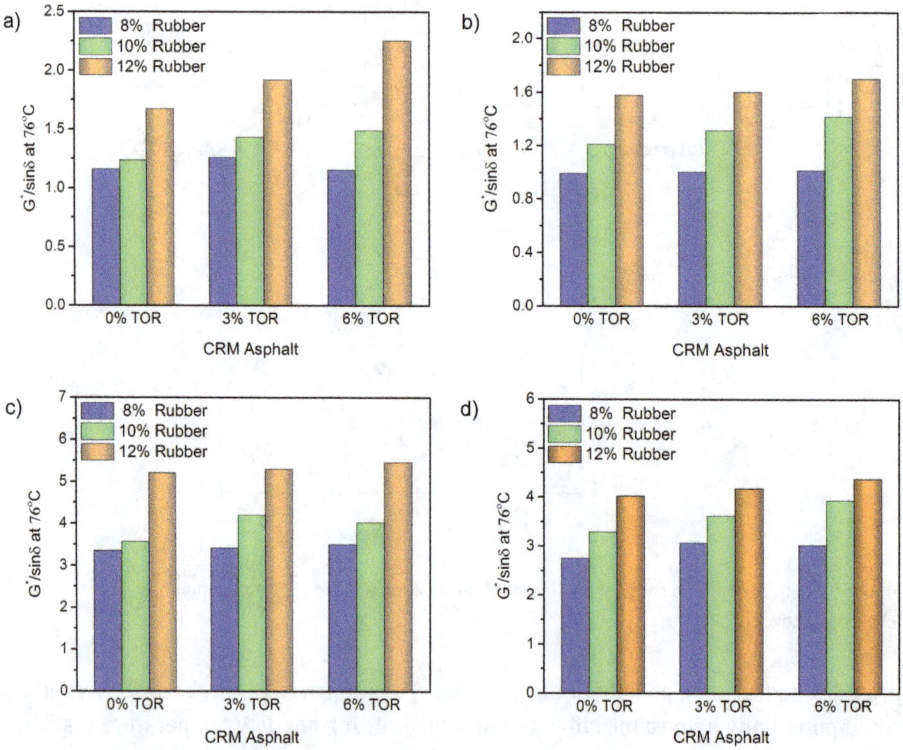


**Figure 9.7:** Effect of crumb rubber-modified asphalt on G*/sin $\delta$ of original binders at 76 °C for (a) asphalt binder of PG 67-22; (b) asphalt binder of PG 64-22; and G*/sin $\delta$ of RTFO residuals at 76 °C for (c) asphalt binder of PG 67-22; and (d) asphalt binder of PG 64-22.

## 9.3.5 Cement-rubber concrete

Rubber from waste tyres used in concrete can be an effective way to reduce the consumption of raw materials and promote sustainable development in the construction industry. Still, as mentioned, CR in concrete can lead to a reduction in stiffness, which may affect the mechanical properties of the concrete [42]. The properties of crumb rubber concrete (CRC) are influenced by various factors related to the CR used in the mixture. These factors include good flexibility, content, particle size, shape, cleanliness, and quality of the surface finish of the CR [43–48]. The incorporation of CR par-

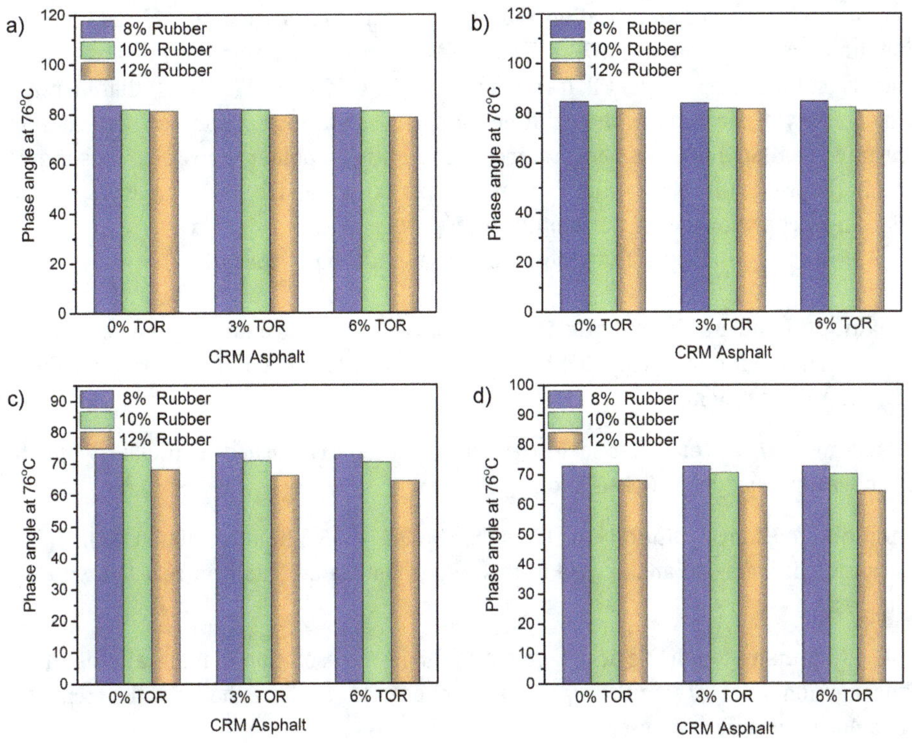


**Figure 9.8:** Effect of crumb rubber-modified asphalt on phase angle of original binders at 76 °C for (a) asphalt binder of PG 67-22; (b) asphalt binder of PG 64-22; and phase angle of RTFO residuals at 76 °C for (c) asphalt binder of PG 67-22; and (d) asphalt binder of PG 64-22.

ticles in concrete can result in a decrease in the fresh density of the mixture, and this decrease is generally proportional to the amount of CR added to the mixture. This is because CR particles are typically less dense than the other materials in the concrete mixture, such as cement, sand, and aggregate.

It is generally observed that the mechanical properties of concrete tend to decrease with increasing CR content. As a result, the addition of CR can reduce the overall strength, stiffness, and durability of the concrete.

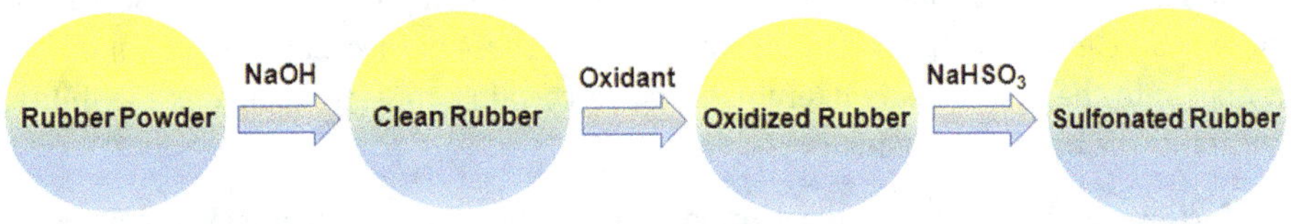


**Figure 9.9:** Schematic presentation of rubber modification process.

In the typical surface modification process, the CR particles are first oxidised with potassium permanganate ($KMnO_4$) solution. $KMnO_4$ reacts with the unsaturated double bonds on the surface of the CR, forming carboxylic acid groups (–COOH) that increase the polarity of the surface; the oxidised CR particles are then sulfonated with sodium bisulfite ($NaHSO_3$) solution, which introduces additional polar groups, such as sulfonic acid groups ($–SO_3H$) on the surface of the CR particles [49]. Figure 9.9 depicts the standard flow chart for the rubber modification process.

Here is a summary of the process and its application in concrete:

**Soaking in sodium hydroxide (NaOH) solution**: The CR block is soaked in a 5% NaOH solution for 24 h. This step is likely performed to clean and prepare the rubber for further treatment.

**Washing with water**: After soaking in NaOH, the CR is washed with clear water to remove any residual NaOH solution.

**Addition to $KMnO_4$ solution**: The CR is then added to a 5% $KMnO_4$ solution. $KMnO_4$ is a strong oxidising agent and is used in this step to initiate an oxidation reaction with the rubber.

**pH adjustment**: Sulphuric acid is added to the $KMnO_4$ solution to maintain the pH of the solution at 2–3. This pH range is likely chosen to facilitate the oxidation reaction and ensure its effectiveness.

**Heating and stirring**: The solution containing the CR, $KMnO_4$, and sulphuric acid is heated at 60 °C with stirring for approximately 2 h. This heating and stirring process helps promote the oxidation reaction and ensures thorough mixing.

**Washing with water**: After completing the oxidation reaction, the modified CR is thoroughly washed with clean water to remove any residual chemicals or by-products.

**Soaking in sodium bisulfite solution**: The modified CR is then soaked in a saturated sodium bisulfite ($NaHSO_3$) solution at 60 °C for 0.5–1 h. This step is performed to induce the sulfonation reaction of rubber, which involves the addition of sulfonic acid groups to the rubber structure.

**Concrete preparation and evaluation**: Two different concretes are prepared using the modified rubber. Details regarding the specific composition and properties of these concretes are not provided. However, the purpose of incorporating the modified rubber into the concretes is likely to enhance certain properties, such as flexibility, durability, or impact resistance.

**Performance evaluation**: The performance of the two different concretes is evaluated, presumably through various tests or measurements. The evaluation may assess properties like compressive strength, tensile strength, water absorption, or other relevant characteristics to determine the impact of the modified rubber on the concretes' performance.

Overall, the described process involves treating CR with NaOH, $KMnO_4$, and sodium bisulfite to modify its properties, followed by incorporating the modified rubber into two different concretes for performance evaluation. Figure 9.10 demonstrates that the impact resistance of modified CRC is 1.9 times greater than that of traditional cement concrete. But, when compared to unmodified rubber-based concrete, the impact energy of modified rubber-based concrete is also higher by 22%. As the content of CR rises, the compressive strength of the rubberised concrete instantly decreases, as shown by the connection between rubber content and compressive strength in Figure 9.11.

Modified rubber concrete's compressive strength, still, declined more gradually. The compressive strength of ordinary CRC (with 4% rubber content) is reduced to 23.6 MPa from the compressive strength of ordinary cement concrete, which is 49.2 MPa. This reduction corresponds to a loss of 52% of the compressive strength when comparing the two types of concrete. When the CR is modified using the described process and incorporated into the concrete at the same 4% content, the compressive strength of the modified CRC is found to be 35.1 MPa. This represents a 48.7% improvement in compressive strength compared to the ordinary CRC. In summary:

- Ordinary cement concrete compressive strength: 49.2 MPa
- Ordinary CRC compressive strength (4% rubber): 23.6 MPa (52% reduction)
- Modified CRC compressive strength (4% rubber): 35.1 MPa (48.7% improvement compared to ordinary crumb rubber concrete).

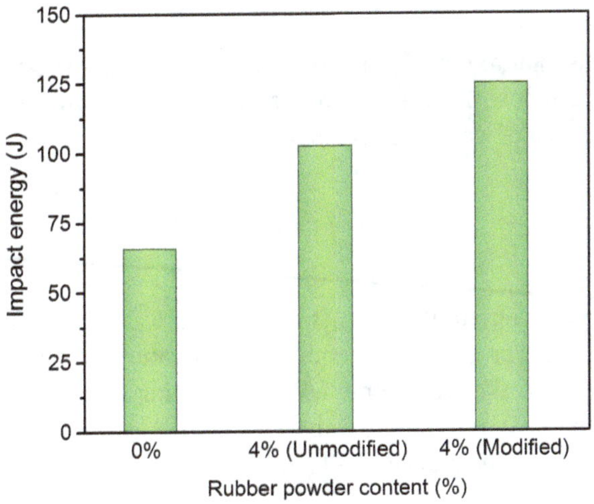


**Figure 9.10:** The effect of modified and unmodified rubber on impact resistance of concrete.

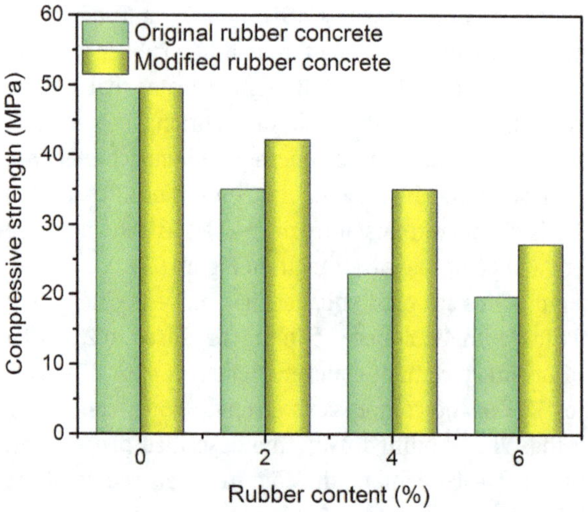


**Figure 9.11:** The effect of rubber content on the compressive strength of the rubber concrete.

# 9.4 Different devulcanisation techniques

## 9.4.1 Preprocessing of waste rubber

In general, first step of devulcanisation to reduce the size of the waste rubber is carried out through the process of grinding. Different grinding processes are adopted, depending on the desired reutilisation of waste rubbers [50].

### 9.4.1.1 Ambient grinding

Ambient grinding of crumb rubber refers to the process of reducing waste tyres or rubber products into smaller particles or granules at ambient temperature, typically through mechanical means. This process is commonly used in recycling facilities to transform discarded tyres or rubber materials into reusable CR.

The ambient grinding process typically involves the following steps:

**Collection and sorting**: Waste tyres or rubber products are collected from various sources and sorted based on their size, type, and quality.

**Pre-shredding**: The collected tyres or rubber materials may undergo pre-shredding to reduce their size and facilitate further processing. This step can involve cutting or shredding the rubber into manageable pieces.

**Grinding**: The pre-shredded rubber is then fed into grinding equipment, such as granulators or hammer mills, where it is further reduced in size. The grinding process breaks down the rubber into smaller particles or granules, typically ranging from a few millimeters to a few centimeters in size.

**Screening and classification**: After grinding, the CR is screened to separate it into different size fractions. This allows for the production of various grades of CR, which can be used for different applications.

**Cleaning and decontamination (optional)**: Depending on the intended use of the CR, additional steps may be taken to remove contaminants, such as steel wires or fibre residues, through processes like magnetic separation or air classification.

The resulting CR can be used in a wide range of applications, including the production of new rubber products, asphalt pavement, sports surfaces, playgrounds, and more. It offers a sustainable and environmentally friendly solution for recycling rubber waste while reducing the demand for virgin rubber materials.

One of the advantages of ambient grinding of CR is that it is a relatively low-cost and energy-efficient method of recycling waste rubber products. The process does not require the use of heat or chemicals, which can help to reduce emissions and minimise environmental impact.

CR produced through ambient grinding can also offer several benefits over traditional materials. It is durable, nontoxic, and provides shock absorption, making it an excellent choice for various applications. Additionally, it can help to reduce the environmental impact of waste tyres by diverting them from landfills and other disposal methods.

Nevertheless, there are two limitations of this process: i) it generates a significant amount of heat, which could cause the CR to oxidise and disintegrate and ii) for production of fine mesh size particles, the cost of production becomes very high [51].

Overall, ambient grinding of CR is an effective and eco-friendly way to recycle waste rubber materials into useful products with a wide range of applications.

#### 9.4.1.2 Cryogenic grinding

Cryogenic grinding is a process used in various industries to reduce the particle size of materials through the application of extremely low temperatures. It is particularly useful for grinding or pulverising heat-sensitive and soft materials that may undergo degradation or lose their properties at higher temperatures. The process produces CR with smaller particles in a cleaner and faster way [52]. The operating cost of cryogenic grinding is generally higher compared to ambient grinding processes. Use of liquid nitrogen or other cryogenic gases in cryogenic grinding requires a continuous supply, which can be expensive. The cost of acquiring and storing these gases adds to the operating expenses; additionally drying step is often necessary to remove the moisture from the CR. This drying process adds extra time, energy, and cost to the overall operation [53].

The process involves freezing the material to a very low temperature, typically below −150 °C (−238 °F), using liquid nitrogen or other cryogenic gases. The frozen material is then ground or pulverised into smaller particles using specialised equipment such as cryogenic grinders or mills.

*The main advantages of cryogenic grinding include:*

- **Preservation of material properties**: Cryogenic grinding helps to retain the original characteristics and properties of heat-sensitive materials. By reducing the temperature during the grinding process, it minimises thermal degradation, oxidation, and volatilisation of essential components.
- **Improved particle size reduction**: The low temperature makes the material more brittle, enabling easier and more efficient grinding. Cryogenic grinding can achieve finer particle sizes compared to conventional grinding methods, resulting in improved product quality and performance.
- **Enhanced process safety**: The use of cryogenic gases for grinding reduces the risk of fire or explosion associated with traditional grinding methods. The extremely low temperatures also help to prevent the generation of dust and the release of volatile components during the grinding process.
- **Wide applicability**: Cryogenic grinding can be applied to a wide range of materials, including plastics, rubber, spices, pharmaceuticals, and food products. It is particularly useful for materials with low melting points or those that are difficult to grind using conventional methods.

*However, cryogenic grinding also has a few limitations and considerations:*

- **Equipment cost**: Cryogenic grinding requires specialised equipment that can handle low temperatures and is designed to prevent material sticking or clogging. The initial investment for such equipment can be higher compared to conventional grinding systems.
- **Energy consumption**: The cryogenic grinding process consumes more energy due to the need for cooling the materials and maintaining low temperatures. The energy costs associated with operating the cryogenic equipment should be taken into account.
- **Material selection**: Not all materials are suitable for cryogenic grinding. Some materials, such as certain elastomers or thermoplastics, may become too hard and difficult to grind when exposed to extremely low temperatures. It is important to assess the material's properties and behaviour at low temperatures before applying cryogenic grinding.

Overall, cryogenic grinding offers unique benefits for certain applications where heat-sensitive materials need to be processed into fine particles. It is widely used in industries such as food processing, pharmaceuticals, plastics, and chemicals to achieve improved product quality and process efficiency.

#### 9.4.1.3 Wet grinding

The wet grinding process involves the mixing of partially refined CR particles with water creating slurry. The slurry is then transferred to a mill for size reduction. When the desired size is achieved, the slurry is then passed in an equipment for removing the major portion of water and then for drying. In an alternative wet grinding technique, water jet is used to mill highly resistant and large truck tyres. The CR generated in this process has high purity and large surface area of particles. The process is eco-friendly, as the water is recycled using a closed-loop system and the energy consumption and noise production in the system is low [54, 55]. The finer particles generated in this process can be used as filler in high quality composites.

Wet grinding refers to a method of particle size reduction or material processing where a liquid (usually water) is added to the grinding process. It is commonly used in various industries such as food processing, pharmaceuticals, and mineral processing. Wet grinding can be considered relatively eco-friendly, because a closed-loop system can be employed, where the water used in the wet grinding process is collected, treated, and reused. This significantly reduces water consumption and minimises the need for fresh water intake [54, 55].

There are different types of equipment used for wet grinding, depending on the specific application. Common examples include ball mills, bead mills, and attritor mills. These machines are designed to handle wet materials and typically involve the rotation or agitation of a grinding media (such as balls or beads) within a container or chamber.

*The process of wet grinding generally involves the following steps:*

**Preparation**: The solid material to be ground is typically pretreated to ensure proper size reduction. This may involve cutting, chopping, or crushing the material to smaller pieces before initiating the wet grinding process.

**Loading:** The solid material and the liquid medium are combined in a grinding mill or similar equipment designed for wet grinding. The ratio of solid to liquid can vary depending on the desired consistency and the properties of the material being ground.

**Grinding**: The grinding mill contains grinding media, such as balls or beads, which help in the comminution process. As the mill rotates, the grinding media collide with the solid particles, causing them to break down into smaller sizes. The liquid medium helps in reducing friction and dissipating heat generated during the grinding process.

**Control**: The wet grinding process may involve controlling various parameters to achieve the desired particle size and distribution. Factors such as the speed of the mill, the size and type of grinding media, the viscosity of the liquid, and the residence time of the material in the mill can be adjusted to optimise the grinding process.

*The advantages of wet grinding include:*
- **Improved particle size distribution**: Wet grinding often produces finer particles compared to dry grinding, which can be beneficial for certain applications that require a narrow particle size distribution.
- **Reduced dust and heat generation**: The presence of a liquid medium in wet grinding helps to suppress the dust formation and dissipate the heat generated during the grinding process, making it a potentially safer and more controlled operation.
- **Enhanced mixing and homogenisation**: Wet grinding enables efficient mixing and blending of ingredients, allowing for better dispersion of additives and achieving a more uniform product.
- **Minimised wear and contamination**: The liquid medium in wet grinding can act as a lubricant, reducing wear on the grinding media and equipment. It can also help to minimise contamination from abrasion or airborne particles.

*However, wet grinding also has a few limitations and considerations:*
- **Higher energy consumption**: Wet grinding typically requires more energy compared to dry grinding due to the need to overcome the viscosity of the liquid medium.
- **Equipment design**: Wet grinding may require specialised equipment that can handle the presence of liquids and operate efficiently in wet environments.
- **Material compatibility**: Not all materials are suitable for wet grinding. Some substances may react with the liquid medium or exhibit poor grinding performance, requiring alternative processing methods.

It is important to note that wet grinding may not be suitable for all materials and applications. Some materials may be sensitive to moisture or require specific drying processes after wet grinding. Additionally, the choice between wet and dry grinding methods depends on various factors, and it is essential to consider the specific requirements and characteristics of the material being processed.

## 9.4.2 Thermochemical devulcanisation

Thermochemical devulcanisation refers to a process that uses heat and chemical reactions to break down vulcanised rubber, such as in used tyres or rubber products, into its constituent components. Thermochemical devulcanisation involves breaking the sulphur cross-links in the vulcanised rubber, allowing the rubber to regain its elastic properties, and making it suitable for reuse. This process typically involves heating the rubber to high temperatures in the presence of a devulcanising agent or additive. The devulcanising agent helps to break the sulphur bonds and promote the reversion of the rubber to its original state.

*The process typically involves the following steps:*
- **Grinding**: The vulcanised rubber is first shredded or ground into smaller pieces to increase the surface area for subsequent reactions.
- **Pyrolysis**: The ground rubber is then subjected to high temperatures in an oxygen-free environment through a process called pyrolysis. Pyrolysis breaks down the rubber into its constituent compounds, including hydrocarbons and other volatile materials. This step is often carried out in a reactor or furnace.
- **Chemical treatment**: Chemical agents, such as solvents or reactive substances, may be added to the pyrolysis process to aid in the devulcanisation process. These chemicals help to break the sulphur cross-links and other chemical bonds within the rubber, facilitating the separation of the rubber into its original components.
- **Recovery and purification**: After devulcanisation, the resulting mixture contains various components, including liquid hydrocarbons, gases, and residual solids. These components are separated and purified using techniques such as distillation, filtration, and solvent extraction.
- **Further processing**: The recovered components can be further processed and used in various applications. The liquid hydrocarbons obtained from devulcanisation can be used as fuels or as feedstock for the production of new rubber or other materials. The residual solids can be used in construction materials or as fillers in other rubber products.

The use of 2-mercaptobenzothiazole disulfide (MBTS) and tetramethyl thiuram disulfide (TMTD) as devulcanising agents in the presence of aromatic and aliphatic oils to devulcanise ethylene-propylene-diene monomer (EPDM) rubber powder from discarded automotive parts in an industrial autoclave is a common approach to rubber recycling [56]. This method involves using high-pressure steam and heat to break down the rubber and then adding chemical agents to facilitate the devulcanisation process.

### 9.4.3 Thermomechanical devulcanisation

Devulcanisation is the reverse process of vulcanisation, and it aims to break the cross-links in the rubber to restore its original properties. Thermomechanical devulcanisation combines heat and mechanical action to achieve this. A general overview of the process is given below:

**Shredding or grinding**: The vulcanised rubber waste is shredded or ground into smaller pieces, increasing its surface area and facilitating subsequent processing.

**Heating**: The shredded rubber is subjected to high temperatures, typically in the range of 150–220 °C (300–430 °F). Heat softens the rubber and weakens the sulphur cross-links, allowing for further mechanical action.

**Mechanical action**: Various mechanical techniques are employed to promote the breakdown of cross-links and dispersion of the rubber components. This can include processes such as extrusion, milling, or shear mixing. The mechanical action exposes the rubber to shear forces and deformation, assisting in the devulcanisation process.

**Chemical additives (optional)**: In some cases, chemical additives may be used during the devulcanisation process to further facilitate the breakdown of cross-links. These additives can include reactive agents or plasticisers that help to disrupt the sulphur bonds and improve the efficiency of devulcanisation.

**Cooling and collection**: Once devulcanisation is complete, the rubber is cooled and collected for further processing. The resulting material can be used as a raw material in the production of new rubber products or mixed with fresh rubber compounds to improve properties like elasticity and strength.

In the thermomechanical devulcanisation process, mechanical milling of scrap waste rubber products is performed at elevated temperatures. This process is employed to break down the cross-linking bonds present in the rubber, which are formed during vulcanisation. During the thermomechanical devulcanisation process, both the main chain and cross-link bonds in the rubber undergo simultaneous breaking. This leads to a significant breakdown in the molecular weight of the rubber.

In the mechanical devulcanisation process of vulcanised natural rubber (NR) sheets, milling was conducted using a two-roll mixing mill at approximately 80 °C. The purpose of this process is to break down the cross-link bonds in the vulcanised NR and reclaim the rubber material for reuse [57]. If the Mooney viscosity of the devulcanised rubber is very high, exceeding 200 (i.e., out of scale), it indicates that the plasticity of the rubber compound is indeed very low. This is typically due to the presence of a higher percentage of cross-linked precursors in the devulcanised rubber.

According to the results shown in Table 9.1, mixing with fresh rubber can improve materials with poor physical attributes and processing characteristics. Furthermore, from the mechanical performance of the vulcanisates from Table 9.1, it is clear that 25% replacement of fresh rubber by mechanically devulcanised rubber showed slightly superior properties than that of the control vulcanisate.

Mechanical devulcanisation of waste rubber is also performed through twin-screw extruder. The rubber network gets shredded down in this process due to the intense shearing action, making it easy to reprocess the resultant material. The benefits of this procedure over the alternative mechanical devulcanisation approach include consistency, high speed, and higher efficiency. Processing parameters like temperature profile, screw rotation speed, and screw configuration mostly determine the quality of devulcanised rubber developed by this procedure [58, 59]. The devulcanisation of the tread section of waste tyres using a corotating twin-screw extruder is a process that aims to break down the cross-linked rubber molecules and restore their elasticity for potential reuse.

The extent of devulcanisation can be influenced by various factors, including barrel temperature and screw speed [60]. The properties of devulcanised rubber, including

**Table 9.1:** Physical and mechanical properties of devulcanised rubber and fresh rubber/devulcanised rubber blend.

| **Mix formulation** | **1** | **2** | **3** | **4** | **5** |
|---|---|---|---|---|---|
| Natural rubber (NR) | 100 | – | 75 | 50 | 25 |
| Devulcanised rubber | – | 100 | 25 | 50 | 75 |
| Zinc oxide | 3 | 2 | 2.3 | 1.5 | 0.8 |
| Stearic acid | 1.5 | 1 | 1.1 | 0.8 | 0.4 |
| Carbon black (N330) | 40 | – | 30 | 20 | 10 |
| Napthenic oil | 4 | – | 3 | 2 | 1 |
| Sulphur | 5 | 2 | 2.8 | 2.5 | 2.3 |
| CBS | 1.6 | 0.8 | 0.8 | 0.8 | 0.8 |
| **Physical properties** | | | | | |
| Mooney viscosity ML (1 + 4) 120 °C | 20 | > 200 | 28 | 29 | 49 |
| Cross-link density (mmol/kg) | 69.1 | 64.4 | 34.7 | 41.1 | 48.0 |
| **Mechanical properties** | | | | | |
| 300% modulus (MPa) | 18.9 | 14.3 | 11.1 | 11.6 | 12.9 |
| Tensile strength (MPa) | 25.4 | 19.2 | 26.7 | 25.6 | 23.5 |
| Elongation at break (%) | 410 | 390 | 550 | 500 | 460 |
| Tear strength (kN/m) | 90 | 48.8 | 79.1 | 75.1 | 58.8 |
| Compression set (%) | 52 | 58 | 69 | 77 | 79 |
| Hardness (Shore A) | 70 | 70 | 65 | 67 | 70 |

CBS, *N*-Cyclohexyl-2-benzothiazole sulfenamide.

the percent of devulcanisation, can be influenced by various factors such as cross-link density, sol fraction, barrel temperature, and screw speed and are depicted in Table 9.2. It appears that Table 9.2 shows the influence of screw speed on the percent of devulcanisation at different barrel temperatures. The data suggests that the screw speed has a significant impact on the devulcanisation process, particularly at low and middle levels of barrel temperature. Furthermore, the highest value of percent devulcanisation achieved in the study was 88%. This maximum value was obtained when the screw speed was set at 12 rpm and the barrel temperature was maintained at 220 °C. When the screw speed is increased at a relatively low barrel temperature, it results in a higher rate of shear and mixing within the rubber.

The increased shear stress facilitates the breaking of sulphur cross-links and promotes the devulcanisation process. As evident from Table 9.2, increase in screw speed at low barrel temperature can also increase the sol content of devulcanised rubber due to the same reason discussed earlier. In a rheometric study, the Δtorque value represents the difference between the maximum torque observed for a revulcanised sample and the maximum torque observed for a devulcanised sample.

Table 9.2 shows the effect of Δtorque versus screw speed at different barrel temperatures, and the values are remarkably comparable. A very low Δtorque value typi-

**Table 9.2:** Devulcanisation conditions and properties of devulcanised samples.

| Devulcanisation condition | | | Properties of devulcanised tyre waste | | | |
|---|---|---|---|---|---|---|
| **Sample** | **Barrel temperature (°C)** | **Screw speed (rpm)** | **Percent of devulcanisation** | **$\Delta_{torque}$ (lb.in)** | **Sol fraction (%)** | **Cross-link density ($\times 10^4$ mol/cm$^3$)** |
| ref | – | – | – | – | – | 1.6 |
| 1 | 220 | 30 | 65 | 48.4 | 26.6 | 3.5 |
| 2 | 220 | 60 | 75 | 43.58 | 35.67 | 3.1 |
| 3 | 220 | 90 | 70 | 49.61 | 33.78 | 6.98 |
| 4 | 220 | 120 | 88 | 37.97 | 44.67 | 4.14 |
| 5 | 250 | 30 | 65 | 36.39 | 37.67 | 2.85 |
| 6 | 250 | 60 | 60 | 25.9 | 33 | 2.91 |
| 7 | 250 | 90 | 65 | 41.2 | 33.23 | 2.21 |
| 8 | 250 | 120 | 83 | 52.4 | 30 | 4.13 |
| 9 | 280 | 30 | 75 | 45.23 | 37.63 | 4.06 |
| 10 | 280 | 60 | 75 | 35 | 37.42 | 5.34 |
| 11 | 280 | 90 | 80 | 102.2 | 41 | 6.73 |
| 12 | 280 | 120 | 85 | 28.2 | 44.15 | 2.1 |

cally indicates that the primary scission mechanism is the cleavage or breakage of polymer chains (main chain scission); on the other hand, a moderate Δtorque value suggests the occurrence of both main chain scission and cross-link scission during devulcanisation. The devulcanisation of ground tyre rubber (GTR) and thermoplastic elastomer (TPE) mixtures can be achieved through a melt extrusion process. In this process, high shear stress is generated by increasing the screw speed and reaction temperature [61].

In the devulcanisation process using a corotating twin-screw extruder, two different types of GTR, namely GTR-A (with a particle size of approximately 30 mesh) and GTR-C (with a particle size of around 20 mesh), are used at the screw speed range of 400–1,200 rpm at temperature 180–270 °C. In the subsequent step of the process, the devulcanised GTR (DGTR) is mixed with EPDM. The gel content of the devulcanised blend is then determined, considering the variation in screw speed and extrusion temperature as depicted in Figure 9.12(a) and (b), respectively. Based on the results obtained, it appears that the gel content of both GTR-A and GTR-C decreases with increasing screw speed at a constant reaction temperature of 240 °C. The minimum value of the gel content is achieved at 1,000 rpm and then marginally increases at 1,200 rpm; furthermore, with a constant screw speed of 1,000 rpm and rising reaction temperature, the gel content of the devulcanised blend continuously decreases.

It appears that the Mooney viscosity of the devulcanised blend is higher when measured at lower screw speeds or lower reaction temperatures compared to measurements taken at higher screw speeds or higher reaction temperatures. Moreover, the Mooney viscosity of the devulcanised blends containing GTR-A was considerably higher than that of GTR-C. Increasing the screw speed in a twin-screw extruder increases the shear stress on the reaction mixture, leading to higher shear rates and a higher degree of mechanical work on the polymer chains. This can facilitate main chain and/or cross-link scission, which can lead to a decrease in the gel content and Mooney viscosity of the devulcanised blends.

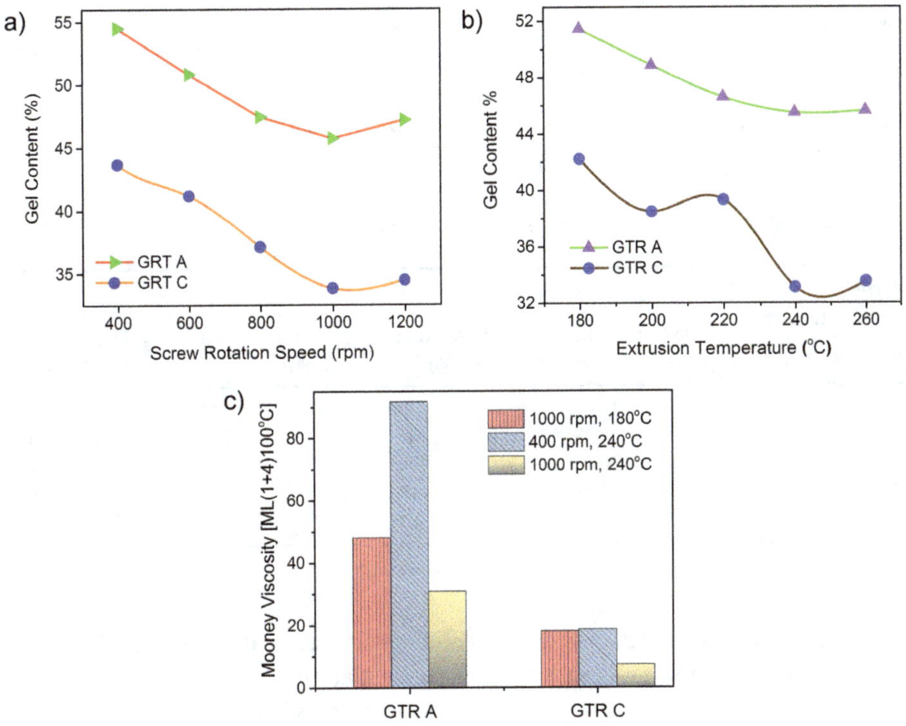


**Figure 9.12:** (a) Influence of screw rotation speed on gel content of devulcanised rubber; (b) influence of extrusion temperature on gel content of devulcanised rubber; and (c) the Mooney viscosity of devulcanised rubber as a function of screw speed and extrusion temperature.

Tables 9.3 and 9.4 respectively, show the effects of screw speed and extrusion temperature on the mechanical characteristics of the styrene–butadiene rubber (SBR)/DGTR/EPDM vulcanisate, such as tensile strength and elongation at break. Investigations showed that at 1,000 rpm and 220 °C, the revulcanised SBR/DGTR-A/EPDM vulcanisate had tensile strength and elongation at break that was significantly higher than those of the revulcanised SBR/DGTR-C/EPDM vulcanisate.

In a twin-screw extruder, SBR-based GTR and EPDM blends were thermomechanically devulcanised using subcritical water as the reaction medium [62]. The ideal processing parameters for devulcanisation in subcritical water were 1,000 rpm screw speed, 200 °C processing temperature, and 1.6 MPa pressure. According to the results,

**Table 9.3:** Effect of screw speed on tensile strength and elongation at break of the revulcanisate.

| Screw speed (rpm) | GTR-A | | GTR-C | |
|---|---|---|---|---|
| | Tensile strength (MPa) | Elongation at break (%) | Tensile strength (MPa) | Elongation at break (%) |
| 400 | 15.77 | 425 | 17.84 | 431 |
| 600 | 17.45 | 443 | 17.63 | 423 |
| 800 | 17.68 | 452 | 17.92 | 425 |
| 1,000 | 18.89 | 488 | 18.58 | 447 |
| 1200 | 17.97 | 444 | 17.13 | 542 |

GTR-A: particle size ~30 meshes; GTR-C: particle size ~20 meshes.

**Table 9.4:** Effect of barrel temperature on tensile strength and elongation at break of the revulcanisate.

| Barrel temperature (°C) | GTR-A | | GTR-C | |
|---|---|---|---|---|
| | Tensile strength (MPa) | Elongation at break (%) | Tensile strength (MPa) | Elongation at break (%) |
| 180 | 19.61 | 386 | 18.07 | 373 |
| 200 | 20.03 | 492 | 18.31 | 441 |
| 220 | 20.18 | 521 | 18.25 | 435 |
| 240 | 18.91 | 487 | 18.53 | 446 |
| 260 | 18.00 | 428 | 17.34 | 485 |

GTR-A: particle size ~30 meshes; GTR-C: particle size ~20 mesh.

subcritical water as a swelling agent promotes the devulcanisation reaction, increases the likelihood of cross-link breakage, protects the material from oxidative deterioration, reduces the size of the gel particles in the devulcanised blends, and significantly improves the mechanical properties of the revulcanisates.

An independently designed twin-screw extruder was used to devulcanise GTR at various barrel temperatures and screw speeds [63]. According to the outcomes, the cross-link density and Mooney viscosity of devulcanised rubber continuously drop as reaction temperature rises from 230–270 °C at a certain screw speed of 30 rpm, while the sol content steadily rises, as shown in Table 9.5.

On the other hand, a completely opposite trend of cross-link density, sol content, and Mooney viscosity of devulcanised GTR was observed when devulcanisation was performed at variable screw speed, 15–35 rpm at a particular barrel temperature, 250 °C. The reason for these different trends of behaviours of devulcanised GTR under dissimilar processing conditions was predominated by the scission of cross-link bonds. Based on the results of monitoring the extent of devulcanisation of GTR and evaluating the mechanical properties of revulcanised DGTR, the optimum processing conditions for the independently developed twin-screw extruder used were determined to be a barrel temperature of 250 °C and a screw speed of 30 rpm.

**Table 9.5:** Effect of processing parameter on sol content, cross-link density, and Mooney viscosity of the devulcanised GTR, respectively.

| Screw speed (rpm) | Barrel temperature (° C) | Sol content (%) | Cross-link density $\times 10^{-5}$ (mol/cm$^3$) | Mooney viscosity ML (1 + 4) 100 °C |
|---|---|---|---|---|
| 30 | 230 | 17.79 | 5.99 | 97.2 |
| | 240 | 18.38 | 5.80 | 85.1 |
| | 250 | 22.01 | 5.40 | 60.2 |
| | 260 | 23.77 | 5.19 | 42.1 |
| | 270 | 26.36 | 4.62 | 30.9 |
| 15 | 250 | 28.73 | 4.89 | 37.8 |
| 20 | | 26.59 | 5.01 | 41.7 |
| 25 | | 25.38 | 5.30 | 47.9 |
| 30 | | 21.88 | 5.41 | 60.9 |
| 35 | | 18.68 | 5.59 | 64.9 |

Under these circumstances, the tensile strength and elongation at break of revulcanised DGTR were 11 MPa and 370%, respectively.

The characteristics and the reprocessing ability of thermomechanically devulcanised GTR revealed that different material structure is obtained depending on the processing condition of devulcanisation in a twin-screw extruder [64]. The highest values of tensile strength (12.9 MPa) and elongation at break (360%) were achieved at specific processing conditions during devulcanisation. These conditions involve a screw speed of 100 rpm and a barrel temperature of 180 °C. The gel permeation chromatography (GPC) analysis of the devulcanised GTR revealed that the extent of devulcanisation had a significant impact on the molecular weight of the resulting material.

The researcher observed that at low levels of devulcanisation, the cross-linked network of the GTR was only partially destroyed, resulting in the formation of high-molecular-weight material. Yet, when the barrel temperature was excessively raised during devulcanisation, both the main chain and cross-linked network were broken down, leading to a decline in molecular weight. The devulcanisation of NR industry wastes in an industrial twin-screw extruder is a promising approach for recycling and reusing rubber waste. The process involves subjecting the rubber waste to various barrel temperatures within the range of 80–220 °C [65]. The effects of input temperature and self-heating on the extent of devulcanisation were studied in the context of the mechanical shearing of rubber during the devulcanisation process.

The input temperature refers to the initial temperature at which the rubber material is introduced into the devulcanisation system, while self-heating refers to the temperature rise that occurs within the rubber due to the friction generated during mechanical shearing. The author's findings suggest that the maximum number of selective S–S bond scissions, a crucial step in devulcanisation, occurred at lower input temperatures of 80 and 100 °C. This result is attributed to the self-heating phenomenon during the devulcanisation process.

The material self-heating phenomenon is indeed an important parameter in thermomechanical devulcanisation processes. It plays a significant role in optimising the processing parameters for devulcanisation. The thermo-mechanical devulcanisation of cryogenically ground EPDM rubber vulcanisate was performed in two stages: initially in a two-roll mixing mill at 210 °C for 25 min, followed by further processing in an internal mixer at 200 °C for 15 min [66]. The study demonstrates that both sulphur- and peroxide-cured EPDM vulcanisate can be devulcanised, with the peroxide-cured sample showing slightly higher efficiency in terms of cross-link density reduction.

Sulphur-cured EPDM sample experienced a decrease in cross-link density of 82.9%, whereas the respective decrease of peroxide-cured sample was 85.7% and the values are presented in Table 9.6. To evaluate the performance of devulcanised EPDM by using different forms of HDPE and EPDM (cured, uncured, and devulcanised), EPDM blend was prepared with varying rubber content such as 20, 40 and 60 wt%. The results of tensile behaviour did not show any systematic trend with various rubber contents.

**Table 9.6:** Cross-link density values of EPDM vulcanising systems.

| Sample | Cross-link density of EPDM vulcanisate (mol/cm$^3$) | Cross-link density of devulcanised EPDM (mol/cm$^3$) | Decrease in cross-link density (%) |
|---|---|---|---|
| Sulphur-cured EPDM | $3.86 \times 10^{-3}$ | $6.61 \times 10^{-4}$ | 82.9 |
| Peroxide-cured EPDM | $3.67 \times 10^{-3}$ | $5.23 \times 10^{-4}$ | 85.7 |

### 9.4.4 Mechanochemical devulcanisation

In mechanochemical devulcanisation along with mechanical energy different types of chemical devulcanising agents are used to enhance the degree of devulcanisation as well as to improve the quality of the devulcanised rubber. The researchers performed mechanochemical devulcanisation of waste rubber powder using thiobisphenols, specifically 4,4′-dithiobis(2,6-di-t-butylphenol). Devulcanisation process was conducted in an internal mixer at two different temperatures, namely 180 and 200 °C [67].

The measurement of extent of devulcanisation such as cross-link density, Mooney viscosity, and sol content of devulcanised rubber clearly specified that sol content increases with increasing thiobisphenol content and this increase was more pronounced at higher processing temperature as shown in Table 9.7.

The increasing proportion of thiobisphenol, decreases both the cross-link density and Mooney viscosity of the devulcanised rubber. This suggests that higher concentrations of thiobisphenol facilitated more efficient devulcanisation, leading to a reduction

**Table 9.7:** The effect of thiobisphenol content on extent of devulcanisation.

| **Mechanical properties** | | **Thiobisphenol content (g)** | | | | | |
|---|---|---|---|---|---|---|---|
| | | **0** | **0.1** | **0.3** | **0.5** | **1.0** | **3.0** |
| Mooney viscosity [ML(1 + 4)100 °C | 180 °C | 200.20 | 188.48 | 178.84 | 169.88 | 158.86 | 107.87 |
| | 200 °C | 109.486 | 97.727 | 93.577 | 85.276 | 76.976 | 59.684 |
| Cross-link density × $10^{-3}$ (mol/cm$^3$) | 180 °C | 1.726 | 1.774 | 1.694 | 1.645 | 1.573 | 1.089 |
| | 200 °C | 1.116 | 0.964 | 0.964 | 0.932 | 0.827 | 0.458 |
| Sol content (%) | 180 °C | 13.023 | 14.608 | 10.436 | 12.515 | 12.998 | 17.711 |
| | 200 °C | 13.730 | 16.508 | 16.012 | 19.286 | 18.194 | 26.131 |

in the cross-link density of the rubber when the processing temperature was 200 °C. In order to see the performance of devulcanised rubber, the curing characteristics, mechanical, thermal, and dynamic mechanical properties of the revulcanised rubber were examined. The degree of vulcanisation (MH-ML) measurement of the devulcanised rubber compound through rheometric study revealed several trends. Researchers found that as the content of the devulcanising agent and the temperature during the devulcanisation process increased, the corresponding degree of vulcanisation decreased.

Additionally, the curing time required for the devulcanised rubber to achieve a certain level of vulcanisation increased. This behaviour may be due to higher amount of main chain scission at high temperature and high concentration of devulcanising agent, leading to the dissipation of functional sites responsible for vulcanisation [68, 69]. The extent of devulcanisation also played a significant role in influencing the mechanical properties of the revulcanised rubber. The results of the study indicated that both tensile strength and elongation at break increased with an enhancement of the devulcanising agent content.

Furthermore, the highest values were achieved when the processing temperature was set at 180 °C. The storage modulus of the revulcanised rubber is an indicator of its stiffness and cross-link density. As a result, the revulcanisate with value at −100 °C is lower (4.37 GPa) in higher devulcanising agent content with respect to the value (4.95 GPa) of no devulcanising agent-containing revulcanisate.

By using ChCl-based DES, the researchers explored the potential of these solvents in enhancing the devulcanisation process. The choice of deep eutectic solvent (DES) compositions, such as ChCl:urea, ChCl:$ZnCl_2$, and $ZnCl_2$:urea, allowed for the investigation of different DES systems and their impact on the devulcanisation efficiency [70].

In the described process, GTR was mixed with each DES at different mass ratios of 1:20, 1:30, and 1:40. The total mixture was fixed at 50 g. Subsequently, the mixtures were subjected to sonication at a frequency of 37 kHz for an hour. After sonication, the mixtures were heated using a hotplate stirrer at 180 °C and 300 rpm with variable heating times of 5, 15, and 30 min. The results of the study indicated that a higher degree of devulcanisation was achieved with increased thermal exposure and higher mass ratios of rubber to DESs. This was attributed to a significant decrease in the cross-link density of the devulcanised rubber.

The Horikx plot, which will give an idea about main chain and cross-link scission suggested that selective scission of cross-link bond occurred, which was further supported by FTIR study through the disappearance of the C–S bond at 669 $cm^{-1}$ in the devulcanised GTR to the GTR. It is seen from FTIR spectrum that the C–S band intensity decreased with increasing heating time and the peak negligibly appeared at 30 min reaction time. Moreover, it is based on the extent of devulcanisation of GTR, ChCl: urea was identified as the most efficient DES. The development of a new low-temperature rubber reclaiming equipment (LTRE) and a devulcanising agent for rubber reclamation highlights the advancements in the field of rubber devulcanisation.

This new approach focuses on low-temperature mechanochemical devulcanisation (LTMD) of vulcanised rubber, and its results are compared to the traditional high-temperature atmospheric devulcanisation (HTAD) method [71]. The equipment consists of two high-shearing units, an internal mixer, and a twin-screw extruder. Indeed, the internal mixer and twin-screw extruder in the LTRE play crucial roles in the devulcanisation process and the efficient processing of waste rubber and devulcanising agent.

In this case, the devulcanising agent consists of three natural plant extracts, namely vegetable oil, vegetable ester, and an anti-re-cross-linking additive, along with some common additives used in rubber processing. Whereas, the twin-screw extruder complements the mixing process by providing extensive shearing and facilitates the devulcanisation, which can steadily extrude from the equipment. It is evident from the study that in the LTMD process with increasing reaction time, $M_n$ decreases gradually, whereas the devulcanisation ratio and Mooney viscosity show more pronounced changes. In contrast to the LTMD process, the study demonstrates that in the high-temperature atmospheric devulcanisation (HTAD) process, the devulcanisation ratio, Mooney viscosity, and Mn of the devulcanised rubber show slower variations with reaction time.

The percent retention of mechanical properties like 50% modulus, 100% modulus, tensile strength, elongation at break, and hardness (Shore A) of LTMD-revulcanised rubber and HTAD-revulcanised rubber were tested with respect to the original rubber vulcanisate and are depicted in Table 9.8. It is also worth mentioning that all mechanical characteristics of LTMD-revulcanised rubber are superior to that of the HTAD-revulcanised rubber. Thus, the performance evaluation and characterisation of LTMD-devulcanised rubber and HTAD-devulcanised rubber indicate that in the LTMD process, the cross-link scission predominates over the main chain scission whereas in the HTAD process, major scission of main chain occurs.

The mechanochemical devulcanisation for industrial waste for NR involves a specialized device that incorporates a triaxial compression reactor into a dynamic hydraulic universal testing machine with a heating chamber [72]. Here devulcanisation

**Table 9.8:** The mechanical properties of LTMD-revulcanised, HTAD-revulcanised, and original rubber vulcanisate.

| | 50% modulus (MPA) | 100% modulus (MPA) | Tensile strength (MPa) | Elongation at break (%) | Hardness (Shore A) |
|---|---|---|---|---|---|
| Original rubber vulcanisate | 2.57 | 5.98 | 23.3 | 388 | 76 |
| LTMD-revulcanised rubber | 1.7 (66%) | 3.4 (57%) | 16.1 (69%) | 336 (86%) | 67 (88%) |
| HTAD-revulcanised rubber | 1.6 (64%) | 3.6 (61%) | 11.6 (50%) | 330 (85%) | 63 (83%) |

LTMD, low-temperature mechanochemical devulcanisation; HTAD, high-temperature atmospheric devulcanisation.

of NR waste was carried out at different temperature in supercritical carbon dioxide ($scCO_2$) in the presence of diphenyl disulfide (DPDS) as devulcanising agent. It is claimed to be a very high degree of devulcanisation (~90%) of NR waste at 160 °C devulcanisation temperature.

Analysis of Horikx plot further suggested the exclusive cross-link scission with almost no polymer degradation and that was supported by the poor thermal behaviour of devulcanised rubber compared to that of the fresh NR vulcanisate. The devulcanisation of waste EPDM rubber using a laboratory intermeshing corotating twin-screw extruder in the presence of a chemical devulcanising agent called disulfide oil (DSO) at different concentrations (5 and 7 phr) and temperatures (220, 250, and 290 °C) is an interesting approach to reclaiming the properties of EPDM rubber waste [73].

The study of extent of devulcanisation clearly specified that increasing the DSO concentration at low temperature (220 °C) leads to cross-link density increase, which ultimately results in decreased sol fraction as well as degree of devulcanisation. The highest decrease in cross-link density and associated highest increase in sol fraction was achieved at high temperature and higher DSO concentration. Thus, highest temperature (290 °C) and maximum concentration of DSO (7 phr) favours devulcanisation reaction and causes maximum breakage of rubber network structure. Horikx plot further suggested that the devulcanisation reactions involve both main chain and cross-link scission. In order to see the performance of devulcanised EPDM, it was blended with virgin EPDM at two different proportions (20 and 40 wt%) and the curing characteristics of the rubber compounds and mechanical behaviour of the vulcanisates were measured.

The data demonstrated that 40 wt% of devulcanised rubber at some specific processing conditions (temp: 290 °C; DSO concentration: 7 phr; and screw speed: 120 rpm) had no adverse effect on the scorch time, optimum cure time, and rate of curing of the rubber compound. At the same time mechanical properties did not deteriorate and amazingly, tensile strength and elongation at break improved in most cases.

Rubber powder was mechanochemically modified via modifier 2,2'-dibenzothiazoledisulfde and high-shearing using a twin-screw extruder at 100 °C [74]. In this process, unmodified rubber powder was dried in a blast drying oven at 100 °C for 2 h and subsequently blended with 2,2′-dibenzothiazoledisulfde on a high-speed blender for 3 min. The mixture then prepared was extruded using a twin-screw extruder at 100 °C to obtain mechanochemically modified rubber powder.

Characterisation of the modified rubber powder clearly indicates that the mechanochemically modified rubber powder had a higher content of oxygen-containing groups on its surface compared to the unmodified rubber powder. The degree of swelling of the modified rubber powder was 1.5 times higher than that of the unmodified rubber powder. Additionally, the modification process resulted in a decrease in the initial degradation temperature and the maximum weight loss rate of the modified rubber powder.

Results suggest that the modified rubber powder exhibited improved surface activity while experiencing a decrease in cross-linking degree due to the combined effect of high shearing and chemical modification. The sol content of untreated rubber powder was only 5%, while that of chemically modified rubber powder increased by 33.1%.

Moreover, the cross-link density of untreated rubber powder was $6.6 \times 10^{-5}$ mol/cm$^3$ while that of modified rubber powder was reduced to $3.1 \times 10^{-5}$ mol/cm$^3$. The desulphurisation efficiency of chemically modified rubber powder was found to be 53.0%. Compared with untreated rubber powder/NR vulcanisate, the tensile strength of chemically modified rubber powder/NR vulcanisate was observed to increase from 12.87 to 16.87 MPa, while the improvement in tear strength and hardness was accompanied by an increase in the elongation at break from 341% to 447%. Thus, during extrusion, the thermal energy and shear forces promote the formation of active sites on the surface of the rubber powder.

These active sites can facilitate chemical reactions or interactions between the rubber powder and the NR matrix, leading to improved interfacial bonding. The presence of a free radical polymerisation reaction during the extrusion process and the subsequent cross-linking activation of the residual modifier during vulcanisation are additional factors that contribute to the formation of a cross-linking network in the system [75–77]. A study of mechanical properties clearly specified the suitability of using chemically modified rubber powder in fresh NR vulcanisate for product application.

## 9.4.5 Devulcanisation by supercritical carbon dioxide

Supercritical fluid has physicochemical characteristics between the states of liquid and gas. Supercritical fluid has benefits for chemical engineering processes, including low viscosity, high diffusivity, and high thermal conductivity [78, 79]. Moreover, by adjusting temperature and pressure, the physical properties can be controlled over a broad range. Since some polymer solids can be quickly penetrated by supercritical fluid, the applications of supercritical water, tetrahydrofuran *n*-butanol, and toluene have been studied for the decomposition of discarded rubbers [80–83]. In these investigations, the rubbers were heated above 300 °C in the presence of the supercritical fluids to break them down into low-molecular-weight hydrocarbon compounds that can be used as fuel.

Indeed, carbon dioxide ($CO_2$) is a commonly used supercritical fluid due to its advantageous properties. Here are some of the key characteristics of $CO_2$ as a supercritical fluid: (i) cost-effective, (ii) nontoxic and environmentally friendly, (iii) nonflammable, (iv) chemically inert, (v) mild critical conditions (critical temperature of 31 °C and a critical pressure of 74 kg/cm$^2$). Its unique properties make it an excellent option for a range of chemical engineering applications, including those related to the decomposition of discarded rubbers. One of the advantages of using $CO_2$ as a swelling solvent is that its removal from the system is relatively easy. At ambient temperature and pressure, $CO_2$ exists in a gaseous state. Furthermore, it was claimed that employing $scCO_2$ as a solvent was comparable to using common hydrocarbon solvents like toluene [84]. Therefore, it is anticipated that $scCO_2$ will be utilised to expand the rubber vulcanisates and to encourage a number of chemical processes in the rubber vulcanisates in order to make the rubber recyclable.

The NR vulcanisate was made by combining zinc oxide and NR (RSS#3): 5 phr, stearic acid: 2 phr, sulphur: 3 phr of *N*-cyclohexyl benzothiazyl sulfonamide (CBS): 1 phr in a Banbury mixer, then heat-pressed at 141 °C for 30 min, and then devulcanised in $scCO_2$ using DPDS at 180 °C under 10 MPa for 60 min [85].

The evaluation of the soluble fraction (sol fraction) of the devulcanised rubber has shown that the amount of reusable linear polymer increases with increasing $scCO_2$ pressure, particularly beyond the critical pressure. The molar mass of the resulting soluble (sol) component is typically in the range of tens of thousands, while the cross-linking density of the gel component decreases.

In the devulcanisation of rubber, it has been observed that the resulting devulcanised rubber may exhibit a slightly higher glass transition temperature ($T_g$) compared to the raw NR. This increase in $T_g$ is attributed to the addition of DPDS onto the polymer main chains during the devulcanisation process. The sol component obtained from the devulcanisation process of rubber, as identified by solid 13C-NMR spectroscopy, was found to have a main structure of *cis*-1,4-polyisoprene with the presence of approximately 7% trans-isomer.

The proposed cross-link cleavage reaction of the devulcanisation is illustrated in Figure 9.13. According to the proposed mechanism of devulcanisation, some DPDS molecules should be dissolved in $scCO_2$. The solvated molecules of $scCO_2$ can penetrate the vulcanised NR. DPDS can also cause cross-link cleavage in NR vulcanisates; this can occur when DPDS attacks the sulphur–sulphur (S–S) cross-link bond that holds the rubber matrix together. During the devulcanisation process, $scCO_2$ acts as a solvent that penetrates the rubber matrix and swells the vulcanisate, facilitating the breakdown of cross-links. The sulphur linkages within the rubber play a crucial role in determining the stability of the cross-links and the susceptibility to devulcanisation. Experimental findings have shown that NR vulcanisates containing predominantly mono-sulfidic linkages tend to exhibit higher resistance to devulcanisation compared to vulcanisates with di- or poly-sulfidic linkages. This observation is supported by theoretical considerations as well.

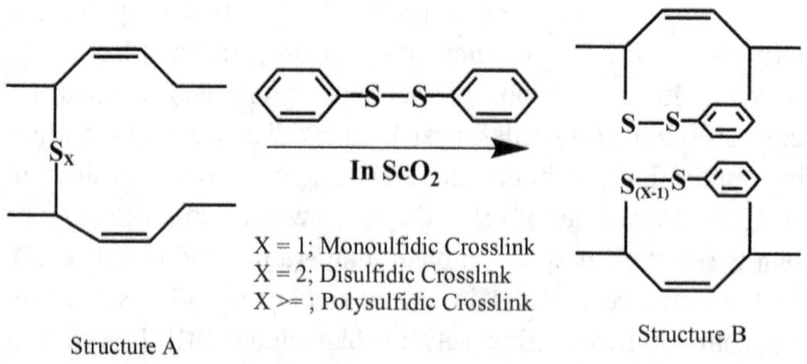


**Figure 9.13:** Possible cross-link cleavage reaction in the NR vulcanisate using diphenyl disulfide.

The recycling of industrial waste of NR in which $scCO_2$ acts as a solvent that facilitates the devulcanisation of the rubber waste in the presence of DPDS that serves as a devulcanising reagent, was carried out in the triaxial compression reactor, a unique equipment integrated into a heating chamber-equipped dynamic hydraulic universal testing machine [72]. Here, unique laboratory equipment as shown in Figure 9.14 was developed, which comprises of a cylinder and a piston that allows for the devulcanisation of rubber by applying triaxial compression to the rubber. This reactor is integrated into a dynamic hydraulic universal testing machine (100 kN) equipped with a heating chamber, which allows it to contain the rubber particles along with all necessary devulcanising agents in a controlled $CO_2$ atmosphere.

The testing device's compression loading raised the pressure of the gas surrounding the rubber particles, which convert $CO_2$ to supercritical state from the gaseous state. Using the Wave Matrix testing machine software, a 38 mm diameter piston moves to control the pressure inside the reactor. The piston's movement during unloading must be reducing to prevent rapid gas decompression, which may damage rubber seals and the investigated rubbers. Controlling parameters during the devulcanisation processes are $CO_2$ pressure, concentration of $CO_2$ content, temperature (typically up to 160 °C), rubber content, quantity of devulcanising agent, time, and compression loading and unloading steps.

The Horikx plot is a graphical representation used to evaluate the quality of devulcanisation of rubber. Figure 9.15(a) shows the results of the swelling test and the Horikx theoretical curves corresponding to the prepared devulcanised rubber. The figure demonstrates that every experimental point is close to the lower curve representing the cross-link bond scission. The selective breakdown of sulphur-sulphur cross-links and little to no polymer degradation indicate that the devulcanisation was

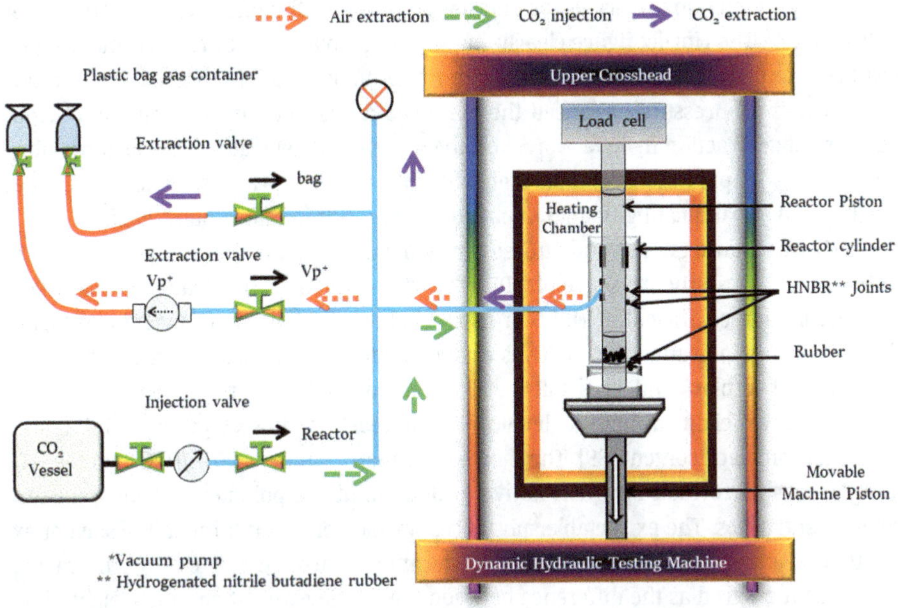


**Figure 9.14:** Devulcanisation device scheme.

successful. Experimental points are moved to the right side of the curve or to a larger amount of devulcanisation, when the temperature rises, as shown in Figure 9.15(b).

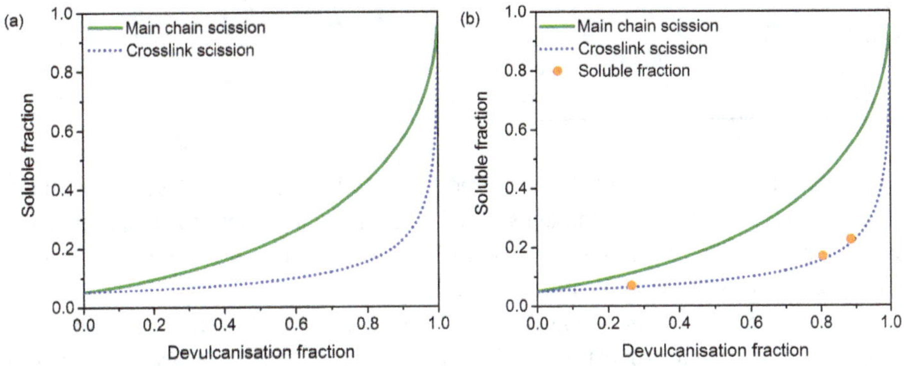


**Figure 9.15:** (a) Theoretical Horikx diagram of the studied NR and (b) theoretical and experimental soluble fractions as a function of the devulcanisation degree for the NR studied.

Waste tyre rubber (WTR) of average particle size 200–300 µm, was efficiently pulverized by $ScCO_2$ jet in the presence of DPDS (10 wt% of WTR) as devulcanising agent [86]. Here, WTR were allowed to swell in $scCO_2$ for 30–60 min before pulverisation followed by immersion in $scCO_2$ for another 60 min for efficient devulcanisation after pulverisation. The Horikx figure clearly showed that moderate jet pressure and a significant reaction time are favourable conditions to promote the selective scission of cross-links. Higher jet pressure enhanced the main chain scission, thus resulting in obvious decrease of gel fraction in GTR. Moreover, the decrease of gel fraction seems more obvious in $scCO_2$ jet pulverisation than that in devulcanisation practices in using $scCO_2$ as reaction solvent, indicating that jet impact tends to result in main chain scission.

GTR was characterised and subsequently devulcanised at 180 °C and pressure of 15 MPa, in $ScCO_2$ in the presence of DPDS (10 wt% of GTR) as devulcanising agent [87]. The distribution coefficient of DPDS in $scCO_2$ is about four orders of magnitude higher than in toluene indicating that $scCO_2$ is a much better swelling agent compared to toluene [88]. Carbon black and other fillers are commonly used in the production of commercial tyres; it was claimed that the devulcanisation reaction via $scCO_2$ was not influenced by the carbon black content [89]. The thermogravimetric analysis (TGA/DTGA) of the GTR samples was performed for quantitative evaluation of the polymer fraction, moisture content, and ashes. The extractable substances were determined through the solvent extraction (acetone followed by chloroform) technique through soxhlet and the carbon black was measured as the difference between the starting weight and the weight of the other fractions. The results of GTR composition are detailed in Table 9.9.

**Table 9.9:** GTR composition.

| Parameter | Value (wt%) |
|---|---|
| Polymer | 53 |
| Carbon black | 30 |
| Extractable | 9.4 |
| Moisture | 0.9 |
| Ashes | 7.2 |

Each of the parameters of the devulcanised GTR (T-GTR), as determined by the analysis of sulphur content, cross-link density, sol and gel fractions, and calculated molecular weight of the GTR and devulcanised GTR (T-GTR), were substantially different from that of the GTR as detailed in Table 9.10.

In order to see the performance of T-GTR, both GTR and T-GTR were mixed with virgin NR in different proportions and the compound formulations are illustrated in Table 9.11.

Each blend was mixed for 30 min in a two-roll mill to avoid the effect of milling on the ultimate properties of the vulcanisates. The stress–strain behaviour of various blend vulcanisates showed that GTR had worse mechanical properties than the reference blend with NR, as measured by moduli at 100% and at 300% of elongation, tensile strength, and elongation at break as a function of the reclaimed rubber content (phr).

**Table 9.10:** Characterisation of the GTR and devulcanised GTR (T-GTR).

| Parameter | GTR | T-GTR |
|---|---|---|
| Sol fraction (wt%) | 1.08 | 8.3 |
| Gel fraction (wt%) | 98.78 | 91.12 |
| Cross-link density × $10^{-3}$ (mmol/cm$^3$) | 0.082 | 0.037 |
| Sulphur % (wt%) | 2.29 | 2.69 |
| $M_n$ (Dalton) | 7,000 | 5,200 |
| PDI | 2.01 | 2.68 |

GTR, ground tyre rubber; T-GTR, devulcanised GTR; $M_n$, number average molecular weight; PDI, polydispersity index.

**Table 9.11:** Compound formulations of different blends.

| Formulation (phr) | Ref. NR | 5-GTR | 5 T-GTR | 10-GTR | 10 T-GTR | 20-GTR | 20 T GTR | 5E-GTR | 5TE-GTR | 10E-GTR | 10TE-GTR | 20E-GTR | 20TE-GTR |
|---|---|---|---|---|---|---|---|---|---|---|---|---|---|
| NR | 100 | 97.5 | 97.5 | 95 | 95 | 90 | 90 | 97.5 | 97.5 | 95 | 95 | 90 | 90 |
| GTR | 0 | 5 | 0 | 10 | 0 | 20 | 0 | 0 | 0 | 0 | 0 | 0 | 0 |
| T-GTR | 0 | 0 | 5 | 0 | 10 | 0 | 20 | 0 | 0 | 0 | 0 | 0 | 0 |
| E-GTR | 0 | 0 | 0 | 0 | 0 | 0 | 0 | 5 | 0 | 10 | 0 | 20 | 0 |
| TE-GTR | 0 | 0 | 0 | 0 | 0 | 0 | 0 | 0 | 5 | 0 | 10 | 0 | 20 |
| Carbon black | 37 | 35.5 | 35.5 | 34 | 34 | 31 | 31 | 35.5 | 35.5 | 34 | 34 | 31 | 31 |

T-GTR, devulcanised ground tyre rubber; E-GTR, acetone-extracted GTR; TE-GTR, acetone-extracted T-GTR; silicon dioxide, 15 phr; zinc oxide, 3.5 phr; stearic acid, 4 phr; waxes, 1.0 phr; sulphur, 1.3 phr; antioxidant-TMQ-6PPD (25–75%), 4.0 phr; CBS, 1.2 phr; resin: 0.3 phr.

Only T-GTR's break elongation was longer than the reference. Contrary to what was shown for moduli at 100% and at 300% of elongation, blends containing T-GTR had higher tensile strength and elongation at break than blends containing GTR. Using sc$CO_2$ as the reaction medium, DPDS was used to devulcanise waste tyre rubber [90]. The effects of various process parameters such as temperature, reaction time, pressure, and the concentration of devulcanising agents were investigated to study their impact on the devulcanisation of rubber. Interactions between these parameters were then demonstrated using a two-level full factorial experiment design.

The concentration of DPDS was identified as the most critical factor for devulcanisation. Experiments showed that when the concentration of DPDS was 12 g/L, a significant decrease in the cross-link density of the gel occurred within a short time (30 min) and at a relatively low temperature of 140 °C. Under such conditions the formation of a small sol fraction suggests that a large number of cross-link bonds in the rubber were broken. Still, at a temperature of 200 °C, the highest sol fraction was achieved in 240 min. This is because more main chains scission occurred as a result of the greater reaction temperature and longer reaction time, which increased the sol fraction.

The results indicated that the highest sol fraction was achieved at a pressure of 0.7 MPa. However, it was observed that under this specific pressure condition, the surface layer of the devulcanised rubber experienced degradation and exhibited a tendency to stick together. The poly dispersity index (PDI), which is shown in Table 9.12, was most significant even though the corresponding molecular mass of the sol was minimum, suggesting that substantial main chain scission took place during heterogeneous devulcanisation. The morphology of the devulcanised rubber remained mostly unchanged when the pressure applied was above 0.7 MPa. At these higher pressures, $scCO_2$ provided a homogeneous reaction environment. As a result of this homogeneous environment, the molecular mass of the sol fraction increased, while the polydispersity index (PDI) decreased.

**Table 9.12:** Molecular weight values of sol at various reaction pressures.

| Temperature (°C) | $M_n$ (g mol) | $M_w$ (g mol) | PDI |
|---|---|---|---|
| 140 | 38,000 | 55,000 | 1.42 |
| 200 | 19,000 | 31,000 | 1.61 |
| 220 | 20,000 | 52,000 | 2.62 |

$M_n$, number average molecular weight; $M_w$, weight average molecular weight; PDI, polydispersity index.

It was interesting to note that when the reaction pressure was lower than the critical value (7.38 MPa), the cross-link density of the gel was higher than that of the vulcanisate. This may be due to the low solubility and penetrability of carbon dioxide ($CO_2$) under certain conditions that can result in a reduced amount of DPDS penetrating into the crosslinked rubber vulcanisate during the devulcanisation process. As a result, the concentration of DPDS inside the vulcanisate remains relatively low, and the cross-link density of the gel fraction may increase. When the pressure was higher than the critical value, yet, more diphenyl disulfide permeated the vulcanisate. The devulcanisation process can occur both on the surface and inside the rubber simultaneously, resulting in a decrease in the cross-link density of the gel fraction. When devulcanising a larger bulk of vulcanisate using $scCO_2$, it is expected that a higher sol fraction can be achieved.

The investigation of devulcanisation in $scCO_2$ for NR-based carbon black-filled truck tyre vulcanisate, which includes additional components such as butadiene rubber (BR), aromatic oil, and antioxidants, is an important step in understanding the devulcanisation process and its potential applications in tyre recycling. The typical compound formulation represented in Table 9.13.

The devulcanisation of the truck tyre vulcanisate was carried out using $scCO_2$ at 180 °C under 10 MPa for 60 min. DPDS was used as the devulcanising agent to break down the sulphur cross-links in the rubber [89]. Then, based on the formulations, the

**Table 9.13:** Recipe for an NR-based truck tyre and recycled rubber vulcanisates made of virgin NR and devulcanised rubber materials in parts per hundred (phr).

| Sample code | Ini-TT | TT-re 20 | TT-re 40 | TT-re 60 |
|---|---|---|---|---|
| NR | 93 | 83.7 | 74.4 | 65.1 |
| BR | 7 | 6.3 | 5.6 | 4.9 |
| DR | 0 | 20 | 40 | 60 |
| HAF carbon black | 70 | 63 | 56 | 49 |
| Aromatic oil | 20 | 18 | 16 | 14 |

Ini-TT, vulcanised truck tyre; TT-re, recycled truck tyre amount (phr); NR, natural rubber; BR, polybutadiene rubber; DR, devulcanised rubber.
**Compound formulation:** Zinc oxide, 2 phr; stearic acid, 2 phr; antioxidant, 2 phr; DG (diphenyl guanidine), 0.3 phr; antioxidant (Sant Flex 6PPD), 2 phr; CBS (*N*-cyclohexyl-2-benzothiazole sulfenamide), 1.5 phr; sulphur, 1.5 phr.
HAF carbon black, high abrasion furnace carbon black.

so-grown devulcanised rubber was blended with virgin rubber in various proportions as shown in Table 9.13 and the compounded rubbers were vulcanised according to their respective optimum cure time ($t_{90}$) generated via rheograph and named as revulcanised rubber.

The tensile properties of revulcanised rubber made from virgin rubber and devulcanised rubber are reported in Table 9.14. The evaluation of tensile properties indicated that no deterioration of tensile properties was observed for 20% devulcanised rubber-containing vulcanisate. The modulus at 100% elongation of 40% and 60% devulcanised rubber-containing vulcanisates was considerably higher than those of the fresh rubber vulcanisates. For 40% devulcanised rubber-containing vulcanisates, 100% modulus was increased by 28% in comparison to that of the fresh rubber vulcanisate.

It is interesting to note that the hardness of all the devulcanised rubber/fresh rubber vulcanisates was higher compared to the fresh rubber vulcanisates. The addition of devulcanised rubber to a blend vulcanisate resulted in a slight decrease in its tensile strength because the devulcanised rubber contains low-molecular-weight rubber. The polymer structural changes that occurred during the devulcanisation process were thought to be the cause of the comparatively low tensile strength of the devulcanised rubber. Nevertheless, for up to 40 phr of devulcanised rubber loading in the vulcanisate, the tensile strength only decreased by around 10%.

In this study, discarded rubber products were modified by devulcanisation using $scCO_2$ and then subjected to cryogenic grinding using a fluidised-bed jet mill [91]. The findings showed that the coarse particles had a high specific breakage rate, which gradually rises with the degree of devulcanisation in each particle size range. Although there is little difference in the specific breaking probability for each particle size range for both original scrap rubber and devulcanised scrap rubber, the fine particles have a comparatively low specific breakage rate.

**Table 9.14:** Mechanical characteristics of the devulcanised NR-based truck tyre rubber/fresh rubber blended vulcanisate.

| Sample code | Ini-TT | TT-re 20 | TT-re 40 | TT-re 60 |
|---|---|---|---|---|
| 100% modulus (MPa) | 3.9 | 4.0 | 5.0 | 4.8 |
| Tensile strength (MPa) | 30.3 | 29.6 | 26.7 | 24.5 |
| Elongation at break (%) | 460 | 460 | 380 | 400 |
| Hardness (JIS K6301) | 62 | 67 | 70 | 69 |

Ini-TT, vulcanised truck tyre; TT-re, recycled truck tyre amount (phr).

Achieving the appropriate devulcanisation degree is essential for upholding efficient particle size reduction. The formation of fine product particles is restricted by the agglomeration of small particles, which rises with increasing devulcanisation degrees. It appears that for the large-size particles (>632 μm), abrasion is identified as the primary breakage mechanism for the original scrap rubber, while cleavage or fracture serves as the secondary mechanism. Cleavage or fracture refers to the breaking or splitting of particles into smaller fragments to achieve the desired particle size range.

In comparatively smaller particles in the size range of 158–224 μm, both abrasion and cleavage/fracture mechanisms are observed for both original scrap rubber and devulcanised scrap rubber during the grinding process. In comparatively smaller particles (158–224 μm), abrasion remains the key grinding mechanism. However, the cleavage or fracture breakage mechanism plays a more vital role in the size reduction of these particles compared to the largest particles. The devulcanisation treatment leads to the transformation of the grinding mechanism from abrasion to cleavage or fracture in the cryogenic grinding process.

In order to find the best devulcanising conditions of GTR by DPDS in $scCO_2$, the analysis of process variables was carried out just by focusing on one variable at a time (OVAT). In the OVAT technique, all of the factors that can impact how well the process performs are held at a fixed level except for one, which is changed until the ideal conditions are attained. It has been identified that temperature ($T$), pressure ($P$), amount of DPDS, and devulcanisation time ($t$) are the key variables that significantly affect the devulcanisation process.

These variables have been found to play a crucial role in reducing the cross-link density and increasing the sol fraction of the devulcanised rubber [92]. It appears that the treatment time has been identified as the least important factor, while the influence of pressure has been determined to be negligible. It has been proven that temperature is the most important factor in the devulcanisation process.

The findings indicate that high temperatures can lead to the degradation of the rubber network. A significant variation of response was observed with increasing temperature and at a high amount of DPDS. Actually, at high temperature the decomposition of DPDS generates more benzene sulfide radicalism leading to the chain scission and cross-link rupture, reducing the cross-link density and gel fraction and increasing the sol fraction [93–95]. These radicals react with the rubber chain and with the cross-linked network increasing the sulphur content of the devulcanised rubber.

### 9.4.6 Low-temperature devulcanisation

Low-temperature devulcanisation involves breaking the cross-links in the rubber using chemical or physical methods at low temperatures, typically around 100–150 °C. This method can be used to recover the rubber from waste tyres and other rubber products, which can be ground into a powder and then subjected to the low-temperature devulcanisation process.

Chemical methods for low-temperature devulcanisation involve the use of various agents, such as peroxides or free radicals, to break the cross-links in the rubber. Physical methods include grinding the rubber into a fine powder and then subjecting it to shear forces, such as through the use of a twin-screw extruder.

There are several low-temperature devulcanisation techniques currently being explored, including:

- Solvent swelling: This method involves the use of solvents to swell the rubber and break the sulphur cross-links. The swollen rubber can then be mechanically or chemically processed to recover the raw materials. The advantage of this method is that it can be performed at lower temperatures, reducing energy requirements.
- Microwave devulcanisation: Microwaves can be used to heat the rubber, selectively breaking the sulphur cross-links. The localised heating effect of microwaves allows for targeted devulcanisation while minimising energy consumption. This method is still under development and requires further optimisation.
- Mechanochemical devulcanisation: This approach involves the use of mechanical forces combined with chemical additives to break the sulphur cross-links. Mechanical energy is applied to the rubber, causing shear and compression forces that promote devulcanisation. Chemical additives facilitate the breakage of cross-links by reacting with the sulphur moieties.
- Enzymatic devulcanisation: Enzymes can be used to selectively break the sulphur cross-links in vulcanised rubber. Specific enzymes with rubber-degrading capabilities are used to catalyse the devulcanisation process. This method shows promise but is still in the early stages of development.

In recent years, there has been a growing focus on low-temperature devulcanisation techniques and the combination of devulcanisation with suitable functionalisation methods for waste rubber treatment. These approaches offer several advantages, including energy efficiency, reduced environmental impact, and the ability to recover high-quality rubber material. The application of hydrogen peroxide ($H_2O_2$) in combination with thermo-oxidative reclaiming and the use of soybean oil have gained attention as a potential method for the treatment of solid SBR vulcanisates at low temperatures [96].

Reclaiming was done in a drying oven in between 0.5 and 4 h. The oxidative reclamation process of solid SBR vulcanisates is represented in Figure 9.16. The results of the study indicate that the combination of soybean oil and hydrogen peroxide ($H_2O_2$) was effective in devulcanising the solid SBR vulcanisates. Specifically, the study found that when the SBR vulcanisates were treated with soybean oil (at 100 parts per hundred rubber, or phr) and a 3 wt% solution of hydrogen peroxide (at 2 phr) at a temperature of 100 °C, the vulcanisates were completely transformed into liquid rubber, with a sol fraction of 100%.

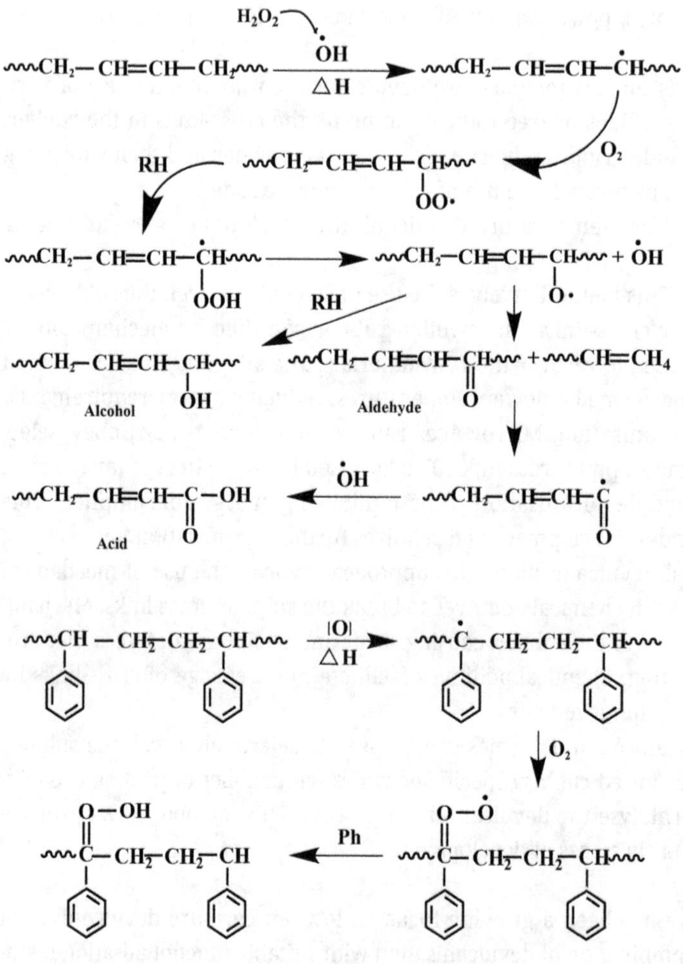

**Figure 9.16:** Schematic representation of oxidative devulcanisation process.

Mechanochemical modification of GTR uses a combination of cryogrinding, chemical treatment with hydrogen peroxide ($H_2O_2$) at room temperature for 3 h, and chemical treatment with inorganic acids ($H_2SO_4$, $HNO_3$, and their mixture) at 100 °C for 3 h [97].

GTR is generally a hydrophobic material due to its predominantly nonpolar rubber matrix. Conversely, through controlled oxidation, new functional groups can be introduced onto the GTR surface, including hydroxyl (–OH) and carboxyl (–COOH) groups. This controlled oxidation leads to the formation of polar functional groups, which enhance the hydrophilicity of the GTR surface. The stability of a water-GTR suspension can depend on several factors, including ambient grinding (Figure 9.17a), cryogenic grinding (Figure 9.17b), $H_2O_2$-treated (Figure 17c); $H_2SO_4$-treated (Figure 17d); $HNO_3$-treated (Figure 9.17e)- and $H_2SO_4/HNO_3$ mixture-treated (Figure 9.17f) and are presented in Figure 9.17. According to the results, only GTR treated with inorganic acids at 100 °C produced a stable suspension in water.

**Figure 9.17:** Hydrophilic characteristics of (a) ambient grinding; (b) cryo-ground; (C) $H_2O_2$-modified; (d) $H_2SO_4$-modified; (e) $HNO_3$-modified; and (f) $H_2SO_4/HNO_3$-modified GTR.

The suitable level of oxidation can indeed enhance the hydrophilicity of GTR, leading to improved dispersion of hydrophilic silica filler into the oxidised GTR-containing blend vulcanisate. This enhanced dispersion can also facilitate the incorporation of GTR into more hydrophilic matrices like polyurethanes, concretes, and other materials.

## 9.5 Microbial desulphurisation

Microbial devulcanisation refers to the process of breaking down or degrading vulcanised rubber using microorganisms. Microbial devulcanisation offers a potential solution by utilising microorganisms to degrade vulcanised rubber. Certain microorganisms, such as bacteria and fungi, have the ability to produce enzymes that can break the cross-linking bonds in vulcanised rubber, resulting in the restoration of its elastic properties.

Microbial devulcanisation of rubber is still a developing technology and faces several challenges that limit its industrial application. One of the main limitations is the low extent of devulcanisation that occurs mainly on the surface of the rubber [98–101]. The typical microbial desulphurisation reactor is displayed in Figure 9.18.

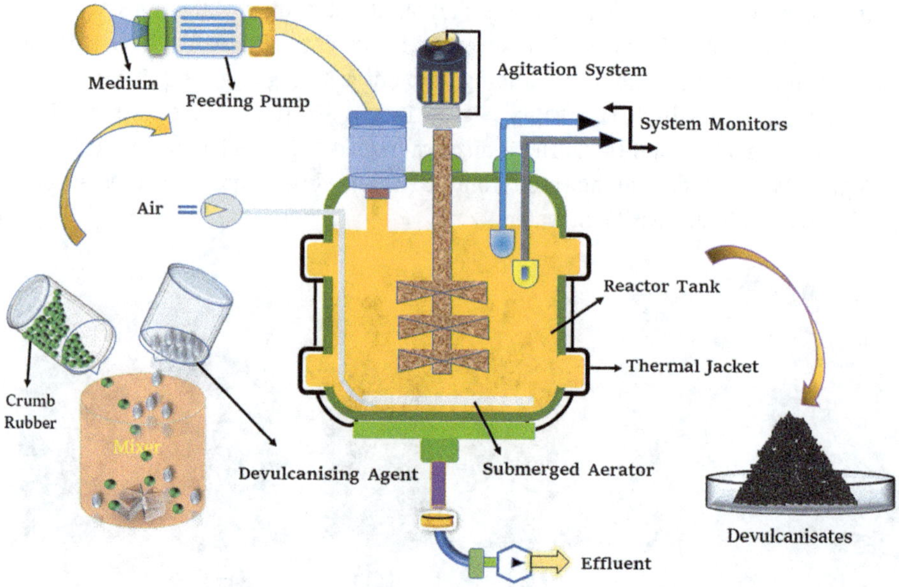


**Figure 9.18:** Microbial desulphurisation reactor.

In a typical microbacterial devulcanisation procedure, the chosen culture medium in the flask was first autoclaved at 115 °C for 30 min [102]. After cooling to room temperature, *Sphingomonas* sp., *Gordonia* sp., and their mixed consortium were placed in different flasks and incubated into the medium, respectively. The inoculums ratio of the single strain and the mixed strain was 10% (V/V), and the ratio for each strain was 5% (V/V). The bacterial growth on the GTR powder and small sheets was quite rapid in the first 24 h. To prepare the GTR material for further use, it was necessary to remove any other bacteria that may have been attached to it. To do this, the GTR powder and small sheets were immersed in 75% ethanol (v/v) for 24 h. The filtration process should be done under sterile conditions to prevent any contamination of the GTR material.

After adequate bacterial growth was achieved and the GTR powder and small sheets were detoxified as described earlier, they can be added to the medium for co-culturing with the bacteria in a rotary shaker under the described conditions – the cultivation temperature of 30 °C and stirring speed of 200 rpm/min, along with continuous monitoring of bacterial biomass. After the 10-day period of desulphurisation, the DGTR (detoxified ground tyre rubber) was filtered out; after filtration, the DGTR material washed thoroughly with distilled water and dried in a room-temperature environment. After the 10-day soaking period, the samples were removed from the culture medium and then treated with the same method as that used for the experimental group.

The results of the study indicate that after treatment by microorganisms, the swelling value of all three types of DGTR increased, while the cross-link density decreased. The study revealed that the $DGTR_M$ sample exhibited a 6.7% increase in swelling value and a 9.5% decrease in cross-link density compared to the GTR sample. The elemental analysis results of the rubber sheets before and after desulphurisation are shown in Table 9.15.

It was observed from the table that after treatment by microorganisms, the carbon (C) content on the surface of DGTR remained relatively unchanged. In contrast, the oxygen (O) content increased, while the sulphur (S) content decreased; the decrease in sulphur content indicates that the microorganisms, specifically *Sphingomonas* sp. and *Gordonia* sp., were capable of breaking the sulphur cross-links present on the GTR surface. When compared to GTR, the sulphur content of $DGTR_S$, $DGTR_G$, and $DGTR_M$ decreased by approximately 24.7%, 19.4%, and 32.4%, respectively. Additionally, the oxygen content is increased by approximately 19.4%, 20.2%, and 33.0%, respectively for $DGTR_S$, $DGTR_G$, and $DGTR_M$.

**Table 9.15:** Elemental analysis results of vulcanised and devulcanised rubber.

| Sample code | Carbon (%) | Oxygen (%) | Sulphur (%) |
|---|---|---|---|
| GTR[a] | 84.91 | 12.1 | 2.99 |
| $DGTR_S$[b] | 83.32 | 14.43 | 2.25 |
| $DGTR_G$[c] | 83.04 | 14.55 | 2.41 |
| $DGTR_M$[d] | 81.88 | 16.1 | 2.02 |

[a]Waste ground tyre rubber sample that immersed in culture conditions without inoculation.
[b]Waste ground tyre rubber sample desulphurized by *Sphingomonas* sp.
[c]Waste ground tyre rubber sample desulphurized by *Gordonia* sp.
[d]Waste ground tyre rubber sample desulphurized by mixed consortium –*Sphingomonas* sp. and *Gordonia* sp.

## 9.6 Ultrasonic devulcanisation

Ultrasonically assisted extrusion is yet another advanced technique [103]. It seems to be the most promising novel approach that applies the most to industrial-scale rubber reclamation. It is possible to explain how ultrasonic devulcanisation works. When rubber is subjected to ultrasonic treatment, the high-frequency vibration from the ultrasonic horn causes quick extension and contraction. This procedure has the potential to pulse bubbles around the rubber's impurities and voids. Therefore, chemical linkages, including cross-links, will be broken when strong stresses and strains are released. A diagrammatic representation of typical ultrasonic devulcanisation reactor is depicted in Figure 9.19.

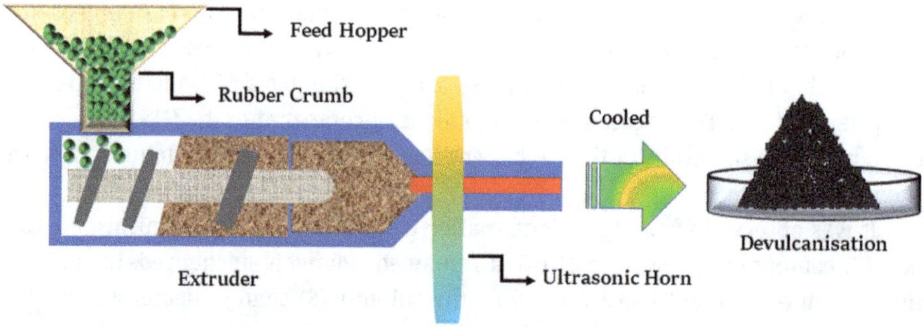


**Figure 9.19:** Ultrasonic devulcanisation reactor.

Ultrasonic waves can lead to cavitations surrounding impurities or voids in cured rubbers. Vibrations generate sufficient localised energy densities to remove cross-links produced by S–S and C–S bonds [104]. Devulcanisation and reprocessability of rubber are achieved in this manner. Since bond energy level of C–C bonds is greater compared to S–S bonds and C–S bonds, fewer C–C bonds are also broken during ultrasonic treatment. As a result, loss of mechanical properties and degradation are reduced. Comparing it to other ways of reclamation, the devulcanisation process just takes a few seconds, which is very efficient.

The insertion of the ultrasonic device in an extruder makes this technique a continuous process, which offers excellent potential for industrial applications. Since it does not require any chemicals, it makes this method clean and environmentally friendly. The mechanical characteristics of the revulcanisates are quite good, sometimes comparable to the virgin vulcanisates. Without the use of any virgin rubber, ultrasonically devulcanised rubber is capable of producing rubber with good mechanical qualities. Extensive studies of ultrasonic devulcanisation have been carried out on various rubbers, including GTR, NR, silicone rubber, SBR, EPDM, polyurethane rubber, and BR [105–111].

In 1995, Isayev et al. [103] published the first article on ultrasonic devulcanisation. The impact of ultrasonic amplitudes and flow rates was investigated, during the devulcanisation of SBR. The degree of SBR devulcanisation was assessed using cross-link density and gel fraction measurements. Increasing ultrasonic amplitudes and lowering flow rate resulted in better devulcanisation, as evidenced by the decreasing cross-link density and gel fraction.

For industrial-scale ultrasonic devulcanisation, GTR is the perfect raw material. Tyre rubber can be fed into an extruder in powder form. Commercially accessible GTR comes in a variety of mesh sizes, from 10 meshes to 200 meshes. Tyre rubber with bigger particle sizes often costs less than tyre powder with finer particle sizes. As a result, it is crucial to identify the powder particle size that is both acceptable and affordable for ultrasonic devulcanisation. Isayev et al. [112] studied the impact of different GTR particle sizes. They discovered that under continuous ultrasonic extrusion,

GTR of 30 meshes experienced higher devulcanisation than 10 meshes. It suggested that improved devulcanisation can result from finer particles.

In a corotating twin-screw extruder, Mangili et al. [113] studied the effects of amplitude, temperature, screw speed, and flow rate on the devulcanisation of GTR using ultrasonic waves. The cross-link density, sol content, gel content, viscosity, and extension strength of ultrasonically devulcanised rubber were measured as a function of devulcanisation time to determine the degree of devulcanisation. The modulus and tensile strength of the revulcanisates were also investigated, and the researchers concluded that the least amount of polymer degradation and the selective cleavage of S–S bonds occur at modest ultrasonic amplitudes.

In order to devulcanise polyurethane foam from seat cushions, an ultrasonic horn oscillation method [114] was recently developed. The outcome showed that after devulcanisation, the gel content, cross-link density, and glass transition temperature decreased by 25% compared to fresh polyurethane foam.

Hong and Isayev worked on combined ground NR and ultrasonically devulcanised NR with virgin carbon black-filled NR [115]. When compared to ground NR/virgin NR blends, the mechanical characteristics of devulcanised NR/virgin NR blends were much better.

This outcome clearly demonstrated the significance of the ultrasound's devulcanising effect is crucial when mixing with virgin rubber.

In order to devulcanise GTR and combine devulcanised GTR with NR, Mangili et al. used a variety of techniques, including supercritical $CO_2$, biological technology, and ultrasonic approaches. In comparison to supercritical $CO_2$ and biological technologies, they discovered that the ultrasonic approach is the most effective technology for controlling the characteristics of the vulcanisates [113].

# 9.7 Microwave devulcanisation

Microwave devulcanisation is a process that utilises microwave energy to break down the cross-linked structure of vulcanised rubber materials. In this process, rubber scraps or vulcanised rubber products are exposed to microwave radiation. The microwaves generate heat and selectively target the sulphur cross-links in the rubber. As a result, the cross-links are weakened or broken, allowing the rubber to regain its elasticity and processability. The application of microwave energy at the particular energy level is crucial for the success of the devulcanisation process. It allows for the selective cleavage of certain bonds in the rubber material. Microwave energy is carried out at a specific energy level.

This energy level is carefully chosen to be sufficient to break carbon–sulphur (C–S) bonds with a bond energy of 302 kJ/mol and sulphur–sulphur (S–S) bonds with a bond energy of 273 kJ/mol. Still, it is inadequate to break carbon–carbon (C–C) bonds with bond energy of 349 KJ/mol.

With this technique, tyre waste may be recycled without depolymerisation, and the material that is created can be recombined, revulcanised, and has physical qualities that are essentially identical to those of new rubber vulcanisate. In the microwave devulcanisation process, the presence of polar groups or components in the polymer is an important requirement for the efficient and effective devulcanisation of the rubber material. The presence of carbon black, which is common filler in rubber formulations, can make nonpolar rubbers more receptive to microwave energy. This characteristic allows for the use of microwave heating in the devulcanisation of waste tyre rubber [116, 117].

In a study, GTR was devulcanised in an adaptive microwave apparatus consisting of a domestic-type microwave oven (Samsung MW71E) and a stirring apparatus. The study involved the devulcanisation of GTR using domestic-type microwave oven. In this case, the Samsung MW71E model, was likely chosen for its availability and convenience [118]. In the mentioned study, the power or magnetron of the domestic-type microwave oven was set to a fixed value of 800 W. Additionally, 100 grams of GTR was taken in a 1,000 ml beaker, and stirring was performed at a speed of 6 rpm. The GTR sample was exposed to microwave irradiation for varying exposure times ranging from 1–5 min.

The devulcanised rubber obtained from the microwave devulcanisation process was mixed with fresh SBR at different proportions to create a new composite material. To compare the properties of the devulcanised rubber-based composite materials with untreated GTR/SBR composites, the curing characteristics, mechanical properties, and morphology of both materials were examined. The extent of devulcanisation was monitored through the measurement of sol content as a function of temperature and time of devulcanisation shown in Table 9.16. It is evident from the table that maximum sol content was achieved for exposure time of 4–5 min and temperature 370 °C.

**Table 9.16:** Sol content vs. time and temperature of devulcanisation.

| Sample code | Time (min) | Temperature (°C) | Sol content (%) |
|---|---|---|---|
| Nil | – | – | 2.3 |
| DV-R1 | 1 | 90 ± 10 | 10.7 |
| DV-R2 | 2 | 160 ± 10 | 13.1 |
| DV-R3 | 3 | 220 ± 10 | 18.5 |
| DV-R4 | 4 | 300 ± 10 | 25.4 |
| DV-R5 | 5 | 370 ± 10 | 34.8 |

Nil, untreated ground tyre rubber (GTR); DV-R, devulcanised rubber.

In order to evaluate the performance of devulcanised rubber, different proportions of devulcanised rubber were mixed with fresh SBR and carbon black along with rubber additives. The compound formulations are presented in Table 9.17. Formulation 1 is the control formulation, which did not contain any devulcanised rubber but formulations 2–4 contained 10–50 phr devulcanised rubber.

**Table 9.17:** Compound formulation of devulcanised rubber/styrene–butadiene rubber (SBR) composites.

| Ingredients (phr*) | 1 | 2 | 3 | 4 |
|---|---|---|---|---|
| SBR 1502 | 100 | 90.9 | 72.7 | 54.5 |
| Devulcanised rubber | 0 | 10 | 30 | 50 |
| Carbon black (N330) | 67.75 | 66.85 | 65.05 | 63.25 |

**Compound formulations:** zinc oxide, 3 phr; stearic acid, 1 phr; sulphur, 1.75 phr; CBS, 1.8 phr added in all formulations.
*phr stands for parts per hundred grams of rubber.

The cure behaviour of the composites is illustrated in Table 9.18. The study found that the addition of devulcanised rubber slightly increased the minimum torque for all the vulcanisates (4 and 5 min). In contrast, the addition of untreated GTR significantly increased the minimum torque of the vulcanisates. The slight increase in minimum torque with the addition of devulcanised rubber could be due to the interaction between the devulcanised rubber and the virgin SBR rubber matrix and the significant increase in minimum torque observed with the addition of untreated GTR could be due to the agglomeration of waste rubber particles in the SBR matrix.

Therefore, the processing of GTR-containing rubber compound is more difficult than the devulcanised rubber-containing compound [119]. The maximum rheometric torque decreased slightly with the addition of devulcanised rubber and the microwave exposure time. This decrease in maximum torque could be attributed to the lower cross-link density of the devulcanised rubber compared to the untreated GTR. The maximum rheometric torque increased significantly with the addition of 30 phr of untreated GTR. On the other hand, further addition of GTR beyond this point resulted in a decrease in the torque.

The existence of short rubber chains and cross-link precursors in the devulcanised rubber may be the cause of the decrease in the maximum rheometric torque value. This indicates that although microwave devulcanisation claimed only breaking of cross-link bonds, depolymerisation also takes place during the process, which results in the presence of short rubber chain in the devulcanised rubber. The scorch time ($t_{s2}$) of the composites decreased slightly either for devulcanised rubber or GTR-containing fresh vulcanisate. However, the $t_{s2}$ of SBR/devulcanised rubber compounds are lower than that of the GTR/devulcanised rubber compound, which indicates the presence of more unsaturation in the devulcanised rubber than that of the GTR.

The presence of active functional sites in the devulcanised rubber may also reduce the $t_{s2}$, which signifies the earlier starting of cross-linking reaction during vulcanisation.

As the concentration of both GTR and devulcanised rubber increased, optimum cure time ($t_{90}$) of the rubber compound decreased.

**Table 9.18:** Curing characteristics of devulcanised rubber/styrene–butadiene rubber (SBR) blend system.

| Sample code | Control | DV-R4 | | | DV-R5 | | | Untreated GTR | | |
|---|---|---|---|---|---|---|---|---|---|---|
| DV-R content | 0 | 10 | 30 | 50 | 10 | 30 | 50 | 10 | 30 | 50 |
| **Curing characteristics** | | | | | | | | | | |
| Maximum torque ($M_H$) dNm | 19.48 | 20.92 | 18.09 | 19.91 | 20.39 | 22.15 | 18.26 | 23.39 | 31.01 | 22.81 |
| Minimum torque ($M_L$) dNm | 10.80 | 11.89 | 12.02 | 13.22 | 11.88 | 12.92 | 11.95 | 13.46 | 18.64 | 16.71 |
| Optimum cure time ($t_{90}$) min | 3.32 | 2.90 | 2.95 | 2.20 | 3.01 | 1.97 | 1.82 | 2.98 | 1.88 | 2.37 |
| Scorch time ($t_{s2}$) min | 1.88 | 1.47 | 1.20 | 0.97 | 1.52 | 0.87 | 0.93 | 1.57 | 1.00 | 1.18 |

DV-R (number), devulcanised rubber (microwave exposure time, min).

Table 9.19 displays the mechanical performance of the SBR-based composites with various GTR and devulcanised rubber concentrations. The results showed that the vulcanisates become weak and brittle with increasing concentrations of GTR and devulcanised rubber. The mechanical properties of SBR/devulcanised rubber were far superior compared to that of the SBR/GTR vulcanisate. It is evident from the table that with the presence of 10 phr devulcanised rubber of GTR to the vulcanisate, a significant increase in the tensile strength and elongation at break was observed.

It is claimed that the GTR or devulcanised rubber's reinforcing effect was responsible for this growing value. Nonetheless, the tensile strength and elongation at break significantly decreased for 30 phr and 50 phr loading of either GTR or devulcanised rubber in the vulcanisate, while the corresponding values were better for devulcanised rubber-containing vulcanisates. Poor dispersion of GTR particles in the SBR matrix can contribute to weak sites for stress transmission, leading to inferior tensile strength and elongation at break in the SBR/GTR vulcanisate. Strong interfacial bonding between SBR and devulcanised rubber and fewer cross-linking sites in the devulcanised rubber, which transmit stress extremely well in the vulcanisate, are the causes of the improved mechanical performance of SBR/devulcanised rubber vulcanisate compared to SBR/GTR vulcanisate. With increasing microwave power exposure time, tensile strength and elongation at break increased.

Modulus at 100% and 200% elongation increased with increasing GTR content in the vulcanisate. The cross-linked network in the GTR cannot be deformed easily and when this material is dispersed in soft rubber material, a local stretching in the vulcanisate is developed, which exceeds the overall strain of the vulcanisate. Conversely, the devulcanised rubber is divided into two parts: a gel part with a greater modulus and a sol part that might be cross-linked with new SBR matrix. The combining effect of both sol and gel fraction of devulcanised rubber enhances 100% and 200% modulus SBR/devulcanised rubber vulcanisate. Longer microwave exposure times and higher

DV-R loading were correlated with increased composite hardness, which is responsible for the increase in the cross-link density of the composites.

**Table 9.19:** Mechanical properties of devulcanised rubber/styrene–butadiene rubber (SBR) blend system.

| Sample code | Control | DV-R4 | | | DV-R5 | | | Untreated GTR | | |
|---|---|---|---|---|---|---|---|---|---|---|
| DV-R content | 0 | 10 | 30 | 50 | 10 | 30 | 50 | 10 | 30 | 50 |
| **Mechanical properties** | | | | | | | | | | |
| 100% modulus (MPa) | 1.96 | 1.48 | 2.56 | 2.13 | 1.80 | 2.18 | 1.98 | 1.78 | 2.14 | 3.67 |
| 200% modulus (MPa) | 3.09 | 2.76 | 3.99 | 3.22 | 2.812 | 3.38 | 2.79 | 2.86 | 3.38 | 6.09 |
| Tensile strength (MPa) | 7.49 | 7.17 | 5.35 | 4.74 | 7.82 | 6.50 | 4.91 | 9.20 | 4.29 | 6.39 |
| Elongation at break (%) | 525 | 595 | 341 | 366 | 614 | 449 | 445 | 618 | 279 | 217 |
| Hardness (Shore A) | 57 | 58 | 64 | 66 | 56 | 67 | 67 | 57 | 64 | 64 |
| Cross-link density (mol/m$^3$) | 54.91 | 69.19 | 79.18 | 94.79 | 74.04 | 93.35 | 106.03 | 77.63 | 87.99 | 119.87 |

DV-R (number), devulcanised rubber (microwave exposure time, min).

Microwave electromagnetic energy was used for devulcanisation of WTR through selective breaking of C–S and S–S bonds of cross-linked rubber network [120]. The effectiveness of the devulcanisation treatment was analysed through FTIR study. The results demonstrated that the material experiences numerous structural changes during devulcanisation, which is confirmed by the formation of new chemical links and the disappearance of several peaks in the devulcanised WTR. The extent of devulcanisation was monitored through the measurement of swelling degree and degree of devulcanisation (Figure 9.20(a)) in the devulcanised rubber as a function of microwave energy.

According to the results obtained, as the treatment of microwave energy increases, the swelling degree of the material exhibits a sharp increase. At a specific energy level of 1.389 kJ/kg, the swelling degree reaches 63.22%. Beyond the critical energy level of 1.389 kJ/kg, the swelling degree of the material tends to stabilise. This means that to achieve a significant increase in the swelling degree beyond this point, a higher increase in the specific microwave energy is required. The figure also showed that degree of devulcanisation follows a similar trend – with the energy employed in the microwave treatment up to the threshold value of 1.389 kJ/kg, there is only a little improvement in the degree of devulcanisation observed. Specifically, the degree of devulcanisation approaches asymptotically to 95%. These results imply that even a relatively weak energy input is sufficient to accomplish the devulcanisation or regeneration process of the material.

In order to see the performance of devulcanised rubber, the GTR and the DGTR were then separately used to prepare epoxy-based composites. The mechanical performances such as stress at break, strain at break, and Young's modulus of epoxy/GTR and epoxy/DGTR are presented in Figure 9.20(b)–(d). The results showed that epoxy composites filled with DGTR have better mechanical properties than those filled with untreated GTR.

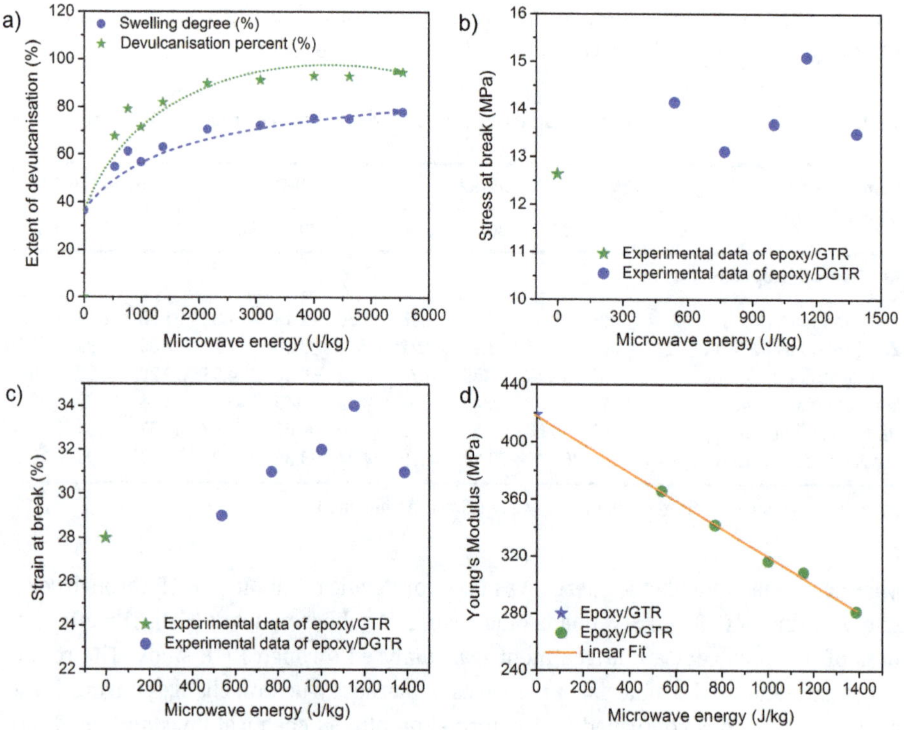


**Figure 9.20:** (a) Swelling degree and devulcanisation degree as a function of microwave energy; and the effect of microwave energy on (b) stress at break; (c) strain at break; and (d) Young's modulus of the composites.

The figure indicates that microwave treatment of GTR leads to an enhancement in the flexural stress and strain at the break of the composites. This suggests that the surface of the DGTR particles is more chemically reactive and exhibits better adhesion with the epoxy matrix compared to untreated GTR particles. This interfacial interaction between the epoxy matrix and DGTR particles is higher for the DGTR powder treated with 1,157 J/kg (750 W for 15 s) compared to untreated GTR particles. This treatment leads to an increase of approximately 20% in the flexural stress and strain at break of the epoxy/DGTR composite, compared to the epoxy/GTR composite. The schematic representation of microwave devulcanisation instrument is depicted in Figure 9.21.

GTR samples were devulcanised using different microwave exposure times to understand the physical and chemical changes occurred in the GTR structure [121]. At this point, GTR samples were recycled in a conventional microwave oven adapted with a motorised speed control stirring system. The devulcanisation process was conducted using different periods in which GTR samples were exposed to microwave for 3, 5, 6, and 7 min followed by homogenisation in an open two roll mill. Table 9.20 demonstrates the percentage of sol and gel content. The results indicated that longer mi-

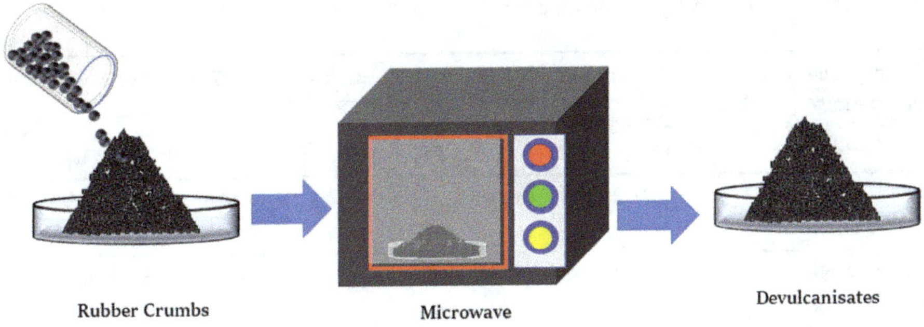


**Figure 9.21:** Schematic representation of microwave devulcanisation instrument.

crowave exposure times led to increase in sol fraction in the devulcanised rubber. The higher soluble fraction with microwave exposure time supports the modification of chemical structure of vulcanised elastomer due to breaking of C–S and S–S bonds, making the vulcanisate more soluble in the solvent.

**Table 9.20:** Percentage of gel and sol content.

| Sample | Microwave exposure time (min) | Gel content (%) | Sol content (%) |
|---|---|---|---|
| $GTR_V$ | 0 | 86.0 | 14.0 |
| GTR 3 | 3 | 83.5 | 16.5 |
| GTR 5 | 5 | 73.0 | 27.0 |
| GTR 6 | 6 | 71.2 | 28.7 |
| GTR 7 | 7 | 69.0 | 31.0 |

$GTR_V$, vulcanised GTR samples; GTR (number), ground tyre rubber (microwave exposure time, min).

A study of thermogravimetry analysis presented in Table 9.21 clearly indicated that the temperature at which the maximum weight loss rate ($T_{max}$) occurs during the devulcanisation process varies for both NR and SBR. Furthermore, the NR peaks of the most devulcanised samples shifted towards higher temperatures. Therefore, the presence of a higher amount of carbon black in the vulcanisate can affect the devulcanisation process by acting as a physical barrier to the release of volatiles.

To determine the glass transition temperature ($T_g$), the differential scanning calorimetry (DSC) method was employed of each devulcanised rubber, shown in Table 9.22. It can be seen from the table that two distinct glass transitions were observed, which were related to NR and SBR in the GTR, respectively. The results indicate that the $T_{g1}$ values, which represent the glass transition temperatures of the polymer samples, were lower for the most degraded samples. This decrease in $T_{g1}$ suggests that the polymer chains experienced a breaking of cross-links. When the cross-links break, the polymer

**Table 9.21:** Thermogravimetry parameters determined from TGA for vulcanised and devulcanised GTR.

| Sample code<br>Parameters | $GTR_V$ | | GTR 3 | | GTR 5 | | GTR 6 | | GTR 7 | |
|---|---|---|---|---|---|---|---|---|---|---|
| | $\Delta m$ (%) | $T_{max}$ (°C) | $\Delta m$ (%) | $T_{max}$ (°C) | $\Delta m$ (%) | $T_{max}$ (°C) | $\Delta m$ (%) | $T_{max}$ (°C) | $\Delta m$ (%) | $T_{max}$ (°C) |
| Oils and additives (~35–320 °C) | 10.3 | ~253 | 10.4 | ~253 | 10.1 | ~256 | 10.0 | ~256 | 8.8 | ~256 |
| Natural rubber (320–410 °C) | 28.4 | 381 | 27.9 | 382 | 25.4 | 384.6 | 22.5 | 387.5 | 19.4 | 391 |
| Synthetic rubber (410–520 °C) | 26.1 | 454.6 | 26.3 | 456.4 | 26.7 | 453.3 | 28.3 | 449.1 | 28.6 | 447.8 |
| Carbon black (520–600 °C) | 29.9 | 569.7 | 30.2 | 565.6 | 32.4 | 567 | 33.6 | 567 | 37 | 568 |
| Residues (600–800 °C) | 5.4 | – | 5.3 | – | 5.5 | – | 5.7 | – | 6.2 | – |

$GTR_V$, vulcanised GTR samples; GTR (number), ground tyre rubber (microwave exposure time, min).

chains become less interconnected, allowing for increased mobility. Although it was anticipated that the GTR7 samples would have the lowest $T_{g1}$ values, the $T_{g1}$ values were actually higher. This could be as a result of carbon black restricting segmental mobility and cyclic sulphur structures development in the polymer chains of the devulcanised SBR [122]. The DSC curves for the GTR5, GTR6, and GTR7 samples exhibited only a single glass transition, indicating that $T_g2$ could not be determined using this technique in these cases.

**Table 9.22:** Calorimetric glass transition ($T_g$) and transition width ($\Delta L$) of the samples.

| Sample code | Glass transition ($T_g$) | | Transition width ($\Delta L$) | |
|---|---|---|---|---|
| | $T_{g1}$ | $T_{g2}$ | $\Delta L_1$ | $\Delta L_2$ |
| $GTR_V$ | −57 | −46 | 5.7 | 8.1 |
| GTR 3 | −56 | −44 | 7.2 | 9.0 |
| GTR 5 | −65 | −44 | 10.5 | 11.4 |
| GTR 6 | −64 | – | 12.4 | – |
| GTR 7 | −62 | – | 15.7 | – |

$GTR_V$, vulcanised GTR samples; GTR (number), ground tyre rubber.

Later, a comprehensive analysis of the GTR's structure modifications through microwave-induced devulcanisation was conducted [123]. The temperature of the GTR samples increased with exposure time after microwave treatment, as indicated by the results presented in Table 9.23. This temperature increase can be attributed to the presence of carbon black in the GTR samples.

**Table 9.23:** The temperature of the GTR samples immediately after their exposure to microwaves.

| Sample | Temperature (°C) |
|---|---|
| GTR 0 | 87.8 |
| GTR 1 | 88.0 |
| GTR 2 | 87.9 |
| GTR 3 | 87.7 |
| GTR 4 | 86.3 |
| GTR 5 | 81.8 |
| GTR 5.5 | 73.8 |

GTR (number), ground tyre rubber (microwave exposure time, min).

According to the findings reported by Garcia et al. [121], the amount of carbon black present in each phase of the GTR material, specifically in NR and SBR, varied with the duration of microwave exposure. In the case of GTR, carbon black is primarily present in the NR phase. According to estimations, it was found that an amount of 45 phr of carbon black was sufficient to improve the devulcanisation efficiency of GTR. Additionally, the preferential location of carbon black in the NR phase raises the temperature more in this phase. The maximum degree of devulcanisation during microwave devulcanisation of GTR can be achieved in the NR phase. Monitoring the extent of devulcanisation can be done by measuring the gel content and cross-link density of the devulcanised rubber as a function of microwave exposure time and is presented in Table 9.24.

**Table 9.24:** Gel content and cross-link density of vulcanised and devulcanised rubber.

| Sample | Gel content (%) | Cross-linking density $v \times 10^{-5}$ mol/cm$^3$ |
|---|---|---|
| GTR 0 | 87.8 | 12.44 |
| GTR 1 | 88.0 | NA |
| GTR 2 | 87.9 | 12.65 |
| GTR 3 | 87.7 | 6.07 |
| GTR 4 | 86.3 | 5.52 |
| GTR 5 | 81.8 | 7.07 |
| GTR 5.5 | 73.8 | 6.07 |

NA: the cross-link density of GTR 1 is unavailable because it broke during the swelling in toluene.
GTR (number): ground tyre rubber (microwave exposure time, min).

Due to the greater amount of energy absorbed by the GTR, the results indicated that the longer the microwave treatment duration, the greater the degree of devulcanisation. Based on an FTIR study, a proposed mechanism for the formation of S–S (sulphur–sulphur) and S–O (sulphur–oxygen) bonds in the devulcanisation process of GTR was presented. This mechanism was proposed considering both the NR and SBR components present in GTR and detailed in Figures 9.22 and 9.23, respectively.

Due to the microwave exposure the S–S (sulphur–sulphur) bond between the rubber main chains can undergo scission, leading to chemical modifications within the material. The chemical modification of the rubber material occurs through the linking of the free radicals formed and the X where X is S or $O_2$.

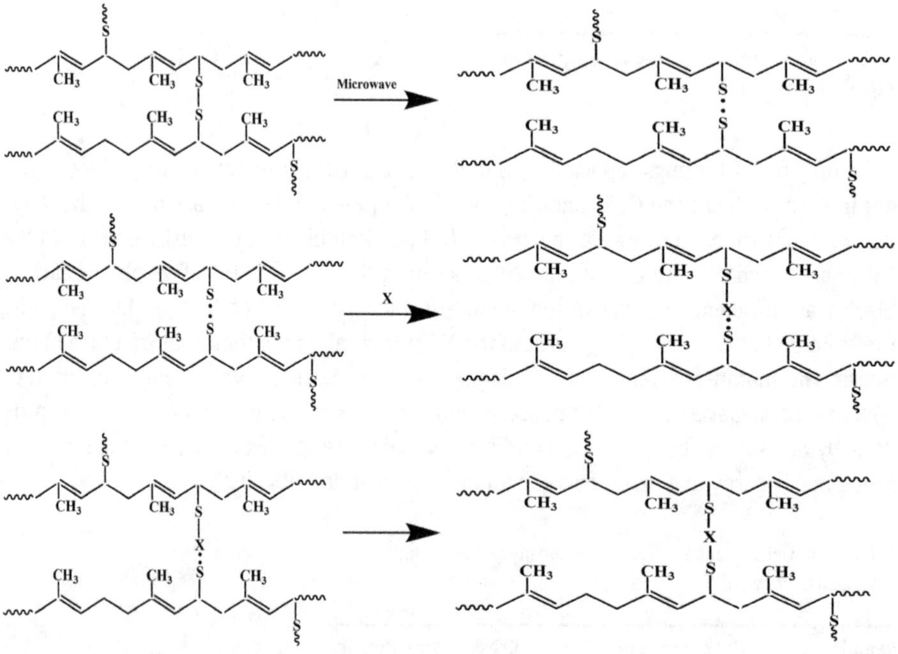


**Figure 9.22:** Simplified mechanism proposed for the formation of S–S and bonds during devulcanisation of NR by microwaves.

## 9.8 Value-added devulcanisation process: a new approach

The value-added devulcanisation process is a new approach to recycling and reusing rubber waste. The majority of tyre recycling techniques used today include civil engineering applications, which involve using whole tyres in various construction projects such as safety barriers, retention tanks, or backfilling. This approach provides an en-

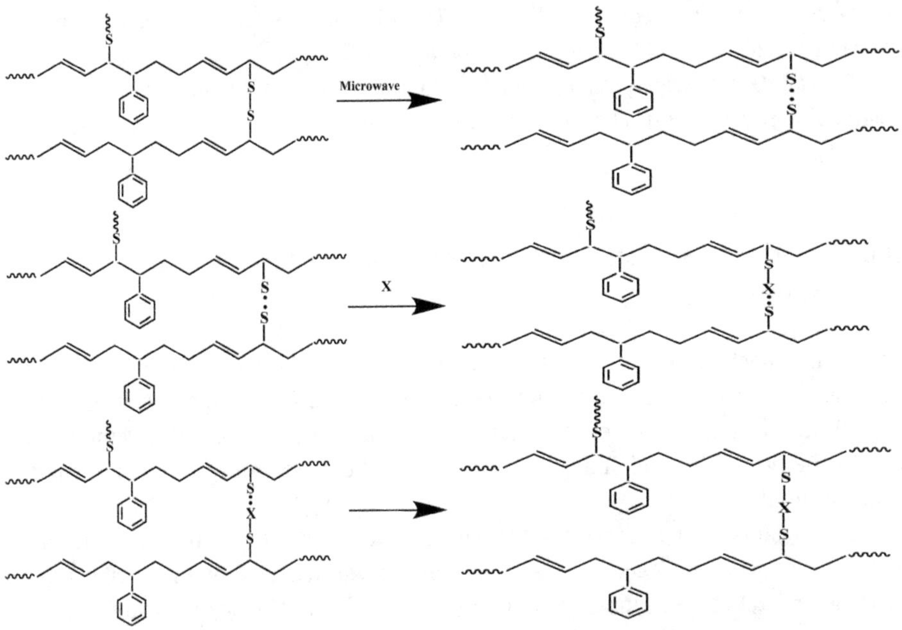


**Figure 9.23:** Simplified mechanism proposed for the formation of S–S and bonds during devulcanisation of SBR by microwaves.

vironmentally friendly solution by reusing tyres in a structural capacity, reducing the demand for other materials and preventing them from ending up in landfills:

- Energy recovery methods, particularly incineration in cement kilns, provide a means of using the energy content of tyres as a fuel source. This approach allows for the recovery of energy from the rubber and other combustible components of the tyres, reducing the reliance on fossil fuels.
- Pyrolysis is another tyre recycling technique that involves heating the tyres in the absence of oxygen to break them down into their constituent components, such as oil, gas, and char. These by-products can then be used as fuels or feed stocks in various industries.
- Material recycling focuses on mechanical disintegration of waste tyres, typically through processes like shredding or grinding, to produce GTR. GTR can then be used in various applications, such as the production of new tyres, rubberised asphalt, playground surfaces, and sports fields. This approach helps conserve resources by reusing the rubber material in a beneficial way.

Currently, it is acknowledged that the main commercially available source of recycled rubber is GTR.

The modification and functionalisation of GTR offer promising avenues for creating value-added products with enhanced properties and performance. This ongoing research and development in the field of GTR utilisation contribute to environmental awareness [124–127] and rubbers or as asphalt modifier [128–134].

Overall, GTR is widely recognised as a commercially viable and environmentally sustainable source of recycled rubber, offering a valuable solution for the recycling of waste tyres.

### 9.8.1 Devulcanisation by multifunctional devulcanising agent TMTD

During the mechanical reclaiming of GTR, tetramethyl thiuram disulfide (TMTD) was used as a multifunctional reclaiming agent. What makes TMTD unique is its dual role in both the reclaiming process and the revulcanisation of the reclaimed rubber sample. In the process of reclaiming GRT (ground rubber tyre), an open two-roll mixing mill was used.

The reclamation was carried out at various time intervals and different concentrations of the reclaiming agent. Degree of reclaiming was assessed by measuring several parameters: gel content, inherent viscosity of sol rubber, Mooney viscosity of the reclaimed rubber, cross-link density, swelling ratio, molecular weight between two cross-link bonds.

Table 9.25 provides an overview of the sol content of reclaimed rubber versus milling time at various reclamation agent concentrations. According to the observations from Table 9.25, it can be seen that, in all cases, the highest sol fraction is obtained at a milling time of 40 min. Nevertheless, it is noteworthy that the maximum increase in sol fraction occurs within the first 20 min of milling. After that, the rate of increase in sol fraction slows down, indicating a significant dependence of the sol fraction on milling time.

**Table 9.25:** Devulcanisation time and concentration of devulcanising agent as a function of sol content.

| Sample code | Concentration of devulcanisation time (mL) | Sol content (%) Time of devulcanisation(min) | | | | | |
|---|---|---|---|---|---|---|---|
| | | 0 | 20 | 25 | 30 | 35 | 40 |
| r-GTR(1.5) | 1.5 | 4.3 | 18.0 | 22.6 | 23.4 | 26.1 | 26.2 |
| r-GTR(2.0) | 2.0 | 4.3 | 19.1 | 21.4 | 21.8 | 25.5 | 25.6 |
| r-GTR(2.75) | 2.75 | 4.3 | 24.4 | 24.9 | 26.1 | 27.5 | 30.3 |
| r-GTR(3.25) | 3.25 | 4.3 | 20.9 | 22.0 | 22.2 | 23.1 | 23.0 |

r-GTR (number): reclaimed GTR (concentration of devulcanising agent, phr).

During the milling process, vulcanised rubber samples are subjected to mechanical shearing forces, which can lead to the breakdown of polymer chains; then due to mechanical shearing temperature rises to break TMTD into radicals [135]. The radicals generated from the breakdown of TMTD can react with the radicals produced from the mechanical shearing-induced breakdown of polymer chains in the vulcanised rubber, thereby preventing the recombination of polymer radicals, which explains the increase in sol fraction with increasing milling time. The cross-link density of devulcanised rubber as a function of devulcanisation time at various concentration of devulcanising agent is shown in Table 9.26. The figure clearly demonstrates that the cross-link density of the sample decreases as the milling time increases. This is due to the breakdown of both cross-links and polymer chains during the milling process.

Additionally, when the concentration of the reclaiming agent is increased, it initially results in a decrease in the cross-link density up to a concentration of 2.75 g, but beyond that concentration, the cross-link density starts to increase again. If the concentration of TMTD is very high in the reclaiming agent, it can indeed lead to the formation of cross-link bonds between polymer chains during the reclaiming process. This cross-linking can result in an increase in the cross-link density of the reclaimed rubber.

**Table 9.26:** Cross-link density as a function of devulcanisation time and concentration of devulcanising agent.

| Sample code | Concentration of devulcanisation time (mL) | Cross-link density x $10^{-4}$ (mol/cm$^3$) Time of devulcanisation (min) | | | | | |
|---|---|---|---|---|---|---|---|
| | | 0 | 20 | 25 | 30 | 35 | 40 |
| r-GTR (1.5) | 1.5 | 12.502 | 9.062 | 8.725 | 8.589 | 7.072 | 7.062 |
| r-GTR (2.0) | 2.0 | 12.502 | 8.368 | 8.624 | 8.610 | 7.074 | 7.021 |
| r-GTR (2.75) | 2.75 | 12.502 | 8.433 | 8.328 | 8.167 | 6.890 | 6.587 |
| r-GTR (3.25) | 3.25 | 12.502 | 8.719 | 8.671 | 8.578 | 8.555 | 8.505 |

r-GTR (number): reclaimed GTR (concentration of devulcanising agent, phr).

After 40 min of milling, the lowest cross-link density is also attained here. The Mooney viscosity of reclaimed rubber is shown in Table 9.27 as a function of milling time at various concentrations.

It demonstrates that as milling time is increased, reclaimed rubber's Mooney viscosity decreases. At 40 min of milling time, the lowest Mooney viscosity was attained for 2.75 g of TMTD. As a function of milling time, Table 9.28 shows the molecular weight between cross-links of reclaimed rubber at various TMTD concentrations. It is observed that as milling time increases, the molecular weight between cross-link bonds also increases. The mechanism used during the process of reclamation can be used to explain why the molecular weight increased with milling time. Reclaiming GRT with TMTD on an open two roll mill generates intense mechanical shearing forces and elevated temperatures, which promote the decomposition of TMTD, leading to the formation of thiocarbamate radicals illustrated in Figure 9.24(a).

**Table 9.27:** Mooney viscosity [ML (1 + 4) 100 °C] as a function of devulcanisation time and concentration of devulcanising agent.

| Sample code | Concentration of devulcanisation time (mL) | Mooney viscosity [ML(1 + 4) 100 °C] Time of devulcanisation (min) | | | | | |
|---|---|---|---|---|---|---|---|
| | | 0 | 20 | 25 | 30 | 35 | 40 |
| r-GTR(1.5) | 1.5 | >200 | 196 | 95.4 | 67 | 54 | 52 |
| r-GTR(2.0) | 2.0 | >200 | 73.4 | 66.6 | 60.7 | 53 | 46.4 |
| r-GTR(2.75) | 2.75 | >200 | 69.3 | 55.1 | 53.3 | 46.6 | 39 |
| r-GTR(3.25) | 3.25 | >200 | 60.2 | 58.4 | 50.8 | 31.8 | 20.1 |

r-GTR (number): reclaimed GTR (concentration of devulcanising agent, phr).

The milling process of GTR on an open two-roll mill involves the simultaneous breaking of polymer chains and cross-link bonds (Figure 9.24(b)). Simple aliphatic disulfides typically have bond strengths of approximately 70 kcal/mol for both the central S–S and the C–S bonds. These bond strengths are relatively high, making thermal dissociation into radicals at moderate temperatures unlikely.

**Table 9.28:** Molecular weight between cross-links as a function of devulcanisation time and concentration of devulcanising agent.

| Sample code | Concentration of devulcanisation time (mL) | Molecular weight between cross-links × $10^3$ Time of devulcanisation (min) | | | | | |
|---|---|---|---|---|---|---|---|
| | | 0 | 20 | 25 | 30 | 35 | 40 |
| r-GTR(1.5) | 1.5 | – | 7.418 | 9.719 | 10.909 | 25.211 | 25.544 |
| r-GTR(2.0) | 2.0 | – | 9.177 | 10.749 | 11.048 | 25.707 | 26.106 |
| r-GTR(2.75) | 2.75 | – | 11.705 | 12.868 | 14.275 | 28.422 | 33.093 |
| r-GTR(3.25) | 3.25 | – | 9.859 | 10.367 | 10.757 | 11.217 | 11.550 |

r-GTR (number): reclaimed GTR (concentration of devulcanising agent, phr).

Potential resonance stabilisation of the radicals in thiuram disulfides should significantly weaken the central bond and enhance the development of the thiocarbamate radical [135]. The molecular weight initially decreases when this thiocarbamate radical combines with a broken polymer radical (shown in Figure 9.24(c)). On the other hand, as the milling process progresses, thiocarbamate radicals are no longer available to couple with fragmented polymer radicals. As a consequence, during the milling process, the end-capping of low-molecular-weight radicals generated by shear forces decreases over time as the availability of thiocarbamate radicals diminishes. Consequently, the fragmented uncapped polymer radicals may have the opportunity to couple with each other, leading to chain extension and an increase in molecular weight during the later stages of milling described in Figure 9.24(d).

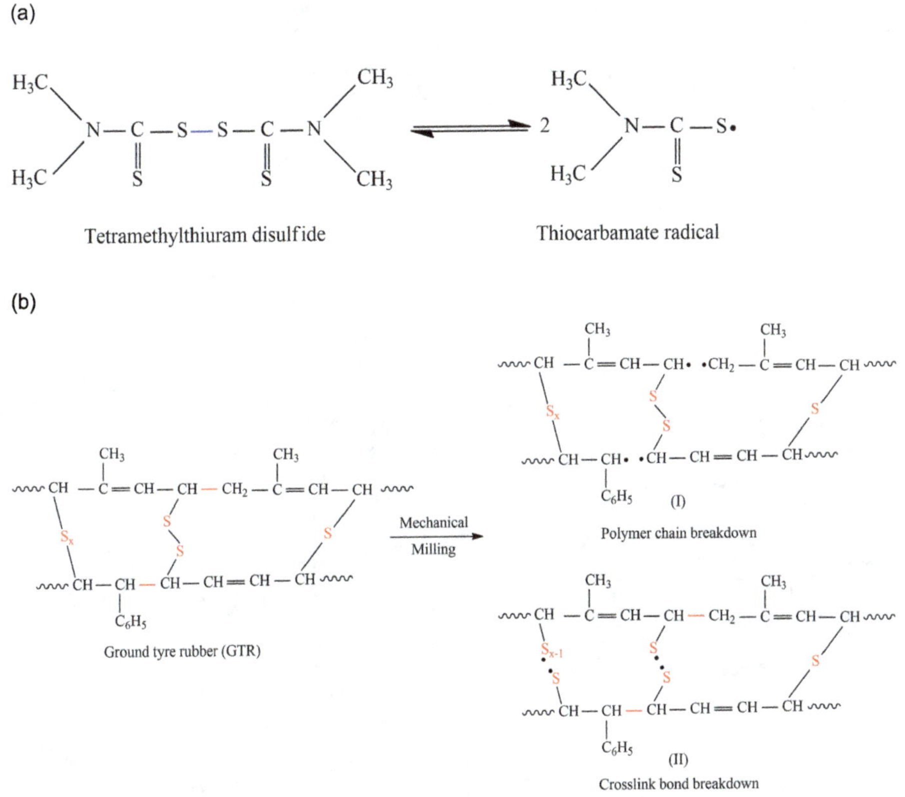


**Figure 9.24:** (a) Decomposition of tetramethylthiuram disulfide. (b) Breakdown of GTR by mechanical milling. (c) Chain capping reaction. (d) Chain extension by radical capping reaction.

In order to see the performance of devulcanised rubber, the mechanical properties of revulcanised rubber were measured. The modulus at 50% and 100% elongation, tensile strength, elongation at break, and hardness of the vulcanised rubber have changed depending on the degree of reclamation, as shown in Table 9.29.

The table shows that for a given concentration of TMTD, moduli at 50% and 100% elongation increase with increasing milling time. The higher cross-link density of the rubber vulcanisates, resulting from the gel present in reclamation rubber may be the cause of the larger 50% and 100% moduli. Cross-linking value data also supports this theory. Chain mobility reduces as cross-link density in the rubber matrix rises, and a higher force is needed for 50% and 100% elongation. With an increase in reclamation rubber's cross-link density, tensile strength rises.

It is possible that a higher degree of reclamation with longer milling times is the cause of the increased tensile strength. Gel content and cross-link density of reclamation rubber decrease with increased milling time, although molecular weight between cross-link bonds and inherent viscosity of sol rubber improve. In the case of reclaimed

(c)

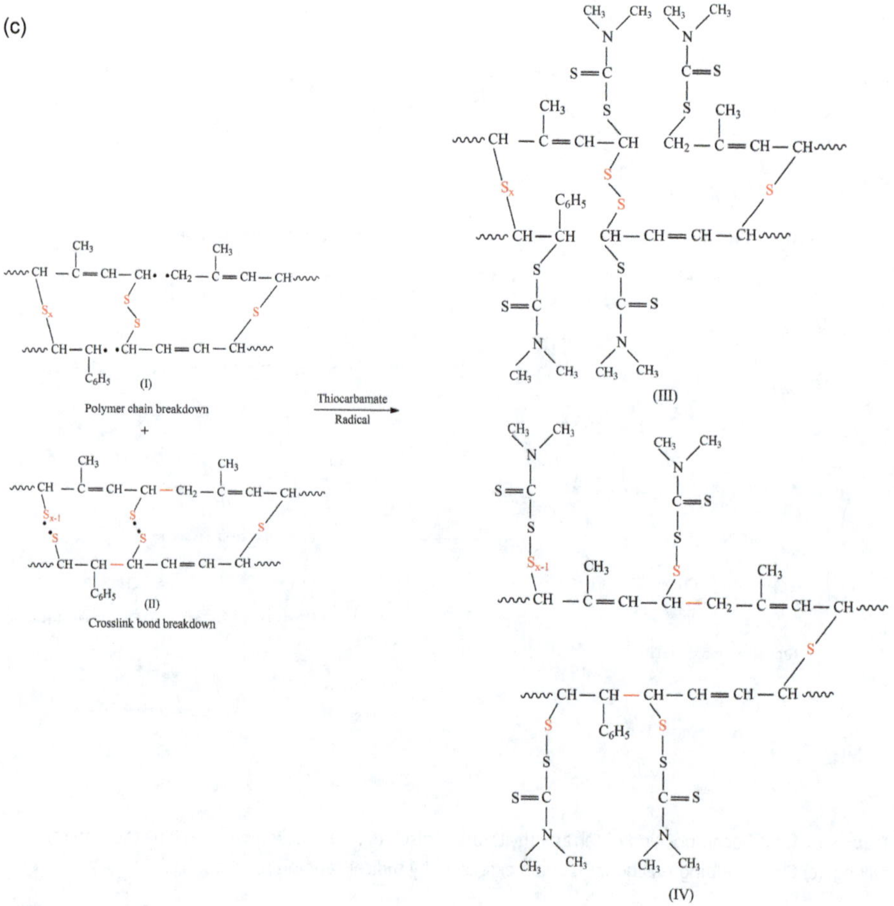


(d)

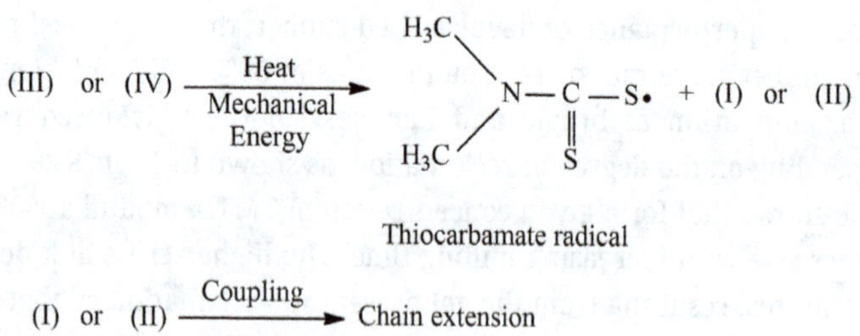


**Figure 9.24** (continued)

rubber, the gel part, which consists of the cross-linked and heavily degraded polymer networks, typically remains intact without dispersing as a continuous matrix.

This gel phase can act as weak sites within the reclaimed rubber, affecting stress transmission and ultimately resulting in lower tensile strength. The tensile strength

**Table 9.29:** Mechanical performance of revulcanised reclaim.

| Properties | Time of reclaiming | | | | | | | | | | | |
|---|---|---|---|---|---|---|---|---|---|---|---|---|
| | r-GTR(1.5) | | | r-GTR(2.0) | | | r-GTR(2.75) | | | r-GTR(3.25) | | |
| | 20 Min. | 30 Min. | 40 Min. | 20 Min. | 30 Min. | 40 Min. | 20 Min. | 30 Min. | 40 Min. | 20 Min. | 30 Min. | 40 Min. |
| 50% modulus (MPa) | 1.458 | 1.545 | 1.580 | 1.507 | 1.522 | 1.523 | 1.957 | 1.959 | 2.005 | 2.037 | 2.013 | 2.149 |
| 100% modulus (MPa) | 2.589 | 2.714 | 2.45 | 2.587 | 2.596 | 2.620 | 3.315 | 3.449 | 3.485 | 3.538 | 3.55 | 3.739 |
| Tensile strength (MPa) | 4.788 | 5.429 | 5.783 | 4.354 | 4.761 | 4.996 | 5.087 | 5.129 | 5.194 | 4.967 | 5.177 | 5.309 |
| Elongation at break (%) | 187 | 200 | 213 | 110 | 174 | 197 | 150 | 155 | 159 | 141 | 143 | 154 |
| Hardness (Shore A) | 74.5 | 75.5 | 76 | 73 | 73.5 | 74 | 75.5 | 76 | 76 | 77 | 77 | 77 |
| Cross-linking value (1/Q) | 0.675 | 0.686 | 0.724 | 0.680 | 0.697 | 0.768 | 0.774 | 0.780 | 0.821 | 0.787 | 0.800 | 0.859 |

r-GTR (number): reclaimed GTR (concentration of devulcanising agent, phr).

value is lower for 20 min milled reclaim rubber and higher for 40 min milled reclaim rubber because 20 min milled reclaim rubber contains more cross-linked gel than 40 min milled reclaim rubber. Elongation at break also rises with increased milling time for the same reason. Additionally, hardness rises as milling time rises.

**Table 9.30:** Mix formulation of NR/RR compounds.

| Ingredients (phr) | Sample code | | | | | |
|---|---|---|---|---|---|---|
| | r-GTR(0) | r-GTR(20) | r-GTR(30) | r-GTR(40) | r-GTR(50) | r-GTR(60) |
| Natural rubber (NR) | 100 | 80 | 70 | 60 | 50 | 40 |
| Reclaim rubber (RR) | – | 20 | 30 | 40 | 50 | 60 |
| Zinc oxide | 5 | 5 | 5 | 5 | 5 | 5 |
| Stearic acid | 2 | 2 | 2 | 2 | 2 | 2 |
| TMTD | 2.16 | 1.61 | 1.335 | 1.06 | 0.785 | 0.51 |
| Sulphur | 0.5 | 0.5 | 0.5 | 0.5 | 0.5 | 0.5 |
| Carbon black (N330) | – | – | – | – | – | – |
| Spindle oil | – | – | – | – | – | – |

r-GTR (number): reclaimed GTR (amount of reclaimed rubber, phr).

The performance of the produced vulcanisates was compared to that of the virgin rubber vulcanisate after the so-grown devulcanised rubber was mixed with fresh NR from 20 to 60 wt%. Table 9.30 shows the compound formulations of fresh NR/devulcanised GRT. On a two-roll mixing mill, virgin NR, various ratios of devulcanised GRT, and compounding additives like zinc oxide, stearic acid, sulphur, and CBS were mixed for 25 min at room temperature. Regardless of the amount of reclaim rubber used in the compound, the additives were employed here in all the vulcanisates based on 100 g of rubber because it was found that the additives in reclaim rubber derived from the parent compound were inactive [136].

Conversely, based on the amount of TMTD utilised during the reclamation of GRT, the amount of TMTD was kept constant at 9 m mol in all vulcanisates. Reclaim rubber is absent from formulation 1, whereas it is present in formulations 2–6, in varying proportions ranging from 20 to 60% by weight. To investigate how carbon black affected the performance of reclaim rubber in the NR/RR (80/20) blend, formulations 7–10 were made. Formulations 8–10 contain varying amounts of carbon black (20, 30, and 40 phr) in an NR/RR (80/20) blend, as compared to formulation 7, which exclusively contains NR with 40 phr. It has been noted that as carbon black is added gradually during compounding, its incorporation and dispersion become progressively more challenging. The stiffness and temperature increase as the carbon black loading is increased because more shearing action is required to achieve better dispersion.

The curing parameters of the rubber compounds including reclaimed rubber are shown in Table 9.31, which demonstrates that as reclaimed rubber content increases, optimal cure time lowers but scorch time does not change in any of the situations.

**Table 9.31:** Curing characteristics of NR/RR compounds.

| Curing characteristics | Sample code | | | | | |
|---|---|---|---|---|---|---|
| | r-GTR(0) | r-GTR(20) | r-GTR(30) | r-GTR(40) | r-GTR(50) | r-GTR(60) |
| Optimum cure time ($t_{90}$, min) | 3.5 | 3.0 | 3.0 | 2.75 | 2.5 | 2.25 |
| Scorch time ($t_{s2}$, min) | 1.5 | 1.0 | 1.0 | 1.0 | 1.0 | 1.0 |
| Extent of cure (dNm) | 40 | 43 | 45.3 | 50.4 | 53.5 | 61 |
| Cure rate index ($min^{-1}$) | 50 | 50 | 50 | 66.7 | 57.14 | 80 |

r-GTR (number): reclaimed GTR (amount of reclaimed rubber, phr).

Low scorch time data indicates that the reclaimed rubber has a tendency to scorch. The higher proportion of reclaimed rubber in NR/RR vulcanisates, which contain cross-linked gel, leads to an increase in the extent of cure. An increase in carbon black loading has been observed to have no effect on the optimum cure time and scorch time, but it does result in an increase in the extent of cure.

**Table 9.32:** Mechanical properties of NR/RR compounds.

| Mechanical properties | Sample code | | | | | |
|---|---|---|---|---|---|---|
| | r-GTR(0) | r-GTR(20) | r-GTR(30) | r-GTR(40) | r-GTR(50) | r-GTR(60) |
| 100% modulus (MPa) | 1.103 | 1.214 | 1.366 | 1.796 | 2.045 | 2.568 |
| 200% modulus (MPa) | 1.474 | 1.737 | 1.980 | 2.639 | 3.026 | 3.966 |
| Tensile strength (MPa) | 14.166 | 13.196 | 11.704 | 10.417 | 9.123 | 7.709 |
| % Elongation at break | 1240.12 | 1140.58 | 908.29 | 805.43 | 671.88 | 501.56 |
| Hardness (Shore A) | 40 | 45 | 49 | 53 | 55 | 60 |
| Cross-linking value (1/Q) | 0.212 | 0.261 | 0.288 | 0.330 | 0.370 | 0.435 |

r-GTR (Number): reclaimed GTR (amount of reclaimed rubber, phr).

Table 9.32 displays the tensile characteristics and cross-linking value of NR/RR blends. According to the data in Table 9.32, 100% and 200% moduli increase as reclaimed rubber (RR) content increases, but tensile strength and elongation at break decrease. The higher 100% and 200% moduli in rubber vulcanisates may be attributed to a higher cross-link density resulting from the presence of gel in the RR. This observation is further supported by the data on cross-linking values. Figure 9.25 demonstrates that as the content of RR increases in the rubber matrix, the cross-link density also increases. This increase in cross-link density leads to reduced chain mobility, requiring higher loads for 100% and 200% elongation. When the cross-link density reaches $5.138 \times 10^{-5}$ mol/cm$^3$ for the control sample (r-GTR(0)), the tensile strength is at its maximum. But, as the cross-link density increases further in other formulations including RR the tensile strength steadily diminishes.

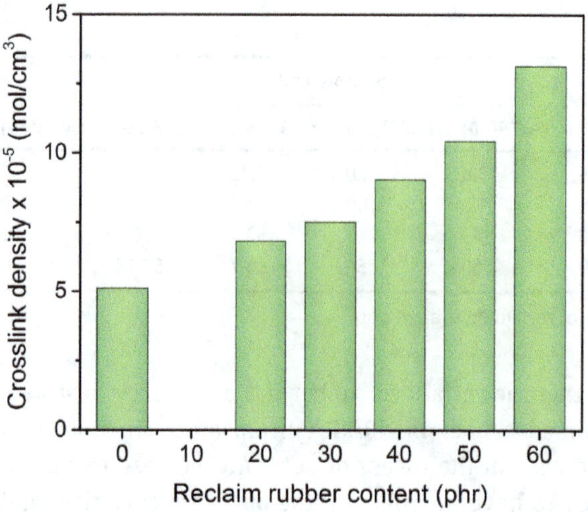


**Figure 9.25:** Effect of reclaim rubber content on cross-link density of reclaim rubber.

Indeed, the presence of cross-linked gel in the RR can contribute to a decrease in the tensile strength of the blended material. When RR is blended with virgin rubber, the cross-linked gel may not disperse uniformly and instead remain as separate entities within the matrix of the virgin rubber. This lack of proper dispersion and integration can lead to weak points or discontinuities in the material, resulting in reduced tensile strength. The presence of the gel in the RR can act as weak sites within the material, affecting stress transmission to the surrounding continuous matrix. This can result in a lower tensile stress and reduced elongations at break as the percentage of RR increases. Additionally, the hardness of the material tends to increase with higher content of RR. Figure 9.26 demonstrates that the Mooney viscosity value increases as the RR content increases.

## 9.8.2 Devulcanisation by multifunctional devulcanising agent TESPT

Mechanochemical devulcanisation of sulphur-cured SBR, NR, guayule natural rubber (GNR) and GRT was carried out in an open roll mixing mill using bis(3-triethoxysilyl propyl) tetrasulfide (TESPT), a dual function devulcanising agent. It is for the first time that TESPT has been used as a devulcanising agent and as-grown devulcanised rubber facilitates the silica dispersion in nonpolar rubber compounds without any coupling agent [137–140].

In a recent work, Ghosh et al. [137] devulcanised and chemically functionalised waste solution SBR simultaneously. Such rubber is used as the most important component in silica-based tread rubber compounds. Ethoxy groups introduced during chem-

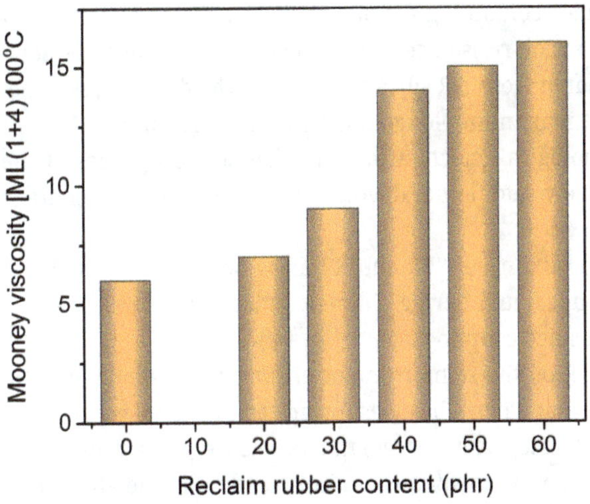


**Figure 9.26:** Effect of reclaim rubber content on Mooney viscosity of reclaim rubber.

ical modification of devulcanised SBR (D-SBR) were used to induce chemical reaction between devulcanised rubber chains and silica. Dynamic mechanical performance of the virgin SBR composites was compared with that of D-SBR. Good covalent bonding between silanol groups of silica and ethoxy groups of D-SBR resulted in its promising performance. Thus this approach could be used for producing rubber products with good elastomeric performance.

The same devulcanisation and chemical modification via thermo-mechanical treatment using a multifunctional devulcanising agent based on TESPT [138] was carried out. The compounds were further strengthened using silica nanoparticles and revulcanised for the fabrication of rubber composites. Tests were conducted to compare the mechanical strength along with curing and processing properties with silica-based green tyre formulations. Evaluation of sol composition, intrinsic viscosity of sol fraction, density of cross-linking and devulcanisation degree as cross-link density function determined the property of devulcanised rubber.

The revulcanised composites had tear strength of 26 and 11 MPa for tensile strength and 450% for elongation at break. The values obtained were similar to that of composites comprising of associated fresh components. Also, the SBR-devulcanised SBR-silica composite had lower tan $\delta$, higher storage modulus, and thus huge dynamic mechanical performance in comparison to SBR-silica composite. This work opened a way for reusing waste tyre for production of tyres.

Ghorai et al. [139] carried out mechanochemical devulcanisation of sulphur-vulcanised natural rubber using silane-based tetrasulfide (TESPT) as a devulcanising agent. This led to formation of silica-based rubber composite. This method helped in rupture of sulphur–sulphur cross-link and subsequent formation of reactive poly (isoprene) chains for interaction with silica. The proposed composite was prepared by 30%

substitution of fresh NR with silica-containing devulcanised NR. The composites had superior mechanical strength, abrasion resistance, tear strength, and dynamic mechanical property in comparison to that of fresh NR silica composites. The devulcanised rubber-based silica composites had maximum elongation strain of 921% and 20 MPa of tensile strength. Further morphological, mechanical, and chemical properties of the composites are studied. Conclusively, addition of devulcanised rubber could encourage its sustainable use.

In incorporating silica into a mixture of SBR and RR using both conventional and sol-gel methods [140], the mechanical and curing characteristics of the SBR/RR vulcanisate were affected by TESPT, a silica coupling agent.

Moreover, the influence of RR on the reinforcement of silica was examined using the prepared blend. It was found that when silica was integrated by the sol-gel reaction of tetra ethoxy silane, it exhibited lesser tensile strength irrespective of the presence of TESPT, when compared to its incorporation by mechanical blending in the absence of TESPT. ATR study confirmed the formation of the bond between the functional groups of RR and silica nanoparticles. A thermogravimetric study confirmed the incorporation of silica via a sol-gel reaction.

It was also a tried-and-true technique for enhancing the reinforcement effect in the silica–graphene oxide-based rubber–filler interaction and reinforcement in SBR/devulcanised NR composites to apply silane coupling agents, TESPT to the silica surface to encourage interactions between the chains of silica and rubber [141]. A significant problem with silica-reinforced nonpolar rubber is the incompatibility of the hydrophilic silica and the hydrophobic rubber since it creates a strong silica-silica network because of the low rubber-filler and high filler-filler interaction. On the other hand, presence of silane coupling agent improves filler–rubber interaction, silica dispersion, and tensile strength.

## 9.9 Proposed mechanism of devulcanisation

During the progress of devulcanisation, the GTR undergoes significant mechanical shearing forces. These forces lead to both random polymer chain breakdown and the scission of cross-links. The devulcanising agent TESPT undergoes homolytic cleavage of its S–S bond, resulting in the formation of fragmented devulcanising agent (FDA) free radicals. This cleavage occurs as a result of mechanical shearing and is further influenced by an increase in temperature. The FDA free radicals formed during the devulcanisation process can react with both broken cross-link bonds and polymer chains. This reaction leads to a process called chain capping, which contributes to the increase in the sol fraction with an increase in devulcanisation time.

The schematic presentation of main chain scission and cross-link scission is represented in Figure 9.27 (a–c). Figure 9.27 suggests that the TESPT concentration of 6 ml

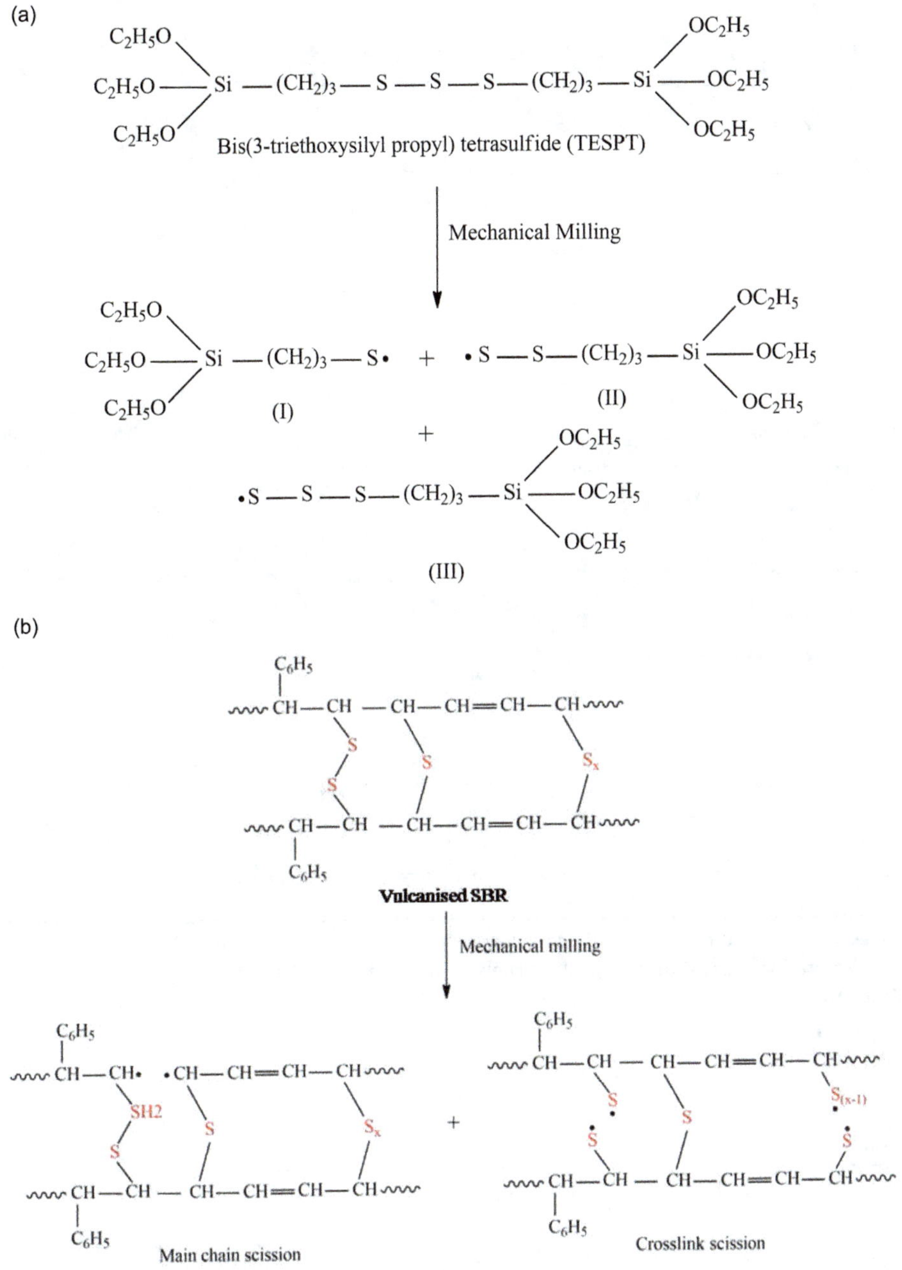


**Figure 9.27:** (a) Fragmented (I) monosulfide, (II) disulfide, and (III) polysulfide radical. (b) Mechanical milling of vulcanised SBR. (c) Chain capping reaction by devulcanising agent.

(c)

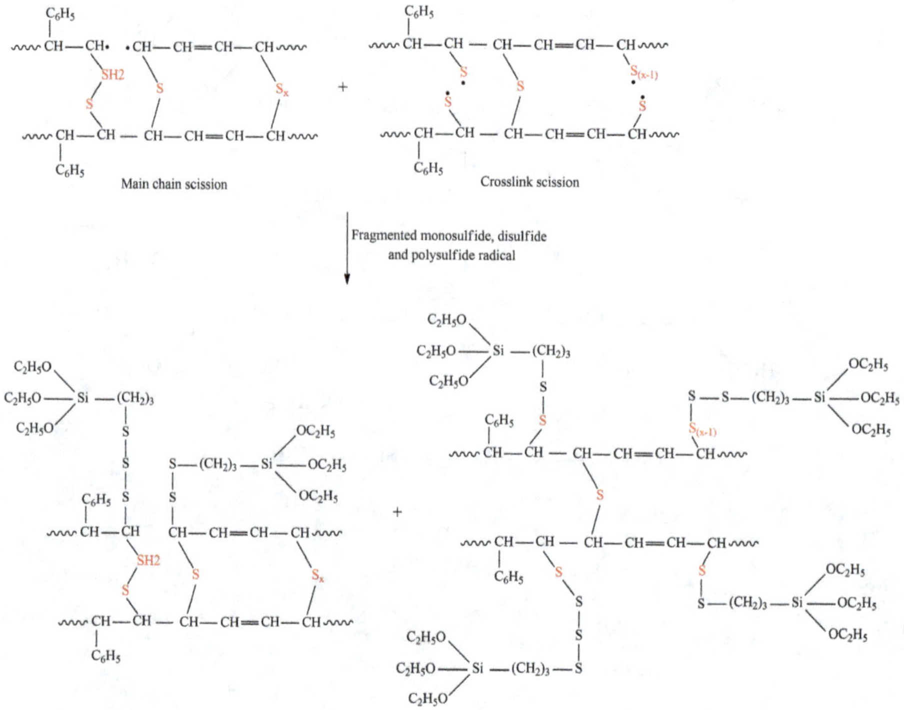


**Figure 9.27** (continued)

**Table 9.33:** Comparison of devulcanisation techniques in recent developments.

| **Types of devulcanisation Processing parameter and characteristics of devulcanised rubber** | **Ultrasonic devulcanisation** | **Microwave devulcanisation** | **Mechanochemical devulcanisation** | **Supercritical $CO_2$-insisted devulcanisation** |
|---|---|---|---|---|
| Instrument required | Ultrasonic reactor | Microwave oven | Extruder/open two-roll mill | Autoclave |
| Raw material for devulcanisation | GTR | GTR | GTR | GTR |

**Table 9.33** (continued)

| **Types of devulcanisation Processing parameter and characteristics of devulcanised rubber** | **Ultrasonic devulcanisation** | **Microwave devulcanisation** | **Mechanochemical devulcanisation** | **Supercritical $CO_2$-insisted devulcanisation** |
|---|---|---|---|---|
| Devulcanising agent | Ultrasonic energy | Microwave energy | Bis(3-triethoxysilyl propyl) tetrasulfide (TESPT) | Supercritical $CO_2$ and diphenyl disulfide |
| Reaction condition | Temperature: 120 °C Screw rotation speed:20 rpm ultrasonic power: 3,000 W Frequency: 20 kHz | Microwave power: 900 W/3 min Microwave frequency:2,000 MHz Temperature: 240 °C | Temperature: 70 °C Time: 10 min TESPT concentration: 6.48 phr | Pressure: 18 MPa, Temperature: 200 °C Time: 2 h Diphenyl disulfide concentration: 12 g/L |
| Environmental concern | Environment-friendly | Environment-friendly | Environment-friendly | Environment-friendly |
| **Physical characteristics** | | | | |
| Sol content (%) | 25 | 7.5 | 25 | 75 |
| Cross-link density (mol/cm$^3$) | $0.105 \times 10^{-3}$ | $0.165 \times 10^{-3}$ | $0.856 \times 10^{-3}$ | $0.84 \times 10^{-4}$ |
| Mooney viscosity [ML (1 + 4) 100 °C] | Not determined | Not determined | 19 | Not determined |
| **Performance of revulcanised rubber** | | | | |
| Tensile strength (MPa) | 10.5 | 10 | 7.27 | – |
| Elongation at break (%) | 250% | 226% | 67% | – |

leads to the highest sol content. However, as the milling process continues, TESPT's disulfide radical fragments are no longer accessible to be combined with the broken polymer chain radicals and cross-link bonds.

The homolytic breakage of the C–C bond by the fragmented disulfide radicals of TESPT causes end-capping of the shear-generated low-molecular-weight DeVulcNRs to decrease gradually with devulcanisation time. In these conditions, chain extension may result from the combination of fragmented, uncapped polymer radicals.

## 9.10 Comparison of recent developments

A comparative evaluation of recent developments in various devulcanisation techniques is shown in Table 9.33. Here, among different devulcanisation processes, ultrasonic devulcanisation, microwave devulcanisation, mechanochemical devulcanisation using TESPT, and supercritical carbon dioxide ($scCO_2$)-insisted devulcanisation using DPDS are critically analysed [142–144].

## 9.11 Conclusion

The tyre recycling technique is continually changing due to developments of tyre production. Among different disposal methods of scrap/discarded rubber products, devulcanisation/reclaiming is the most promising approach to solve the disposal problem of waste rubber products because this not only solves the disposal problem of scrap tyre products but also saves valuable petroleum resources.

Of late, mechanochemical devulcanisation using different disulfide chemicals and supercritical carbon dioxide-insisted devulcanisations using various disulfide chemicals are being widely used to get superior quality devulcanised rubber. The mechanochemical devulcanisation by silane tetrasulfide produces high quality devulcanised rubber, which can be incorporated into the fresh rubber compound up to 30 phr without compromising the mechanical properties of the vulcanisate.

Moreover, the added advantage of the developed material is that it can facilitate the silica to disperse in the rubber matrix without the use of any coupling agent. Thus, the material has potential application in green tyre production.

# References

[1] Kaza, S., Yao, L., Bhada-Tata, P., Van Woerden, F. What a Waste 2.0. A Global Snapshot of Solid Waste Management to 2050. Washington, DC World Bank: World Bank (2018).
[2] Hoornweg, D., Bhada-Tata, P. What A Waste – A Global Review of Solid Waste Management. Urban development series, the World Bank, (2012).
[3] Subulan, K., Taran, A. S., Baykasoglu, A. Designing an environmentally conscious tyre closed-loop supply chain network with multiple recovery options using interactive fuzzy goal programming. Applied Mathematical Modelling 39:2661–2702 (2015).
[4] Martinez, J. D., Puy, N., Murillo, R., Garcia, T., Navarro, M. V., Mastral, A. M. Waste tyre pyrolysis – A review. Renewable Sustainable Energy Reviews 23:179–213 (2013).
[5] Wang, W. C., Bai, C. J., Lin, C. T., Prakash, S. Alternative fuel produced from thermal pyrolysis of waste tyres and its use in a DI diesel engine. Applied Thermal Engineering 93:330–338 (2016).
[6] Zhang, L., Zhou, B., Duan, P., Wang, F., Xu, Y. Hydrothermal conversion of scrap tyre to liquid fuel. Chemical Engineering Journal 285:157–163 (2016).
[7] Rowhani, A., Rainey, T. J. Scrap tyre management pathways and their use as a fuel – A review. Energies 9:1–26 (2016).
[8] Parthasarathy, P., Choi, H. S., Park, H. C., Hwang, J. G., Yoo, H. S., Lee, B. K., Upadhyay, M. Influence of process conditions on product yield of waste tyre pyrolysis – A review. Korean Journal of Chemical Engineering 33:2268–2286 (2016).
[9] Matade, S. P., 2016. Need to regulate the end-of-life tyres market in India. Rubber Asia. http://www.rubberasia.com/2016/06/11/need-to-regulate-end-of-life- (accessed 11 June 2016).
[10] Serumgard, J. R., Eastman, A. L. Scrap tyre recycling: Regulatory and market development progress. Plastics, Rubber, and Paper Recycling-A Pragmatic Approach, ACS Symposium Series; American Chemical Society 609:237–244 (1995). doi: 10.1021/bk-1995-0609.ch020.
[11] Akiba, M. The status of recycling of waste rubber. Porima Daijesuto 48(2):79–93 (1996).
[12] Smith, F. G. Rubber ACS Division Meeting. IRC, Orlando, FL (26–28 October 1993).
[13] Amari, T., Themelis, N. J., Wernick, I. K. Resource recovery from used rubber tyres. Resources Policy 25:179–188 (1999).
[14] Roy, C., Chaala, A., Darmstadt, H. The vacuum pyrolysis of used tyres: End-uses for oil and carbon black products. Journal of Analytical and Applied Pyrolysis 51:201–221 (1999).
[15] Mark, H. F., Bikales, N. M., Overberger, C. G., Menges, G. Encyclopedia of Polymer Science and Engineering. NewYork: Wiley, Vol. 14, 787–804 (1988).
[16] Manar, E., Abdul, R., Nermine, E., Azim, A. Thermo chemical recycling of mixture of scrap tyres and waste lubricating oil into high caloric value products. Energy Conversion and Management 51:1304–1310 (2010).
[17] Environmental Protection Agency 40 CFR Part 241: Identification of non-hazardous secondary materials that are solid waste. Federal Register 76:15456–15551 (2011).
[18] Samolada, M., Zabaniotou, A. Potential application of pyrolysis for the effective valorization of the end-of-life tyres in Greece. Environmental Development 4:73–78 (2011).
[19] Cantanhedes, A., Monge, G. State of the art of tire management in the Americas. Publication of Pan American Center for Sanitary Engineering and Environmental Sciences. Lima, Peru: World Health Organization (2002).
[20] Williams, P. T., Brindle, A. J. Fluidised bed catalytic pyrolysis of scrap tyres: Influence of catalyst: Tyre ratio and catalyst temperature. Waste Management and Research 20:546–555 (2002).
[21] Abinasa, F., Daud, W. M. A. W. Optimization of fuel recovery through the stepwise co-pyrolysis of palm shell and scrap tyre. Energy Conversion and Management 99:334–345 (2015).

[22] Martinez, J. D., Veses, A., Mastral, A. M., Murillo, R., Navarro, M. V., Puy, N., Artigues, A., Bartroli, J., Garcia, T. Co-pyrolysis of biomass with waste tyres: Upgrading of liquid bio-fuel. Fuel Processing Technology 119:263–271 (2014).
[23] Choi, G. G., Oh, S. J., Kim, S. J. Clean pyrolysis oil from a continuous two-stage pyrolysis of scrap tyres using in-situ and ex-situ desulfurization. Energy 141:2234–2241 (2017).
[24] Ouyang, S., Xiong, D., Li, Y., Zou, L., Chen, J. Pyrolysis of scrap tyres pretreated by waste coal tar. Carbon Resources Conversion 1:218–227 (2018).
[25] Onay, Ö. The catalytic co-pyrolysis of waste tyres and pistachio seeds. Energy Sources, Part A 36:2070–2077 (2014).
[26] Lo Presti, D. Recycled tyre rubber modified bitumens for road asphalt mixtures: A literature review. Construction and Building Materials 49:863–881 (2013).
[27] Huang, B., Mohammad, L. N., Graves, P. S., Abadie, C. Louisiana experience with crumb rubber modified hot-mix asphalt pavement. Transportation Research Record 1789:1–13 (2002).
[28] Liang, R. Y., Lee, S. Short-term and long-term aging behavior of rubber modified asphalt paving mixtures. Transportation Research Record 1530:11–17 (1996).
[29] Shen, J., Amirkhanian, S. The influence of crumb rubber modifier (CRM) microstructures on the high temperature properties of CRM binders. The International Journal of Pavement Engineering 6:265–271 (2005).
[30] Hicks, R. G. Asphalt Rubber Design and Construction Guidelines: Volume I-Design Guidelines, Northern California Rubberized Asphalt Concrete Technology Centre – California Interated Waste Management Board. Sacramento, CA, USA (2002).
[31] Santagata, E., Zanetti, M. C. The Use of Products from End-of-life Tyres in Road Pavements. Milan, Italy: ECOPNEUS (2012).
[32] Santagata, E., Lanotte, M., Dalmazzo, D., Zanetti, M. C. Potential performance related properties of rubberized bituminous mixtures produced with dry technology. In: Proceedings, 12th International Conference on Sustainable Construction Materials, Pavement Engineering and Infrastructure, Liverpool, UK (27–28 February 2013)
[33] Amirkhanian, S. N. Utilization of crumb rubber in asphalt concrete mixtures – South Carolina's Experience. Research Report, Clemson University, Clemson, SC, USA (2008).
[34] Zanetti, M. C., Fiore, S., Ruffino, B., Santagata, E., Lanotte, M. Assessment of gaseous emissions produced on site of bituminous mixtures containing crumb rubber. Construction and Building Materials 67:291–296 (2014).
[35] Doh, Y. S., Kim, H. H., Kim, K. W. Estimation of properties of CRM-modified asphalt mixtures by addition of hydrated lime. Journal of Advanced Mineral Aggr Compos 13:9–17 (2008).
[36] Kim, S., Lee, S. J., Yun, Y. B., Kim, K. W. The use of CRM-modified asphalt mixtures in Korea: Evaluation of high and ambient temperature performance. Construction and Building Materials 67:244–248 (2014).
[37] Zhou, X., Wang, F., Yuan, X., Kuang, M., Song, Y., Li, C. Usage of slurry oil for the preparation of crumb-rubber-modified asphalt emulsion. Construction and Building Materials 76:279–285 (2015).
[38] Huang, S. C., Xu, J., Quing, Y. C. Modifying Emulsifying Asphalt and Micro-surfacing Technique. Beijing: China Communication Press, (2010).
[39] Liu, H., Chen, Z., Wang, W., Wang, H., Hao, P. Investigation of the rheological modification mechanism of crumb rubber modified asphalt (CRMA) containing TOR additive. Construction and Building Materials 67:225–233 (2014).
[40] Cong, P., Xun, P., Xing, M., Chen, S. Investigation of asphalt binder containing various crumb rubbers and asphalts. Construction and Building Materials 40:632–641 (2013).
[41] Xie, Z., Shen, J. Effect of cross-linking agent on the properties of asphalt rubber. Construction and Building Materials 67:234–238 (2014).

[42] Elchalakani, M. High strength rubberized concrete containing silica fume for the construction of sustainable road side barriers. Structures 1:20–38 (2015).
[43] Thomas, B. S., Gupta, R. C. Properties of high strength concrete containing scrap tyre rubber. Journal of Cleaner Production 113:86–92 (2016).
[44] Meddah, A., Bensaci, H., Beddar, M., Bali, A. Study of the effects of mechanical and chemical treatment of rubber on the performance of rubberized roller-compacted concrete pavement. Innovative Infrastructure Solutions 2:1–5 (2017).
[45] Pacheco-Torres, R., Cerro-Prada, E., Escolano, F., Varela, F. Fatigue performance of waste rubber concrete for rigid road pavements. Construction and Building Materials 176:539–548 (2018).
[46] Angelin, A. F., Miranda, E. J. P., Jr, Dos Santos, J. M. C., Lintz, R. C. C., Gachet-Barbosa, L. A. Rubberized mortar: The influence of aggregate granulometry in mechanical resistances and acoustic behavior. Construction and Building Materials 200:248–254 (2019).
[47] Mohammadi, I., Khabbaz, H. Shrinkage performance of crumb rubber concrete(CRC) prepared by water-soaking treatment method for rigid pavements. Cement and Concrete Composites 62:106–116 (2015).
[48] Raffoul, S., Garcia, R., Pilakoutas, K., Guadagnini, M., Medina, N. F. Optimisation of rubberised concrete with high rubber content: An experimental investigation. Construction and Building Materials 124:391–404 (2016).
[49] He, L., Ma, Y., Liu, Q., Mu, Y. Surface modification of crumb rubber and its influence on the mechanical properties of rubber-cement concrete. Construction and Building Materials 120:403–407 (2016).
[50] Pehlken, A., Muller, D. H. Using information of the separation process of recycling scrap tyres for process modeling. Resources, Conservation and Recycling 54:140–148 (2009).
[51] Meyasami, M. A study of scrap rubber devulcanisation and incorporation of devulcanised rubber into virgin rubber compounds. Ph. D Thesis. University of Waterloo, Ontario, Canada.
[52] Blumenthal, M. Scrap tyre market in the United States: An update. Rubber Manufactures Association Border Meeting. Nogales, Arizona, USA 528 (2012).
[53] De, S., Isayev, A. I., Khait, K. Rubber Recycling. CRC Press. Boca Raton, FL, USA (2005). ISBN 0-8493-1527-1
[54] Gyorgy, M., Apparatus for regular grinding rubber vehicle tyres and other elastic materials by ultra-high pressure fluid. Patent HU 0800444.
[55] Sienkiewicz, M., Kucinska-Lipka, J., Janik, H., Balas, A. Progress in used tyres management in the European Union: A review. Waste Management 32:1742–1751 (2012).
[56] Mohaved, S. O., Ansarifar, A., Nezhad, S. K., Atharyfar, S. A novel industrial technique for recycling ethylene-propylene-diene waste rubber. Polymer Degradation and Stability 111:114–123 (2015).
[57] Phadke, A. A., Bhattacharya, A. K., Chakraborty, S. K., De, S. K. Studies of vulcanization of reclaimed rubber. Rubber Chemistry and Technology 56:726–736 (1983).
[58] Sutanto, P., Picchioni, F., Janssen, L. P. B. M. The use of experimental design to study the responses of continuous devulcanisation processes. Journal of Applied Polymer Science 102:5028–5038 (2006).
[59] Sutanto, P., Picchioni, F., Janssen, L. P. B. M. Modelling a continuous devulcanisation in an extruder. Chemical Engineering Science 61:7077–7086 (2006).
[60] Yazdani, H., Karrabi, M., Ghasmi, I., Azizi, H., Bakhshandeh, G. R. Devulcanisation of waste tyres using a twin-screw extruder: The effects of processing conditions. Journal of Vinyl and Additive Technology 17:64–69 (2011).
[61] Si, H., Chen, T., Zhang, Y. Effects of high shear stress on the devulcanisation of ground tyre rubber in a twin-screw extruder. Journal of Applied Polymer Science 128:2307–2318 (2013).
[62] Wang, X., Shi, C., Zhang, L., Zhang, Y. Effects of shear stress and subcritical water on devulcanisation of styrene–butadiene rubber-based ground tyre rubber in a twin-screw extruder. Journal of Applied Polymer Science 130:1845–1854 (2013).
[63] Lv, X. L., Huang, X. H., Lv, Y. B. Balancing mechanical properties and processability for devulcanised ground tyre rubber using industrially sized single-screw extruder. Journal of Applied Polymer Science 133:43761 (2016).

[64] Tao, G., He, Q., Xia, Y., Jia, G., Yang, H., Ma, W. The effect of devulcanisation level on mechanical properties of reclaimed rubber by thermal-mechanical shearing devulcanisation. Journal of Applied Polymer Science 129:2598–2605 (2013).
[65] Seghar, S., Asaro, L., Rolland-Monnet, M., Hocine, N. A. Thermo-mechanical devulcanisation and recycling of rubber industry waste. Resources, Conservation and Recycling 144:180–186 (2019).
[66] Pirityi, Z. D., Pölöskei, K. Thermomechanical devulcanisation of ethylene propylene diene monomer rubber and its application in blends with high-density polyethylene. Journal of Applied Polymer Science 138:50090 (2021).
[67] Zhang, X., Saha, P., Cao, L., Li, H., Kim, J. Devulcanisation of waste rubber powder using thiobisphenols as novel reclaiming agent. Waste Management 78:980–991 (2018).
[68] Shi, J., Jiang, K., Ren, D., Zou, H., Wang, Y., Lv, X., Zhang, L. Structure and performance of reclaimed rubber obtained by different methods. Journal of Applied Polymer Science 129:999–1007 (2013).
[69] Levin, V. Y., Kim, S., Isayev, A. Vulcanization of ultrasonically devulcanised SBR elastomers. Rubber Chemistry and Technology 70:120–128 (1997).
[70] Saputra, R., Walvekar, R., Khalid, M., Shahbaz, K., Ramarad, S. Effective devulcanisation of ground tyre rubber using choline chloride-based deep eutectic solvents. Journal of Environmental Chemical Engineering 7:103151 (2019).
[71] Guo, L., Wang, C., Lv, D., Ren, D., Zhai, T., Sun, C., Liu, H. Rubber reclamation with high bond-breaking selectivity using a low temperature mechano-chemical devulcanisation method. Journal of Cleaner Production 279:123266 (2021).
[72] Asaro, L., Gratton, M., Poirot, N., Seghar, S., Hocine, N. A. Devulcanisation of natural rubber industry waste in supercritical carbondioxide combined with diphenyl disulfide. Waste Mangement 118:647–654 (2020).
[73] Sabzekar, M., Zohuri, G., Chenar, M. P., Mortazavi, S. M., Kariminejad, M., Asadi, S. A new approach for reclaiming of waste automotive EPDM rubber using waste oil. Polymer Degradation and Stability 129:56–62 (2016).
[74] Liu, H. L., Wang, X. P., Jia, D. M. Recycling of waste rubber powder by mechano-chemical modification. Journal of Cleaner Production 245:118716 (2020).
[75] Jia, S., Dong, Z., Zhenxing, Z., et al. Influence of screw speed ratios of twin screw extruder on properties of rubber powder and its natural rubber blends. Synthetic Rubber Industry 39 (5):396–399 (2016).
[76] Magini, M., Cavalieri, F., Padella, F. Mechanochemical treatment of scrap tyre rubber. Journal of Metastable and Nanocrystalline Materials 13:263–268 (2002).
[77] Luo, M. C., Liao, X. X., Liao, S. Q., et al. Review on the broken three-dimensional network modification methods of waste rubber powder. Advanced Materials Research 554–556:181–186 (2012).
[78] Kajimoto, O., Tucker, S. C., Peters, C. J., Gauter, K., Brennecke, J. F., Chauteauneuf, J. E., Baiker, A., Jessop, P. G., Ikariya, T., Noyori, R., Darr, J. A., Poliakoff, M., Kendall, J. L., Canelas, D. A., Young, J. L., DeSimone, J. M., Kirby, C. F., McHugh, M. A., Savage, P. E., Mesiano, A. J., Beckman, E. J., Russell, A. J. Super critical fluids. Chemical Reviews 99:353–634 (1999).
[79] Perrut, M. Supercritical fluid applications: Industrial developments and economic issues. Industrial and Engineering Chemistry Research 39:4531–4535 (2000).
[80] Chang, S. H., Park, S. C., Shim, J. J. Phase equilibria of supercritical fluid–polymer systems. Journal of Supercritical Fluids 13:113–119 (1998).
[81] Funazukuri, T., Ogasawara, S., Wakao, N., Smith, J. M. Subcritical and supercritical extraction of oil from used automotive tyre samples. Journal of Chemical Engineering of Japan 18:455–460 (1985).
[82] Lee, S. B., Hong, I. K. Depolymerization behavior for *cis*-polyisoprene rubber in supercritical tetrahydrofuran. Journal of Industrial and Engineering Chemistry 4:26–30 (1998).
[83] Kershaw, J. R. Supercritical fluid extraction of scrap tyres. Fuel 77:1113–1115 (1998).
[84] Hyatt, J. A. Liquid and supercritical carbon dioxide as organic solvents. Journal of Organic Chemistry 49:5097–5101 (1984).

[85] Kojima, M., Tosaka, M., Ikeda, Y. Chemical recycling of sulfur-cured natural rubber using supercritical carbon dioxide. Green Chemistry 6:84–89 (2004).
[86] Wang, Z., Zeng, D. Preparation of devulcanised ground tyre rubber with supercritical carbondioxide jet pulverization. Materials Letters 282(1–5):128878 (2021).
[87] Mangili, I., Collina, E., Anzano, M., Pitea, D., Lasagni, M. Characterization and supercritical $CO_2$ devulcanisation of cryo-ground tyre rubber: Influence of devulcanisation process on reclaimed material. Polymer degradation and stability 102:15–24 (2014).
[88] Kojima, M., Kohjiya, S., Ikeda, Y. Role of supercritical carbon dioxide for selective impregnation of decrosslinking reagent into isoprene rubber vulcanizate. Polymer 46:2016–2019 (2005).
[89] Kojima, M., Tosaka, M., Ikeda, Y., Kohjiya, S. Devulcanisation of carbon black filled natural rubber using supercritical carbon dioxide. Journal of Applied Polymer Science 95:137–143 (2005).
[90] Liu, Z., Li, X., Xu, X., Wang, X., Dong, C., Liu, F., Wei, W. Devulcanisation of waste tread rubber in supercritical carbon dioxide: Operating parameters and product characterization. Polymer Degradation and Stability 119:198–207 (2015).
[91] Li, X., Xu, X., Liu, Z. Cryogenic grinding performance of scrap tyre rubber by devulcanisation treatment with $ScCO_2$. Powder Technology 374:609–617 (2020).
[92] Mangili, I., Oliveri, M., Anzano, M., Collina, E., Pitea, D., Lasagni, M. Full factorial experimental design to study the devulcanisation of ground tyre rubber in supercritical carbon dioxide. The Journal of Supercritical Fluids 92:249–256 (2014).
[93] Jiang, K., Shi, J., Ge, Y., Zou, R., Yao, P., Li, X., Zhang, L. Complete devulcanisation of sulfur-cured butyl rubber by using supercritical carbon dioxide. Journal of Applied Polymer Science 127:2397–2406 (2012).
[94] Rajan, V., Dierkes, W. K., Joseph, R., Noordermeer, J. W. M. Science and technology of rubber reclamation with special attention to NR-based waste latex products. Progress in Polymer Science 31:811–834 (2006).
[95] Jana, G. K., Mahaling, R. N., Rath, T., Kozolowska, A., Kozolowski, M., Das, C. K. Mechano-chemical recycling of sulphur cured natural rubber. Polimery 52:131–136 (2007).
[96] Zhang, Z., Li, J., Wan, C., Zhang, Y., Wang, S. Understanding $H_2O_2$-inducedthermo-oxidative reclamation of vulcanised styrene butadiene rubber at low temperatures. ACS Sustainable Chemistry and Engineering 9:2378–2387 (2021).
[97] Araujo-Morera, J., Verdugo-Manzanares, R., Gonzalez, S., Verdejo, R., Lopez-Manchado, M. A., Hernandez Santana, M. On the use of mechanochemically modified ground tyre rubber (GTR) as recycled and sustainable filler in styrene-butadiene rubber (SBR) composites. Journal of Composites Science 5:68 (2021).
[98] Tsuchii, A., Takeda, K. Rubber-Degrading enzyme from a bacterial culture. Applied and Environmental Microbiology 56(1):26–274 (1990). PMID. 16348100.
[99] Tsuchii, A., Tokiwa, Y. Microbial degradation of tyre rubber particles. Biotechnol Letters 23(12):963–969 (2001).
[100] Tsuchii, A., Suzuki, T., Takeda, K. Microbial degradation of natural rubber vulcanizates. Applied and Environmental Microbiology 50(4):965–970 (1985).
[101] Tsuchii, A., Takeda, K., Tokiwa, Y. Degradation of the rubber in truck tyres by a strain of Nocardia. Biodegradation 7(5):405–413 (1996). PMID.9144970.
[102] Cui, X., Zhao, S., Wang, B. Microbial desulfurization for ground tyre rubber by nixed consortium-Sphingomonas sp. and Gordonia sp. Polymer Degradation and Stability 128:165–171 (2016).
[103] Isayev, A. I., Chen, J., Tukachinsky, A. Novel ultrasonic technology for devulcanisation of waste rubbers. Rubber Chemistry and Technology 68(2):267–280 (1995).
[104] Isayev, A. I., Yushanov, S. P., Chen, J. Ultrasonic devulcanization of rubber vulcanizates. I. Process model. Journal of Applied Polymer Science 59:803 (1996).

[105] Yun, J., Oh, J. S., Isayev, A. I. Ultrasonic devulcanization reactors for recycling of GRT: Comparative study. Rubber Chemistry and Technology 74:317 (2001).

[106] Hong, C. K., Isayev, A. I. Continuous ultrasonic devulcanization of carbon black-filled NR vulcanizates. Journal of Applied Polymer Science 79:2340 (2001)

[107] Shim, S. E., Isayev, A. I. Ultrasonic Devulcanization of Precipitated Silica-Filled Silicone Rubber. Rubber Chemistry and Technology 74:303 (2001).

[108] Levin, Y. V., Kim, S. H., Isayev, A. I. Vulcanization of Ultrasonically Devulcanized SBR Elastomers. Rubber Chemistry and Technology 70:120 (1997).

[109] Yun, J., Isayev, A. I., Überlegene mechanische Eigenschaften von ultraschallrecycliertem EPDM-Vulkanisat. Fasern Kunstst, G. 55:628 (2002).

[110] Ghose, S., Isayev, A. I. Recycling of unfilled polyurethane rubber using high-power ultrasound. Journal of Applied Polymer Science 88:980 (2003).

[111] Oh, J. S., Isayev, A. I. Continuous ultrasonic devulcanization of unfilled butadiene rubber. Journal of Applied Polymer Science 93:1166 (2004).

[112] Isayev, A. I., Liang, T., Lewis, T. M. Effect of particle size on ultrasonic devulcanisation of tyre rubber in twin-screw extruder. Rubber Chemistry and Technology 87(1):86–102 (2014).

[113] Mangili, I., Lasagni, M., Anzano, M., Collina, E., Tatangelo, V., Franzetti, A., Caracino, P., Isayev, A. I. Mechanical and rheological properties of natural rubber compounds containing devulcanised ground tyre rubber from several methods. Polymer Degradation and Stability 121:369–377 (2015).

[114] Moon, J., Kwak, S. B., Lee, J. Y., Kim, D., Ha, J. U., Oh, J. S. Synthesis of polyurethane foam from ultrasonically decrosslinked automotive seat cushions. Waste Management 85:557–562 (2019).

[115] Hong, C. K., Isayev, A. I. Blends of ultrasonically devulcanised and virgin carbon black filled NR. Journal of Materials Science 37(2):385–388 (2002).

[116] Scuraccio, C. H., Waki, D. A., Silva, M. L. C. P. Thermal analysis of ground tire rubber devulcanized by microwaves. Journal of Thermal Analysis and Calorimetry 87:893–897 (2007).

[117] Sun, X., The Devulcanisation of Unfilled and Carbon Black Filled Isoprene Rubber Vulcanizates by High Power Ultrasound, Ph.D. Thesis. University of Akron, USA (2007).

[118] Karabork, F., Pehlivan, E., Akdemir, A. Characterization of styrene butadiene rubber and microwave devulcanised ground tyre rubber composites. Journal of Polymer Engineering 34(6):543–554 (2014).

[119] Li, S., Lamminmaki, J., Hanhi, K. Effect of ground rubber powder and devulcanisates on the properties of natural rubber compounds. Journal of Applied Polymer Science 97:208–217 (2005).

[120] Aoudia, K., Azem, S., Hocine, N. A., Gratton, M., Pettarin, V., Seghar, S. Recycling of waste tyre rubber: Microwave devulcanisation and incorporation in a thermoset resin. Waste Management 60:471–481 (2017).

[121] Garcia, P. S., de Sousa, F. D. B., de Lima, J. A., Cruz, S. A., Scuracchio, C. H. Devulcanisation of ground tyre rubber: Physical and chemical changes after different microwave exposure times. Express Polymer Letters 11:1015–1026 (2015).

[122] Morera, J. A., Santana, M. H., Verdejo, R., Manchado, M. A. L. Giving a second opportunity to tire waste: An alternative path for the development of sustainable self-healing styrene-butadiene rubber compounds overcoming the magic triangle of tires. Polymers 11:2122 (2019).

[123] de Sousa, F. D. B., Carlos, H. S., Guo-Hua, H., Sandrine, H. Devulcanisation of waste tyre rubber by microwaves. Polymer Degradation and Stability (2017).

[124] Kakroodi, A. R., Rodrigue, D. Highly filled thermoplastic elastomers from ground tyre rubber, maleated polyethylene and high-density polyethylene. Plastics, Rubber and Composites 42(3):115–122 (2013).

[125] Mujal-Rosas, R., Orrit-Prat, J., Ramis-Juan, X., Marin-Genesca, M., Rahhali, A. Study on dielectric, mechanical and thermal properties of polypropylene (PP) composites with ground tyre rubber (GTR)Polym. Polymer Composites 20(9):797–808 (2012).

[126] Wang, Y.-H., Chen, Y.-K., Rodrigue, D. Production of thermoplastic elastomers based on recycled pe and ground tyre rubber: Morphology, mechanical properties and effect of compatibilizer addition. International Polymer Processing 33(4):525–534 (2018).

[127] Wiśniewska, P., Zedler, Ł., Formela, K. Processing, performance properties, and storage stability of ground tyre rubber modified by dicumyl peroxide and ethylene-vinyl acetate copolymers. Polymers 13(22):4014 (2021).

[128] Formela, K., Haponiuk, J. Use of tall oil in chemical industry. Przem Chem 91(6):1160–1163 (2012).

[129] Zedler, Ł., Colom, X., Saeb, M. R., Formela, K. Preparation and characterization of natural rubber composites highly filled with brewers' spent grain/ground tyre rubber hybrid reinforcement. Composites Part B: Engineering 145:182–188 (2018).

[130] Zhang, X., Zhu, X., Liang, M., Lu, C. Improvement of the properties of ground tyre rubber (GTR)-filled nitrile rubber vulcanizates through plasma surface modification of GTR powder. Journal of Applied Polymer Science 114(2):1118–1125 (2009).

[131] Zhao, X., Hu, H., Zhang, D., Zhang, Z., Peng, S., Sun, Y. Curing behaviors, mechanical properties, dynamic mechanical analysis and morphologies of natural rubber vulcanizates containing reclaimed rubber. e-Polymers 19(1):482–488 (2019).

[132] Li, Y., Shen, A., Lyu, Z., Wang, S., Formela, K., Zhang, G. Ground tyre rubber thermo-mechanically devulcanised in the presence of waste engine oil as asphalt modifier. Construction and Building Materials 222:588–600 (2019).

[133] Liang, M., Xin, X., Fan, W., Sun, H., Yao, Y., Xing, B. Viscous properties, storage stability and their relationships with microstructure of tyre scrap rubber modified asphalt. Construction and Building Materials 74:124–131 (2015).

[134] Rath, P., Love, J., Buttlar, W., Reis, H. Performance analysis of asphalt mixtures modified with ground tyre rubber modifiers and recycled materials. Sustainability 11(6):1792 (2019).

[135] Bevilacqua, E. M. Vulcanization with TMTD. Rubber Chemistry and Technology 32:721–738 (1959).

[136] De, D., De, D., Singharoy, G. M. Reclaiming of ground rubber tyre by a novel reclaiming agent. I. virgin natural rubber/reclaimed GRT vulcanisates. Polymer Engineering and Science 47:1091–1100 (2007).

[137] Ghorai, S., Mondal, D., Hait, S., Ghosh, K. A., Wiessner, S., Das, A., De, D. Devulcanization of waste rubber and generation of active sites for silica reinforcement. ACS Omega 18: 17623–17633 (2019).

[138] Ghosh, J., Hait, S., Ghorai, S., Mondal, D., Wießner, S., Das, A., De, D. Cradle-to-cradle approach to waste tyres and development of silica-based green tyre composites. Resources, Conservation and Recycling 154:104629 (2020). Mar 1

[139] Ghorai, S., Mondal, D., Hait, S., Ghosh, A. K., Wiessner, S., Das, A., De, D. Devulcanisation of waste rubber and generation of active sites for silica reinforcement. ACS Omega 4(18):17623–17633 (2019 Oct 18).

[140] De, D., Das, A., De, D., Panda, P. K., Dey, B., Roy, B. C. Reinforcing effect of silica on the properties of styrene butadiene rubber–reclaim rubber blend system. Journal of Applied Polymer Science 99(3):957–968 (2006).

[141] Mondal, D., Ghorai, S., Rana, D., De, D., Chattopadhyay, D. The rubber–filler interaction and reinforcement in styrene butadiene rubber/devulcanize natural rubber composites with silica–graphene oxide. Polymer Composites 40(S2):E1559–E1572 (2018).

[142] Mondal, D., Hait, S., Ghorai, S., Wießner, S., Das, A., De, D., Chattopadhyay, D. Back to the Origin: A Spick-and-span Sustainable Approach for the Devulcanization of Ground Tire Rubber. Journal of Vinyl & Additive Technology 29:240–258 (2023). https://doi.org/10.1002/vnl.21974

[143] Isayev, A. I., Yushanov, S. P., Kim, S. H., Yu Levin, V. Ultrasonic devulcanization of waste rubbers: Experimentation and modeling, Rheol. Acta 35:616–630 (1996).

[144] Molanorouzi, M., Mohaved, S. O. Reclaiming waste tire rubber by an irradiation technique. Polymer Degradation and Stability 128:115–125 (2016).

Kasilingam Rajkumar and Santosh C Jagadale

# Chapter 10
# Cost of quality in rubber processing

**Abstract:** New Age Economics is driven by sustainability development goals and resource optimisation. Quality plays a pivotal role in such demanding situations that are backed by consumer sentiments. Industry focuses on segregation of materials based upon its quality. Lower rejection and rework rates not only lead to increased bottom-line, customer loyalty, brand enhancement, and increased overall competitiveness but also assist the industry in achieving the sustainable development goals and social harmony. Typically, in an industrial scenario, rejection rate/rework rate may vary from 5% to 30% and at times, it may be even more. There are various hidden costs associated where there is low quality. It is just a tip of the iceberg that is visible. Reasonable inferences can be drawn from the fact that many of the costs of quality are hidden and are difficult to be estimated/measured by prescribed measurement systems. Further, few costs of poor and good quality are obvious and are tangible. Such things present immense potential opportunities for identification of hidden quality costs and their reduction over time period. It is thus a necessity for competitively placed businesses to discover the hidden part of the quality orientation. Most of the small and micro small rubber industries are unorganised in their process of manufacturing and hence, are unfamiliar with the concept of cost of poor quality (COPQ). Further, their supply chains are driven by price sensitivities, neglecting prescribed quality standards and are therefore weak in responsiveness. Such a supply chain creates competitiveness driven by constraints. Introduction to the COPQ and the application of techniques towards its reduction or elimination enhances not just the competitiveness but also results in increased profitability. This chapter discusses at depth the COPQ and provides remedies/alternatives to deal with the same for increased competitiveness of the industry.

## 10.1 Introduction

Cost discovery of quality is unfamiliar in most of the rubber product manufacturing units, especially in small and microbusinesses. Typically, quality failure during work processes result in material and labor wastages, leading to reworks and additional operational time, and increased material and labor costs, resulting in reduced profitability and operational effectiveness. This quality failure costs are generally not measured and are not reflected in a company's financials separately, and thus go unnoticed. They remain buried in the financial numbers and its consequence being management being unaware about the loss estimation and the gravity of its significance. The cost of poor quality (COPQ) cannot be traced or identified using the existing financial reporting and

https://doi.org/10.1515/9783110668537-010

auditing systems. Identification of cost figure on quality is a difficult job and financial reporting systems are ill-equipped to discover the 'true' cost of quality (COQ). The top leadership is mainly concerned with the overall cost and is lesser aware of the extent of their hidden losses on account of poor quality.

Campanella and Rao et al. state that it was Juran who gave rise to the concept of quality costs in his first *Quality Control Handbook*, wherein he tells his famous analogy of 'gold in a mine'. Still, Barbara et al. [1] and Evans et al. [2] contend that the concept of 'COQ' and 'cost of non-quality' was developed by Frank Gryna in the 1950s, with the objective of presenting to top executives the language of quality translated into monetary value [4, 5]. However, it is largely accepted that the traditional COQ concept was developed by W.J. Masser in his 1957 article, 'The quality manager and quality costs', where he subdivided the quality costs into prevention, appraisal, and failure. Lending further validity to the COQ concept, the American Society of Quality Control formed the Quality Cost Committee in 1961 to make the business community aware of quality costs so that businesses might improve their quality through the measurement of quality costs [5]. Two years later, the US Department of Defense adopted the quality cost program in 1963.

Finally, Feigenbaum [6] further developed the COQ model in his classic book, *Total Quality Control*. The concept of COQ was introduced by Juran in the year 1951 [7]. COQ theories have become important economic measures on quality issues for several decades. As mentioned by Juran [7], a company will not survive if its quality level is low, except where there is a monopoly. For these very reasons, the industry spends significant resources on improving the product or service quality in order to survive and remain competitive. Atkinson et al. studied how effectively the cost appraisal system proposed measures for the COPQ in a construction project [8]. The results showed that for the 60-day study period, COPQ decreased by about 24% while labor productivity and profitability increased by about 17 and 11%, respectively, after the implementation of the COPQ measuring system.

The process involved in making of finished rubber product involves conversion of raw materials into finished product without any rejection at any of its stages. In such an instance, the COQ will be only the cost of prevention/appraisal cost. Various uncertainties and variations in processes lead to process deviations. Internal processes fluctuations and failure to comply with the specified requirements result in COPQ, which is added to the cost of product. These failures and its associated costs can be of internal or external nature. Therefore, total quality cost includes prevention cost, appraisal cost, internal failure, and external failure costs.

## 10.2 Costs of quality

COQ is deemed to be understood as the arithmetic sum of the conformance and nonconformance costs, where the cost of conformance is the cost incurred for prevention of poor quality (e.g., inspection and quality appraisal) and cost of nonconformance is the COPQ caused by the product and service failure (e.g., rework and returns) [9]. Juran [7] has suggested that the COQ can be understood in terms of the economics of the end-product quality or in terms of the economics of the conformance to standards.

According to research in 1997, six primary theories are existent related to the COQ: (i) Juran's model, (ii) loser's classification, (iii) prevention-appraisal-failure model, (iv) the economics of quality, (v) business management and the COQ, and (vi) Juran's model revised [10]. Among the various proposed theories on COQ, Juran was the one who proposed the concept of COQ and the value derivatives of quality; he further demonstrated relationship between the optimum cost and perfection of the product. Feigenbaum, on other hand, presented that COQ shall include cost categories of total quality control such as prevention cost, appraisal cost, and failure cost; it is widely accepted as the prevention–appraisal–failure model. The work of Faigenbaum is a 'milestone' in the development of COQ theories [6].

On deployment of huge funds on quality-related efforts, the management seeks to measure quality in measured management terms – on its returns on these investments, and the demands for further improvements, and risk mitigation. Traditionally, the solution to these needs has been the COQ. Shahid Mahmood studied the method of determination of COPQ and its impact on productivity and profitability in a civil construction project [11].

The COQ is defined as the monetary measurement for quality loss of a business [7]. Shahid and Sajid [12] reported the different aspects of COPQ in public sector projects. COQ analysis empowers organisations to identify gaps, work out remedies, and control the consequences of poor quality. The major goal of a COQ approach is to improve the bottom line by eliminating/minimising poor quality [13]. The COQ concept is further strengthened by the implementation of Sig Sigma, Kaizen, and Total Quality Management.

Arithmetically, the discovery of quality costs are too complex using basic mathematical functions; instead they are heavily dependent on the support processes like maintenance and human resources, which majorly contribute to the summation of COPQ. The major quality costs are aggravated by incapable support processes. COQ, after its recognition, can be reduced through structured approaches [3].

After price and delivery time, questions are raised about a product's durability. This is usually a very difficult question to answer for rubber products because

- Product life cycle is in approximation of decades
- Complexity of service conditions
- Lack of definitive data on durability
- Complex matrix of degradation agents, service conditions, properties of different rubbers

Data, data analytics, and its monitoring are thus priority areas to answer such questions. Such acts add to the cost of the product.

COQ is defined as a methodology that allows a company to determine the extent to which its resources are used for activities that prevent poor quality, that appraise the quality of the organisation's products or services, and that result from internal and external failures. Usually, rejections occur due to poor quality of raw materials, wrong formulation design, mixing error, processing errors, etc. The quality of raw materials can be accessed through an established system of good quality control.

Raw material rejections and added testing add an additional cost to the COQ. Supply chain associations/partnerships need to be strengthened and quality verified for its correctness as per quality control plan to ensure the right quality of raw materials. Quality assurance plan of the overall product manufacturing processes should possess quality control plan of the raw materials as an essential ingredient. Effective human capability development program for technicians and operators, and regularly monitoring the quality of work in process (WIP) can minimise the consequential effect. WIP materials get rejected if the process is not well established and implemented.

It is therefore important to assess the essential COQ and the avoidable COPQ for taking precautionary action to minimalise or eliminate such problems. Procedures and systems, along with their implementation, would help businesses minimise/eradicate unwanted or value-deficit processes. Such implementations will reduce the COPQ to some extent. Quality control, resulting in the rejection of any in-process materials at any stage of the production process, is classified as appraisal cost, leading to corrective/preventive action to rectify the same.

In his book, *Quality Is Free*, Crosby defined COQ as having two main components: the cost of good quality (or the cost of conformance – prevention and appraisal costs) and the COPQ (or the cost of nonconformance – internal and external failure costs). Furthermore, Crosby (1983/1987) stated that no subject has received more attention from quality professionals over the past years than COQ [14].

The COQ can be divided into four categories, namely, (i) prevention cost, (ii) appraisal cost, (iii) internal failure cost, and (iv) external failure cost. They are briefly explained below.

## 10.2.1 Prevention cost

The cost accrues and originates from defect prevention efforts at different stages of the rubber production process. This is the cost of related activities of quality assurance mechanism involved at the different stages of the manufacturing processes. Prevention costs are mainly incurred to avoid or evade quality problems during the product manufacturing and life cycles. These costs are linked with the design, implementation, and maintenance of the quality management system. They are planned and incurred before the actual operation, and these could include capability building training that includes design, development, preparation, and maintenance. A customer perceives a product as having a high level of quality if it conforms to customer expectations. Establishing a quality management system and certification of the company for compliance with ISO 9001 will help in preventing COQ.

In the rubber industry, there are different key processes that are critical in the manufacturing of a rubber product. Few of these processes or stages includes testing and verification of inward material quality, mixing process, vulcanisation process by using different vulcanisation techniques, product finishing, product testing, inspection, packaging, and dispatch. In each of these stages, there is added value in terms of material or labor.

These stages are considered as critical stages. Data collection, in terms of defects, has to be done for all these stages so that the net production rejection can be calculated for the required period.

### 10.2.2 Appraisal cost

This cost accrues and originates from defects detection efforts. This will appear when the prevention of quality is not properly implemented. This is the cost associated as a part of quality control. The testing and segregation of costs are included in this type of appraisal cost. Quality appraisal activities are the most traditional quality practices and these activities are quite visible incidentals. These can be easily seen in the balance sheet under asset schedules, recurring manpower costs, etc. Appraisal is imperfect and an expensive way of organisational quest for achieving quality. Appraisal is a confirmatory mechanism to validate that the production processes and preventive measures are working.

Appraisal is about segregation of good from the bad product, counting defects, scrapping and calculating yield, while this is an essential ingredient of the quality program. Preventive methods to reduce cost of failures are thus pivotal to reduce the ultimate impact on the COPQ.

### 10.2.3 Internal failure cost

This is the cost incurred due to deviation or variations in different processes, rejections, rework, scrap, etc. This is incurred due to ineffective QC and QA mechanisms. This cost can be minimised through effective implementation of systems and procedures within the factory, for example, cost of rework (fixing of internal defects and re-testing). This could be achieved by deploying supervision of processes, establishing statistical data, and analysing the same. The process improvement programme should be regularly conducted and documented. Regular review should be conducted to monitor the improvement towards reduction of internal costs. Pictorial display methods of such failures will help workers to improve the system.

## 10.2.4 External failure cost

This cost is associated with various possible sources that includes wrong design, misuse, etc. Failure investigations are generally undertaken to quantify such costs and to identify and investigate the causes of failure. This cost accrues from client/end-user identified defects and the corrective efforts associated with it. There are several quality tools applied to identify such causes and factor the associated cost. Estimation of external failure costs or costs that occur after delivery of a defective rubber product is highly challenging, and quantification of it is thus difficult. This cost includes cost of rework (fixing of external defects and re-testing) and any other costs associated with external defects (product service/liability/recall, etc.). Intangible costs of external quality failures are hard to calculate but are easy to simulate/visualise it's huge negative impact on the future performance of the business as it results in loss of brand image, dissatisfied customers, customer attrition, etc., with added costs on the ground for penal provisions on sustainability. The only possible way to avoid external quality failure costs is to have a clear focus on improving the balance of the three COQs.

In the rubber sector, mostly automotive rubber components are the best example to have a system to monitor the COQ. Customer-originated and monitored demand by original equipment manufacturers (OEMs) has compelled the rubber industry to implement such a vibrant quality system. This has helped it reduce the COQ, including COPQ. It is mandatory to demonstrate the improvement in systems related to the improvement of quality, improved productivity, reduction in rejection, rework and replacement, etc. Quality Dashboards are widely used for constant monitoring and periodic reviews are conducted.

There are numerous possible sources within each of four categories that are cost-related, on basis of quality performance of rubber products. Therefore, to summarise, the COQ is a measure that quantifies the cost of control towards conformance to processes to make rubber products and the failure costs associated with failure of control systems to check nonconformance involved in the product lifecycle. One can thus easily infer it as a summation of the costs related to prevention and detection of defects and the costs due to occurrences of defects.

Organisation spending on preventive actions versus failure fixing thus being derivative trade-off costs. Additionally, through the conventional way of reducing the COQ by defect reduction, it is possible to tackle the efficiency of quality management system, which in itself results in reduced COQ.

## 10.3 Quality in the rubber industry and its associated costs

It is thus expedient in the interest of better understanding to examine the different COQs involved in the rubber product manufacturing processes. Conventionally, the rubber product manufacturing involves the following major steps.

### 10.3.1 Structural design of the rubber product (engineering drawing of the product)

The product, mould, or assembly drawing contains numerical aspects that define the size and location of the part. Important attributes, including the material, surface finishes, fabrication methods and other industrial processes necessary to create the part are communicated in it. These variables add to the overall production cost. It is the designer's responsibility to determine every dimension's tolerance and should be determined based on how the parts fit together [15].

If the product fits properly and functions as per the requirement by *allowing a greater amount of tolerance*, then a larger tolerance should be allowed. Dimensional tolerances should be as large as possible without impacting the assembly or the performance of the part from the prism of cost. Also, the tolerance defines the acceptable amount of deviation from the dimension's nominal value. The 'allowable tolerance' window can have a dramatic effect on the manufacturing method and total cost of producing the part. As a general rule, smaller the allowable deviation, the larger the amount of money it takes to manufacture.

This phenomenon is due to the fact that extremely precise dimensions are more difficult to achieve and increase the chance of rejection, rework, and scrap [16]. For 'tight tolerance" there is a need to use much more expensive production methods, machine tools, inspection devices, and a significantly greater amount of total processing time. This can add up to a significant amount of money as it based on product fitment and optimum performance; hence, tolerance needs to be assigned in design. More stringent the tolerance, more are the chances of rejection.

Structural design should be well documented to represent the actual product and its assembly. Most rubber industries do not have the facility and competency to document the design control of product. In fact, design changes in the rubber product lead to confusion and to rejection. Therefore, one has to properly document the design of a rubber product. Product design is different from the mould design and hence, it is essential to properly document and control the same to avoid confusion. Similarly, the assembly drawing of a product function should be documented.

Changes in design should be properly documented and replaced. Old design should be marked as 'Obsolete' and should be removed from the point of use.

### 10.3.2 Development of suitable rubber compound materials

Development of a rubber compound involve several steps, which include the selection of rubber(s) and other rubber compounding ingredients such as fillers, plasticisers, curatives, and any other special additives; suitable mixing processes; and then testing and confirmation of meeting the specifications. The right selection of materials of right quality, and controlled parameters during the mixing process will result in a better consistent rubber compound for use in the industry. Any variation in the quality of raw materials and variation in the mixing parameters will lead to rejection. Industries do not properly document the process control parameters required for mixing a rubber compound. Adequate care should be taken to prepare the process flowchart, with a process information log sheet. Batch to batch variation needs to be analysed and controlled to reduce the cost of (poor) quality.

### 10.3.3 Process to shape the products through mixing, extrusion, moulding, etc

In the rubber industry, various stages are involved in the manufacturing processes. The key steps being raw materials stores management, weighing, mixing, intermittent processes like extrusion or calendaring, etc. as per the product types, and finally moulding – compression, transfer, or injection techniques. Raw materials should be stored properly with lid containers in a controlled environment. Weighing errors should be minimised. The mixing sequence should be properly maintained during mixing. The temperature should be controlled during mixing to avoid variation in the compound properties. Proper process conditions should be maintained during the intermittent processes, like extrusion, calendering, moulding, etc.

Moulding temperature and timing need to be controlled and monitored during the production time. Most of the time variations in temperatures of the top and bottom platens lead to lots of variation in the quality of the products. A digital monitoring system with a computerised data logger can assist in data generation and analysis for monitoring the process precisely.

## 10.4 Reducing the cost of quality failures

Reducing the COPQ is achievable by reducing the COQ failures. COPQ is the cost associated with providing poor quality products or services due to failure to conform to the quality standards of customer requirements. Harrington defines COPQ as all the costs incurred by the company and the customer because the output did not meet the specifications and/or customer expectations [17].

Production/supplier site inspections conducted at production, receiving or supplier site, allows one to examine/investigate some aspects of products, materials, and components that are nonconformant to specifications. This nonconformance could lead to rework, scrapping, returns, and recalls; all of which should be documented in nonconformance or discrepancy reports and classified in a database in such a way that the organisation can use the data analytics to determine the costs and areas for improvement. Rubber compound gets rejected due to noncompliance to specification, and scrapped. Such rejection and scraps on the shop floor are very common. The absence of a rheometer in any factory is leading to wrong judgment of rubber mixing, and leading to continued production without ensuring the quality at the mixing stage. The products get rejected at the final stage, leading to increase in COPQ.

## 10.5 Role of suppliers in the control of cost of quality

Supply chains play a key role in the COQ. A right supplier can play a big role in minimising the COPQ. The suppliers of raw materials and assembly need to be treated as part of the business model and hence, enough awareness needs to be created about the criticality of the quality of supplies made at any time. Periodic audit and feedback mechanism should be in place for the verification of the requirement for the actual production processes. Proper assessment of supplier, supplier ratings, etc. will help in implementing quality culture at the supplier end as well. This will minimise the COPQ to a greater extent. Ongoing practices focus on supplier certification and receiving inspection for supplier appraisal.

Minimising inventory levels and fast mobilisation of the product to the end-user consumer has been a recent focus of the industry. At times, there is the possibility of a technical issue being discovered during receiving inspection that will adversely affect the production schedule. Thus, there is renewed demand by companies and focus on additional inspection requirements into supply chain, using more source inspections. Inspection templates allow requirements to be effectively managed for both source and receiving locations.

## 10.6 Role of maintenance of equipment in the control of cost of quality

Financial reporting must include maintenance-related costs. Analysis of costs of maintenance helps management evaluate the comparative impact of its maintenance problems on the organisation's bottom line. This results in a communication synergy between the top management and the maintenance function to address areas where the management should allocate resources to rectify quality problems and make improvements. The system facilitates new avenues for communication to share best practices and coordinated efforts of experts in quality management with the maintenance team. This will lead to improved understanding of the statistical capability and reliability of the equipment [18].

Improved quality and reliability of products and processes may be achieved through early stage detection and elimination of common and special causes. Based upon machine condition and pattern of stoppages, one can decide upon the need for overhaul. It is easy to establish a common database that facilitates the monitoring of deviations in the condition of complex and highly modernised equipment. Maintenance of such equipment can be achieved through training of the maintenance staff in the interpretation of vibration spectra for fault identification at an early stage and of the operator on how to respond to program's warning signals [19].

According to a study reported by Mobley, 15–40% (with an average of 28%) of the total cost of finished goods can be attributed to maintenance activities in the factory [20, 21].

Equipment resources must be maintained to assure their optimal performance to capabilities, especially measurement equipment used for product verification purposes. Preventive maintenance and calibration processes for equipment and tools must be standardised and documented. Tools can be calibrated more efficiently, based on actual usage, instead of using the traditional data-driven expiration scheme, which ends up in more calibration work than is needed.

The absence of preventive maintenance mechanism leads to machine breakdowns and results in losses on account of productivity. Sometimes, improper maintenance of equipment and tools can cause the rejection of products, which will ultimately add to the COQ.

## 10.7 Role of quality documentation

Several quality documentations help in improving the control of COPQ. Few such documents are described below for the necessary implementation and benefits.

### 10.7.1 Quality plan

A quality plan is a document, or a set of several documents that together specify quality standards, practices, resources, specifications, and the sequence of activities relevant to a particular product, service, project, or contract. Quality plans should clearly define a method for measuring the achievement of the quality objectives. The word quality is often used broadly for many different meanings. Quality can be defined as 'readiness for use', 'customer delight', 'doing things right the first time' or 'zero defects'.

These definitions are acceptable because quality can refer to degrees of excellence. In the rubber industry, a quality plan is often wrongly understood or defined. It is highly essential to make the workforce read and understand the quality plan implanted at the level of various departments/processes in the rubber industry.

It is important to plan which activities will be used to measure the quality of the product. And it needs to think about the cost of all the quality-related activities it wants to do. Then, it is needed to set some guidelines for what it will measure. Finally, it is essential to design tests that will run when the product is ready to be tested.

In the rubber industry, the quality plan is implemented for the following, but not limited to:

1. Raw materials
2. In-process materials
3. Finished products

The quality plan for raw materials may involve various tests like ash content, moisture content, heating loss, melting point, etc. In-process materials like mixed rubber compound is tested for its Mooney Viscosity, Rheometric test, specific gravity and, if required, the physical properties and other critical tests specified in the product standards and specifications. An internal control mechanism must be implemented to ensure right quality of mixing and that there are also upper limit and lower limits of the rheometric curve so that they are statistically studied and analysed to narrow down the gaps in the process while implementing continual improvement in the system. The finished products are tested for its dimensions as per the drawing, and any other critical/functional tests that are specified in the technical contract/agreement.

The following are few parameters that are usually specified for its minimum requirements/ASRTU specification for tread rubber [22]:

- Polymer content
- Tensile strength
- Elongation at break
- Modulus
- Tear strength
- Adhesion strength

The following are few parameters that are usually specified for its maximum requirements as specification [23]. Specification for automotive rubbers:

- Ash content
- Specific gravity
- Compression set
- Tension set
- Relative vol. loss
- % Volume swell

The following are few parameters that are usually specified as a range with certain tolerance as requirements in the specification. Sample specification for bonding gum for retreading application are [24]:
- Mooney viscosity
- Mooney scorch time
- Hardness ± 5
- % change in physical properties.

There are also many inherent difficulties in designing tests for some products to ensure a good quality assurance plan. Whilst large amounts of durability data are generated by accelerated methods, much of it is only useful for quality control purposes and relatively little has been validated as being realistically capable of representing service. Lifetime is not necessarily measured in time. For example, for some rubber products, it will be mentioned as number of cycles of use. There has been increasing pressure to improve performance so that rubbers can be used at higher temperatures and in harsher environments.

### 10.7.2 Quality checklist

A quality checklist is a tool used to assess the quality of a product or service. Essentially, a quality checklist comprises a list of items that are relevant to quality assurance and for the quality control process of a company. The quality checklist is an integral part of quality management systems. Quality checklists are properly numbered and the version of such document is controlled. The rubber industry uses such checklists for quality control of various critical processes involved in rubber product manufacturing. Such critical processes involve quality of the raw material, raw material weighing, mixing, extrusion or calendaring or moulding, etc. Such checklists would contain check parameters that will focus on continuous monitoring of a set of parameters such as temperature, pressure, time, etc. Since such quality checklists come with the user's concurrence, the question of dispute is negligible.

This checklist may be used for analysis during the process of customer feedback/complaint evaluation or at later stages for process evaluation and for improvement purposes.

## 10.8 Standard operating procedure (SOP)

Standard operating procedures (SOPs) are the documented processes that allow a company to ensure that services and products are delivered consistently every time. SOPs are tools used to ensure compliance to regulation or operational practices as well as to document how particular tasks must be performed and completed within a company. Procedures are not one-stop solution; they may not guarantee excellent performance or good results. However, SOPs facilitate that structured management systems and processes are in place for qualified/trained/skilled employees, and inculcate quality culture in the company. SOPs, in their purest form, empower people to perform effectively and efficiently in a robust environment.

SOP facilitates not just ongoing activities within companies but can also be used to further decisions regarding operational processes, moving forward. There are many additional benefits that can be realised by bringing standardisation in processes and documenting formal SOPs of the company. They assist people on the aspect of safety, health, and environmental issues as also relevant operational information in connection with the same so as to perform a job/task effectively and efficiently, with operational consistency and quality control of processes and products. They also allow process flow in an uninterrupted and consistent manner for the timely completion of the production schedule. Adherence to SOPs allows company to avoid process shut downs caused by equipment failure or other facility damage [26]. Frequent equipment failures can be averted through implementation of preventive and predictive maintenance activities of equipment through proper SOPs. Such SOPs also mitigate the risks of equipment/process failure during manufacturing activities that could possibly harm the surrounding community or the environment. They also allow the company to function and perform in conformity to relevant government or regulatory compliance framework.

SOPs serve as ready reckoner for employee training, enabling users to learn operational parts about the process for which the SOP was written, to serve as a checklist for coworkers who monitor the job performance, to serve as a checklist for auditors, etc.

An SOP is a set of step-by-step instructions compiled by any rubber factory to help workers carry out the routine production activities. Defining the process of rubber product manufacturing is again subjective in nature and one has to be very careful in writing SOPs for such processes. SOPs aim to achieve efficiency, quality output, and uniformity of performance within the factory.

An SOP is a unique template designed by the management and implemented by it to ensure that the people in the company can easily recognise the document and implement it through its strict compliance. Version control allows proper identification of the latest document and eliminate older documents from the operational environment.

## 10.9 Quality assurance plan

Quality assurance plan is a series of measures that cover all technological stages of the development, release, and operation to guarantee the quality of the rubber product under manufacture. It is a systematic process that ensures product and service excellence. A robust QA team examines the requirements to design, develop, and manufacture reliable products, thereby increasing client confidence, company credibility, and the ability to sustain and thrive in a competitive market. Many a times people get confused between quality control and quality assurance. The difference is that QA is process-focused and QC is product-focused. Testing being product oriented, it comes under the QC domain.

The quality control of raw material procurement, raw material storage, weighing of raw materials, mixing and its subsidiary processes such as extrusion/calendaring/moulding, etc. are extremely essential to implement and the monitoring of their quality control is extremely important. Any deviation from the prescribed quality parameters should be addressed before it results in increased COPQ. A quality assurance plan is illustrated in Figure 10.1.

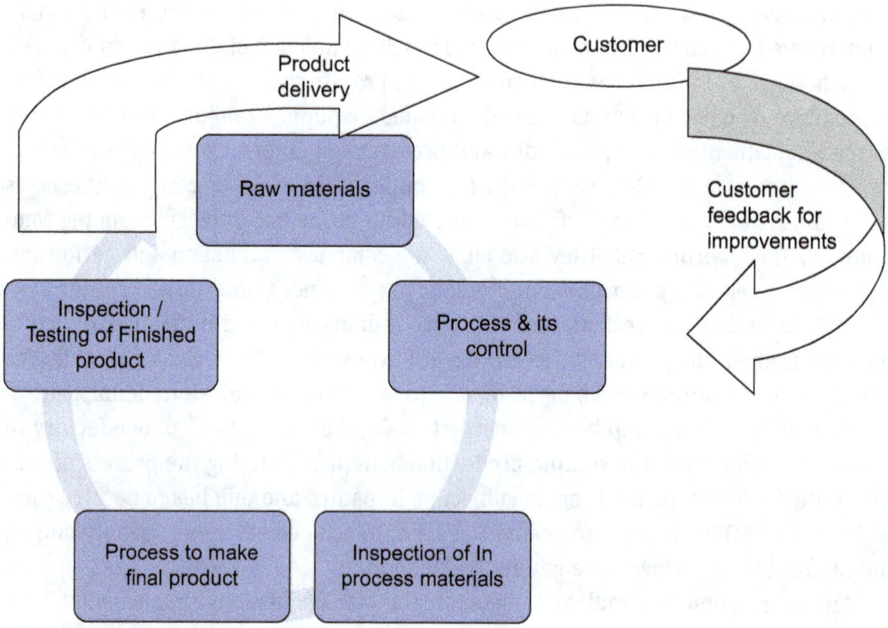


**Figure 10.1:** Quality assurance plan.

Testing for quality is not assuring quality; it is about controlling it. Quality assurance makes sure that people are doing the right things in a right way:

1. Effect of poor quality and the associated problems
2. Types of COPQs

COPQ can be broadly categorised into three costs:

1) Due to nonconformities
2) Due to lost sales
3) Due inefficient processes

When companies refer to these costs, they do not stick to these categories or look at four categories: (i) internal failure cost, (ii) external failure cost, (iii) appraisal cost, and (iv) prevention cost.

## 10.10 Methods to identify the cost of poor quality

In an operational environment, a rubber technologist should act like a detective – gathering evidences, understanding the material type, and using deductive reasoning. There are various methods for identifying the COPQ. By doing proper due diligence of the below listed methods, one can discover the COPQ:

- Establishing proper product and process traceability.
- Closed loop nonconformance and corrective action mechanisms
- Systematic preventive maintenance procedures
- Periodic effective internal quality audit
- Seamless change management process.
- Aggregated customer complaints and timely redressing mechanism.
- Streamlined supplier quality program

## 10.11 Different techniques to deal with cost of poor quality

### 10.11.1 Training and skill development

Human resource is an important constituent of quality. One cannot underestimate or undermine the important role human element plays in addressing quality problems. Despite sophisticated production systems and state-of-the-art facility, one can still have quality problems if people associated with it fail to comply with SOPs or do not implement QC or QA guidelines or quality checklists.

Certification of personnel as competent should be based on education, training, skills, and experience. Human resource qualification processes must be standardised and documented. Supervisory certification of a subordinate for the task is absolutely essential. System for validating each employee's skills and certifications and their latest training records are absolutely essential before they can be assigned to a particular task/job.

Human resource qualification processes must be standardised and documented. Trained and qualified manpower is the need of the hour for the rubber industry. The rubber industry workforce lack necessary qualifications and mostly are trained in-house and used for the processes, particularly in small and micro small companies.

They are mostly unaware about the criticality of the processes. The supervisor should sensitise them of the criticality of the processes and their significance in the final quality of the finished product.

The workforce need to be well trained, certified, and monitored for their competency to make them fit/suitable for the production/associated processes in the rubber industry. As part of Skill India Movement, National Occupational Standards are formulated for every industry by the National Skill Development Corporation (NSDC) and rubber industry-oriented job roles are already defined by Rubber and Chemicals, Petrochemicals Skill Development Council (RCPSDC) and certification schemes are available. Industry can very well avail the certification program offered by RCPSDC and get benefited [25].

## 10.11.2 Reducing reliance on 100% inspection

Implementation of a comprehensive inspection plan, based on production process validation or first article inspection rules, when there is (a) a change in design, affecting fit, form, or function of the product, or (b) a change of manufacturing sources, processes, inspection methods, location, tooling or materials, which affect the fit, form, or function would facilitate the transition from 100% inspection to 100% of the time.

In an ideal environment, one would like to move away from 100% inspection to more inspection by production personnel; leaving only a small percentage of random over inspection for quality management personnel. A manufacturing execution system (MES) enables these types of strategies. One way to reduce the inspection levels from 100% inspection, 100% of the time, is to trigger comprehensive inspection, based on production process validation or first article inspection rules when there is

- a change in design, affecting the fit, form, or function of the product, or
- a change of manufacturing sources, processes, inspection methods, location, or tooling, or
- materials, which affect the fit, form, or function. In the rubber industry, it is often that 100% inspection is not carried out.

They are randomly sampled and tested and in case the reference samples are not passed, then the entire lot is rejected or the lot is subjected to re-sampling processes before rejection. The cost of rejection is very huge and the rubber industry does suffer a lot in this regard. This is all because of the absence of proper internal quality control process. The 100% inspection, as deemed fit, are not carried out at the factory level before they are offered to an inspection agency or to the customer. Inspection and testing are not taken with a serious note, which leads to huge rejection at the final stage. In order to reduce the cost of inspection and testing process, scientific sampling is introduced.

### 10.11.3 Scientific sampling

Acceptance sampling is the process of evaluating a portion of the product in a lot for the purpose of accepting or rejecting the entire lot.

Advantage of sampling is 'ECONOMY': If the past history shows that the quality level is much better than the breakeven point and is stable from lot to lot, little if any inspection may be needed. If the level is much worse than the breakeven point and consistently so, it will usually be cheaper to use 100% inspection rather than sampling.

If the quality is at neither of these levels, a detailed economic comparison of no inspection is required.

### 10.11.4 Sampling risks

Neither sampling nor 100% inspection can assure that every defective item in a lot can be traced. The following risks are always associated during the quality assurance of a rubber product:

Sampling risks

α Risk β Risk.

α-Producer risk: Good lots can be rejected

β-Consumer risk: Bad lots can be accepted

Quality plan should include a sample plan to mitigate risks associated, by making the producer secured and safeguarded against rejection of good lots and the consumer safeguarded and secured against acceptance of bad lots. The costs associated with such sampling risks need to be captured.

## 10.12 Automated inspection and automated statistical process control (SPC)

Automation allows data to be collected quickly and efficiently and 100% inspection becomes feasible.

For example, in the volume-driven hose industry, a laser system can be used to measure the diameter of tubing coming out from an extruder in an automated environment at high speeds. Additionally, a vision system may inspect the multiple dimensions and surface properties. Such types of multicharacteristic testing occurring in an automated way on the production line is a major advancement in inspection and measurement methods as it eliminates the need for an operator by performing manual testing with hand held gages. It also helps one collect data quickly and efficiently. Wherever possible, such systems can be introduced to minimise human errors, improve the product quality, and the mitigate risks associated with rejection.

In general, process fluctuation is often observed in the rubber industry. The processes, in general are never stable; they appear to be stable if:

a) Observation period is too short
b) Measuring precession is inadequate
c) Such fluctuations are caused by

- Raw material variation from batch to batch
- Temperature and humidity/other environmental condition may vary from time to time/day to day
- Operator changes
- Electrical voltage fluctuation and so on

Therefore, the statistical process control (SPC) should help one identify, quantify, and reduce them to keep them within permissible limits that are compatible with the specification or objective. Automation in process has been introduced to reduce the variations in process and mitigate the associated risk.

Lot of data needs to be collected wherever inspection is needed so as to explore the possibility of reducing manual efforts using automated inspection methods, such as coordinate measurement machines or visual inspection machines that are integrated directly into the MES inspection data collection records. Inspection and test results coming out of measurement and inspection machines can be imported directly into the MES. Critical measures and results can be tied to data collection points and to SPC run charts to monitor the control levels.

Automation and its integration allow increased levels of operational efficiency over the traditional processes of inspectors using spreadsheets and a separate SPC software on the side. In the rubber industry, the scope for automation is very limited and only large sized companies like the tyre industry can afford to implement such automation. Mostly, it is identified that the mixing process can be automated in any rubber industry and other processes can be automated, based on requirements. The production processes like moulding, etc., operate in a semiautomated environment and human intervention is essential to ensure proper production. On the bright side, the semiautomatic process improves the quality of product with reduction in the rejection/rework opportunities.

Small and medium-scale industries have not yet realised the significance of automation and its role in the reduction of COPQ. Thus, creating awareness is the pressing need among such SME-based rubber industry. Low-cost automation systems should be implemented in the MSME rubber sectors to improve the productivity and quality.

Wherever possible, the industry should plan such low-cost automation system so that the risks associated with human error can be mitigated. This approach will enormously assist in reducing the poor quality material that goes into waste or rejection and increases the COPQ.

## 10.13 Rework, re-inspection, and retesting

Management information systems can integrate the data pertaining to rework, re-inspection and retesting and analyse the same to find opportunities to reduce them, so that overall quality improvement is achieved. The resulting rework instructions can be issued and integrated with the same job list along with the planned work so that technicians at the shop floor do not have to learn a different process for rework, re-inspection, and retesting.

However, these activities do get highlighted and integrated with financial reporting as COPQ and thus its price cannot be measured for financial consideration. In the rubber industry, it is very difficult to do re-processing of the rejected rubber-vulcanised product. In case any defective products are manufactured, they cannot be recycled, unlike plastic products, as the rubber vulcanisation reaction is an irreversible reaction. The curing or vulcanisation process involved in rubber product manufacturing are thermoset in nature and hence, once it is cured or vulcanised, it cannot be reprocessed/reshaped.

For such reasons, wastage would be significant and hence, the COPQ will be very high in such cases. The only way to reduce the COPQ in the rubber industry is by designing the processes carefully before implementing them. Critical processes involved in the rubber industry are selection of ingredients, and identification and control of suitable processes, including inspection.

## 10.14 Housekeeping

Proper storage of various chemicals such as polymers, plasticisers, fillers, additives, metal parts, textiles, fuels, and different type of oils is an essential part of a good housekeeping and for good product quality. Contamination of the same with each other or improper segregation can be the one of the sources that may cause defects in the process and hence the product. This results in degradation of quality parameters. Figure 10.2 is the example of poor housekeeping.

Poor housekeeping results in increased risks associated with fire and might result in explosions also. Good housekeeping is necessary to ensure safety for building occupants in the event of an emergency.

For good housekeeping, specific information regarding storage may be found on the product container label, material safety data sheet, or technical data sheets. Un-

**Figure 10.2:** Scattered material – poor housekeeping.

less otherwise specified by the manufacturer, store chemicals in a cool, dry, well-ventilated location, and away from direct sunlight.

Examples of proper and dedicated clean workshops are portrayed in Figures 10.3 and 10.4.

**Figure 10.3:** Clean and dedicated area for each working activity.

**Figure 10.4:** Clean processing area to avoid contamination.

## 10.15 Maintenance

In the rubber industry, for operational effectiveness and defect-less products production, proper working of each process is critical. Running processes related to maintenance of different activities to keep them in working condition, which will reduce rise in poor quality products is very challenging task. Thus, proper periodical maintenance of machineries, different types of tooling, optimisation for proper process is very much critical. Following are few examples to concentrate.

- Continuous and uninterrupted electrical energy and fuel energy with economic viability. For Diesel Generator (DG) sets, there is a need to evaluate the average cost of power generation, specific energy generation, and subsequently identify areas where energy savings could be achieved by optimising different machineries and tooling, accessories, etc.
- Proper power factor, load factor, no-load losses of machineries, installation of capacitor banks, ac line reactors, and variable frequency drives
- Study of motors in terms of measurement of voltage ($V$), current ($I$), power (kW) and power factor; if required, reduction in the size of motors or installation of energy-saving devices in existing motors Study of temperature rise and voltage imbalance; choosing devices with proper rating for energy saving
- Study of pumps and their flow, and whether the same is required or not or for energy-saving, reducing the size of motors and pumps suitable for that particular application – so, it is optimisation of existing motors, pumps, and such type of other equipment
- In preventive maintenance, proper maintenance of bearings, seals, lubrication, and alignment; wear monitoring and vibration analysis, reduction of throttling losses, pressure/flow monitoring; reduction of pipe sizing, etc. will lead to considerable savings.
- By incorporation of variable frequency drives for different types of internal mixers and open mill mixers, energy can be saved; there is a need to concentrate on vulcanising machineries by the implementation of correct heat generation and transfer technologies. In the case of boilers, to reduce loss of heat, boiler efficiency needs to be focused, apart from thermal insulation, flue gas analysis, condensate recovery, scale removal, boiler and blow down heat recovery.
- In lots of applications, there is need of continuous air during or after the completion of a specific task and before starting a new job, such as mould cleaning. Hence, such systems should give good performance without any interruption. Thus, there is a need to concentrate here on plugging of leakages, proper maintenance to ensure proper gradient in line, installation of electronic condensate drain traps for removal of condensate, reduction of inlet air temperature, choice of proper dryer, heat recovery unit, etc.
- For observing all activities involved in the different process, illumination system of the surrounding areas should be proper and it also needs to be checked at the same time for improvements and for energy conservation, wherever feasible.

These factors are directly related to the quality of products for which we are using these mechanisms, which, if not implemented well, could gives rise to poor quality product as well as high production cost.

## 10.16 Conclusion

Quality is an essential ingredient that needs to be embedded into the processes to ensure an approach to improvements with a quality mindset. It is possible to achieve this in every company with a defined quality system without needing to support an entire quality department. To implement an effective improvement program, one needs to factor in the customer's quality requirements, understand how a quality approach impacts the business, map our processes, use the appropriate tools to help in seeing what is really happening, and committing to fixing the quality as soon as an error is found.

One needs to be acquainted with the sensitivity of inculcating quality and its significance to be sustainable. In the rubber industry, the rubber compound mixing and associate processes for manufacturing a rubber product are complex in nature and by nature, lots of process variations take place. Thus, with proper control in processes and system, the desired level of quality can be achieved and thus, reduction in the COPQ.

## References


[1] Barbará, C. E., de Souza, E. C., Catunda, R. Modeling the cost of poor quality. In: Simulation Conference, 2008. WSC 2008, 1437–1441, Winter (2008)

[2] Evans, J. R., Lindsay, W. M. The Management and Control of Quality, 6th edition, Ohio, OH: THOMSON-South Western, 398–415 (2005).

[3] Retnari, D. M., Rapi, A., Nilda, A. The measurement of quality performance with sigma measurement and cost of poor quality as a basis for selection process of quality improvement. In: Proceedings, IMECS, Hong Kong (March 17–19 2010).

[4] Rao, A., Carr, L. P., Dambolena, I., Kopp, R. J., Martin, J., Rafii, F., Schlesinger, P. F. Total Quality Management: A Cross Functional Perspective, New York, NY: John Wiley & Sons, 119–163 (2010).

[5] Campanella, J. Principles of Quality Costs: Principles, Implementation, and Use, 2nd edition, Milwaukee: ASQC Quality Cost Committee (1990).

[6] Feigenbaum, A. V. Quality and productivity. Quality Progress 10(11):18–21 (1977).

[7] Juran, J. M. Quality Control Handbook, New York, NY: McGraw-Hill (1951).

[8] Atkinson, J. H., Jr, Hohner, G., Mundt, B., Troxel, R. B., Winchell, W. Current Trends in Cost of Quality: Linking the Cost of Quality and Continuous Improvement, National Associations of Accountants (1991).

[9] Schiffauerova, A., Thomson, V. A review of research on cost of quality models and best practices. International Journal of Quality and Reliability Management 23(4):647–669 (2006).

[10] David Chiu, Y.-F. PhD Thesis on A Study on the Economics of Quality in a Technology Management Environment, Industrial Engineering Department, Texas Tech University, (2002).
[11] Mahmood, S. Determining the cost of poor quality and its impact on productivity and profitability Built Environment Project and Asset Management. 4(3):296–311 (2014).
[12] Shahid, M., Sajid, A. Cost of poor quality in public sector projects. Journal of Marketing and Management 1(1):70–93 (2010).
[13] Mohandas, V. P., Raman, S. Cost of quality analysis: Driving bottom-line performance. International Journal of Strategic Cost Management 3(2):1–8 (2008).
[14] Crosby, P. B. Quality Is Free: The Art of Making Quality Certain, New York, NY: McGraw-Hill, (1979).
[15] https://www.kellertechnology.com/blog/how-dimensional-tolerances-impact-part-production.
[16] https://www.kellertechnology.com/blog/how-dimensional-tolerances-impact-part-production.
[17] Harrington, J. H. Poor Quality Cost, New York, NY: Marcel Dekker Inc., ASQC Quality Press, (1987).
[18] Weinstein, L., Vokurka, R. J., Graman, G. A. Costs of quality and maintenance: Improvement approaches. Total Quality Management & Business Excellence 20(5):497–507 (2009). doi: 10.1080/14783360902863648.
[19] Al-Najjar, B. Total quality maintenance: An approach for continuous reduction in costs of quality products. Journal of Quality in Maintenance Engineering 2(3):4–20 (1996). https://doi.org/10.1108/13552519610130413).
[20] Mobley, R. K. An Introduction to Predictive Maintenance, New York, NY: Van Nostrand Reinhold (1990).
[21] Blanchard, B. S. An enhanced approach for implementing total productive maintenance in the manufacturing environment. Logistics Spectrum 35–41 Winter (1994).
[22] ASRTU specification for precured Tread Rubber, AS:272:83:Septt.:2016.
[23] ASRTU Specification for inner tubes made of butyl rubber for radial Tyre application, AS:310:86:Jan2020.
[24] ASRTU Specification for tyre retreading material (cold process): vulcanizing cement/solution; bonding gum/cushion gum AS:163:74:Aug:2019.
[25] http://rsdcindia.in/ssc/training_assessments/#page-content
[26] Shivananda, N. K. World Class Maintenance Management, India: $M_c$ Graw Hill Education (2017).

Bireswar Banerjee

# Chapter 11
# Lean productivity and cost optimisation for rubber processing industries

## 11.1 Introduction

'Lean' is a recognisable word in the present industrial world; it arrived from the ability to accomplish more with fewer resources by the continuous elimination of waste. It calls for eliminating waste in order to reduce costs and to increase competitiveness. Lean management has gathered success throughout most sectors of the industry and the rubber industry is just one environment where lean can methodically promote business operations and profitability.

Globally intensified business confrontations are designed to increase emphasis on operational functioning in rubber-based product manufacturing industries.

The perception of lean manufacturing was engineered to make best use of the resources through reduction of waste. Later, lean was formulated as a rejoinder to the variable and competitive business scenario.

Lean manufacturing, on the whole, has proved to be an effective control technique in the present manufacturing scene. In order to make existing manufacturing practices better, rubber industries have to concentrate on their process stream by programming to improve their productivity and quality.

The methodology of lean was derived from Toyota's 1930 operating model, 'The Toyota Way' TPS (Toyota Production System).

Subsequently, the lean production system was industrialised and introduced to automobile manufacturers after the Second World War. In the 1950s, the then executive of Toyota – Taiichi Ohno, who is called the father of the lean concept, introduced the Lean perception in the production system of Toyota in Japan, as a successful response to competition from larger car manufacturers.

Procedure of lean method is to eradicate waste in its manufacturing operations and improve the efficiency of the operators. Its application became so successful that the process has been taken up in manufacturing sectors around the world. In the USA, lean methodology has been adopted for competing against lower cost countries.

The thinking of lean management set the focus on the product and its needs, relatively, than the organisation or the equipment, as all the procedures that necessitate the designing, ordering, and producing a product materialise in a continuous flow. Lean thinking also redefines the work of functions, departments, and the company; consequently on the positive employee contributions to value.

If a study is conducted in a rubber products manufacturing company, it can be observed that there are quite a lot of wastes during the production processes. The

https://doi.org/10.1515/9783110668537-011

major reasons are (i) breakdown of machines, (ii) damaged equipment, and (iii) delays in production due to lack of proper maintenance on machines and equipment. In general, wastes are created during the processes of production owing to (i) overproduction, (ii) delays, and (iii) defects. It is essential to eliminate those wastes to maintain economics of production. On implementation of the lean production system, the wastes during manufacturing of a product can be eliminated [1].

## 11.2 The lean management

Manufacturers of Japan rebuilding their production units after the Second World War had a hard time; they retained a limited amount of people, a limited amount of raw material, and money. These predicaments led to the growth of lean manufacturing practices.

It can be defines as a long-term philosophy of growth by generating value for the customer, society, and the economy, with the objectives of reducing costs, improving delivery times, improving productivity, and thus the quality through the total elimination of wastes. The word 'lean' has become an often–used term in the milieu of the economic downturn; it simply indicates 'doing more with less'.

Besides the rubber-based products manufacturers, a study done by Wong et al., directed to investigate the adoption of lean manufacturing in the Malaysian electrical and electronics industries. From his study, it is established that most of the Malaysian manufacturing industries have implemented lean manufacturing system to some extent.

The primary goals of lean manufacturing are reduction or elimination of waste in all segments of the production process, and emphasis on continuous quality improvement. In simple terms, it can be defined as doing more with less, as efficiently and economically as possible, without compromising quality and being receptive to the demands of the customer.

'Lean' necessitates shedding waste in order to reduce costs and increase competitiveness [2].

## 11.3 Lean principles

The term 'lean' was coined in 1988 by John Krafcik. In 1996, James Womack and Daniel Jones defined it to consist of five key principles; 'Specify value, identify value, make value flow, customer pull value, and pursue perfection', as shown in Figure 11.1.

The philosophy of lean can be recapitulated in five following key principles,

- Precisely specify 'value' for a specific product
- Identify the 'value stream' for each product
- Make the value 'flow' without disruption

- Let the customer 'pull' value from the producer
- Pursue 'perfection'.

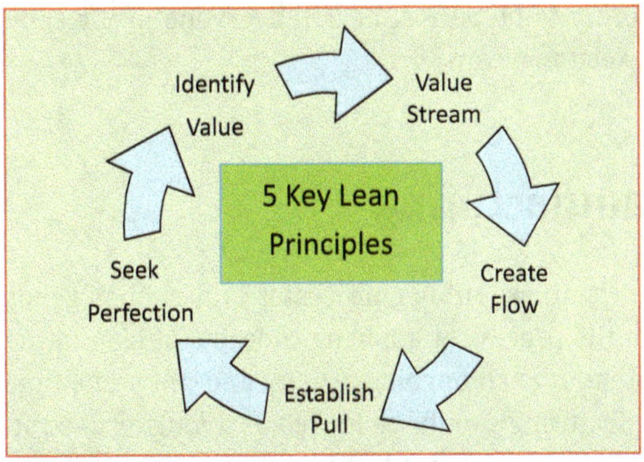


**Figure 11.1:** Five key lean principles.

**Identify value:** The customer's prerequisite is delineating for a specific product; it illustrates its value. Significant provisions, manufacturing timeline, delivery and the price point, are the essential information for defining value.

**Value stream:** On ascertaining the 'value' the subsequent phase is value stream mapping that includes raw materials, processes, and the delivery of final product to the customer. On illustrating this visual tool, manufacturers can understand and analyse the flow of materials from the supplier to the ultimate customer, in addition to information about the flow within the company. The manufacturer can easily identify operational information about the production bottlenecks and can achieve innovative solution for the impediments.

**Create flow:** On removing waste from value stream, the next step is to assure maintaining a smooth flow with no interruptions, delays, or impediments. This may require innovative thinking and attempt cross-functional handovers to all the processing departments, which can be one of the challenges for lean programs to overcome. Studies revealed that this will lead to enormous increase in productivity and efficiency, sometimes enhanced up to 50%.

**Pull:** In accordance with the flow advancement, the time to customer can be improved considerably. Providing products becomes simple as it requires resembling to 'just in time' manufacturing or delivery. It signifies that as per the requirement, the customer can 'pull' the product in a much shorter duration.

Therefore, products need not be produced in advance or stock may be built up. Provide the stock just when the customer pulls; thus is no waste of resources. Appropriately, creating expensive inventory needs to be managed, saving money equally to the manufacturer and to the customer.

**Perfection:** Getting the above 4 steps done is an inspiring initiation. The fifth step is believably imperative – introduce lean thinking and process upgrading as a culture of the organisation. It is essential to keep in mind that lean is not an immobile process; it needs persistent attempts and scrutiny to be perfect [3, 26]. Everyone in the company is required to be concerned in executing lean.

## 11.4 Designing lean manufacturing

Design of lean manufacturing for the natural rubber processes and rubber goods manufacturing industries is based on the process of applying concepts of lean to the design phase of a manufacturing system. It is an ongoing activity and not a one-time activity; so, the design for lean manufacturing should be looked as a long-term strategy for a manufacturing company. Lean manufacturing technique in product and process development: In the region of process, skilled people, technology, and machinery are needed. Documentation in lean manufacturing designing is every person's activity in an organisation, making sure that everyone is aware about the lean design.

Major lean manufacturing tools can exclusively be employed by a design-for-lean manufacturing team.

According to Rinehart et al., lean manufacturing will be the standard manufacturing mode of the twenty-first century. It has been established that lean manufacturing methodology can be implemented in all types of commercial enterprises, including in rubber processing industries. The methods contained by lean tools within the design for lean manufacturing include the following:

1) Prevention of waste
2) Improve productivity

The lean manufacturing and lean production process pay attention to minimising waste within the process of product manufacturing systems and at the same time, take advantage of productivity. The benefits of lean manufacturing can entail lessening in lead times, reduced operating costs, and improved product quality:

- Prevention of waste
- Accordance with methodical measurement
- Innovation and problem solving
- Provide leadership by system designer
- Sustain support by the senior management.

Design for lean manufacturing is a method for applying lean concepts to the design phase of a system, such as a complex product and a process like rubber products manufacturing. It is operational in decreasing the use of time, resources, and the number of company events. It is revealed that there are three main characteristics of companies that put into operation lean manufacturing design.

a) An effective design is one that concurrently reduces waste and provides value.
b) Design for lean manufacturing is articulated out as four value issues: (i) customer, (ii) product design, (iii) test production, and (iv) knowledge.
c) Innovative product manufacturing processes with the objectives of flexibility and ease of implementation.

Lean management may be transmissible; company's improvements in performance will also come to the notice of customers. They may perceive and to expect to be part of implementing the process. Like this way, shared thinking will extend to the suppliers as well, who may desire to employ lean technique themselves to generate their own improvements [15, 21].

According to Anvar and Irannejad [32], there are several methods in lean manufacturing that are used to reduce waste. One of the lean manufacturing

methods used to understand current conditions and find potential improvements to reduce and eliminate waste is VSM can be viewed in Figure 11.2.

According to Hines and Taylor [33], there are seven types of waste, namely, overproduction, defects, unnecessary inventory, inaccurate processes, the ineffectiveness of transportation, waiting, and unnecessary movements.

The activities in a company are divided into three types, namely, value-added activities (VAs), nonvalue-added activities (NVAs), and activities that are not value-added but required (NNVA) [4, 18].

**Figure 11.2:** Lean natural rubber processing unit.

## 11.5 The eight wastes

Lean was initially introduced by Toyota to eradicate waste and improve inefficiency in its manufacturing operations. The process became thriving; therefore, it has been taken up by the manufacturing industries all over the world. An American company determined lean as a decisive process for competing against lower cost countries.

Rubber processing units, as we all know, are highly labor-intensive and power-consuming where there are 7 + 1 = 8 categories of wastes; those can be identified in a lean manufacturing study.

Those categories are (1) overproduction, (2) unnecessary stock, (3) inefficient transportation, (4) unnecessary motion, (5) waiting times, (6) rejects and defects, (7) inappropriate processing. They are portrayed in Figure 11.3.

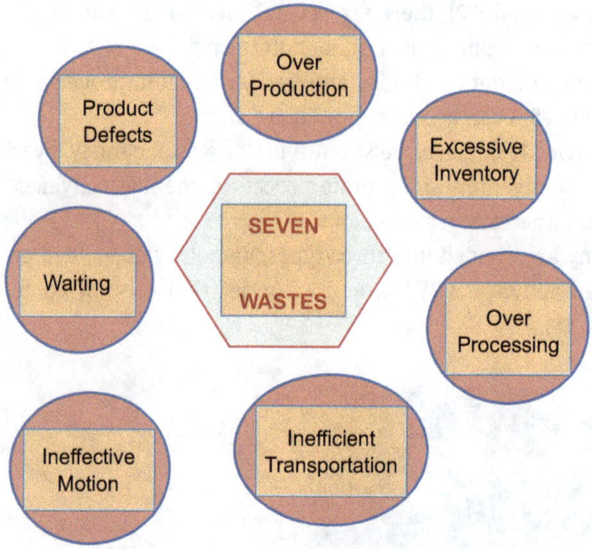


**Figure 11.3:** Seven wastes.

In addition, the eighth waste is underutilisation of employees.

Employees who are underutilised due to lack of proper training for work and not motivated will be considered as an additional waste – number eight, in addition to the original seven wastes.

A manufacturing company, on capitalising the employees' capability and creativity, can eliminate/reduce the other seven wastes and can improve their functioning continuously.

Process or product defects are the result of production of materials that do not conform to the specifications; these defects may be seen by the downstream internal or external customer. Overprocessing affects the material when it is processed to a greater degree than is required by the downstream customer.

All of these add to costs and can be reduced and/or eliminated through the application of these methodologies [5, 20].

### 11.5.1 Lean implementation

The focus of lean implementation is on getting the right things to the right place at the right time in the right quantity to attain a perfect flow of work, concurrently minimising wastes, in addition to being flexible and being able to change, as described in the management of inventory that is illustrated with a cause-and-effect diagram depicted in Figure 11.4.

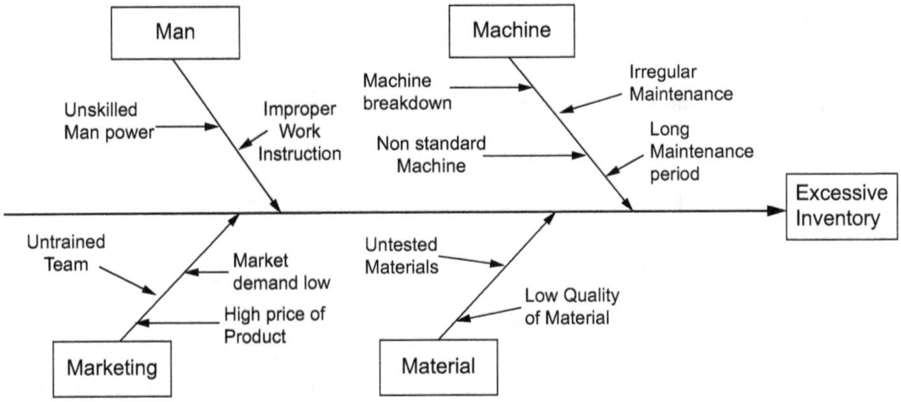


**Figure 11.4:** Cause-and-effect diagram of management of inventory.

A fishbone diagram is to manage and identify any probable cause of impediments.

On conducting several brain-storming sessions of machine operators, processors, and production managers, a fishbone diagram can be developed to identify the problems on manufacturing rubber items [15, 6]; it is described in Figure 11.5.

The diagram helps to identify the causes of a defective product in the process of manufacturing in a rubber industry – explained in a cause-and-effect diagram illustrated in Figure 11.6.

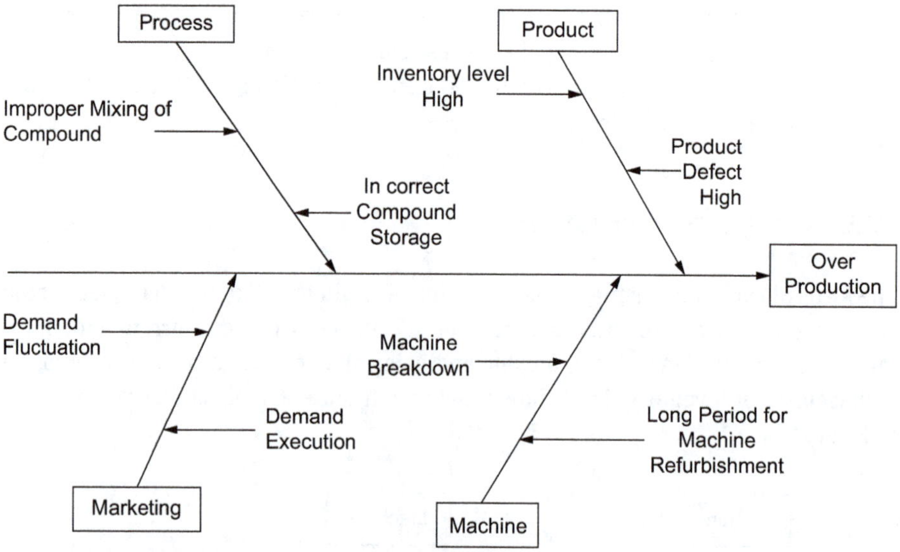


**Figure 11.5:** Cause-and-effect diagram of overproduction.

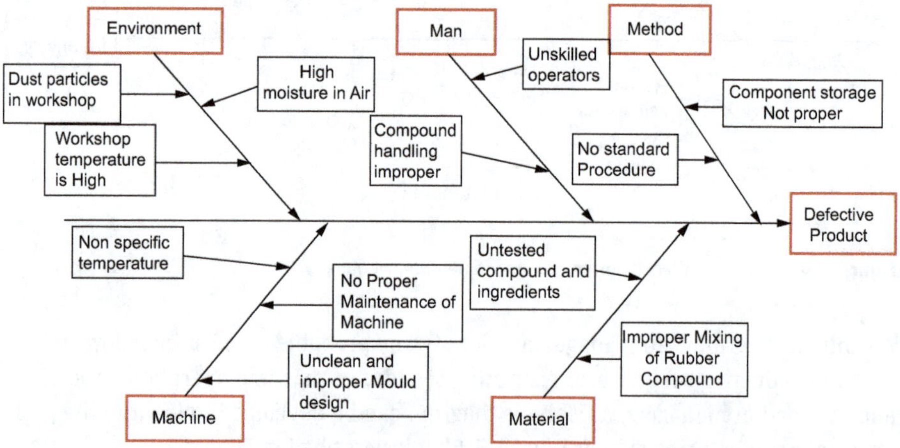


**Figure 11.6:** Cause-and-effect diagram of a defective product.

## 11.6 Lean Six Sigma methodology

Six Sigma is principally a methodology to make a better manufacturing process. The main objective of Six Sigma is to implement a process, which systematically gets divested of inefficiency and product defects.

Lean Six Sigma is a system that relies on a combined team effort to improve performance by methodically removing waste and reducing variation.

Lean Six Sigma principle was introduced by an engineer at Motorola in 1986 and was motivated by Japan's Kaizen model. Mikel Harry and Bill Smith developed Six Sigma to create improvements on the manufacturing shop floor.

Its objective is to make a better processes by identifying and eliminating the causes of defects and variations in the manufacturing methods.

Lean Six Sigma, in combination with Lean manufacturing, helps to build a comprehensive method that eliminates the eight kinds of waste and brings down process variation to make a more efficient manufacturing procedure and to achieve creating the best possible products.

The system of lean manufacturing is an organised approach to eradicating waste and actuate flow in the production process, while Six Sigma is a set of techniques that makes an effort to greatly reduce the percentage of defects.

A tool room of a rubber moulding unit, before and after the application of lean manufacturing system, is represented in Figure 11.7.

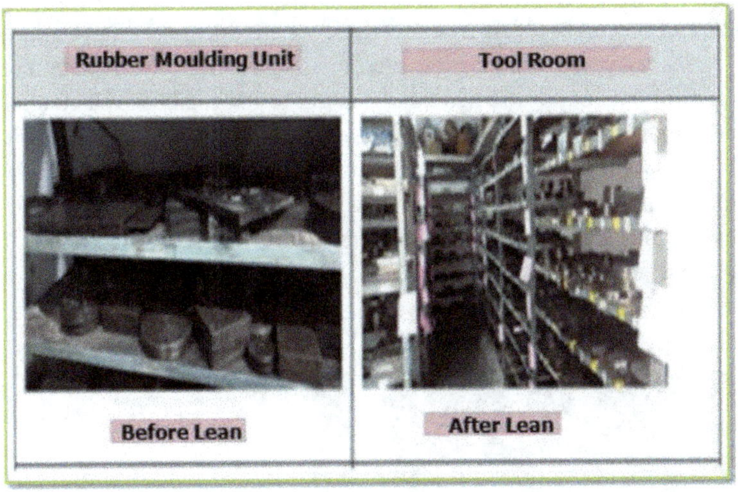


**Figure 11.7:** Lean application in a rubber moulding unit.

## 11.6.1 The elemental Six Sigma program

- Make customer the focal point
- Create a competent process flow
- Drive down waste and systematically improve value
- Prevent defects by way of eliminating variation

- Organise the team through collaboration
- Establish the efforts methodologically and scientifically

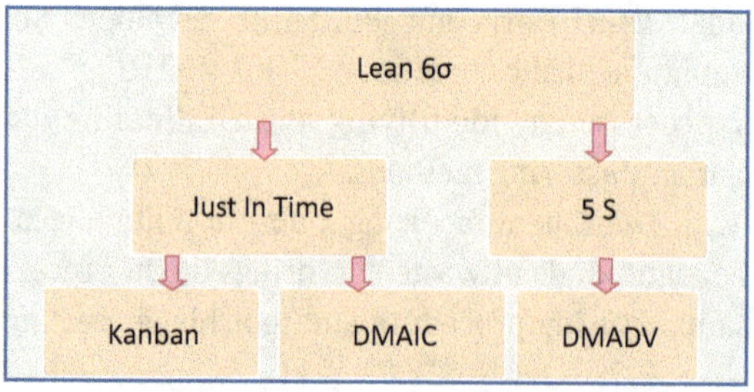


**Figure 11.8:** Six Sigma problem-solving program.

DMAIC stands for define, measure, analyse, improve, and control. It is the five- phase method for problem-solving.

DMADV signifies the process phases of define, measure, analyse, design, and verify for is a Six Sigma agenda.

Implementation of 'Lean Six Sigma' not merely for the reduction of defects in a process and for the management of waste, but additionally facilitates the formation of change in the organisational culture in general. Lean Six Sigma establishes a state of mind in employees that translates to growth and continuous improvement in the course of process development. This transformation in culture and the approach of a company amplifies the morale of the employees and enhances profitability. The Lean Six Sigma problem-solving program is portrayed in Figure 11.8.

In order to successfully execute Lean Six Sigma, a combination of tools from lean manufacturing and Six Sigma are to be employed. Some of these tools include kaizen,

**Figure 11.9:** Lean rubber workshop.

value-stream mapping, line balancing, and visual management [7]. An example of Lean Six Sigma in a rubber workshop is shown in Figure 11.9.

## 11.7 Kaizen methodology

Kaizen comes from two Japanese words: 'kai' means 'change' and 'zen' means 'good'. Masaaki Imai made the term famous in his book Kaizen: The Key to Japan's Competitive Success. Kaizen philosophy was first introduced by Toyota in the 1980s and has since been functioning in the global industrial sectors.

Lean manufacturing includes a philosophy that lean intellectuals use to accomplish improvements in productivity, quality, and lead time by eliminating waste through kaizen. **Kaizen** is a Japanese word that essentially means 'change for the better'. The objective is to offer the customer a defect-free product when it is desired and in the quantity it is necessary.

Kaizen is basically a system to put efforts for continuous improvement across seven different sectors. The five constituents of Kaizen are depicted in Figure 11.10.

**Figure 11.10:** Five constituents of Kaizen.

### 11.7.1 The Kaizen events

- Set goals and provide the essential background
- Review the existing situation and develop a plan for development
- Execute improvements

- Evaluate and resolve those that do not work
- Recount results and establish any follow-up items

The approach in moving towards Kaizen is that an employee and his place of work are inclusive, irrespective of his position in the company, and all can provide ideas for the improvement of processes.

Everybody's ideas may not effectively result in transformation within a company. Nevertheless, in due course of time, small improvements will combine to gather and lead to considerable reduction lessens in wasted resources [12]. An ideal Kaizen pyramidal representation is shown in Figure 11.11.

**Figure 11.11:** The Kaizen pyramid.

## 11.8 Lean productivity improvement techniques using the five S program

It is a methodology adopted for lean quality implementation. The simple, widely practiced tool is a derivative of the five Japanese words **SEIRI, SEITON, SEISO, SEIKETSUE,** and **SHITSUKE**. It is found to be very effective and successful for the management of housekeeping, productivity, and ultimately the quality of the product, as signified in Figure 11.12.

**Seiri** – This word indicates organising or reorganising – organise the work area, leaving only the tools and materials necessary to perform the daily activities, distinguishing the necessary and unnecessary things. When this is implemented properly, communication between workers is improved, and productivity and product quality are increased.

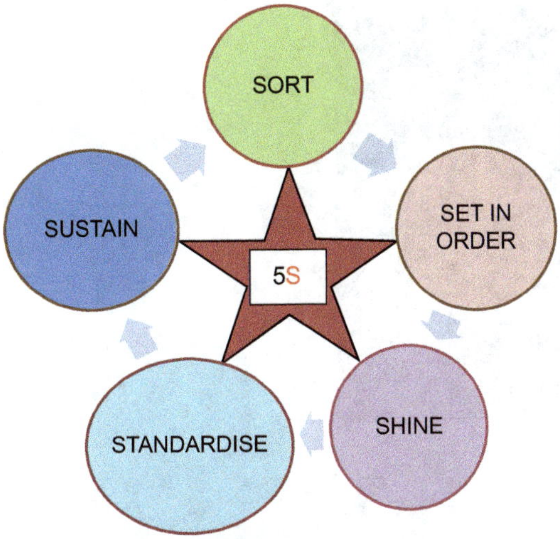


**Figure 11.12:** 5 S implementation.

**Seiton** – It means neatness – arrange everything thing in order; it involves the orderly arrangement of the needed items so they are easy to use and they are accessible for 'anyone' to find. The method eliminates waste in production.

**Seiso** – It suggests job cleaning or shining. Keep the work place and environment always clean. This maintains a safer work area and the problem areas are quickly identified. Do not allow litter, scrap, cuttings, etc., to land on the floor, in the first place.

**Seiketsu** – This word calls for systematising the practices. Standardise or maintain what has been achieved by the above three good practices – those that can be introduced but difficult to maintain.

**Shitsuke** – It is the practice of self–discipline. Be disciplined; follow the procedure [16, 23]. A five 'S'-implemented rubber company's tyre workshop is shown in Figure 11.13.

## 11.9 Effective cost saving tips for rubber manufacturing companies

Manufacturers are always in the quest for saving the cost of products that can help move up the business and the rubber industry, whether small or large – there is no exception. This target can be achieved by small cost-saving efforts, which in due course, lead to a sizeable amount of savings.

There is a compilation of nine suggested cost saving aspects that can be applied by all categories of industries, including rubber goods manufacturers.

**Figure 11.13:** Lean workshop of a tyre plant.

The following strategies can be applied to help in saving costs.

### 11.9.1 Comprehensive assessment

Start with a systematic and pragmatic review. A detailed assessment is the first step in tackling a problem. Allot time to observe the full picture and make objective assessments of each component.

### 11.9.2 Preference to return on investment (ROI)

Subsequent to assessment, prefer the findings. Normally, manufacturers go a number of times without a comprehensive audit which results in out-of-date processes, procedures, and technologies. So, it is essential to take a factual step towards prioritisation.

### 11.9.3 Look for improvement from within

The factory workforce can be an impressive source of ideas for improving the different stages of processes. It is effective to talk with the shop floor workmen pertaining to process improvement at the different stages. Since process operators on the shop floor are all the time carrying out the processes, they can propose valuable ideas for improvement and simplify the existing process, which can result in value. Reward employees as an appreciation of their ideas for contributing to cost savings; this will motivate them further to save.

## 11.9.4 Revaluate the old ideas

Looking at the preceding cost – saving view points, it is proposed to look for cost – saving clues. It may locate one or more indication that make signification to retain at present or in the forthcoming evaluation.

## 11.9.5 Consult ISO 9001 standards

The ISO 9001:2015 standards are guiding principles for any manufacturing organisation focused on consistently providing quality products, improving customer satisfaction and improving system processes.

## 11.9.6 Perceive cost savings using energy audit

Energy consumption is one of the many substantial expenses and reducing it is a challenge to rubber manufacturing companies. By shifting the focus of production decisions away from capacity-utilisation and towards stress-driven mode, make decisions to scale back production during a slow-down phase without disrupting the output.

On the shop floor, there are ways to gain energy savings as it has sizeable impact on your company's cost savings. Conducting preventive maintenance, repairing the defective parts and equipment of processing machinery, and decisively setting up equipment operations can add to the energy conservation efforts.

## 11.9.7 Work in a skillful way

By automating or consolidating repetitive manual processes, you can increase the product quality, improve throughput, and potentially cut down costs. Technology is always evolving and offers new ways of making tasks much easier to complete. Both on the shop floor and in the office, eliminating monotonous tasks through the use of technology is a great way to optimise the labor spends.

## 11.9.8 Scrap disposal at regular interval

Scrap is the unproductive material from a manufacturing process and rubber industries are no different. Collect discarded rubber products due to wrongly processed end products, including the in-process unvulcanised rubber compounds that are likely to be recycled. In most cases, this is usually in the form of a tyre, the major rubber product.

In some cases, send rubber compounds, textile wastes, vulcanised rubber trimmings etc. for recycling. And at times, broken, discarded machine parts, hand tools, electrical components, etc. can also be recycled. Scrap are trash to the company, but for some other person it is considered a treasure.

### 11.9.9 Collaborate with suppliers and customers

Development of ongoing relationships with suppliers and customers is indispensable to a company's success. Take advantage of the position as a manufacturer and try to renegotiate a better offer with the supplier.

In employing a few of the cost-saving features delineated above, the company can be on the way to cost saving modalities.

Before going on the cost savings expedition, it is imperative not to affect the quality of the product and the company's image. Curtailment of costs indiscriminately may ultimately result in imperfect products or uninspiring service [13, 22]. The best cost efficiency with quality is reproduced in Figure 11.14.

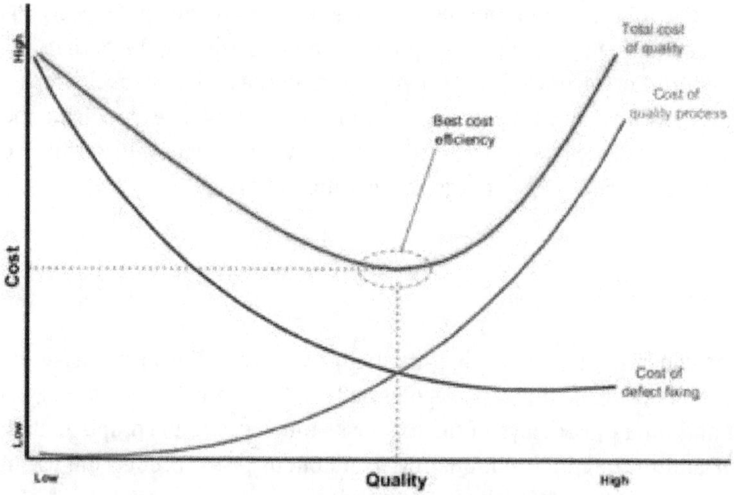


**Figure 11.14:** Best cost efficiency and quality.

## 11.10 Karakuri: a predominant constituent of the lean philosophy

Karakuri is an automation mechanism that was invented in Japan between the seventeenth and nineteenth centuries. This simple, intelligent mechanical trickery is an automation that been a central element of the lean philosophy followed in Japan for decades. Karakuri is the low – cost intelligent automation of processes, based on physical principles with no drives, sensors, electricity, or compressed air.

The objective of Karakuri is to use resources that are already in place – available plants, apparatus, and equipment on the one hand and the inventiveness of the working place on the other. In practice, it ought to be amusing, but Karakuri has been a central element of the lean philosophy, where it refers to the simple but skillful system adopted by the manufacturing industry for decades.

Shop floor operators use levers, pulleys, counterweights, and gravity to increase productivity, quality, and safety [25].

The system of Karakuri is the utilisation of simple mechanisation to develop a lean manufacturing solution; it increasingly forms part of production equipment design and safety. An example is presented in Figures 11.15.

**The Karakuri makeup:**
- Optimise processes
- Relieve strain
- Improve efficiency
- Swift operation
- Save space
- Simplify work
- Ensure flexibility and compatibility
- Work delightfully using simple mechanical devices.

**Figure 11.15:** Karakuri mechanisation.

The Karakuri concept system is appropriate for rubber processing units. A schematic graphic is depicted in Figure 11.16(a) and (b).

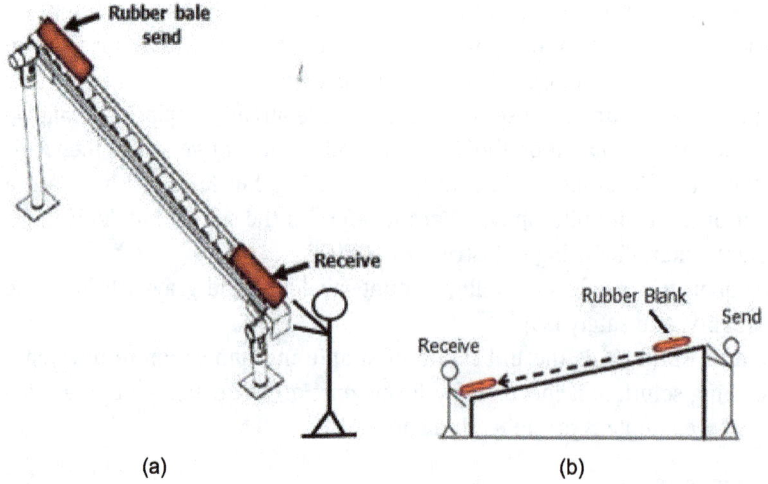


**Figure 11.16:** (a) and (b) Karkuri system for material handling.

## 11.11 The 3 M practice in manufacturing

The 3 M – Muda, Mura, and Muri are three Japanese terms that signify a management philosophy. It was developed by Taiichi Ohno of Toyota Production System. It plays a vital role in reducing wasteful actions that unconstructively impact overflow, productivity in manufacturing process, and finally to customer satisfaction.

Mura can be kept away by means of just-in-time (JIT), 'Kanban' systems, and other pull-based method that limit overproduction and excess inventory. The basic concept of a JIT procedure is delivering and producing the right part at the right amount and at the right time.

Muda, Mura, and Muri are interrelated; removing one of them will affect the other two.

Muri signifies overstrain – disproportionate burden. Muri is a consequence of Mura. It may be the reason for excessive deletion of Muda (waste) from the manufacturing process. For example, the 3 M signifying workload versus time is shown in Figure 11.17.

The practice of the three Japanese magic words Muda, Muri, and Mura at the work place will bring down the cost of production without affecting the quality of the product.

**Muda** – means waste, an activity that does not add any value but is included in the total cost of the product.

**Muri** – indicates excessive strain or unreasonable approach to any field of operation.

**Mura** – depicts discrepancy – the actions that are irregular, uneven, or inconsistent.

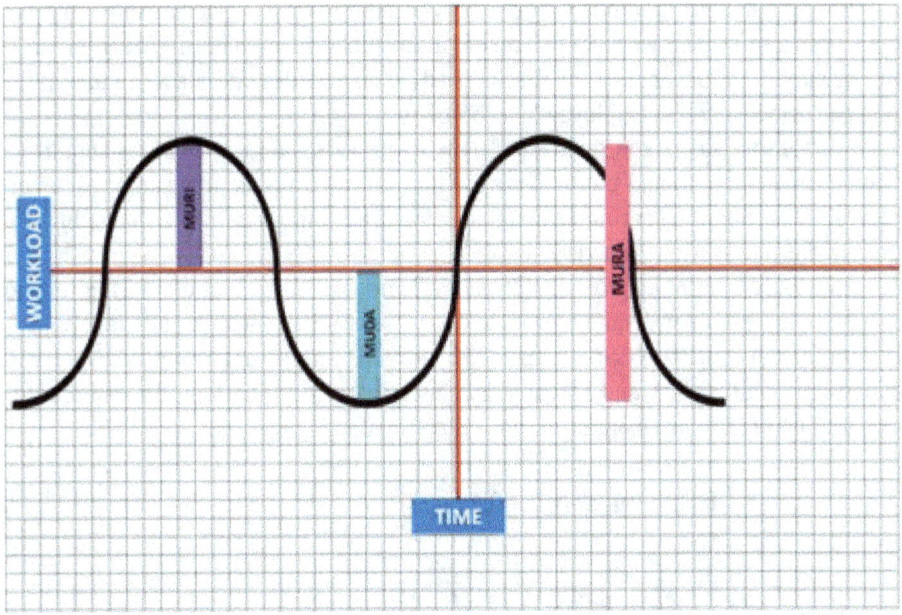


**Figure 11.17:** 3 M work load versus time.

It can be observed that the main method of lean is not the tools, but the reduction of the three types of waste: Muda – the nonvalue-adding work, Muri – the overburden, and Mura – the unevenness, to depict problems systematically and to use tools where the ideal cannot be achieved. From this perception, tools are workarounds tailored to different situations, which explains any apparent inconsistency of the viewpoints stated [1, 2].

## 11.12 Standard operating procedure

**Standard operating procedure (SOP)** is a written, approved, and controlled procedure detailing the required actions or activities for a specific function. All steps and activities of a process or procedure should be carried out without any deviation or modification to guarantee the expected outcome. SOP is an established written directive that documents scheduled or repetitive functioning to be abided by a manufacturing company.

It is a document that portrays the methodically established procedures step by step, significant to the quality of the exploration. The objective of an SOP is to carry out the operations accurately and always in the same mode. An SOP should be accessible at the place where the work is performed.

The manufacturing standard operating procedure is a set of documented instructions prepared to help workers to execute a scheduled manufacturing assignment. It can be employed for manual and mechanised tasks, and also operate as a manual for safe performance to the employee in a company.

All manufacturing and quality influencing processes and procedures should be laid out in SOPs. It should be the basis for the routine training program of each employee. SOPs should be regularly updated to assure compliance to the controlling requirements and working practices [8].

An SOP is a set of step-by-step directives composed by a manufacturing unit to assist employees in performing the assigned task. It aspires to achieve efficiency, quality output, and consistency in all stages of operations, while minimising miscommunication and malfunction and to abide by the industry's regulations. The method of preparation of SOP for an organisation is illustrated in Figure 11.18.

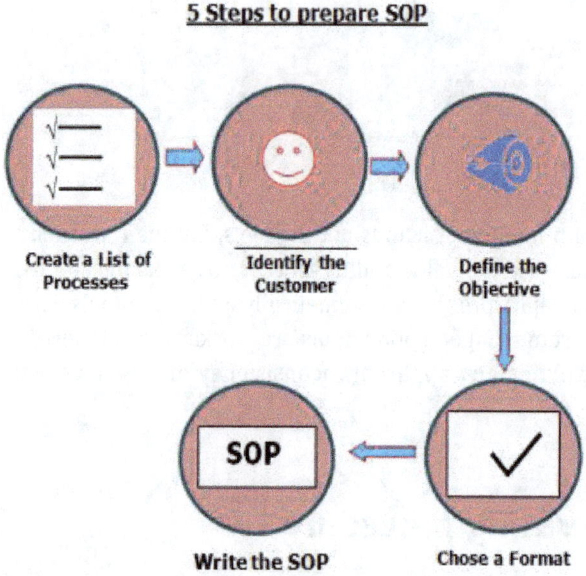


**Figure 11.18:** Documentation of standard operating procedure.

On implementing well-structured SOPs in a rubber product manufacturing company, the following seven tangible benefits are possible [9].

- Enforcing best practices
- Making processes perceptible
- On-boarding employees
- Ensuring regulatory and standards conformity
- Reducing process failures

- Diminishing process errors and corrective actions
- Developing exchange of ideas

## 11.13 Just-in-time production (JIT)

JIT originated in Japan. It is a management philosophy that has been in practice since the early 1970s in various Japanese manufacturing industries. It was first developed and perfected within the Toyota manufacturing plants by Taiichi Ohno to meet consumer requirements with minimum delay.

Taiichi Ohno and Shigeo Shingo at the Toyota Motor Company incorporated the Ford production and other techniques into an approach called the Toyota Production System or JIT.

JIT is a management philosophy and not a technique. The initial proposition is to the making of goods to meet customer demand precisely and in time.

The JIT inventory system is a management approach that brings raw-material orders from suppliers in line with production planning.

The JIT inventory management system reduces wastage, gets better efficiency and productivity, and provides effective production flows. A shorter production cycle can minimise financial costs, inventory costs, and operational costs.

JIT is a mode of inventory management that necessitates working closely with the supplier's raw materials arrivals, whilst the production is scheduled to start. One of the most significant steps in the implementation of lean manufacturing is JIT because the Just-in-Time production is the backbone of lean manufacturing.

JIT production is about not having more raw materials, work in process, or products than what are required for an uninterrupted production operation. The main intent of JIT is to produce the goods when it is needed; the requirement is generated by the customer's demand. Next, performance is on the shop floor and production should be performed with no disruptions and without work in progress.

This will help to carry out an uninterrupted process of manufacturing. There are tools and techniques available to accomplish JIT. On implementing lean, reduction in process timing is exemplified in Figure 11.19.

This method promotes improved quality products and incessantly improves competence. It lays emphasis on constant communication with the customer to modify processes, meet varying needs, and elevate customer satisfaction [21, 22].

The JIT advantages are:

i) Improve production efficiency and competitiveness
ii) Eliminate overproduction
iii) Minimising waiting times
iv) Decreasing transport costs
v) Saving in stock, on modification of production systems

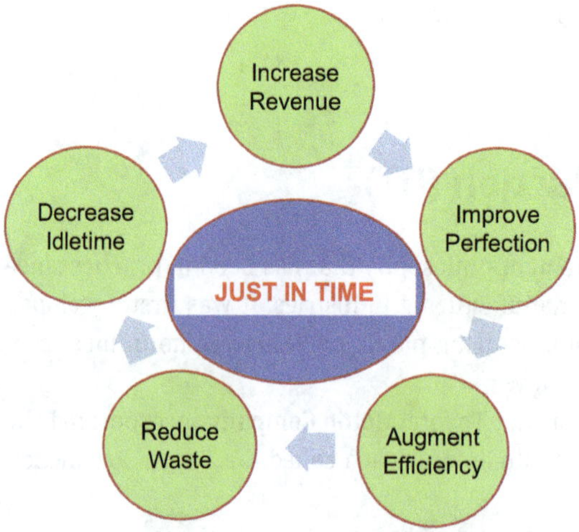


**Figure 11.19:** Just-in-time achievement and success.

## 11.14 The Kanban system

**Kanban** is a framework that falls under the functional methodology. The configuration was developed in the late 1940s by an industrial engineer of Toyota, Taiichi Ohno. The implementation of the Kanban framework focuses on visualising the whole project on boards in a sequence to augment project transparency and enhance collaboration between team members.

The Kanban name comes from two Japanese words, 'Kan' denotes sign and 'Ban' denotes a board. It is a visual system that works to manage the trail of work as it moves in the course of a process.

It was created as a simple scheduling system, intended to restrain and manage work and inventory at every stage of the production to the extent possible.

Kanban is contemplated as a 'lean production' system, or one that eradicates labor and inventory waste. One of the ways Kanban decreases waste is by means of the 'pull production' mode that standardises product manufacture on the basis of customer's supply and demand.

It is an inventory control system employed in JIT manufacturing to track production and order new consignment of parts and materials [29].

### 11.14.1 Implementing Kanban

In implementing a Kanban pull system effectively, the team needs to integrate the six core procedures of the system:

1. Visualise the workflow
2. Eradicate disruptions
3. Administer flow
4. Make process procedures specific
5. Maintain open feedback circle
6. Progress collaboratively

Kanban is an established lean workflow management system to efficiently manage and upgrade services that provide knowledge work. It helps to visualise work, maximise efficiency, and improve continuously. Work is designated on Kanban boards, allowing optimising of work delivery across several teams and handling even the most diverse projects in a single location [30]. It is presented in Figure 11.20.

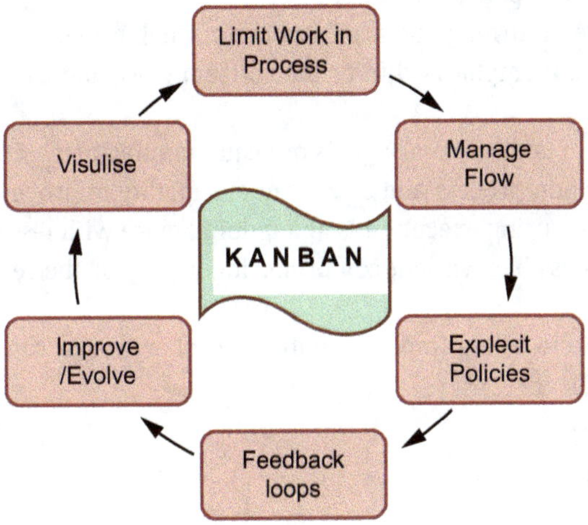


**Figure 11.20:** Kanban process.

## 11.14.2 The six rules of Kanban

- By no means, pass defective products
- Procure only what is essential
- Produce the exact necessary quantity
- Level the production
- Modify the production or process optimisation
- Establish and modify the process

## 11.15 Quality circle for the success of lean quality execution

The quality circle is a small group activity. It consists of a group of workmen from the same working area. It is for self and mutual development and has learned to analyse the problem and solve them. **QC**, when implemented, can be successful in any industry anywhere in the world.

Quality circle concept was first illustrated by W. Edwards Deming in the 1950s; later, the objective of QC spread across Japan in 1962; Kaoru Ishikawa and others enhanced the idea. The QC movement was communicated in Japan, in association with the Japanese Union of Scientists and Engineers (JUSE).

It was prevalent during the 1980s, and maintained its existence in the form of Kaizen groups and similar employees' participation schemes.

Lean quality circle is appropriate in turning out large number of small projects to meet the company's tangible benefits. It emphasised interrelated team work and continuous improvement activities.

Reduction or elimination of waste is the primary goals of Lean manufacturing. It can be in all segments of the production process and as continuous quality improvement. In simple terms, a lean manufacturing execution is about doing more with less, as proficiently and economically as possible, without compromising quality or the requirements of the customer [3, 16].

The effectiveness of a quality circle, when implemented in a rubber works, can result in what is shown in Figure 11.21.

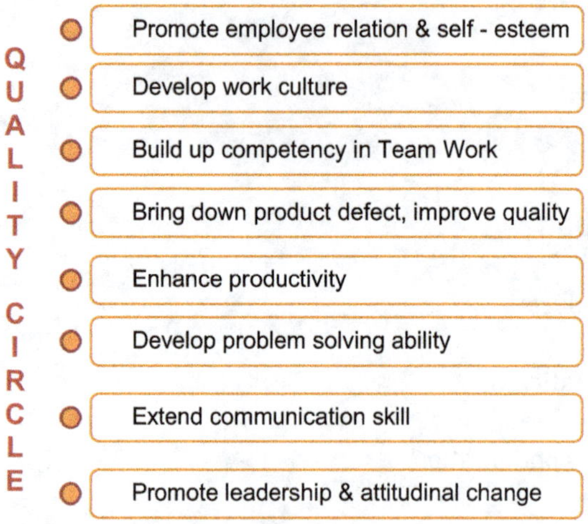


**Figure 11.21:** Quality circle effectiveness.

## 11.16 Lean versus mass production

When implementing lean manufacturing, it is essential to understand the difference between the conventional manufacturing system and the lean manufacturing concepts to achieve success [2]; an example is given in Table 11.1.

**Table 11.1:** Mass production versus lean production.

| Mass production | Lean Production |
|---|---|
| Large batches | Small Batches |
| Big, Fast Machines, inter changeable parts | Right – Size Machines |
| Economics of scale | Economics of speed |
| Changeover not important | Quick change over |
| Functional Silos | Production Cells |
| Efficient | Adaptive & Flexible |
| Push | Pull |
| Slow to change | Fast to respond |
| Produce & Sell | Build to order |
| Specialized knowledge | General knowledge |

## 11.17 Zero defects in lean manufacturing

Zero defects is a method of thinking that underlines the concept that imperfections are not acceptable, and that everyone in the manufacturing industry should, 'do things right the first time'.

According to Philip Crosby, the principles of zero defects is 'Quality is the conformance to Requirements'. Every product has a requirement. It is a portrayal of what the customer is expecting. Therefore, a particular product is said to have accomplished quality if it meets those requirements.

Everything that is improvident and does not add value to a product should be obliterated, which is identified as the procedure of removal of wastes. Eliminating wastes generate a process of development and likewise lowers the costs of the product. Therefore, the zero defects in a scheme means to execute necessities at that point in time.

The philosophy of zero defects was movement promoted by Philip B. Crosby, a quality control manager of an American company. It is a management-led program to eliminate defects before they reach the customer. It was recognised by the American industries between 1964 and 1970 as a means to increase the profit margin, which can be achieved by eliminating the cost of failure and by increasing revenues through increased customer satisfaction.

Zero-defect manufacturing is a perception of the quality of a product that is defect-free and to eradicate waste linked with defects. The objective of zero defects is the foundation on the principle that defects are precluded by the method of the stages of manufacturing processes. Zero Defects is a very effective, inexpensive management tool when enthusiastically sustained by the management [10, 11].

The total quality management system in a manufacturing establishment is the basis of zero-defect planning; it relates to the tools and techniques and is portrayed in Figure 11.22.

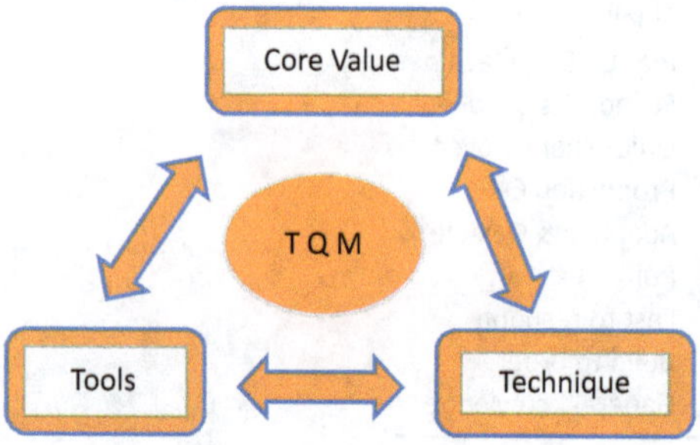


**Figure 11.22:** Zero-defect planning.

### 11.17.1 Principles of zero defects

Zero defects in quality management of an organisation intends to ensure that all products leaving the company are defect-free. The industry can do this through comprehensive planning and with observant employees who are constantly looking for errors:
- Quality is conformance to requirements
- Defect prevention
- Standard quality means zero defects
- Quality is measured in financial terms

It is a motivational program where the workmen/employees are made more responsible, more achievement oriented, and are prouder of their work. To monitor the progress of quality improvement on a consistent basis, it is essential to establish a zero defects council in the organisation [17, 24]. A model is depicted in Figure 11.23:
- Quality council
- Quality awareness
- Plans for zero defects
- Supervisor training
- Zero-defect day
- Quality measurement
- Corrective actions
- Cost of quality evaluation

## 11.17.2 Zero waste in the rubber industry

The secret of zero waste is all about employing the rules of reducing, reusing, and recycling – the effective methods of a cost saving system in producing general rubber-based products.

Reusing, for example, 'feed the tail' in rubber compounding system, such as crumbs of vulcanised scraps/wastes crushed into powder in a rubber crusher mill and reused. The 8–10% crumbs can be added appropriately with a fresh rubber compound for thicker rubber moulding products without any deterioration of the major physical properties.

In a rubber crusher mill, vulcanised scraps and even rubber blocks can be made into 5- to 60-mesh rubber crumbs that can be reused in fresh rubber compounding stock of various types.

Commercial applications of crumb rubber include: foundation waterproofing, rubberised asphalt, shock absorption, and kindergarten play grounds. The recycled tyre rubber crumb, as an additive in hot mix asphalt mixture, is considered a sustainable construction method. It is considered as a smart solution for workable development.

The most significant and established recycling example of rubber product is the tyre retreading process on serviceable tyre casing. This can be conceivable to recycle all categories of light, heavy duty, and unfluctuating aircraft tyres.

Reclaimed or regenerated rubber from dry rubber tyres and rubber latex product wastes are widely used in the rubber industry for reducing cost and improving processability.

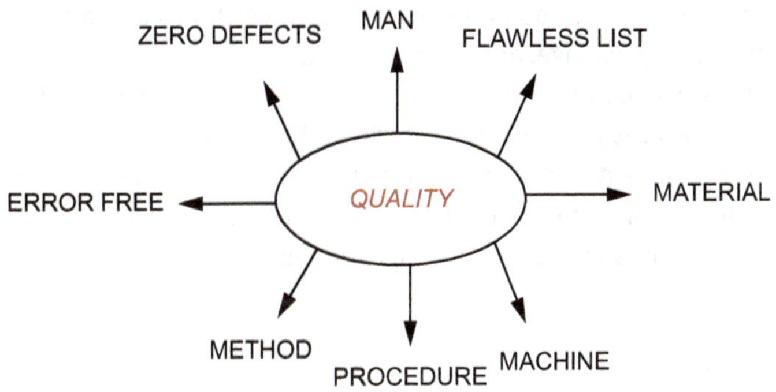


**Figure 11.23:** Controlled areas of quality assurance.

### 11.17.3 Zero accidents in workshops

**Importance in housekeeping**

Industrial safety implies safety management practices in plant machinery, operators, facilities and other installations in good order. Usually, the rubber goods manufacturing units, particularly in the small sectors, are unorganised and the working areas are muddled with dusting powders, earth dust, carbon black, cutting wastes, and wet with oily and watery floors; this is the scene of the walkways too. It is imperative to maintain the workstations spotless on regular basis to avoid accidents, for example,

Installation of proper safety guarding mechanisms to machineries such as open-roll mixing mills, calender machine, and extruders to avoid accidents in the operation stages; installation of appropriate fencing, guards, railings to electrical equipment, and installation of danger signals to prevent accidents; using proper personal protective gears like hand gloves, facemask, helmets, etc.

These are prerequisites for the safety of operators. These safety measure correlates to get defect-free output [10, 11].

### 11.17.4 Zero pollution

Zero pollution is demarcated as reducing emission to air, soil, and water up to levels that are no longer perilous to human health and ecology.

It is imperative to perceive that the rubber goods manufacturing industry burn considerable quantities of coal and petroleum-based fuels to produce heat or generate electricity. Fossil fuels release harmful pollutants into the air, which has a damaging effect on the environment, climate change, and human health.

Specific improvement is needed in those areas in the company, where reduction of pollution is noticeable. Measures for pollution control and to follow pollution control rules of local municipal administration directly relate to get a defect-free product. Lean and environment-friendly manufacturing activities both focus on waste reduction, and as a result have a synergistic effect on the environment [31].

### 11.17.5 Zero breakdowns

Zero-breakdown approach is projected to provide cost effective solutions to the operation of machinery and equipment that will allow the companies to achieve the best return on investment.

Unpredicted breakdown losses result in equipment downtime for repairs. Cost can include downtime and loss of production, labor, and spare parts [19, 28].

The hidden defects of the processing plant and machinery of a rubber goods manufacturing industry are:

- Normal wear
- Dirt
- Deformation
- Surface damage
- Vibration and noise
- Leakage
- Corrosion
- Fatigue

In establishing zero breakdown measures, the following are the suggested strategies for zero breakdown events:
- Breakdown meeting
- Breakdown presentation
- Breakdown analysis
- Breakdown countermeasure
- Zero breakdown course of action
- Maintaining basic conditions
- Maintaining operating standards
- Restoring or preventing deterioration
- Improving or eliminating design weaknesses
- Preventing human error

Implementation of standards is necessary for the company's machinery and equipment to work appropriately. Standards are to be for the following categories [14, 27]. The challenges in implementing a zero-defect methodology in a manufacturing company are depicted in Figure 11.24.
- Cleaning standard
- Lubrication standard
- Inspection standard
- Preventive maintenance standard

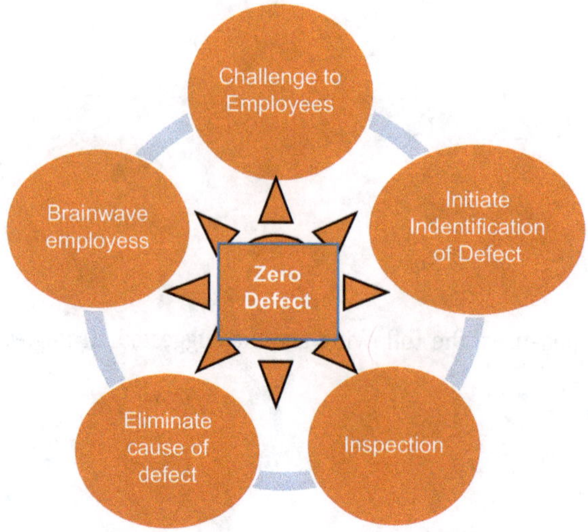


**Figure 11.24:** Achieving zero defects.

## 11.18 Conclusion

In today's rubber products' global market, staying competitive comes down to two key fundamentals: speed and quality. The task involves eradicating all forms of waste on all fronts – everything from supplier networks and customer relations to product design, factory management, inventory handling, time management, and all non-value-added dealings. One of the major drivers behind polymer product manufacturer's implementation of lean manufacturing techniques as a means to eliminate waste is globalisation.

The manufacturing units' survey study revealed that nearly 70% of global manufacturing companies are operating in a way to benefit from lean manufacturing practices. Application of continuous improvement is the most common premeditated practice in the global manufacturing scenario.

Lean manufacturing methodology can facilitate organisations to give rise to smart factories that operate with innovative sensors, robotics and software, to enable superior and prognostic investigative ideas that may initiate advanced decision-making. The decisive result is end-to-end digitisation across the supply chain and method of production.

## References

[1] Banerjee, B. Pursuit of organizational excellence through TQM. In: International Conference on Rubber & Rubber – like Materials –Indian Institute of Technology, Kharagpur, India (2008).
[2] Banerjee, B. Application of lean quality management to meet global competitiveness for chemical processing & pharmaceutical industries. In: National Seminar on Chemical & Pharmaceutical Industries in India, Present Scenario & Future Prospects, Calcutta, India: University of Calcutta & Jadavpur (2011).
[3] 5 Lean Principles Every Engineer should know – Mark Crawford – The American Society of Mechanical Engineers, (2016).
[4] National Manufacturing Competitiveness Programme, Lecture Notes – Lean manufacturing Competitiveness Scheme – Govt. of India, (2009).
[5] Dhiraidamani, P., et. al. Application of lean tools for production of cycle time in rubber moulding process cell – an industrial case study. International Journal of Advanced Engineering Technology, II:967–977 (2016).
[6] The 5 Tools that Make Lean Manufacturing Thrive – Technology Lean Manufacturing Supply Chain, Oliver Freeman, (2020)
[7] Six Sigma Methodology, www.leansixsigmaprowess.com.
[8] SOP Manual for Rubber Product Manufacturing for Mechanical Use, September 2020, https://www.powerdms.com.
[9] The 8 steps Guide to Building a Standard Operating Procedure, (2017) http://log.lern.org.
[10] https://www.industr.com, Zero defect Manufacturing.
[11] https://www.erp-information.com, Zero defect.
[12] http://www.researchgate.net.
[13] 9 Effective Cost Saving Tips for Manufacturing Companies, Posted by:Marjorie Dunn on Sept. 2016.
[14] Industrial Safety & Hygiene News (ISHAN), (2022) www.Ishan.com.
[15] Quality circle to improve productivity. International Journal of Engineering Research and Application 3: 814–819.
[16] Lean Quality Circle, http://qcfi.in.
[17] https://www.erp-information.com>, Zero defects
[18] Genc, R. Lean Manufacturing Practices and Environmental Performance, (2021) https://www.intechopen.com.
[19] Sheikh, H., et.al. The application of 7 zeros in improvement of lean agility manufacture, Iran. Interdisciplinary Journal of Contemporary Research in Business, 5(8): (2013).
[20] United States Environmental Protection Agency, Lean Thinking and Methods – TPM – for Machine Breakdown.
[21] Juran, J. M. Juran on Planning for Quality, New York: Maxwell Macmillan International, 151 (1988).
[22] Productivity Management – Joseph Prokopenko, New Delhi: Oxford & IBH Publishing Co. Pvt. Ltd., 133–135 (1990).
[23] TQC Wisdom of Japan – Managing for Total Quality Control- Hajime Karatsu, Tokyo: JUSE Press Ltd., (1988).
[24] Crosby, P. B., Quality Is Free, New York: Penguin Books USA Inc (1980).
[25] Implement Karakuri as a Material Handling in Production Sealer Line, Green Technology & Engineering Seminar, Eliza Shamsudin Malaysia, (2019).
[26] Productivity Improvement in the Rubber Production Process Using Value Stream Mapping Method to Eliminate Waste – Evan Haviana – Operations Excellence, Jakarta, Indonesia, 2019.
[27] The Concept of Zero Defects in Quality Management – Simplilearn, (2022) https://www.simplilearn.com.

[28] Zero Defects in Quality Management – ZQC, analysis, plan, example, https://www.erp-information.com.

[29] Kanban Methodology: The Simplest Agile Framework, (2022) https://kissflow.com.

[30] Six Rules of Kanban: How to Better Implement Kanban. (2019) https://kanbanzone.com.

[31] Jagadle, S. C., et al. Environmental concern of pollution in rubber industry. International Journal of Research in Engineering and Technology 4(11): (2015).

[32] Anvar, M. M. and Irannejad, P.P. Value stream mapping in chemical processes: A case study in Akzonobel Surface Chemistry, Stenungsud, Sweden. Proceedings of the Lean Advancement Initiative, A seminar paper Florida; Daytona Beach. (2010).

[33] Hines, P. and Taylor, D. Going Lean. Lean Enterprise Research Centre Cardiff Business School, Cardiff, UK, 3–43. (2000).

# Index

2-mercaptobenzothiazole disulfide (MBTS) 385
3 M 486
5 S 481

AAS 293
abrasion 158, 167, 173, 184, 327–328
abrasion resistance 8–9, 10, 27, 34, 41, 68, 96, 99, 110–111
abrasion resistant 213, 240, 255
accelerator 6, 8–9, 14–15, 20, 25, 204, 221, 223, 286–288, 291, 293, 308, 310
ACM 282–283, 286
acrylic rubber 282
acrylonitrile butadiene rubber 160, 170, 284
activation energy 339
activator 287
adhesion 7, 11, 13, 17, 21, 24, 28, 33–34, 35, 36, 44, 161, 170, 174, 180, 197, 210, 213, 219, 221, 223–224, 232, 234–236, 256, 260
adhesion strength 197, 213, 221, 223–224
adhesive 21, 208, 269
ageing 176
aggregate 56, 70–71, 73, 77–78, 87
aggregate size distribution 87
aging 13, 31, 33, 36–37, 39, 176, 208, 220, 324, 357–358
Agriculture 199
ambient temperature 203
amorphous 73
antidegradants 18, 24, 279, 286–287, 290, 293
antioxidant 6, 8–9, 18–19, 31, 284, 310
antiozonants 8–9, 18–19, 284
application 199, 202, 206, 469, 475, 477
appraisal 446–449, 459
aramid 21, 35, 45, 213, 231–232, 237
ASD 88–89
ash content 116
asphalt 374, 381
asphalt binders 373
asphalt concrete 373
assessment 482
ASTM 9, 12–13, 206, 211, 224
ASTM D1765 59, 284
atomic absorption spectroscopy 293
automated 462
automobile 153, 199, 206
automotive 37, 41, 44, 184, 197, 199, 201, 208–209, 211
automotive industries 209
average dispersed phase size 125

bale pickup time 131
ball bladder 43–44
belt velocity 202
belts 1, 8, 23, 25, 28, 34, 197–200, 202–204, 208–209, 215–221
biomass 372
bituminous mixtures 372
bladder 42–43
blend homogenisation 129, 132, 139, 141–142, 145
blend homogenisation times 133–134, 146
bonding 233, 235–236, 238–239, 261, 266–267, 270
bottlenecks 471
bottom-up process 153–154
bound rubber 330–331
BR 121–122, 126–128, 132, 134, 147
build up 267, 269
building process 23, 215, 216
business 469, 481
business operations 469
butadiene rubber (BR) 410
butyl rubber 282

carbon black 151–152, 167, 169, 171–173, 176, 183–184, 211–212, 222–224, 284, 299–300, 302–304, 306, 310, 329, 345, 359, 412, 419, 429
carbon dioxide ($CO_2$) 397
carboxylic acid groups 378
cause-and-effect 475
cellulose 214–215
centre of the pulley 203
channel black process 61–62
chemical additives 386
chemical bonding 233–235, 270
chemical interactions 159, 163
chemical modification 158, 180–181
chemical resistance 10–11, 37, 41
chemical treatment 385
chemically reduced 156
chemical-resistant 37, 213
chemicals 382
chlorinated isobutylene isoprene copolymer 121
chlorinated polyethylene 121, 136

https://doi.org/10.1515/9783110668537-012

chlorobutyl rubber 170
chloroprene rubber 284
chlorosulfonated polyethylene 121
chlorosulfonated polyethylene rubber 283
CIIR 121, 125–128, 132, 134, 139–142, 145–148
civil engineering application 367, 420
classical V-belt 198
climetic conditions 101
CM 121, 126–128, 132, 134, 136–137, 139–142, 145–148
coagulation 169, 182
COAN 59, 110
coefficient of friction 200, 202–203
cold solution 270
compatabilisation 122
compatibility 121–122, 124
compatibiliser 121–122, 126, 132–148
compatibilising agent 121–123, 126, 128, 141, 145–147
competitiveness 445
components of tyre 100
composite 231
compound 1–2, 5–13, 15–17, 19–28, 31–32, 34–38, 40–42, 45–46, 155, 162, 167–169, 171, 173–175, 178–179, 184, 197, 203–204, 207, 210–212, 215, 219–226, 231–234, 258, 263, 267, 269, 276
compounding 180
concrete 379
concrete preparation and evaluation 378
conformance 447–448, 450
construction 23, 202–203, 205
construction industry 376
construction materials 385
contact angle 158, 166
continuous belt 199
conveyance technology 202
conveyor belt 1–2, 6, 21, 33–37
co-pyrolysis 372
cord 7, 23–24, 28, 30, 35, 197, 205, 213, 215, 217–221, 221, 223–224
cost 1, 10–11, 12, 15–16, 18, 20–22, 26, 38, 151–153, 180, 182, 184, 197, 210–211, 226
cost of quality 446–454
cost optimisation 469
cost savings 483
cotton 35, 45, 200, 214–215, 219
cover grade 255
cover rubber 232, 241, 256
CR 283, 286
crack 11, 28, 36, 203–204, 208
create flow 471
CRGO 156, 169–170
CRM mixture 373
CRMA 373–375
cross-link density 324, 346, 352, 355–356, 358
crumb rubber 376, 379, 381
crumb rubber content 374
crumb rubber modified asphalt 374
crumb rubber modifier (CRM) 372
cryogenic grinding 381–382
CSM 121, 126–128, 132, 134, 139–142, 145–147, 283
curatives 3, 8–9, 25, 39
cure 158, 166
curing bladder 42–43
curing system 204

day-light press 218
decontamination 381
deep eutectic solvent (DES) 393
defects 448–450, 454, 474, 493–494, 497–498
degradation 447, 463
degree of polymerisation 214
density 252, 272–273
design criteria 219
desulfurisation 371, 396, 409
deterioration 176, 203
devulcanisation 367, 380, 385–387, 391–392, 394, 396–398, 402–404, 408, 410–411, 415, 419, 431–432, 436
devulcanisation efficiency 393
devulcanisation process 385
devulcanisation reactions 395
devulcanisation techniques 436
devulcanised natural rubber 432
devulcanised rubber 386–387, 397–398, 403, 405, 413–415, 423, 425, 428, 431
devulcanised rubber blend 387
devulcanised rubber-based silica composites 432
devulcanised styrene–butadiene rubber (D-SBR) 431
devulcanising agent 394, 423, 430–431, 433
devulcanising reagent 398
dielectric 178, 182
dielectric property 178
die-swell resistance 96
differential scanning calorimeter 293–294

dimension 198, 205–206
diphenyl disulfide (DPDS) 398, 402
dipped fabric 236
discarded rubbers 397
dispatch 449
dispersed phase size 122, 124, 132, 144, 146–147
dispersion 96, 109–112, 114
disposal methods 366
distillation 385
DMA 173–175, 197, 225
drive mechanism 202
driven 197
drum friction 256–259
dry grinding 384
DSC 294–295, 297–306
dumbbell 260
durability 184, 203
dynamic applications 41, 151, 203
dynamic mechanical 151, 158, 173–174
dynamic properties 346
dynamic shear rheometer (DSR) 375
dynamic stress-strain 203

EBT model 172
efficiency 471, 484–485, 488–489, 491
eight wastes 474
elastic modulus 337–339, 345, 347
elasticity 386
elastomer blends 121, 125–126
elastomeric 151, 159, 167, 181–182
elastomeric compounds 39, 197
elastomers 382
electric motors 200
electrical conductivity 151, 177–178, 183
electrical property 177
electrical resistivity 256, 258
electrically conductive 78
Elemental Analyser 294
elevator belt 243, 251
elongation 234–238, 256, 260
emissions 381
emulsion SBR 281
energy audit 483
engineering industries 199
ENR 121, 125–128, 141–148
enthalpy 84
entropy 84
enzymatic devulcanisation 405
EPDM 122–124, 126, 166, 178, 181, 283, 286, 304
EPDM rubber powder 385
EPM 121, 126–128, 130–148
epoxidation 127, 141–142, 145–146, 148
epoxidised natural rubber 121
equipment 454, 457, 465, 470
eradicate 469, 474, 477, 494, 498
ESBR 281
ethylene-propylene copolymer 121
ethylene-propylene-diene monomer (EPDM) 388
ethylene-propylene-diene rubber (EPDM) 410
European tyre labeling 96
expanded graphite 155–157, 161
expensive 449, 451
external failure 446, 448, 450
extractable material 291
extrusion 452

fabric 6–7, 21–22, 24–25, 34–36, 45, 197–200, 204–205, 213–217, 219–221, 223–224, 230–237, 267, 270
fabrication 153
failure cost 445, 447–448
failure modes 204
failures 203
fastener 265–266
fatigue crack 20, 203
fatigue life 203, 346, 353–355
fatigue life of rubber 204
fatigue resistance 31, 33, 35, 203–204, 220
fatigue to failure 327, 353, 355
FEF 221–223
fibres 21, 213–214
filament 238
filler 1, 8–9, 10, 15–17, 25, 27, 39, 151–152, 160–161, 166–167, 170, 172, 174–175, 177–183, 210, 284, 385
filtration 385
fines content 112
fishbone 475
flat belts 200
flex 324, 352–354
flexibility 159, 176, 201, 236–237, 248, 256
flow visualisation in internal mixer 121
Flow visualisation studies 125, 147
food processing 382
formulation 253–254
Fourier-transform infrared spectroscopy 159
fractal theory 79

fresh rubber 387
friction 253, 258
FTIR 158–159, 296–306, 308
fuel economy 102
fuel efficiency 101
functionalised graphene 152, 156, 169
furnace black process 64

gas barrier property 151, 179
Gas chromatography Mass spectroscopy 296
GC-MS 293, 296–297, 305, 307–309
Gibbs free energy 84
Glass transition temperature 281, 294, 298–299, 302, 304–305, 324, 347, 360
good quality 448, 463
granules 380
graphene 151–163, 165–184
graphene nanoplates 166, 176
graphene oxide 152, 156, 158–161, 163, 166, 168–171, 173, 176, 179, 182
graphene–rubber nanocomposite 152, 158, 167, 177
graphite 153–158, 170, 176, 179–180, 183
graphite oxide 155, 157
grinding 381, 385, 421
grinding mill 383
grinding processes 380
grit 105, 108–109
groove 200–202
ground rubber tyre (GRT) 430
ground tyre rubber (GTR) 388, 393, 407, 410, 415, 421, 424
guayule natural rubber (GNR) 430

HAF 221–223
hardness 7–8, 11, 15–17, 36, 38, 63, 90–91, 102, 105, 110–111, 210, 220–222, 224, 324, 328, 345
heat buildup 197, 203–204, 209, 221–223, 225
heat generation 151–152
heat loss 116–117
heat resistance 36–37, 203, 235
heat resistant 36, 240, 255
heating rate 157
Hevea brasiliensis 280
hidden 446
hidden quality 445
high-temperature atmospheric devulcanisation (HTAD) 394
history of transmission belts 200
HNBR 184
Horikx plot 395, 398
hose 21, 44–46, 210
human 447, 459, 462
hybridisation 76
hydrated lime 373
hydrodynamic effect 76
hydrogenated carboxylated nitrile–butadiene rubber 160
hysteresis 8, 10, 24, 27, 63, 96, 167, 204, 324, 347, 349, 351

ICP-OES 293, 295, 308
identify value 471
idler 199
IIR 282, 286
immiscible blends 147
implementation 478, 497
improvement 449–450, 453, 455, 463, 466
in situ polymerisation 152, 180, 182
incineration 367–368, 421
inductive coupled plasma 293, 295
industrial machinery 200
industrial rubber products 102–103
industrial rubber products (IRP) 102
industrial waste 398
industry 445–446, 449–452, 461–463
inner liner 282, 301
inspection 447, 449, 451, 453, 460–461, 463
internal friction 251
internal mixer 121–122, 125–128, 131, 135, 141, 145–147
Iodine Adsorption No 115
IR 121, 124, 126–128, 130–131, 135, 138–142, 144, 146–148
IRP 102

Japan 469–470, 477, 479, 484, 489, 492

Kaizen 477, 479–480, 492
Kaizen methodology 479
Kaizen pyramid 480
Kanban 486, 490–491
Kanban system 490
Karakuri 484–485

labor 445–446, 449
laboratory equipment 398

landfilling 366–367
latex 152, 159, 161, 168–170, 175–177, 179–182, 184, 207, 234
latex blending 152, 161, 170, 177, 180–182
latex co-coagulation 168
layer of components 205
lean 469–470, 472–480, 482, 487, 489, 491–493, 496, 498
lean management 469–470
Lean Manufacturing 472, 477, 493, 498
lean principles 470
leather 200
liquid nitrogen 381
low-temperature devulcanisation 405
low-temperature devulcanisation techniques 405

machineries 199
macromolecule 167, 169, 175, 182, 207, 214
maintenance 447–448, 453–454, 457, 465
Malaysian 470
management 447–448, 450, 452–453, 456–457, 460, 480, 494
manufacturing 1, 25, 33, 151–153, 184, 197, 199, 212, 481, 486, 494
manufacturing defects 219
mass production 493
MBTS 221, 223
mechanical bonding 233
mechanical devulcanisation 386
mechanical exfoliation 151, 157
mechanical fastener 265
mechanical milling 386, 433
mechanical performance 414
mechanical property 152
mechanochemical devulcanisation 392, 394, 405, 430–431, 436
mechanochemically modified rubber powder 396
melt mixing 152, 168, 170, 177, 180–181
metal oxide 286, 290, 293, 295
metal pulley 202
methodology 476, 480
microbacterial devulcanisation 408
microbial desulfurisation reactor 408
microbial devulcanisation 407
microorganisms 409
microwave devulcanisation 405, 411, 413, 436
milling process 424
mineral processing 383
miscibility 122, 133–134, 141–142, 147
mixing 323–326, 328–329, 448–449, 452–453, 455, 462, 466
modulus 8, 10–11, 15, 21, 34, 151, 160, 167–170, 173–175, 184, 210, 220–222, 224
molded poly-V-belt 201
molding 5, 24, 42, 217, 325–326, 328–329, 344, 357, 452, 456, 462
monomer 152, 182, 207–210
Mooney scorch 332, 334
Mooney viscosity 323, 326, 331–334, 386, 388, 390, 392, 394, 422–423, 430
Muda 486–487
Mura 486–487
Muri 486–487

natural rubber 3, 11–12, 25, 28, 40, 159, 161, 163, 166, 168–169, 171, 174–177, 179, 182, 184, 207–208, 279–281, 299, 323, 349, 353, 357, 398, 472–473
natural rubber (NR) 410, 417, 430
natural rubber (NR) vulcanisate 397
NBR 121, 123–127, 131–137, 141, 147–148, 159, 161, 166, 178, 283–286
nitrile rubber 121
nitrogen surface area 115
noise level 98
nomenclature 205
non-conformance 447–448, 450, 453
NR 207–210, 219, 221–223, 281, 285–286, 298, 306, 310
nylon 24, 35, 45, 213

oil absorption no. 110
oil resistant 240, 255–256
optical and scanning electron microscopy 121
optical microscopy 124
organisation 469, 472, 488, 494
O-ring 41, 283, 303–304

packaging 449
particle size 56, 59, 62, 68, 73, 111
Payne effect 81–82, 96, 338
PBR 281
pellet hardness 114
perfection 472
performance of devulcanised rubber 412
performances 469
PET 213–214
pH 114

pharmaceutical stoppers 39
pharmaceuticals 382–383
phase morphology 121, 123–124, 126, 129–130, 132, 134–135, 137, 139–142, 144–147
philosophy 470, 479, 484–486, 489, 493
physical properties 15, 31, 41, 197, 204, 210, 222
plastic applications 117–118
plasticiser 8, 17, 285–286, 293
plastics 382
ply 267, 275
polarity 121, 126, 134, 141, 147–148
poly dispersity index (PDI) 402
polybutadiene 121–122, 147
polybutadiene rubber 281
polychloroprene 14, 210
polychloroprene rubber 283–284
polydimethylsiloxane 166
polyester 23–24, 35, 45, 213–215
polyisoprene 6, 121, 127, 159, 207
polymer 197, 207–209, 213–214, 374
polymer degradation 398
polymerisation 152, 182, 208–210, 214
polyurethane foam 411
polyurethane rubber 410
poly-V-belts 199
poor quality 445–450, 452–454, 458–459, 462–463, 465–466
porosity 56, 70–71, 115, 117
potassium permanganate 378
pour density 105, 113
power generation automobile 202
power law index 338
power transmission 197
power transmission capacity 200
pre-shredding 380
prevention 446–450
principles 470
pristine 156–157, 179
process aids 284
processability 323, 335, 340
processing 469, 473–474
processing aids 8, 17, 20, 286
production 448–449, 451–453, 457, 460, 462, 465–466
production of tyre 431
productivity 446–447, 450, 454, 469–472, 479–480, 485–486, 489
profitability 445–447, 453
pull 471
pulley 197–198, 202
pulley diameter 203
pulley groove 202, 204
pulverisation 400
pyrolysis 367, 370, 385, 421
pyrolysis efficiency 371
pyrolysis of tyres 370
pyrolysis process 385
pyrolytic oil 372

quality assurance 448, 456, 461
quality circle 492
quality control 447–448, 456–457, 460
quality plan 454–455

Raman spectroscopy 158, 162
raw edge belts 199–200
raw edge cogged belts 199
raw edge laminated belts 199
raw edge plain belts 199
raw material 448, 452–453, 455, 462
rayon 213
REC 199
reclaim rubber (RR) 432
recycling 366–367, 372, 398, 420–421
reduced graphene oxide 152, 156, 168–169, 175
reinforced 197–198, 202
reinforcement 59, 68, 70, 73, 76–77, 98–99, 111–112, 116, 151, 167, 169, 172, 178, 181–183, 197, 205, 210–211, 213–214
reinforcing member 204
reinforcing Unit 213
re-inspection 463
rejection 446, 448–454, 460–463
REL 199
REP 199, 205
residual rubber production 366
resilience 324, 347, 349–350
resin 221, 223
retesting 449–450, 463
return on investment 482
reusable crumb rubber 380
reusable linear polymer 397
revulcanisates 411
revulcanisation 422
revulcanised rubber 403, 425
rework 447, 449–451, 453, 462–463
RFL 213, 215, 217
rheograph 335

rheology 166
rice polisher 40
road pavement construction 372
rolling resistance 96, 99, 101–102, 152–153, 173–174, 184
rolling thin film oven (RTFO) 375
rubber 197, 200–208, 210–211, 213, 215, 217, 220, 222, 230, 232, 234, 236, 238–240, 250–251, 256, 258, 261, 263, 266–267, 269–270, 273–274, 276, 386, 469–470, 472, 474–475, 477–478, 481, 483–485, 488, 492, 495
rubber compound 160, 168, 171, 177, 179, 184, 204
rubber goods 472
rubber hose 44–45
rubber industry 469, 495
rubber molecules 386
rubber network structure 395
rubber powder 395–396
rubber product 203, 211, 381, 445, 447–451, 456–457, 466, 472
rubber roller 302
rubber vulcanisate 425, 428
rubber waste 420
rubberised fabric 216
rubber-modified bitumen 372

SBR 161, 169, 172–173, 175–177, 179, 181–182, 208–209, 219, 221–223, 281, 285–286, 306, 310
scanning electron microscopy 124, 158, 161
scrap 483
section 198, 201, 205–206, 217–218
SEIKETSUE 480
SEIRI 480
SEISO 480
SEITON 480
SEM 154–155, 158, 161, 164
separation 290–291, 305
service life 203, 211, 220
sheaves 202
SHITSUKE 480
Shore A Durometer 90
shredding 385, 421
sidewall 202, 209
silane 167, 173, 176, 182, 212, 223
silica 58, 99, 151–153, 167, 169, 173, 183–184, 211–212, 219, 221–224, 291, 293, 296, 309
silica nanoparticles 432
silica-based rubber composite 431
silicone rubber 410
Six Sigma 476–478
skim compound 25, 28, 31, 35, 37–38
slippage of the belt 200
slippage problem 200
slurry oil 373
snap back velocity 85
snapping of belt 204
sodium silicate 212
solubility parameter 121, 123, 134
solution blending 152, 161, 181
solution intercalation 152, 158, 180–182
solution SBR 281
solvent etching 130–131
solvent extraction 385
solvent swelling 405
soot 54–55, 60
SOP 487–488
space saver 205
special black for tyre 99
specific gravity 323, 329
specific heat 324, 358
SRF 221–223
SSBR 281
statistical 449, 454
statistical thickness surface area 98, 105, 115
stopper 39–40
strain-induced crystallisation 163
stress 237, 248
stress induced crystallisation 141
stress relaxation 332–333, 340–341
stress–strain analysis 171
styrene–butadiene rubber (SBR) 170, 172, 208, 281, 410, 414, 417, 432
sulfur 204, 207, 219, 221, 223
sulphur-cured styrene–butadiene rubber (SBR) 430
sulphur-vulcanised natural rubber 431
supercritical carbon dioxide ($scCO_2$) 402–403
suppliers 453
surface area 151, 155, 169, 171, 177
synthetic rubbers 151, 208

TD 105, 111
tear strength 169–170
tensile properties 403
tensile strength 10, 34, 36, 38, 68, 110–111, 167–172, 176, 210, 222, 260, 341–342, 357
tensioner 199

TESPT 182
testing 448–449, 452, 460–461, 463
tetramethyl thiuram disulfide (TMTD) 385, 422
textile machines 202
TGA 293–294, 297–306
thermal conductivity 324
thermal exfoliation 151, 157
thermal expansion 324, 358, 361
thermal stability 152, 175–176
thermochemical devulcanisation 384
thermodynamics of rubber elasticity 84
thermogravimetric analyser 293
thermogravimetry analysis 417
Thermo-mechanical devulcanisation 385
thermomechanical devulcanisation process 386
thermo-mechanically devulcanised GTR 391
thermoplastic elastomer (TPE) 388
thermoplastics 382
timing belt 304–305
tint strength 111
tire 208–209
TMTD 14, 31
tolerance 221, 223, 451, 456
toluene discoloration 105, 111
TOR 121, 125–128, 132, 134, 139–142, 145–147, 374–375
Toyota 469, 474, 479, 486, 489–490
traction 153, 174, 183–184
training 454, 459
transmission belt 38, 197, 200, 203
transmission electron microscopy 124, 158, 161–162
transmission engineering 202
*trans*-polyoctenamer rubber (TOR) 374
*trans*-polyoctenylene 121
treatment by microorganisms 409
TRGO 152, 156, 168–170, 174, 176, 179
triaxial compression reactor 398
truck tyre vulcanisate 402
tyre classification 99
tyre construction 99
tyre demand 365
tyre production 436
tyre recycling technique 420, 436
tyre scrap 365
tyres tread 298–300, 305, 309

ultrasonic devulcanisation 409–410, 436
ultrasonic horn oscillation method 411
ultrasonically assisted extrusion 409
used tyres management 367
UV 238
UV-resistant 213

v-belt 21, 197–199, 202, 205–206, 213, 219–221
V-belts developed 200
viscosity 456
viscous modulus 337, 339, 347, 350
volume fraction 76–78
vulcanisate 403, 411, 413–414, 432
vulcanisation 231, 267, 269, 286–287, 290–291, 293, 386, 449, 463
vulcanised natural rubber (NR) 386
vulcanised rubber 384, 422
vulcanising agent 284, 286–287

wastage 445, 463
waste 469–474, 477–479, 481, 486–487, 490, 492, 494–496, 498
waste management 368
waste rubber pyrolysis 370
waste rubber treatment 405
waste tyre rubber (WTR) 399, 415
waste tyres 421
WAXD 171
wear resistance 152, 184
wearing 204
wedge V-belt 198
wet grinding 383–384
wet grinding process 383
wet grip 97
wrapping 199, 217

XNBR 159–160, 162–163, 170, 176, 182

yarn 214

zero defects 493–494
zero pollution 496